POWER PLANT ENGINEERS GUIDE

by Frank D. Graham
revised by Charlie Buffington

Macmillan Publishing Company
New York

Collier Macmillan Publishers
London

THIRD EDITION

Copyright © 1974 by Howard W. Sams & Co., Inc.
Copyright © 1983 by The Bobbs-Merrill Co., Inc.
Copyright © 1987 by Macmillan Publishing Company, a division of Macmillan, Inc.

All rights reserved. No part of this book may be reproduced or transmitted in any form or by any means, electronic or mechanical, including photocopying, recording or by any information storage and retrieval system, without permission in writing from the Publisher.

Macmillan Publishing Company
866 Third Avenue, New York, N.Y. 10022
Collier Macmillan Canada, Inc.

Library of Congress Cataloging in Publication Data

Graham, Frank Duncan, 1875-
   Power plant engineers guide.

   Includes index.
   1. Steam power-plants—Handbooks, manuals, etc.
I. Buffington, Charlie. II. Title.

TJ400.G72 1983   621.1′9   82-17779
ISBN 0-672-23329-0

Macmillan books are available at special discounts for bulk purchases for sales promotions, premiums, fund-raising, or educational use. For details, contact:

   Special Sales Director
   Macmillan Publishing Company
   866 Third Avenue
   New York, N.Y. 10022

10 9 8 7 6 5 4 3 2

Printed in the United States of America

Foreword

The purpose of this book is to include in one volume a complete library of practical and theoretical steam engineering. An attempt is made to present the basics in numerous subjects in plain language and in terms that any serious student can understand.

The information contained in this book is designed to prepare a beginner for the first exam and to help those who want to upgrade an existing license. Numerous illustrations are included along with typical exam questions and answers to further simplify the wide range of subject matter involved in the field.

An expression of thanks and appreciation is given to various manufacturers and individuals for their cooperation in furnishing complete information covering their equipment and for encouragement and suggestions during the preparation of this book.

ACKNOWLEDGMENTS

Sincere thanks are due the following companies for their assistance and cooperation in supplying information for the revision of this book. Some supplied photos and drawings as well.

ABCO Industries
A.O. Smith Corporation
Babcock and Wilcox Co., Power Gen. Division
CAM Industries, Inc.
Cleaver Brooks, Division of Aqua-Chem Inc.
CNB Div., Combustion Service and Equipment Co.
Econo-Therm
Elliot Co.
Industrial Steam Div., Kewanee Boiler Corp.
Ingersoll-Rand Co.
ITT Domestic Pump
JOY Machinery Co.
O'Connor Combustor Corp.
Oswego Package Boiler Co., Cyclotherm Division
Pennsylvania Separator Corp.
Ray Burner Co.
Reimers Electra Steam, Inc.
Sentry Equipment Corp.
The Trane Company
VaPower Products
William Schank, ret., Babcock and Wilcox Co., Tube Division

Special thanks to the National Association of Power Engineers and Executive Director Dorian Dvorak for allowing the use of their educational materials, particularly those of the Minnesota State Association (Refrigeration and Basic Electricity) and the Stationary Engineering Home Study Course. Thanks also to the National Institute for the Uniform Licensing of Power Engineers, Inc., and the *National Engineer* Magazine.

Appreciated also has been the willing assistance of Bill Everhart, chief engineer with the State of Maryland at a high-rise office facility.

CHARLIE BUFFINGTON

CONTENTS

Foreword 3

Chapter 1
BASIC PRINCIPLES 13

Matter—heat—temperature—work—steam—friction—efficiency—characteristics of metals—refrigeration

Chapter 2
FUELS 31

Solid fuels—liquid fuels—gaseous fuels—waste heat—nuclear energy—geothermal energy—solar energy

Chapter 3
HEAT 53

The effect of heat—specific heat—experiments with heat

Chapter 4
COMBUSTION 65

Combustible material—combustion products—draft—smoke and flue gases—ash—experiments with combustion

Chapter 5

STEAM AND HOW A BOILER MAKES IT........................85

Nature of steam—types of steam—the formation of steam—condensation—steam above atmospheric pressure—the steam tables—superheated steam—evaporation—how a boiler makes steam

Chapter 6

TYPES OF BOILERS 109

Boiler classifications—steam generators—waste-heat recovery boilers—packaged boilers

Chapter 7

SHELL OR FIRE-TUBE BOILER CONSTRUCTION 131

Welded joints—riveted joints—stays—boiler tubes—vertical tubular boilers—horizontal fire-tube boilers—shell boiler openings

Chapter 8

WATER-TUBE BOILER CONSTRUCTION 165

Boiler types and construction—water tubes—water-tube boiler parts—sectional and nonsectional construction—advantages of various tube types—miscellaneous boiler types

Chapter 9

SOME BIG-BOILER FEATURES 193

Classifications—main advantages of straight tubes over bent tubes—water-cooled furnaces—forced-circulation boilers

Chapter 10
STRENGTH OF BOILER MATERIALS 205
   Metals used—characteristices of boiler metals—metal testing

Chapter 11
BOILER CALCULATIONS 219
   Shell pressure and stress—working and bursting pressure—shell thickness—riveted joints—welded construction—ligaments—reinforcement of flat surfaces—boiler output-furnace—heating surface

Chapter 12
BOILER FIXTURES, FITTINGS AND ATTACHMENTS 261
   Safety valves—stop and check valves—blow-off and blow-down valves—cocks—water gauge—feed-water connection—steam gauge—fusible plugs—steam traps—separators—collectors—steam loop—furnace

Chapter 13
BOILER FEED PUMPS 297
   Pump types—the simplex system—the duplex system—centrifugal pumps—maintenance—variable stroke power pumps

Chapter 14
FEED-WATER HEATERS AND ECONOMIZERS 349
   Types of heaters—open heaters—closed heaters—economizers

Chapter 15

FEED-WATER REGULATORS, INJECTORS, DEAERATORS,
AND EJECTORS 369

Displacement regulators—evaporation regulators—expansion regulators—injectors—deaeration—steam-jet air ejectors

Chapter 16

FEED-WATER TESTING AND TREATMENT................... 405

Scale and mud—hardness of feed water—saturation point—precipitation—chemical terms relating to feed water—characteristics of chemical substances—simple tests—feed-water treatment

Chapter 17

CONDENSERS..439

Economy of condensing—air evacuation pump—direct-contact (class 1) condensers—surface (class 2) condensers—condenser operation—typical modern condensate systems

Chapter 18

COOLING PONDS AND COOLING TOWERS 465

Cooling pond types—cooling pond operation—cooling towers—cooling-water problems and treatment

Chapter 19

BOILER INSTALLATION, STARTUP AND OPERATION 485

Installation—running—firing—boiler water—routine operation and care

Chapter 20
BOILER MAINTENANCE AND REPAIRS 533

Shall repairs—boiler patches—gasket repairs—tube repairs—tube cleaners—lubricants—lubrication

Chapter 21
MECHANICAL STOKERS 575

Overfeed stokers—underfeed stokers—rotary or sprinkler stokers—traveling-grate stokers—spreader or sprinkler stokers—stoker operation

Chapter 22
PULVERIZED COAL BURNERS 595

Points on combustion—firing pulverized coal—pulverizers—burners—automatic control

Chapter 23
OIL, GAS, AND WASTE-FUEL BURNERS 613

Fuel oils—gravity-feed oil burners—pressurized oil burning—classifying oil burners—vaporizing burners—atomizing burners—rotary burners—spray burners—low-pressure proportioning burners—pot-type burners—gun-type burners—dual- and multi-fuel systems—waste-fuel burners

Chapter 24
STEAM TURBINES..639

Classifications—velocity ("impulse") turbines—compound turbines—multistage turbines—unequal-pressure (reaction) turbines—miscellaneous turbines—erection of turbines—turbine operation—turbine maintenance—turbine trouble-shooting

Chapter 25
AIR COMPRESSORS 677

Using Boyle's and Charles' Laws—effect of Boyle's and Charles' Laws combined—types of compressors—cooling—compressor valves—packaged compressor systems—intercoolers and aftercoolers—speed and pressure controls—unloaders—compressor operation

Chapter 26
REFRIGERATION BASICS 721

Refrigeration system components—refrigerants—compressors—other refrigeration methods—heat pumps—controls—defrosting systems—maintenance and repair

Chapter 27
BASICS OF ELECTRICITY 751

Magnetism—current, voltage, and resistance—generators—alternating current—measuring voltage, current, and resistance—motors

Chapter 28
DIESEL ENGINE PRINCIPLES 779

Four-cycle diesel engines—compression pressures—fuel-injection methods—combustion—fuel-injection system—service—two-cycle diesel engines

Chapter 29
PLANT SAFETY 805

Safety valves—low-water cut-off—flame detectors—steam gauge test—safety guidelines

Chapter 30

TABLES AND DATA..................................823

Basic math calculations—square, rectangle, and parallelogram calculations—triangle, trapezoid, and cylindrical calculations—boiler horsepower—engine horsepower—circle calculations—inscribed circle calculations—spherical calculations—water calculations—boiler conversion factors—metric measurement—metric-to-English conversions—English-to-metric conversions—English-to-English conversions—Miscellaneous conversions—Fahrenheit/Celsius equivalents—circumferences and areas of circles—properties of saturated steam—properties of superheated steam—specific heat of superheated steam—factors of evaporation—specific heat of various substances—maximum allowable stress for carbon and alloy steel—efficiency of riveted joints—properties of standard lap-welded boiler tubes—circumference area and volume—standard wrought pipe—properties of extra strong pipe—double extra strong pipe—American standard bolt sizes, course thread series—physical properties of common refrigerants—pressure-temperature relations for refrigerants—pressure

Appendix A

RULES AND REGULATIONS OF THE ENVIRONMENTAL PROTECTION AGENCY (EPA) 867

Summary of standards—Part 60—standards of performance for new stationary sources—EPA regional offices

Appendix B

CRITERIA FOR REACTOR OPERATOR TRAINING AND LICENSING AS ESTABLISHED BY THE NUCLEAR REGULATORY COMMISSION (NRC) 917

Training and licensing—training in heat transfer, fluid flow, and thermodynamics—training criteria for mitigating core damage—control manipulations—Part 55: rules and regulations: operator's licenses

CHAPTER 1

Basic Principles

Define the term "basic principles."
Answer: The fundamental laws that govern the behavior of substances when acted upon by any external agency or agencies.

Of what value is this in taking an examination for an engineer's license?
Answer: With a knowledge of these fundamental laws, the applicant is better equipped to reason out the problems presented and answer the questions.

Why are engineers required to take an examination before obtaining a license?
Answer: Because a boiler is a dangerous thing, especially in the hands of a person who is ignorant of the operation of a boiler and its related equipment.

Power Plant Engineers Guide

Which type of person makes the best engineer: one who knows *how* and *why* equipment works? Or, one who knows the *way* it works?

Answer: Neither. The best engineer is the one who has both *practical* and *theoretical* knowledge and who can put the two to work.

Is it important for an engineer to keep up with new developments?

Answer: Yes. The most successful engineer is one who makes it his business to stay current with new developments in theory and how they are being or will be used.

MATTER

What is a molecule?

Answer: The smallest particle of a substance or chemical combination that can exist by itself and still retain the properties of that substance and which is composed of two or more atoms.

What is matter?

Answer: Any material substance having mass (occupying space and having weight) but not a specific, formed body.

What is an element?

Answer: A substance in nature, one of the 104 known to exist, that cannot be divided or broken up into any simpler or more basic form.

What is an atom?

Answer: The smallest particle of an element which has all the properties of the element.

What are the three states of matter?

Answer: (1) Solid, (2) liquid, and (3) gas.

How are the three states distinguished with respect to molecules?

Answer: By their amount of motion.

How do the molecules move in a solid?

Answer: Back and forth like tiny pendulums. Molecules in a solid are held together by the force of *cohesion* which enables a solid to hold its shape.

How do they move in a liquid?

Answer: They wander at random without any definite path. Heat energy can overcome the force of cohesion between molecules, allowing them to move freely.

How do they move in a gas?

Answer: In straight lines. Molecules in a material in gaseous form are in a state of repulsion. The hotter they get, the more they repulse each other.

HEAT

What is heat?

Answer: A basic form of energy known by its effects.

How are these effects indicated?

Answer: Through *touching* and *feeling*, as well as by the *expansion, fusion, combustion,* or *evaporation* of the matter upon which it acts.

What is temperature?

Answer: A measure of how hot or cold a substance is, a measure of *sensible* heat.

What is sensible heat?

Answer: That heat which produces a measurable rise of temperature as distinguished from *latent* heat, which produces a change of state.

What is latent heat?

Answer: The quantity of heat required to change the *state* or condition under which a substance exists without changing its temperature.

What is latent heat of sublimation?

Answer: The heat needed to change a solid to a vapor without a change in temperature.

What is specific heat?

Answer: The ratio of the quantity of heat required to raise the temperature of a given weight of any substance 1°F. (or 0.6° C.) to the quantity of heat required to raise the temperature of the same weight of water from 62° to 63°F. (16.6° to 17.2°C.). The capacity of any substance for receiving heat as compared with another which is taken as a standard, this being generally water. Thus, the same quantity of heat which will raise one pound of water 1°F. will raise about 4¼ pounds of cast iron 1°F. So, the specific heat of water is taken as 1.000; that of cast iron is 0.241.

When does a transfer of heat take place?

Answer: When bodies of unequal temperatures are placed near each other, heat leaves the hot body and is absorbed by the cold body until the temperature of each is equal.

How fast does the transfer take place?

Answer: The rate by which heat from the warmer body is absorbed by the colder body is proportional to the difference of temperatures of the two bodies.

How does a transfer of heat take place?

Answer: By radiation, conduction, or convection.

How is the quantity of heat measured?

Answer: By a standard unit called the *British thermal unit*, abbreviated Btu.

What does the term "heat release per cubic foot per hour" (cubic centimeter per second), mean?

Basic Principles

Answer: A measure of the amount of heat produced by burning fuel. It is measured Btu (calories) per hour.

The term, "heat transfer per square foot" (square centimeter) of heating surface tells us what?

Answer: A measure (in Btu or calories per hour) of the amount of heat released in the furnace that is transferred to the water side of the boiler.

Explain how heat passes through a steel plate.

Answer: Heat is conducted from molecule to molecule within the steel.

What does thermal conductivity mean?

Answer: A measure of how well a substance can conduct heat. It is a measure of the amount of heat passing per Btu per hour (or calorie per second) per sq. ft. (sq. cm.) per inch (cm.) of thickness per degree F. (or C.) difference in temperature on opposite sides of the substance.

On what four factors does thermal conductivity depend?

Answer: (1) The surface area of the material (larger surfaces conduct more heat); (2) the thickness of the material (more heat will flow through a thin material); (3) the temperature difference on each side; and (4) the type of material (steel is a better conductor than some metals).

Explain the meaning of thermal flywheel effect.

Answer: It's the heat-storage capacity of high-temperature water.

What is thermal shock?

Answer: The sudden cooling of overheated steel in a boiler, usually the result of adding water during a low-water condition.

What usually causes thermal shock?

Answer: A sudden heating or cooling of a part of a boiler changes the pattern of stress distribution in the steel.

As water is heated in a boiler, how does it naturally circulate?
Answer: It rises from bottom to top.

What is the relation between heat and work?
Answer: Heat develops mechanical force and motion, hence it is convertible into mechanical work.

What is the "mean Btu"?
Answer: Technically it is 1/180 of the heat required to raise the temperature of one pound of water from 32° to 212°F. (0° to 100°C.).

How much heat is a therm?
Answer: 100,000 Btu.

What happens to solids, liquids, and gases as they heat and cool?
Answer: They all expand in volume when heated and contract or decrease in volume when cooled. Water obeys the same rule, but it begins to expand again at 39°F. (3.9°C.). There is further expansion when it freezes. Gases expand much more rapidly than solids or liquids.

TEMPERATURE

What is a thermometer?
Answer: A device to measure temperature.

Of what does it consist?
Answer: It consists of a glass tube terminating in a bulb which is usually charged with mercury.

How does it measure the temperature?
Answer: By the contraction or expansion of the liquid with temperature changes, causing the liquid to rise or recede in the tube.

How is the degree of heat indicated?
Answer: An arbitrary scale divided into "degrees."

Basic Principles

Good quality thermometers are usually filled with mercury, alcohol, or nitrogen. Why is this, and in which temperature range is each used?

Answer: Mercury is used from –40°F. (°C.) to 200°F. (93°C.) because mercury either solidifies or vaporizes beyond these extremes. If nitrogen is used to fill the space above the mercury, which reduces the tendency of mercury to vaporize or separate, the thermometer will work to approximately 500°F. (260°C.). Nitrogen-filled stock thermometers are accurate to approximately 1000°F. (538°C.). *Alcohol* thermometers are used in the range from 158°F. (70°C.) to –112°F. (–80°C.).

What is a pyrometer?

Answer: An instrument used to measure temperatures above 750°F. (399°C.)

What are the Fahrenheit and Celsius scales?

Answer: The Fahrenheit scale was once generally used in English-speaking countries. The Celsius scale has replaced it in many cases and is considered the world standard. On the Fahrenheit scale, the freezing point of water is 32° and the boiling point, 212°. At Celsius water freezes at 0° and boils at 100°.

What is absolute zero?

Answer: A point on the thermometer scale beyond which a further decrease in temperature is inconceivable. It is that temperature at which the volume of a gas would have become zero or it would have lost all the molecular vibration which manifests itself as heat.

What is the absolute zero temperature?

Answer: Minus 459.6°F. or –273°C.

WORK

What is work?

Answer: The overcoming of resistance through a certain distance by the expenditure of energy.

How is work measured?
Answer: By a standard unit called the foot-pound (ft. lb.).

What is a foot-pound?
Answer: The amount of work done in raising 1 pound 1 foot, or in overcoming a pressure of 1 pound through a distance of 1 foot.

What is the relation between the unit of heat and the unit of work?
Answer: One Btu equals 777.52 foot-pounds.

ENERGY

What is energy?
Answer: Stored work; that is, the ability to do work.

Name two kinds of energy.
Answer: Potential energy and *kinetic* energy.

What is potential energy?
Answer: Energy due to position.

What is kinetic energy?
Answer: Energy possessed by a moving body due to its momentum.

Power

What is power?
Answer: The rate at which work is done.

What is a unit of power?
Answer: The amount of work done in a given period of time or the *rate* of doing work measured in *horsepower* (hp).

Is there a relationship between the unit of work and the unit of power?
Answer: Yes, the unit of power (hp) tells how much work has been or is being done.

Basic Principles

What is one horsepower?
Answer: 33,000 foot-pounds per minute.

Pressure and Vacuum

What is pressure?
Answer: As defined by Rankine, a force of the nature of a thrust, distributed over a surface.

How is pressure usually measured?
Answer: As pounds per square inch (psi).

What is atmospheric pressure?
Answer: The force exerted by the weight of the atmosphere on every point with which it is in contact.

How great is the atmospheric pressure?
Answer: It is generally taken at 14.7 psi at sea level.

Why do we not feel this pressure?
Answer: Because air presses the body both externally and internally so that the pressures in different directions balance.

What effect has atmospheric pressure on steam engine operation?
Answer: It acts as a back pressure on the engine piston and so reduces the power or adds load to the engine.

How is most of the atmospheric pressure gotten rid of?
Answer: By the use of a condenser.

How is the pressure of the atmosphere measured?
Answer: By an instrument called a *barometer*.

How is a barometer constructed?
Answer: Essentially it consists of a glass tube 33 or 34 inches long, sealed at one end, filled with mercury, and inverted in an open cup of mercury. A device called an aneroid barometer measures atmospheric pressure mechanically.

How does it measure the pressure of the atmosphere?

Answer: By the height of the column of mercury in the tube above the level of the mercury in the cup. The reading gives the pressure in terms of *inches of mercury.*

What is a vacuum?

Answer: A space devoid of matter; a space in which the pressure is zero.

Define absolute pressure.

Answer: The pressure registered on a gauge plus the pressure of the atmosphere.

What is gauge pressure?

Answer: The pressure indicated by a gauge and read from a gauge scale.

What causes the pointer on a pressure gauge to move?

Answer: Expansion and contraction of a Bourdon tube.

What is bursting pressure?

Answer: Accumulated pressure beyond the strength of a tube, boiler, or any confining vessel. It's determined by the thickness of the material, tensile strength, the quality of any joints, and the size of the vessel.

What is the difference between gauge pressure and absolute pressure?

Answer: Gauge pressure is pressure measured above atmospheric pressure; absolute pressure is pressure measured above zero; that is, gauge pressure plus pressure of the atmosphere.

What are the applications of these pressures?

Answer: Gauge pressure (psig) is used for measuring the pressure in a boiler, automobile tire, etc. Absolute pressure (psia) is used in all calculations relating to the expansion of steam.

How is absolute pressure expressed as gauge pressure?

Answer: By subtracting 14.7.

BASIC PRINCIPLES

How are pressures below that of the atmosphere usually expressed?

Answer: As psia in making calculations or the equivalent in "inches of mercury" in practice.

What is the meaning of the term "referred to a 30-inch barometer"?

Answer: It means that the variable pressure of the atmosphere is such that it will cause the mercury in the barometer to rise to a level of 30 inches.

What can be said about such expressions as a 24-inch vacuum?

Answer: Ridiculous in that it is not a vacuum strictly speaking, but only a *partial* vacuum; yet, nothing can be done about it.

How is the pressure in psia obtained from the barometer reading?

Answer: Multiply the barometer reading in inches of mercury (inHg) by 0.49116 to get atmospheric pressure in psia. Various readings are given in Table 1-1, which is based on the standard atmosphere of 29.921 inches of mercury, or 14.696 psia.

Table 1-1. Psia Equivalents of Barometric Pressure

Barometer (inHg)	Pressure (psia)	Barometer (inHg)	Pressure (psia)
28.00	13.75	29.921	14.696
28.25	13.88	30.00	14.74
28.50	14.00	30.25	14.86
28.75	14.12	30.50	14.98
29.00	14.24	30.75	15.10
29.25	14.37	31.00	15.23
29.50	14.49		
29.75	14.61		

STEAM

What is steam?

Answer: The vapor of water, a colorless, expansive, invisible fluid.

What is the white cloud seen issuing from an exhaust pipe mistakenly called "steam"?

Answer: This is not steam, but in reality a fog of minute liquid particles produced by the condensation of steam.

Under what conditions does steam exist?
Answer: When there is the proper relation between the temperature of the water and the external pressure.

Under atmospheric pressure at what temperature does water boil?
Answer: 212°F. or 100°C.

What does factor of evaporation mean?
Answer: It's the amount of heat needed to produce steam under pressure above atmospheric, divided by the amount of heat it takes to generate steam at atmospheric pressure. The number is usually more than one.

Which liquid expands more when heated, water or alcohol?
Answer: Alcohol.

Condensation

What is condensation?
Answer: The reduction in bulk of any substance accompanied by an increase in density.

What is dew point?
Answer: It's the temperature at which the water vapor (moisture) in the atmosphere starts to condense.

Describe the condensation of steam.
Answer: When the temperature of steam becomes less than that corresponding to its pressure, condensation takes place. That is, it ceases to exist as steam and becomes water.

When steam condenses what is the liquid called?
Answer: The *condensate*.

Basic Principles

What happens during the condensation of steam?
Answer: Air that was originally mechanically mixed in the water is liberated.

What must be provided for condensers to maintain a vacuum?
Answer: The liberated air coming in with the condensate must be removed from the condenser.

How is this done?
Answer: By an air pump, mistakenly called a vacuum pump.

Expansion

What is the law of expansion and contraction?
Answer: Practially all substances expand with an increase in temperature and contract with a decrease of temperature.

What is the difference between linear and volumetric expansion?
Answer: Linear expansion is the expansion of solid bodies in a longitudinal direction; the expansion in volume is called volumetric expansion.

As a material expands during a change in state, what will happen if the material is subjected to a high pressure?
Answer: High pressure (above normal atmosphere) *opposes* an increase in volume and *aids* a decrease in volume.

What happens to the boiling point as pressure is increased?
Answer: It rises; the boiling point increases with an increase in temperature.

FRICTION

Define friction.
Answer: It is that force which acts between two bodies at their surface of contact so as to resist their sliding on each other.

EFFICIENCY

Define efficiency.

Answer: It is the ratio of the useful work performed by a prime mover to the energy expended. In other words, the output divided by the input.

The term, thermal efficiency of a boiler, refers to what?

Answer: The percentage of heat produced by the fuel burned that is actually used to convert water in the boiler into steam.

CHARACTERISTICS OF METALS

How would you define the term "stress" as related to a steam boiler?

Answer: During the welding of boiler seams, the force applied to the metal surrounding the weld area that is not heated to the weld temperature is called stress. It is the result of uneven expansion and contraction in the same piece of metal and differing thicknesses of the metals joined.

How is this stress relieved?

Answer: The entire boiler is slowly heated by firing or heating uniformly to at least 1100°F. (593.4°C.) to 1200°F. (649°C.) or higher, if distortion won't result, and held for a period of time based on one hour per inch of thickness. The boiler should then be allowed to cool slowly at normal air temperatures to around 600°F. This slow heating and cooling equalizes stress throughout the metal. Stress around small weld areas can be relieved by several sharp blows with a ballpeen hammer.

When applied to metals, what do these terms mean: hardness, ductility, elasticity, plasticity, brittleness and malleability?

Answer: Hardness refers to the surface of the metal and its resistance to machining. *Ductility* describes how easily a metal can be drawn into a new shape (the easier it is to draw, the more ductile it's said to be). *Elasticity* of metal is its ability to return to its original shape after being bent or deformed. When a material is

Basic Principles

plastic, it does not return to its original shape after bending. *Brittleness* is the characteristic of a metal that causes it to break rather than bend; the more brittle, the less it'll bend before breaking. A *malleable* material can be bent without breaking.

What do these terms mean: load, stress, strain?

Answer: Load is any external force applied to an object or material; it can be measured in pounds, tons, grams, kilograms, etc. *Stress* is the amount of resistance the molecules of a material exert in an attempt to separate them; it is measured in the same units as load. *Strain* describes the amount a material can be stretched and is usually measured in inches or centimeters.

Define these terms: tensile stress, comprehensive stress, shear stress.

Answer: Tensile stress is a material's resistance to stretching. *Compressive stress* is a material's resistance to crushing. *Shear stress* is a material's resistance to cutting across it.

Explain the meaning of these terms: ultimate strength and safe working strength.

Answer: The *ultimate strength* of a material is the point at which it will break. *Safe working strength* is the percentage of ultimate strength that experience has shown to be safe.

Describe what is meant by the limit of elasticity.

Answer: The point where any further load on a metal would cause it to become permanently deformed; that is, if the limit of elasticity is exceeded, a metal would not return to its original shape after the load is removed.

What is the factor of safety?

Answer: The difference between the *safe working strength* and the *ultimate working strength*.

Very often materials used to build boilers are subjected to three different tests. Name them.

Answer: Tensile, homogeneity (uniform structure or composition throughout) and hydrostatic (resistance to pressures exerted by liquids).

What are three purposes of threads in engineering work?

Answer: Feed screws on machine tools—*squares* and *acme*—and threads to apply pressure to force rolls together—buttress.

Would there be any advantage in having one standard and form of thread and one standard for thread pitch?

Answer: Yes, it would be easier to interchange equipment produced by various manufacturers.

Describe Whitworth and American standard threads.

Answer: The *Whitworth* standard is formed at a 55° angle and has one-sixth of the thread depth rounded off at the top and bottom. *American standard* threads are formed at a 60° angle and one-eighth of the thread depth is cut off square with the axis of the thread at the top and bottom.

Holes to be tapped (threaded) are usually drilled a little larger than the diameter of the bolt at the base or root of its thread. Why?

Answer: Tapping is easier and quicker, and broken taps are avoided.

If bolt stress is to be calculated, why is the diameter at the base of the thread always used?

Answer: The bolt is only as strong as its smallest diameter, which is that found at the bottom of the thread.

REFRIGERATION

What is refrigeration?

Answer: The removal of heat, transferring heat and the means of controlling the transfer.

What type of change do most liquids used in refrigeration systems undergo?

Answer: Compression and expansion.

BASIC PRINCIPLES

What is the most basic, elementary refrigerant?
Answer: Ice.

What will happen to a liquid that boils at 0°F. (−19°C.) at atmospheric pressure if it is put into an open container at 0°F. (−19°C.)?
Answer: It will boil as it extracts heat from the surrounding air.

Why does a liquid refrigerant boil or vaporize in the evaporator?
Answer: Because it absorbs heat from the substance to be cooled.

More questions and answers on the subject of refrigeration appear in Chapter 26.

CHAPTER 2

Fuels

Define the term "fuel."
Answer: A fuel is any substance, such as coal, oil, gas, wood, or other product, that will produce heat by burning.

How are fuels broadly classed?
Answer: Solid, liquid, and gaseous. Waste heat is a fourth classification, although it is not a fuel by the preceding definition.

Are there alternate energy sources that produce heat without combining with oxygen?
Answer: Yes, they are solar and nuclear energy.

What other energy sources should be considered?
Answer: Geothermal, and possibly wind energy as an electricity-producing source.

Two different types of analysis can be made of fuel. What are they called and which fuel type is each suited for?

Answer: The ultimate or complete analysis can be made on any fuel. An approximate analysis can be made on solid fuels.

SOLID FUELS
Coal

Explain these terms: volatile matter, fixed carbon, total combustible, and ash.

Answer: In the language of the chemist, *volatile matter* is that part of coal (moisture excepted) which is driven off when a sample is subjected to a temperature up to about 1750°F. (955°C.); the solid carbon is the *fixed carbon*; the sum of volatile matter and fixed carbon is the *total combustible*; the part that does not burn is *ash*.

What are the chemical constituents of coal?

Answer: Carbon, hydrogen, oxygen, nitrogen, and the inorganic matter that constitutes the ash. Sulfur in the free state is sometimes present in coal.

Why is the sulfur content of coal and fuel oil getting more attention?

Answer: During combustion, sulfur produces sulfur dioxide, which pollutes air and has adverse effects on iron and steel. (See Appendix A for more information on air-quality regulations.)

What causes the different heating values of the mining grades of coal?

Answer: The varying quantities of the chemical constituents and their combinations.

Name two general divisions of coal.

Answer: Hard coal, or *anthracite*, and soft coal, or *bituminous*. (Table 2-1 lists the classifications of American coal and Table 2-2 the sizes of anthracite that are popular. Some may not be generally available in some areas.)

Table 2-1. Classification of American Coal

Class	Volatile matter in % of combustible	Oxygen in combustible %	Btu per pound of combustible
Sub-bituminous and lignite	27 to 60	10 to 33	9,600 to 13,250
Bituminous, low grade	32 to 50	7 to 14	12,400 to 14,600
Bituminous, medium grade	32 to 50	6 to 14	13,800 to 15,100
Anthracite	less than 10	1 to 4	14,800 to 15,400
Semi-anthracite	10 to 15	1 to 5	15,400 to 15,500
Bituminous, high grade	30 to 45	5 to 14	14,800 to 15,600
Semibituminous	15 to 30	1 to 6	15,400 to 16,050
Eastern cannel	45 to 60	5 to 8	15,700 to 16,200

Fixed Carbon Content
Anthracite, 92.5 to 97%
Semianthracite, 87.5 to 92.5%
Semibituminous, 75 to 87.5%
Eastern bituminous, 60 to 75%
Western bituminous, 50 to 65%
Lignite, under 50%

Table 2-2. Sizes of Anthracite Coal

Size	Diameter of opening through or over which coal will pass, in inches	
	Through	Over
Broken	$4\frac{1}{2}$	$3\frac{1}{4}$
Egg	$3\frac{1}{4}$	$2\frac{5}{16}$
Stove	$2\frac{5}{16}$	$1\frac{5}{8}$
Chestnut	$1\frac{5}{8}$	$\frac{7}{8}$
Pea	$\frac{7}{8}$	$\frac{9}{16}$
No. 1 Buckwheat	$\frac{9}{16}$	$\frac{5}{16}$
No. 2 Buckwheat	$\frac{5}{16}$	$\frac{3}{16}$
No. 3 Buckwheat	$\frac{3}{16}$	$\frac{3}{32}$
Culm	$\frac{3}{32}$	...

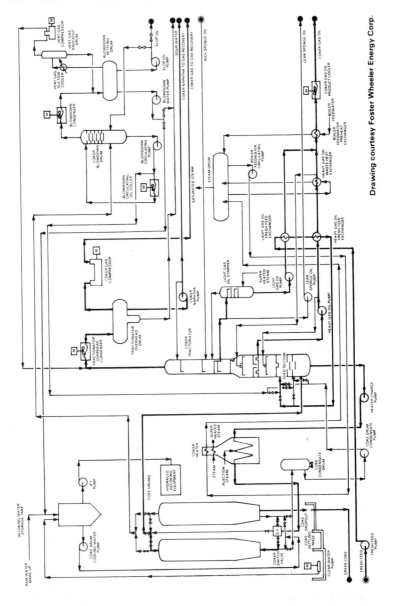

Fig. 2-1. Schematic diagram of a typical Foster Wheeler delayed coking system.

FUELS

What is the heating value of coal?

Answer: Anthracite delivers 14,000 to 15,000 Btu, while bituminous delivers between 12,000 and 15,000 Btu (see Table 2-1). Table 2-3 is comparison of the amounts of coal and oil needed to evaporate water.

What is coke?

Answer: The solid substance remaining after the partial burning of coal (the most volatile gases) in an oven or after distillation in a retort, a heating vessel designed for distilling coke.

How is gas-retort coke produced?

Answer: By the application of high temperature to the outside of the retort for a short time.

What are its principal uses?

Answer: For domestic firing and sometimes in firing steam boilers.

What is the heating value of coke?

Answer: Roughly, about 14,000 Btu.

What is delayed coke?

Answer: A solid hydrocarbon fuel produced from residues of petroleum after lighter fuels have been removed. See Fig. 2-1.

Why is lump size important in choosing coal?

Answer: In general, the smaller the size, the greater the amount of impurities present. Also, the heat value is lower, more coal sifts through the grate, and other objectionable results are increased.

What is a consequence of this?

Answer: The larger sizes usually command higher prices, especially with anthracite.

How is coal graded as to size?

Answer: By screening through standard openings, but these openings differ in size and shape in different localities and do not

35

relate to pulverized coal blown into a furnace by an air blast. See Table 2-2.

What is pulverized coal?

Answer: Any type of coal crushed or pulverized to a fine powder.

What are the advantages of pulverized coal?

Answer: Any grade of coal can be burned, according to the fuel's supporters, and burned more efficiently. Obviously, firing rates are easier to control, very much like oil or gas. Fig. 2-2 shows a modern coal pulverizer.

How is pulverized coal burned?

Answer: It is blown into the boiler, usually, and burned in suspension.

Table 2-3. Comparative Evaporation of Coal and Oil

(1) Coal	(2) Water Evaporated	(3) Petroleum Equivalent
Pittsburgh lump and nut	10.0	4.0
Pittsburgh nut and slack	8.0	3.2
Anthracite, Pennsylvania	9.8	3.9
Indiana block	9.5	3.8
Georges Creek lump, Maryland	10.0	4.0
New River, West Virginia	9.7	3.8
Pocahontas lump, West Virginia	10.5	4.2
Cardiff lump, Wales	10.0	4.0
Cape Breton, Canada	9.2	3.7
Nanaimo, British Columbia	7.3	2.9
Co-operative, British Columbia	8.9	3.6
Greta, Washington	7.6	3.0
Carbon Hill, Washington	7.6	3.0

Source: U.S. Geological Report on Petroleum

Column One shows the type of coal under consideration. Column Two shows the number of pounds of water evaporated at 212°F. per pound of this coal. Column Three shows the number of barrels of petroleum at 18° to 40° Baume required to evaporate the same amount of water as will one ton of coal from each location.

FUELS

Fig. 2-2. This coal pulverizer renders coal to a powder-like consistency and delivers it directly to a furnace where it is burned in suspension (not on a grate).

Wood

What is the heating value of wood?

Answer: In dry wood, roughly 7800 Btu; in ordinary firewood, about 5800 Btu. Sawdust produces the same heat value as the wood from which it was made.

What is the relative heating value of wood as compared with coal?

Answer: The heating value of thoroughly dried wood is about 40% that of coal. The approximate composition of dry wood is: carbon, 49%; oxygen, 44%; hydrogen, 6%; and ash, 1%.

As a fuel, what do we mean here when we speak of "wood"?

Answer: The term designates the limbs and trunks of trees as they are felled.

What is the effect of moisture in wood?

Answer: It causes a loss of economy, although wood with a 20% to 35% moisture content can be burned satisfactorily in a waste-fuel boiler. In fact; material having in excess of 35% moisture can be burned in these devices under special grating conditions.

What problems are to be encountered in wood-burning units?

Answer: Creosote buildup in flues and stack.

Why is this a problem?

Answer: The creosote will eventually catch fire, causing dangerous stack heat.

What causes creosote formation?

Answer: Incomplete combustion during low firing periods, which allows the distilled oils and pitch to condense on the stack walls.

Is there a solution?

Answer: Yes. Use only woods that are sufficiently dried; use low-pitch wood, and adjust the air supply so that efficient combustion takes place.

Waste Fuel

What is a waste-fuel boiler?

Answer: A boiler designed to use combustible materials that were formerly destroyed or discarded.

What specific materials are these?

Answer: Wood scraps, such as byproducts of furniture manufacturing, sawdust, cotton hulls, bagasse (dry refuse of sugar cane after juice extraction), municipal waste, and similar combustibles.

Is there an optimum size for these waste fuels?

Answer: Generally, they should be in cubes measuring 2 inches or less.

Why would these materials have to be so small?

Answer: For the sake of efficient combustion, conveyance, and storage. Also, this makes it possible to meter the fuel rate.

How would waste wood be reduced to the proper size?

Answer: The process is called "hogging" and is a mechanical operation.

How are the materials delivered to the firebox?

Answer: Either by gravity feed, by a low-pressure air system through pipes, or by stoker-type devices.

Can material with a relatively high moisture content be used in a waste-fuel boiler?

Answer: Wood with a 20% to 35% moisture content can be burned satisfactorily, as can material with an excess of 35%; but the latter must be burned under special grating conditions. The boiler shown in Fig. 2-3 is a wood-burning unit.

LIQUID FUELS

In general, what is the composition of crude oil?

Answer: It consists of carbon and hydrogen, although it also

POWER PLANT ENGINEERS GUIDE

Courtesy York-Shipley, Inc.

Fig. 2-3. Fluidized bed boiler designed to burn virtually any combustible product.

contains varying quantities of moisture, sulfur, nitrogen, arsenic, phosphorus, and silt.

What is the percentage of moisture?
Answer: It varies from 1% to more than 30%.

What is the heating value of petroleum?
Answer: Usually in the range 18,000 to 22,000 Btu.

What are the relative values of oil and coal as fuel?
Answer: Under favorable conditions, 1 pound of oil will evaporate from 14 to 16 pounds of water at 212°F. (100°C.). One pound of coal will evaporate from 7 to 10 pounds of water at 212°F. See Table 2-3 for more detailed information.

Which type of fuel oil is most commonly used in power plants?

Answer: Residue oil left after lighter products have been distilled off during refining.

Why would residue oil have a more uniform and higher Btu value per pound?

Answer: Because the lighter oils have been removed.

What does the term "flashpoint" mean?

Answer: The flashpoint of oil is the temperature where enough inflammable vapor is given off to flash (ignite) when brought into contact with a flame. Fuel oil has to be heated to the flashpoint.

GASEOUS FUELS

What kinds of gaseous fuels are used in steam boilers?

Answer: Natural gas, waste gas from blast furnaces, coke-oven gas, and producer gas. It is possible to use methane or similar gases produced from biomass products and wastes. See Fig. 2-4.

How do gas fuels compare with liquid fuels?

Answer: Gas fuels offer all the advantages of liquid fuels and very few of the disadvantages.

What is "natural" gas?

Answer: A fuel obtained by drilling into subterranean gas domes. It is processed by dehydration and, with some additives, is piped under pressure for domestic and commercial use. Natural gas contains methane and ethane (basically carbon and hydrogen compounds).

POWER PLANT ENGINEERS GUIDE

What is the heating value of natural gas?

Answer: It varies from 800 to 1100 Btu per cubic foot.

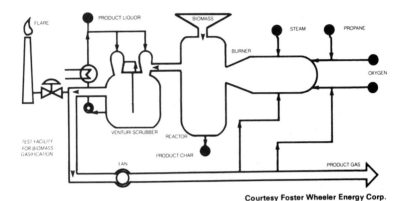

Courtesy Foster Wheeler Energy Corp.

Fig. 2-4. Developed by Foster Wheeler Power Products Ltd., the United Kingdom's Department of Energy and the Commission on European Communities, this system is being used to investigate practical modes for gasifying biomass before the gas is converted to fuel methanol.

What is the relative value of natural gas as compared to coal?

Answer: One thousand cubic feet of natural gas is approximately equivalent to 57.25 pounds of coal.

What are the products of gas combustion?

Answer: Carbon dioxide, water, oxygen, nitrogen, carbon monoxide, hydrocarbons, nitric oxides, and sulfur dioxide.

Which of these are considered pollutant emissions and are therefore regulated in many communities?

Answer: Carbon monoxide, hydrocarbons, nitric oxides, and sulfur dioxide. See Appendix A for more information on environmental regulations.

How is this pollutant data expressed?

Answer: In parts per million, or ppm.

FUELS

How does gas compare with light oil and heavy oil in producing smoke and particulates?

Answer: Assuming proper burner setup in each case, gas will produce no smoke and no particulates; light oil produces maximum smoke No. 2 on the Bacharach scale with negligible particulates; heavy oil produces maximum smoke No. 4 on the Bacharach scale, with particulates being primarily a function of the oil composition. (See Chapter 4 for an explanation of the Bacharach scale.)

Are there any alternatives to natural gas as a fuel?

Answer: Yes, hydrogen is receiving attention because natural gas reserves seem to be limited.

What is the source of hydrogen?

Answer: It is made by water decomposition, using electrical energy to separate the oxygen and hydrogen.

Are there any pollutants formed in its manufacture?

Answer: No, the oxygen can be vented to the atmosphere or compressed for commercial uses.

Are there any pollutants formed in its use?

Answer: Hydrogen burns to produce only water, so the usual products of combustion such as CO, hydrocarbons, sulfur, and particulates are not formed.

Does hydrogen have the same heating value?

Answer: Compared to natural gas, it produces about one-third the value under equal conditions.

Does hydrogen have any special hazards?

Answer: Its flammability limits are much greater than methane, and because of its low viscosity, it will escape more rapidly

43

through a given leak. It is colorless and odorless, so it would have to be supplemented with odorizers to make it suitable for household use.

WASTE HEAT

What is a waste-heat-recovery boiler?
Answer: A boiler designed to recover heat that would otherwise be wasted to the atmosphere.

What sources of heat does such a boiler have?
Answer: Heat is recovered from fume or solid-waste incinerators, industrial furnaces, drying ovens, and internal-combustion engines.

What heat range is necessary to make this boiler function?
Answer: Exhaust gases in the 600-2000°F. range.

What is the tube style?
Answer: Generally, fire-tube construction.

Does a waste-heat-recovery boiler need refractories?
Answer: Yes, the same as with direct-heat units.

Can these units supply steam as well as hot water?
Answer: Yes. Their application depends on the heat source.

Do these boilers need unique control systems?
Answer: No, the same controls and safety devices found on a direct-fired boiler are found on this unit.

NUCLEAR ENERGY

How long have scientists been working to harness nuclear energy safely?
Answer: About 50 years.

FUELS

What is fissioning?

Answer: The splitting apart of atoms. Neutrons were first used to split uranium atoms into lighter elements of barium and krypton, releasing energy in the process.

Does this process have a practical application?

Answer: Yes. A controlled, nuclear chain reaction produces heat energy, converting water into steam to turn electric power generators.

What is radioactivity?

Answer: The spontaneous breakup of atoms releases three types of rays: alpha, beta, and gamma. The alpha ray has very slight penetrating power; the beta ray can penetrate thin sheets of metal; gamma rays will penetrate relatively thick layers of metal.

Where does the heat come from?

Answer: Radioactive rays. Measurements show that if one ounce of radium were to release all of the rays it is capable of releasing, the energy produced would be equivalent to 10 tons of coal.

What is a nuclear reactor?

Answer: A type of housing or device that will contain a controlled nuclear chain reaction. A commercial reactor has a means for transferring the heat generated during fission.

What is a chain reaction?

Answer: There are two types: controlled and explosive. Explosion occurs when the fission of one atom effects fission in two, then four, etc., in rapid sequence. A controlled chain reaction is achieved by limiting the number of atoms that are allowed to fission or split.

What element is used in contemporary nuclear reactors?

Answer: Nuclear fuel for generating electricity is made from uranium.

45

Power Plant Engineers Guide

What part of the atom produces the energy?
Answer: U-235 is the fissionable isotope in uranium that produces most of the energy.

In what form is the nuclear fuel prepared?
Answer: Usually, the uranium is compressed into pellets.

How large are the pellets?
Answer: Less than one-half inch in diameter. However, each pellet contains the equivalent energy of approximately 100 gallons of oil.

How are the pellets introduced as a fuel?
Answer: They are stacked in metal tubes, 200 of which make up the core of the reactor.

Is the nuclear reactor a feasible means of producing electricity?
Answer: Electrical production from one reactor can supply the average needs of 750,000 homes.

How does this compare with fossil-fuel usage?
Answer: It is equal to 10 million barrels of oil or 3 million tons of coal.

What is the future of nuclear energy?
Answer: It has been predicted that nuclear fuels could provide about 33% of all U.S. electrical demand by 1990.

How is steam produced in a nuclear reactor?
Answer: By transferring the heat produced by fission through a heat exchanger. In some cases, water is sprayed into the chamber where a nuclear reaction is producing heat and is immediately converted to steam. It passes to a drying area and then to the turbine generators. See Fig. 2-5.

How is heat controlled?
Answer: Control rods are moved in and out of the reaction zone

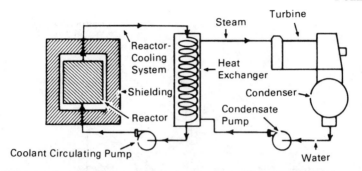

Fig. 2-5. Simple diagram of a nuclear power plant. The heat generated in the reactor is released from the heat exchanger to produce steam to drive a turbine.

containing the fuel, thus increasing or slowing the rate of fission and the consequent volume of heat. See Fig. 2-6.

Why doesn't the unit explode spontaneously like the atomic bomb?

Answer: The interaction of atoms in a nuclear reactor is under strict control, thus prohibiting a "runaway" or uncontrolled fission.

How does the heat-producing capability of nuclear energy compare with the heat generated by coal?

Answer: It has been estimated that 1 pound of U-235 can produce the same amount of heat as 1400 pounds of coal.

What type of shielding surrounds a reactor?

Answer: Usually 5 to 8 feet (1.52 to 20.32 m) of solid concrete, necessary to contain gamma rays.

How are nuclear reactors classified?

Answer: Power, breeder, research. The power type is the only type used commercially in the U.S. Reactors are also classified by fast, intermediate, and slow or thermal. Power plant reactors are usually the slow type.

Is a special license needed to operate a nuclear reactor?

Answer: Yes. Nuclear plant operators are licensed by the U.S.

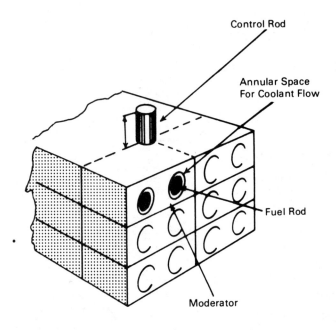

Fig. 2-6. Very basic concept of a nuclear reactor core showing the general relationship of the fuel rods, control rods, space for coolant flow, and moderator material (usually carbon).

Nuclear Regulatory Commission. See Appendix B for a summary of the NRC licensing requirements.

GEOTHERMAL ENERGY

What is geothermal energy?
Answer: Heat that comes from the earth.

How can this natural resource be harnessed?
Answer: Hot-water springs or geysers can be capped, and the resulting heat-energy applied to practical usage.

What about geographical areas that do not have hot springs or geysers?

Answer: Scientists are drilling to "hot rocks," some of which can be found at less than 9000 feet below the surface of the earth.

Could a volcano be a source of heat energy?

Answer: Yes. The temperatures in a volcano are hot enough to melt rocks, so experiments aimed at utilizing this natural heat source are currently being conducted in the state of Hawaii.

What are the possible applications of geothermal energy?

Answer: Low-cost electricity, energy for heating, and energy to operate air conditioning could be available for industrial and commercial processes as well as for home usage.

Why are these alternate sources under serious consideration?

Answer: As fossil-fuel supplies are depleted, new applications of known sources need to be developed for man's survival.

SOLAR ENERGY

What is the fourth form of matter?

Answer: It is called *plasma*, a very hot, glowing gas existing in the solar system. Interstellar matter and stars are plasmas, and 99% of all matter in the universe is believed to be in this form.

What are the rays of the sun called?

Answer: Low-energy cosmic rays.

Are there practical uses for this form of energy?

Answer: The first large-scale use is now directed toward home heating and cooling.

Why so?

Answer: There is a concern for the supply of fossil fuel because 25% of the current fuel consumption is for home heating and cooling.

What principle is involved with using the sun's energy?

Answer: It is based on the absorption of solar radiation.

Does any form of combustion take place?
Answer: No.

How can the energy be used?
Answer: Some collector units are covered with a glass that allows solar radiation to pass through but is opaque to reflected or emitted heat from the liquid-carrying tubing inside the collector. In effect, the glass acts like a one-way valve.

What takes place after the collector unit is heated?
Answer: The fluid is pumped into the storage system on a continuous basis. Transfer units then pick up the heat and send it through the heating circuit.

What is one disadvantage of the solar hot-water system?
Answer: A very large storage tank is necessary because as the sun stops supplying heat in the afternoon, a large reserve of heat is necessary at night to carry over. The possibility of the sun not shining for several days also adds to this disadvantage and creates the need for a back-up, or supplemental, source of heat.

What are the chief advantages of solar energy?
Answer: The "fuel"—in this case, the sun's rays—is free, and scientists estimate the sun will last approximately five billion years before burning out. Second, because there is no combustion, there are no polluting wastes emitted.

If there are several advantages to using solar energy, why aren't more such systems in use?
Answer: Widespread adoption has been limited because of the lack of well-engineered, reasonably priced systems.

Is hot water the only useful form of solar energy available?
Answer: No. Electricity can be generated from the sun's rays by solar or photovoltaic cells.

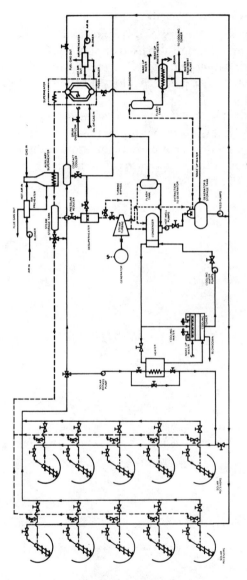

Fig. 2-7. Schematic of a solar-fossil fuel 5 MWe power plant designed for the town of Crosbyton, Texas, with the assistance of Foster Wheeler Energy Applications.

POWER PLANT ENGINEERS GUIDE

Where are they being used?

Answer: Nearly every NASA spacecraft has been fitted with solar cells, and they are also used on some navigational buoys and automatic weather stations. Commercial supplementary use is beginning to increase as technology advances. See Fig. 2-7.

What disadvantages do the electrical systems have?

Answer: Nightfall again requires the use of stored, supplemental energy, and the cost of manufacture makes them too expensive for industrial, commercial, or residential use.

CHAPTER 3

Heat

What is heat?
Answer: A form of energy.

What is the unit of heat?
Answer: The British thermal unit (Btu).

THE EFFECT OF HEAT

Heat Measurement

A substance is said to be *hot* or *cold* according to its physical or *sensible* effect when touched.

Define temperature.
Answer: The thermal state of a body with respect to its ability to pick up heat from or pass heat to another body.

When is a body at a higher temperature than another body?

Answer: When its molecules move faster than those of the other body.

How is temperature measured?

Answer: Ordinary temperatures by a *thermometer*. Very high temperatures in a combustion chamber are measured by a *pyrometer*.

What is the basic principle of thermometers?

Answer: Expansion and contraction of substances due to the effect of heat.

What are the two fixed points on the thermometer scale?

Answer: The freezing point and the boiling point of water at sea level atmospheric pressure.

How is the scale graduated?

Answer: By graduating the distance between the two fixed points into the proper numbers of degrees corresponding to the particular scale used.

What thermometer scales are in general use?

Answer: Most common are the Fahrenheit and Celsius scales (the latter was formerly known as "centigrade," a term now obsolete). Water boils at 212° and freezes at 32° on the Fahrenheit scale; each degree F. is 1/180 of the difference between these two fixed points, and zero is located 32 degrees below the freezing point of water. On the Celsius scale, water boils at 100° and freezes at 0°, with each degree being 1/100 of the difference between freezing and boiling. Other scales include Reaumur, Kelvin, and Rankine, but these are in limited or specialized use.

How do you convert from Fahrenheit degrees to Celsius degrees and vice-versa?

Answer: Subtract 32 from the Fahrenheit reading and take 5/9 of the remainder. The resulting figure will be the equivalent Celsius temperature. To convert Celsius to Fahrenheit, take 9/5 of the

HEAT

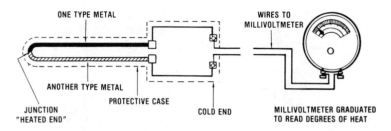

Fig. 3-1. Elementary thermocouple thermometer used for measuring high temperatures.

Celsius reading and add 32. The resulting figure will be the equivalent Fahrenheit reading. One degree Fahrenheit is equal to 5/9 of one degree Celsius; one degree Celsius equals 9/5 of one degree Fahrenheit (both are approximate).

What is a pyrometer?

Answer: An instrument usually used for measuring temperatures beyond the upper limit of a mercury thermometer. Their heat-sensing units can operate by means of several different principles, among which are the contraction of clay by heat, the curling of a bimetallic strip, changes of electrical resistance according to heat, sensing the amount of light given off by a heated body, and generation of an electric current by two dissimilar metals when their junction is heated. The latter method is used in a *thermocouple,* which generates an amount of electricity in proportion to the amount of heat applied; this current is brought to a galvanometer or a voltmeter calibrated in degrees of temperature (Fig. 3-1). These principles can also be used to construct thermometers operating within conventional temperature ranges whenever the use of a mercury instrument such as that of Fig. 3-2 or another type would be inconvenient or impossible.

Can the temperature of steel materials be judged by eye, according to their color?

Answer: Approximately, yes, but usually only by the experienced eye. See Table 3-1. Optical pyrometers are used to measure temperature according to the color of the furnace brickwork.

55

POWER PLANT ENGINEERS GUIDE

What effect does heat have on the molecules in a material?

Answer: An increase in heat increases the rate of molecular vibration, gradually weakening the force of cohesion holding

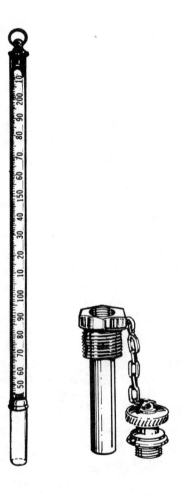

Fig. 3-2. Tagliabue mercury light well temporary thermometer connection and type of thermometer used with it. The mercury well is designed for use with a solid glass thermometer for test work or for application where only an occasional reading is required. There is a seating plug provided with a gasket for confining the mercury.

HEAT

molecules together. When enough heat is applied to break down the force of cohesion, the material turns to a liquid. If more heat is added, molecular movement becomes so violent that molecules break away from the surface (evaporate) of the liquid into the surrounding space. Adding enough heat to a solid, therefore, can turn the entire quantity of material into a vapor or gas and fill the space that confines it.

Table 3-1. Steel Temperature by Color

Color	°C.	°F.	Color	°C.	°F.
Incipient red	525	977	Deep orange	1100	2021
Dull red	700	1292	Clear orange	1200	2192
Incipient cherry red	800	1472	White	1300	2372
Cherry red	900	1652	Bright white	1400	2552
Clear cherry red	1000	1832	Dazzling white	1500+	2732+

Source: From a table by Pouillet.

Expansion of Metal by Heat

What is linear expansion?

Answer: For example, the amount by which a metal rod or pipe increases in length when subjected to a rise in temperature. The amount of expansion varies with the amount of heat and the type of metal. When the temperature of the metal in a pipeline is increased by steam, adequate provision must be made for the resulting expansion. If members of a structure are rigidly fixed, expansion will cause compression and contraction which eventually damage the metal and joints.

Define the coefficient of linear expansion.

Answer: It is the ratio of the increase in length produced by a rise of temperature of 1° to the original length (Fig. 3-3).

What provision must be made in boilers because of the expansion of the metal due to the heat?

Answer: In setting horizontal shell boilers by the old method, one end is supported on rollers. Old methods are shown in Figs. 3-4 and 3-5. A better method is shown in Fig. 3-6.

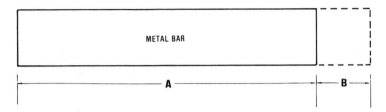

Fig. 3-3. The coefficient of expansion is the ratio of B ÷ L.

What provision is made in water-tube boilers?

Answer: The tubes are arranged so that they are free to expand and contract.

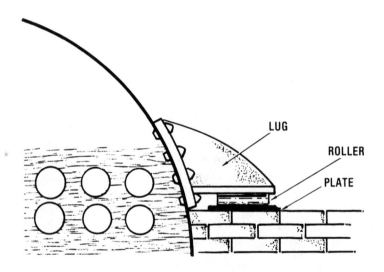

Fig. 3-4. The usual but objectionable method of providing for expansion on lug-supported boilers. Note that no provision is made for transverse expansion of the shell; consequently, the side walls must "breathe" with the boiler, which tends to produce air leaks by cracking and loosening the mortar on brick settings.

Give an advantage of expansion and contraction by heat.

Answer: Boiler plates are fastened by red-hot rivets. When the rivets cool, they contract and bind the plates together with great force.

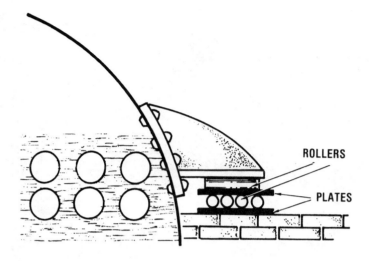

Fig. 3-5. Approved method of providing for expansion on lug-supported boilers. The two sets of rollers placed at right angles to each other form a universal joint, allowing free movement both endwise or crosswise.

Do liquids expand when heated and contract when cooled?

Answer: Yes, with the exception of water. Water contracts as it cools until it reaches about 39°F. (4°C.), then begins to expand. As it freezes, further expansion occurs.

Do liquids usually expand at a greater rate than solids?

Answer: Yes, and alcohol expands at a greater rate than water.

At what rate do gases expand?

Answer: Gases expand much more quickly and nearly all gases expand by the same amount when heated, depending on pressure, temperature, and volume.

Give a disadvantage of expansion and contraction by heat.

Answer: A short space must be left at railroad rail-joints to permit this expansion and contraction.

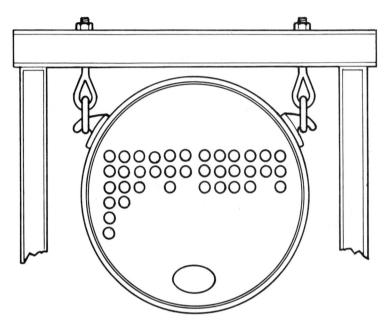

Fig. 3-6. Method of providing for expansion by link suspension. As may be seen, the weight of the boiler is carried by the steel work, thus relieving the brickwork of this duty. The side columns may be either of solid channel iron or built up from angles and lattice work, and channel bars are carried across the top of these columns as shown. The boiler is suspended from these channels by suspension links or rods arranged with nuts and washers, permitting easy leveling and adjustment of the height of the boiler.

Describe radiation, conduction, and convection in boiler operation.

Answer: Heat from the burning fuel passes to the metal of the heating surface by *radiation*; through the metal by *conduction*; and is transferred to the water by *convection* (circulation). See Fig. 3-7.

What is conductivity?

Answer: The relative value of a material, as compared with a standard, in affording a passage through itself or over its surface for heat.

HEAT

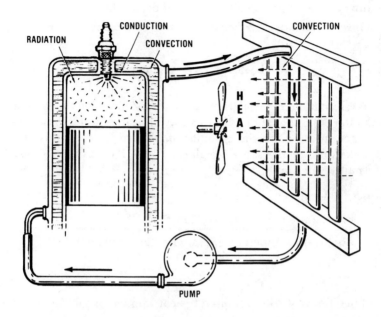

Fig. 3-7. In this internal-combustion engine cylinder, heat from the burning fuel-charge transfers to the metal cylinder wall by radiation, passes through the wall by conduction, and into the surrounding water by convection. Air drawn through the radiator at right cools the water (also by convection), which then returns to the cylinder water passages to begin a new cycle. The term "water-cooled engine" is technically a misnomer, since it is air that performs the ultimate cooling; the water serves only as a transfer medium.

What is a very bad conductor called?

Answer: An insulator.

Name a good conductor.

Answer: Copper. Any substance that is a good conductor of electricity is a good conductor of heat.

What does heat transfer per square foot of heating surface mean?

Answer: The amount of furnace-produced heat that is transferred to the water side of the boiler. It is measured in Btu per square foot per hour.

How does temperature affect thermal conductivity?
Answer: In most substances it varies.

SPECIFIC HEAT

By definition, the specific heat of a substance (see the table in Chapter 30) is the ratio of the quantity of heat required to raise its temperature 1°F. to the quantity of heat required to raise the temperature of the same weight of water 1°F. Expressed as a formula, it becomes:

Specific Heat =

$$\frac{\text{Btu to raise temperature of substance 1°F.}}{\text{Btu to raise temperature of same weight of water 1°F.}}$$

or

Specific Heat = Btu required to heat 1 pound of substance 1°F.

For example, the same quantity of heat that will raise 1 pound of water 1°F. will raise about 7.6 pounds of cast iron 1°F. Accordingly, if the specific heat of water is taken as 1, that of cast iron would be 1/7.6, or only 0.131. The standard is usually water raised from 62°F. to 63°F.

EXPERIMENTS WITH HEAT

Experiment 1

To check the accuracy of an existing thermometer or to determine the freezing point on a blank thermometer, arrange the apparatus as shown in Fig. 3-8. Wash some ice, break it small, and pack it around the bulb of the thermometer in a glass or metal funnel. Place a vessel beneath the funnel to receive the water that forms as the ice melts. The ice should be heaped up around the thermometer tube until only the top of the column is visible, and the thermometer is left this way for 15 minutes. At the end of that

HEAT

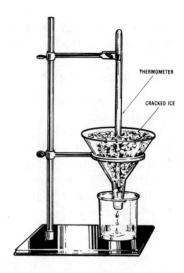

Fig. 3-8. Apparatus for determining the freezing point on a thermometer.

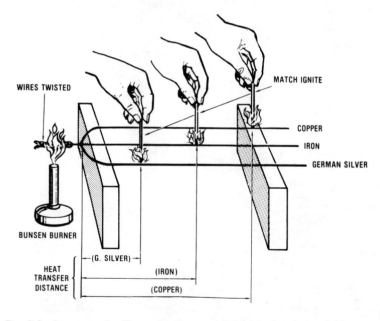

Fig. 3-9. Apparatus for illustrating heat conductivity of various metals.

period, a calibrated thermometer should indicate the freezing point according to the scale used. A blank thermometer can be marked with a fine file, opposite the top of the mercury column.

Experiment 2

To show the properties for heat-transfer possessed by different metals, secure identical lengths of copper, iron, and german (nickel) silver wire and twist their ends together to form a pigtail. Separate the strands and mount on insulating blocks as shown in Fig. 3-9. Place a lighted bunsen burner or alcohol lamp under the pigtail and, after a short wait, slide the tip of a kitchen match along one wire, starting at the end opposite the heat source. Note and mark the point at which the match ignites as a result of heat being conducted by the wire. Repeat for each wire and note the spread of distances involved.

CHAPTER 4

Combustion

What is combustion?

Answer: Rapid oxidation of a combustible material to produce heat.

What is oxidation?

Answer: The act of combining with oxygen, or subjecting a material to the action of oxygen or of an oxidizing agent.

What is this oxygen called?

Answer: The supporter of combustion. Without oxygen, combustion is impossible.

Where is it obtained?

Answer: From the air.

How much oxygen is contained in the air?
Answer: About 20.91% by volume; 23.15% by weight.

What is combustion rate?
Answer: The rate heat is released (in Btu per hour) per cubic foot of combustion space.

COMBUSTIBLE MATERIAL

What is material called which is capable of combustion?
Answer: A combustible. In steam engineering practice the term "combustible" is applied to that portion of the material which is dry and free from ash.

What is a fuel?
Answer: Any material that produces heat by combustion, such as coal, coke, or oil.

Give another definition for combustible.
Answer: It is that portion of the fuel which burns.

What are the principle combustibles in coal and other fuels?
Answer: Carbon, hydrogen, and sulfur.

What is sulfur?
Answer: An elementary mineral substance, yellow in color, brittle, insoluble in water, easily fusible, and flammable. It burns with a blue flame and gives off a peculiar, suffocating odor.

Why is the presence of sulfur in fuel objectionable?
Answer: It has a tendency to aid in the formation of clinkers, and the gases from its combustion, when in the presence of moisture, may cause corrosion.

What is carbon?
Answer: A combustible element, nonmetallic in nature, which

is present in most organic compounds. As a combustible, it forms the base of lampblack and charcoal and enters largely into mineral coals.

How much carbon is contained in bituminous coal?
Answer: It is about 50% carbon.

How is carbon obtained from wood?
Answer: It is separated from wood in the form of charcoal by distilling off the more volatile elements.

What is hydrogen?
Answer: A colorless, odorless, tasteless gas, the lightest element known. This element is combustible, burning with an almost invisible flame. It is a nonsupporter of combustion.

What are the steps in the combustion of coal?
Answer: Coal absorbs heat, which distills off volatile components. Those that are combustible will burn, while others are heated and go up the chimney. The fixed carbon left on the grate is burned, leaving only noncombustible ash.

What conditions are necessary in the furnace to burn coal efficiently and economically?
Answer: Furnace heat must be kept at 1800°F. (982°C.) to burn carbon. Volatile gases burn at 220°F. (104°C.). To support the combustion, enough air must be drawn or forced in to supply oxygen; the air supply must be controlled so oxygen is supplied to all parts of the firebed.

What conditions are necessary for efficient combustion of oil?
Answer: The temperature of the oil must be at the fire point of the fuel and must be properly atomized or broken up into particles surrounded by air.

What do combustion controls regulate?
Answer: (1) Fuel supply; (2) air supply, primary and secondary; (3) mixture of fuel and air.

COMBUSTION PRODUCTS

What is the ignition or kindling point?

Answer: The temperature at which a combustible will unite with oxygen and allow combustion to take place.

Describe what happens during combustion.

Answer: The two principal elements of coal, carbon and hydrogen, have an affinity for oxygen. When they unite, chemical heat is produced. The oxygen, having the stronger affinity for hydrogen, unites with it first and sets the carbon free. A multiplicity of solid particles of carbon are scattered in the midst of burning hydrogen and raised to a state of incandescence. The carbon in due time unites with the oxygen, forming carbon dioxide or carbon monoxide.

What happens to the hydrogen during combustion?

Answer: The hydrogen unites with the oxygen in the proportion of two atoms of hydrogen to one atom of oxygen, forming water (H_2O).

Mention an important feature in the process of combustion.

Answer: Chemical compounds are formed by the combination of carbon and hydrogen.

What are these compounds called?

Answer: Hydrocarbons.

What are the most important hydrocarbons?

Answer: Methane or marsh gas, ethylene gas, acetylene, and penzole.

What are the products of complete combustion?

Answer: Carbon dioxide (CO_2), and water (H_2O).

What other names are given to carbon dioxide and carbon monoxide?

Answer: Carbonic acid and carbonic oxide, respectively.

COMBUSTION

When is combustion complete?

Answer: When the combustible unites with the greatest possible amount of oxygen, as when one atom of carbon unites with two atoms of oxygen to form carbon dioxide—CO_2.

When is combustion incomplete and why?

Answer: When the combustible does not unite with the maximum amount of oxygen, as when one atom of carbon unites with one atom of oxygen to form carbon monoxide (CO). The reason is that the carbon monoxide (CO) may be further burned to form carbon dioxide (CO_2).

What causes incomplete combustion?

Answer: Insufficient supply of air.

DRAFT

What happens when too little air is admitted to the fire?

Answer: There will not be enough oxygen present to supply two atoms of oxygen to each atom of carbon liberated. Hence, carbon monoxide will be formed, which has a heating value of only 4450 Btu, instead of carbon dioxide, which has a heating value of 14,500 Btu.

What is primary air? secondary air?

Answer: Primary air is the air introduced with fuel at the burners. Secondary air is added, by natural draft or forced, to supplement primary air.

How much air does it take to burn a pound (0.45 kg) of coal?

Answer: About 12 pounds (5.4 kg), theoretically. In practice, the amount needed is almost double the theoretical.

What is natural draft?

Answer: Convection flow of air through the furnace and up the stack, taking with it unburned combustion products. The heat of the furnace causes air to rise and maintains the draft (see Fig. 4-1). Control over natural draft is limited.

POWER PLANT ENGINEERS GUIDE

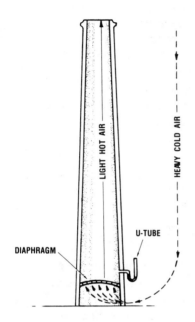

Fig. 4-1. Natural chimney draft. The U-tube (out of proportion here) measures the amount of draft. See Fig. 4-3.

What is forced draft?

Answer: Draft caused by a blower or fan, either *forced* into and through the firebox by a blower or *drawn* through by a stack-mounted blower. In some cases, a combination of the two methods is used (Fig. 4-2). Much more control is possible with mechanical draft systems. A good chimney is needed to carry away gases from a forced draft, but induced draft will compensate for a poor-drawing chimney, since the gases are blown into the chimney.

How is draft measured?

Answer: Draft gauge or manometer, a U-shaped glass tube partially filled with water. The level of the water in the tube indicates the amount of draft. A flexible hose is used to connect the tube to the point in the stack system where draft is to be measured.

Draft pulls air out of the tube, which allows the water in the tube to move toward the stack. The drop in water level in the calibrated side of the tube indicates draft pressure. See Fig. 4-3.

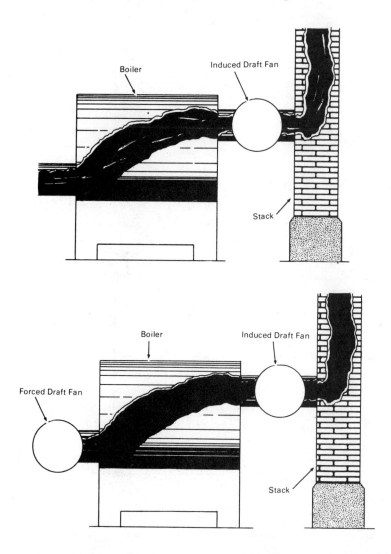

Fig. 4-2. Locations of forced and induced draft blowers.

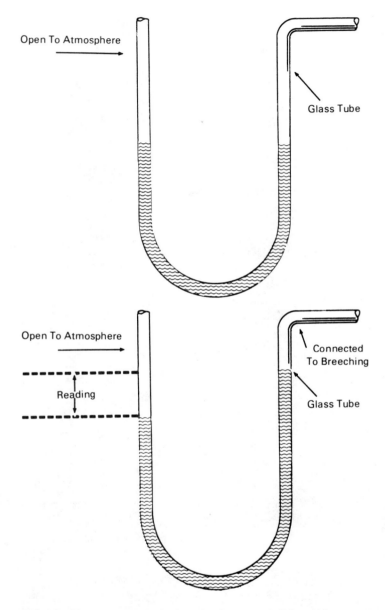

Fig. 4-3. Manometer, attached to the furnace stack breeching, measures draft.

COMBUSTION

How is draft controlled?

Answer: It varies with the furnace. The manufacturer's instructions should be followed.

What results when too much air is supplied?

Answer: Since carbon cannot combine with oxygen in any greater ratio than two atoms of oxygen to one atom of carbon, any excess air supply simply dilutes the gases and cools the furnace. As a result fuel is wasted, usually, in the formation of smoke.

Are steam boilers usually operated with too much air supply?

Answer: Yes, an excess supply as large as 150% is not uncommon, too much draft being employed as a rule.

What is the effect of heating the air supply?

Answer: It increases the rate of combustion.

What are the objectionable effects of the nitrogen contained in the air supply?

Answer: In passing through the furnace without change, it dilutes the air, absorbs heat, reduces the temperature of the product for combustion, and is the chief cause of heat loss in furnaces.

What is the useful effect of nitrogen?

Answer: It prevents too rapid combustion. Without the large proportion of nitrogen in the atmosphere, the latter would be so rich in oxygen that the resulting high rate of combustion would burn out the grates.

Is it possible in practice to obtain perfect combustion with the theoretical amount of air?

Answer: No. An excess is required, amounting to sometimes double the theoretical supply, depending upon the nature of the fuel to be burned and the method of burning it. The reason for this is that it is impossible to bring each particle of oxygen in the air into intimate contact with the particles in the fuel that are to be oxidized, due not only to the dilution of the oxygen in the air by nitrogen, but because of such factors as the irregular thickness of the fire and the varying resistance to the passage of the air through the fire in separate parts on account of ash and clinker.

Is as large an excess of air required for oil as for coal?

Answer: No.

SMOKE AND FLUE GASES

What is flame?

Answer: Visible flame is a combustible gas heated to an intense heat (Fig. 4-4). The form of the candle flame is common to all

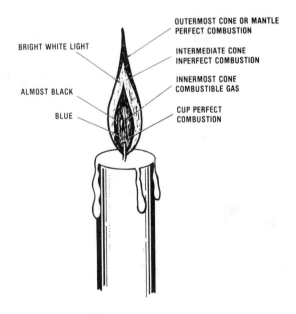

Fig. 4-4. Elements of a candle flame.

flames which consist of gas issuing from a small, circular jet, like the wick of a candle. The gas issues from the jet in the form of a cylinder which immediately becomes a diverging cone by diffus-

Combustion

ing into the surrounding air. When the cone is kindled, its margin (where interruption with the surrounding air is nearly complete) will be perfectly burned; however, the gases in the interior of the diverging cone cannot burn until they have ascended sufficiently to meet with fresh air. Since these unburned gases are continually diminishing in quantity, the successive circles of combustion must diminish in diameter, thus resulting in the typical conical shape.

What is the product of the perfect combustion of carbon?

Answer: Carbon dioxide. The product of perfect combustion of hydrogen is invisible water vapor, or steam.

How is the state of the combustion in a furnace determined and why?

Answer: It is determined accurately by finding the amount of carbon dioxide in the flue gases, because carbon is the principal constituent of the fuel.

What happens when fresh coal is fired into a hot furnace?

Answer: Incomplete combustion.

What is the visible indication of incomplete combustion?

Answer: Smoke.

What is smoke?

Answer: The term is applied to *all* the products of combustion escaping from the furnace, whether visible or invisible (See Fig. 4-5.).

What are the black particles in smoke?

Answer: Solid carbon; therefore, wasted fuel.

What does colored smoke indicate?

Answer: Incomplete combustion.

What is volatile matter?

Answer: Substances that distill at low temperatures and are released when the coal is heated. The amount and nature of these distillates determines the amount and nature of the smoke produced.

Upon what does the smoke-producing tendency of the coal depend?

Answer: Upon the nature, rather than the volume, of the volatile content.

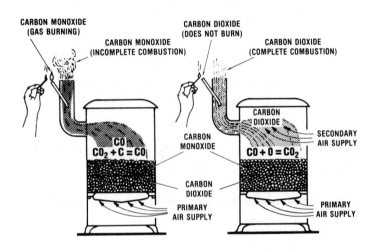

Fig. 4-5. Incomplete combustion (left) and complete combustion in an ordinary stove.

What is the effect when the air supply does not thoroughly mix with the gases from the fuel?

Answer: It causes slower combustion, resulting in a longer flame.

How can the hydrocarbon gases be completely and smokelessly burned?

Answer: By admitting and thoroughly mixing sufficient air before the gases are cooled below a certain temperature.

Are there local, state, and federal air-quality regulations dealing with furnace emissions?
Answer: Yes. See Appendix A for general guidelines.

How should the combustion chamber be proportioned for burning bituminous coals?
Answer: It should be extra large.

How is smoke classified with respect to density or analyzed?
Answer: By dividing it into several shades or comparing it with a smoke chart. The Ringelmann scale (Figs. 4-6 and 4-7) consists of

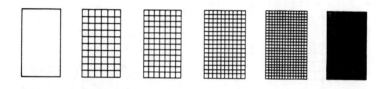

Fig. 4-6. Ringlemann scale for grading smoke density.

four large cards ruled into progressively darker shades of gray. Together with a white and a solid black card, they are placed 50 feet from the observer, in line with the chimney. The observer glances quickly from the chimney to the cards and judges which one corresponds with the color and density of the smoke. Ringelmann readings are taken at 30- to 60-second intervals during an hour or more and plotted in a log. They give a good general idea of the manner and regularity of smoke emission but are very unsatisfactory for ordinary stacks.

Simple Bacharach testers draw a sample of flue gas through a piece of white filter paper to produce a stain on the paper which corresponds with the amount of combustion smoke in the gas. The stain is compared with a scale of ten stains of varying density and the closest one indicates the amount of smoke being generated.

Other testers make use of photoelectric methods to measure the density of emitted smoke. One type measures the electrical con-

POWER PLANT ENGINEERS GUIDE

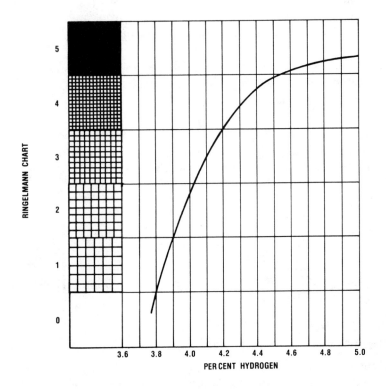

Fig. 4-7. Ringlemann chart showing how the density of smoke varies with the percentage of hydrogen in the coal.

ductivity of the gas and compares each instantaneous value with a standard. The result of the comparison is recorded on a chart.

How does a CO_2 recorder work?

Answer: Most of them work on the principle of absorption of CO_2 from flue gases by a solution of potassium hydroxide. A measured sample of flue gas is passed through the potassium hydroxide solution, which absorbs the CO_2 and decreases the total volume of gas in proportion. The size of the reduced volume compared with the original volume causes a recording mechanism to draw an ink line on graph paper, the length of the line depending on the amount of CO_2. Readings are taken automatically, 10 to 20 times per hour.

What does the CO_2 percentage indicate?

Answer: The volume of excess air flowing through the furnace and the power of the boiler.

Just what is this percentage?

Answer: It is the ratio between the air that is taken for a useful purpose (burning the fuel) and for a wasteful purpose (cooling the furnace gases). That is *all* it indicates, and its indications are only approximate.

Describe the operation of a flue-gas analyzer.

Answer: The Orsat apparatus in Fig. 4-8 works on the same basic principle as just described, except the gas sample is passed through additional reagents which absorb its content of oxygen and carbon monoxide. The percentage of each gas is determined in the same way. Pipette B contains the potassium hydroxide

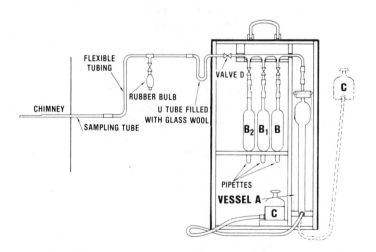

Fig. 4-8. Orsat apparatus for accurate flue-gas analysis.

solution, consisting of one part commercial caustic potash in two parts of water, and used to absorb CO_2. Pipette B1 contains a solution of 5 grams pyrogallic acid in 15cc of water, mixed with 120 grams of caustic potash dissolved in 80cc of water, and used to absorb oxygen. Pipette B2 contains a saturated solution of copper chloride in hydrochloric acid, used to absorb carbon monoxide. Each pipette is filled a little over halfway with its respective solution, and the flue-gas sample tested first for CO_2, then for O, then for CO.

The analyzer is connected to the sampling tube and the rubber bulb operated several times to get rid of air. Bottle C, about two-thirds full of water, is placed on top of the case and three-way valve D is opened, allowing the water to flow into vessel A. Place bottle C below the level of the analyzer and open valve D to the sampling tube; as the water flows out of vessel A, flue gas will replace it. Repeat this operation several times, venting the gas to the atmosphere through valve D, in order to get an undiluted sample. Draw the final sample well below the zero mark, close valve D, hold bottle C so that its water level coincides with the zero mark on the measuring tube, and open valve D to the atmosphere to allow surplus gas to escape. This provides tube A with a full measure of gas at atmospheric pressure.

Close valve D, open the valve above pipette B, and raise bottle C. Water entering vessel A will force the gas sample into pipette B, but the water level in A should be watched closely so that it never rises above the mark at the top of the measuring tube. Draw the gas back into vessel A and repeat the operation three or four times to ensure thorough absorption of the CO_2. Measure the quantity of gas returned to A by holding the water bottle so that the water level in it is exactly the same as the level in the graduated tube (otherwise, the gas will be compressed or expanded by the difference between the two columns of water). Read the percentage directly off of the graduations on the measuring tube. Two identical readings in succession show the CO_2 to be absorbed.

Follow the same procedure with the remaining two pipettes, taking the reading of the reduced quantity of gas in vessel A after each operation. The CO_2 will be absorbed after three passes through the pipette, but the gas must be passed through the pyrogallic acid at least five or six times to absorb all oxygen. If this is not

COMBUSTION

done, any oxygen remaining will be absorbed by the copper chloride and be mistaken for carbon monoxide.

The potassium hydroxide solution should absorb about 40 times its own volume before it becomes exhausted; the potassium pyrogallate, twice its own volume; the copper chloride, only its own volume. Care should be taken to keep the pyrogallic solution from air, since it absorbs oxygen rapidly—it is best to mix the two solutions together in the pipette.

What largely determines the temperature of combustion?
Answer: The design of the furnace.

ASH

Give a definition of the term "ashes."
Answer: All the mineral matter left after the complete combustion of fuel.

Why do fuels contain noncombustible matter?
Answer: Because the plants of which the coal was formed contained inorganic matter and because of the earthy matter in the drift of the coal period.

What are the principal constituents of ash?
Answer: Silica, alumina, lime, oxide and bisulfide of iron.

What is a clinker?
Answer: A product formed in the furnace by fusing together impurities in the coal, such as oxides of iron, silica, and lime.

Which coals clinker least under high temperature as judged by the color of the ashes?
Answer: Those whose ashes are nearly pure white.

What substance in ashes causes clinker?
Answer: Iron oxide.

With complete combustion of coal, what percentage of ash remains?

Answer: It varies considerably for different coals, but average values will be 5% to 10%.

EXPERIMENTS WITH COMBUSTION

Experiment 1

Using a bunsen burner and a piece of wire gauze, light the gas first above, then below the gauze, as shown in Fig. 4-9. Note that the flame will not pass through the gauze. This is because the latter

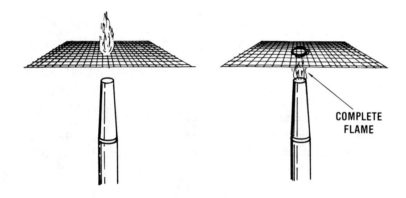

Fig. 4-9. Principle of the Davy safety lamp.

conducts the heat away from the flame so rapidly that the gas on the opposite side cannot be raised to ignition temperature. The principle forms the basis for the Davy safety lamp.

Experiment 2

Wind a thin copper wire into a helix as shown in Fig. 4-10 and carefully lower it over the wick of a burning candle. The heat of the flame will be rapidly transmitted along the wire, lowering the temperature below that at which combustible gases combine with

COMBUSTION

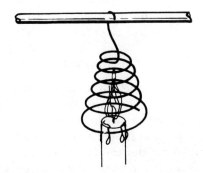

Fig. 4-10. Experiment illustrating the cooling of flame below the igniting temperature.

oxygen and therefore extinguishing the flame. This cooling effect is noticeable in the operation of an internal-firebox boiler where the heat of the fuel lying next to the furnace walls is transmitted through the walls to the water so rapidly that the fire becomes "dead" along the walls. Perform the experiment again, but heat the helix red-hot before you lower it over the candle wick. Can you explain what happens?

Experiment 3

Insert one end of a small, open tube into the body of a candle flame as in Fig. 4-11. Wait a few seconds and then light the

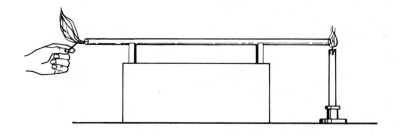

Fig. 4-11. Experiment showing that combustion occurs only at the surface of an ordinary flame.

83

combustible gas escaping from the opposite end of the tube, using a match. Note the candle flame and observe that combustion occurs only at the surface of an ordinary flame.

Experiment 4

Remove the glass chimney from an ordinary kerosene lamp, light the wick, and note the dark, smokey flame. Replace the chimney and note the difference in the character of the flame. The effect of the chimney is to produce a draft of air, which is thrown against the flame by the deflector surrounding the wick and supplies sufficient oxygen to consume all particles of carbon. Without the draft produced by the chimney, the supply of oxygen is insufficient to ignite all the carbon, which then escapes in the form of smoke or soot. Leave the chimney in place, turn the wick higher, and note that once again the flame begins to smoke. Can you explain why? Again, is there enough oxygen to completely burn all fuel? Not likely if it's smoking.

CHAPTER 5

Steam and How a Boiler Makes It

It is important that those who install or have charge of boilers should have some knowledge of the *nature* of steam, its *formation* and *behavior* under various conditions. Unfortunately, this knowledge is usually sadly lacking. It should be of a higher order than that possessed by some individuals who call British *thermal* units, British "*terminal*" units.

NATURE OF STEAM

What is steam?
Answer: The vapor of water.

What are its characteristics?
Answer: It is a colorless, expansive, invisible gas.

What is the white cloud seen issuing from an exhaust pipe incorrectly called "steam"?

Answer: It is not steam at all but in reality a fog of minute liquid particles formed by condensation. In other words, it is finely divided condensate.

What causes the steam to change into a white cloud?

Answer: Exposure to a temperature lower than that corresponding to its pressure. For instance, steam exhausting into the atmosphere encounters a temperature ordinarily between 60° and 80° and condensation immediately takes place, forming the white cloud. If the outside temperature were 212°, no condensation would take place and the exhaust would remain steam, which is invisible.

What is generally meant when the term "steam" is used?

Answer: It is generally understood to mean *saturated* steam.

With respect to power, what is steam?

Answer: The medium or working substance by which some of the heat energy liberated from the fuel by combustion is transmitted to the engine and partly converted into mechanical work.

TYPES OF STEAM

How is steam classified according to its quality?

Answer: As saturated, dry, wet, superheated, highly superheated, or gaseous, as shown in Fig. 5-1.

What is saturated steam?

Answer: Steam at a temperature corresponding to its pressure. This is the important definition, and the one expected by the examiner in an examination.

Why is the last answer important in an examination?

Answer: Because, strictly speaking, it specifies the only condition in which true steam can exist.

Steam and How a Boiler Makes It

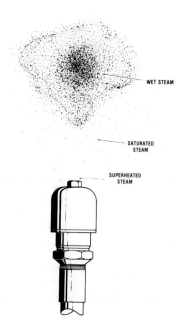

Fig. 5-1. Safety valve blowing, illustrating the three kinds of steam.

What is wet steam?

Answer: Steam containing intermingled moisture, mist, or spray.

What is dry steam?

Answer: Steam containing no moisture. It may be either saturated or superheated steam.

What is superheated steam?

Answer: Steam having a temperature higher than that corresponding to its pressure.

What is gaseous steam?

Answer: A poor, inaccurate classification for highly super-

87

heated steam. This is a poor definition because both superheated steam and gaseous steam are gaseous.

How is steam classified with respect to pressure?
Answer: (1) Vapor, (2) atmospheric, (3) low-pressure, (4) medium-pressure, (5) high-pressure, (6) extra-high-pressure.

What is the application of atmospheric-pressure steam?
Answer: In steam-heating systems in which the pressure is only a few ounces above or below atmosphere pressure.

What is the application of low-pressure steam?
Answer: In ordinary steam-heating systems working at approximately 5 to 10 psi.

What is the application of medium-pressure?
Answer: For power plants working up to 150 psi.

What is the standard "stationary" working pressure?
Answer: Eighty psi.

What is the application of high-pressure steam?
Answer: For plants operating at pressures from 150 psi to 300 psi, approximately.

What is the application of extra-high-pressure steam?
Answer: For plants operating from 300 psi to over 2000 psi.

THE FORMATION OF STEAM

How much heat does it take to generate steam?
Answer: The sensible heat plus the internal latent heat plus the external latent heat.

What is sensible heat and why is it so-called?
Answer: That part of heat which produces a rise in tempera-

ture, as shown by the thermometer, in distinction from *latent heat*. It is so-called because it is sensible to the touch. See Fig. 5-2.

What is the British thermal unit (Btu)?

Answer: The quantity of heat that may be added to a substance during a change of state without causing a temperature change.

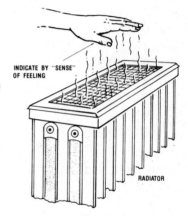

Fig. 5-2. Familiar radiator example of sensible heat.

What is latent heat?

Answer: The amount of heat required to raise the temperature of one pound of water one degree Farenheit.

What is internal latent heat?

Answer: The amount of heat that water will absorb at the boiling point without a change in temperature. In other words, the amount of heat which must be absorbed by water at the boiling point before vaporization will begin. See Figs. 5-3 and 5-5.

What is external latent heat?

Answer: When vaporization takes place, the amount of heat required because of the work in pushing back the atmosphere to make room for the steam. Those who disagree with this calculation hold that since the water already existed at the beginning of vaporization, the atmosphere was already displaced to the extent

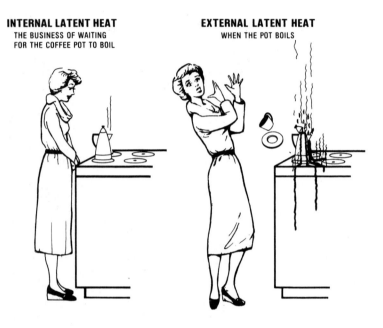

Fig. 5-3. Domestic illustrations of internal and external latent heats.

of the volume occupied by the water. Therefore, they say, this displacement must not be considered as contributing to the external work done by the steam during its formation.

What is the meaning of "from and at 212°F."?

Answer: In boiler operation, it is an evaporation that would be the equivalent of the actual evaporation when the feed water enters the boiler at 212°F. and steam is formed at standard atmospheric pressure (Fig. 5-4).

Describe the process of boiling.

Answer: When heat is applied to a liquid such as a quantity of water in a boiler, the lower layers are the first warmed. These expand and rise to the top, their place being taken by the colder layers from above. By this process the mass is warmed throughout.

Steam and How a Boiler Makes It

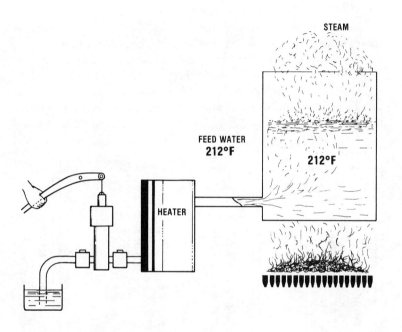

Fig. 5-4. Evaporation "from and at 212°F."

The air which is contained in the water expands as the temperature is raised, and rises to the top. The temperature of the lower layers in time increases to slightly above the atmospheric boiling point, 212°F. (100°C.), and steam is formed as bubbles adhering to the heating surface; these bubbles, by expansion, become large enough to detach themselves and rise into the colder layers above. On reaching the colder layers, they condense and their sudden collapse sets up vibration in the water which is communicated to the metal of the containing vessel, causing the familiar "singing" heard at this stage. The steam which composes the bubbles gives up its latent heat, thus warming the water until the whole mass is at the boiling point.

What is another way to describe saturated steam?

Answer: Steam that is just at the dividing point between water and steam.

Power Plant Engineers Guide

Describe the total heat of saturated steam.

Answer: In transforming one pound of water into saturated steam at atmospheric pressure (Fig. 5-5), the amount of heat to be supplied may be tabulated by stages. Stage 1 requires 180 Btu of sensible heat to raise the temperature of the water to the boiling point. Stage 2 is the internal latent heat absorbed by the water at 212°F. (100°C.) before a change of state takes place, and totals 897.51 Btu. Stage 3 requires 72.89 Btu of external latent heat to do

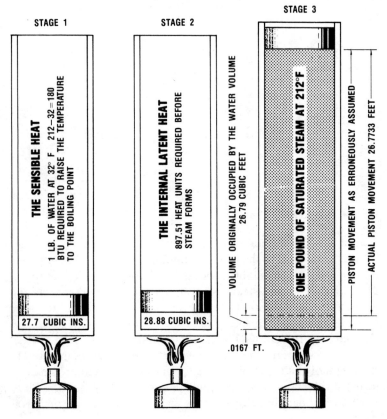

Fig. 5-5. Three stages in the formation of steam.

the work on the atmosphere. The sum of these three items, 1150.4 Btu, is known as the *total heat above 32*°F. (0°C.), the temperature taken as the starting point.

Steam and How a Boiler Makes It

What is the nature of superheated steam?

Answer: Dry steam containing no water. It is steam heated to a temperature beyond the evaporation point. Superheat is the amount of heat steam must lose before condensation can begin. (More later.)

In the foregoing, where is the sensible heat?

Answer: In the water.

Where is the total heat?

Answer: In the steam.

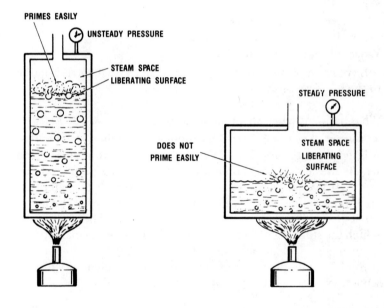

Fig. 5-6. Priming.

What is priming?

Answer: Violent agitation at the surface of boiling water, caused by inadequate steam space or a too small water surface area. See Fig. 5-6.

What is foaming?

Answer: Water surface agitation caused by impurities in the water.

CONDENSATION

What is condensation?

Answer: The change of state of a substance from the gaseous to the liquid form.

What is the liquid called?

Answer: The *condensate*.

What causes condensation?

Answer: A reduction of temperature below that corresponding to the pressure of the gas.

What happens when steam condenses?

Answer: See Fig. 5-7. Steam produced by boiling water in flask A is led through pipe C into a coiled section surrounded by cold water. In the coil, it is cooled below the boiling point and condenses, the condensate passing into receptacle B as water. The cooling or "circulating" water enters the condenser at the lowest point (D) and leaves at the highest point (E). The water from which the steam was originally formed contained a small percentage of air mechanically mixed with it, and this air does not recombine with the water of condensation; it remains liberated in a condenser or in the pipes of a steam heating plant.

How is this air removed?

Answer: By air pumps and air valves, respectively.

How was the principle of condensation first utilized?

Answer: The early engineers discovered that by condensation the pressure of the atmosphere is made available for doing work, resulting in the introduction of so-called atmospheric engines.

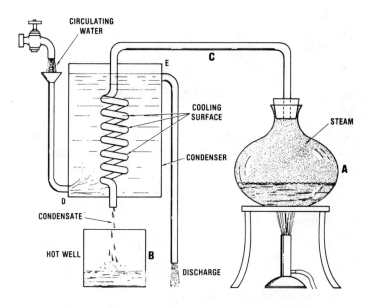

Fig. 5-7. The condensation of steam.

STEAM ABOVE ATMOSPHERIC PRESSURE

When vaporization takes place in a closed vessel, what happens due to rising temperature?

Answer: The pressure rises until equilibrium between temperature and pressure is reestablished.

When equilibrium is established, what is the temperature of equilibrium called?

Answer: The boiling point.

How does the boiling point vary?

Answer: The higher the pressure, the higher the boiling point, as shown in Fig. 5-8.

Why?

Answer: More heat must be added to the water because of the increasing amount of work that must be done to push back the air to make room for the steam. That is, because of the increasing external latent heat in establishing "thermal equilibrium."

THE STEAM TABLES

The amount of study of the steam table and knowledge absorbed are not always in proportion. Sometimes after considerable study, one gets "tripped up" in using the steam table in solving problems.

What is the steam table?
Answer: By definition, the properties of steam for various pressures given in tabulated form.

Name two kinds of steam tables.
Answer: "Properties of Saturated Steam" and "Properties of Superheated Steam."

As water evaporates, does it expand or contract?
Answer: At atmospheric pressure, the volume of space occupied by water expands by roughly 1700 times as the water turns into steam. If the steam is confined, the volume occupied by the steam decreases as the pressure in the confining vessel increases.

Give some examples of how to use the "Properties of Saturated Steam" table.
Answer: You will find this table in Chapter 30. Suppose the question came up as to how many heat units are saved in heating 25 pounds (11.3 kg) of feed water from 90°F. (32°C.) to 212°F. (100°C.). Go to column 4 and note that the total heat in the water at 212°F. (100°C.) is 180 Btu, while at 90°F. (32°C.) there are 58 Btu; the difference between the two represents a saving of 122 Btu per pound. With 25 pounds (11.3 kg) of feed water, the saving would be 3050 Btu.

Steam and How a Boiler Makes It

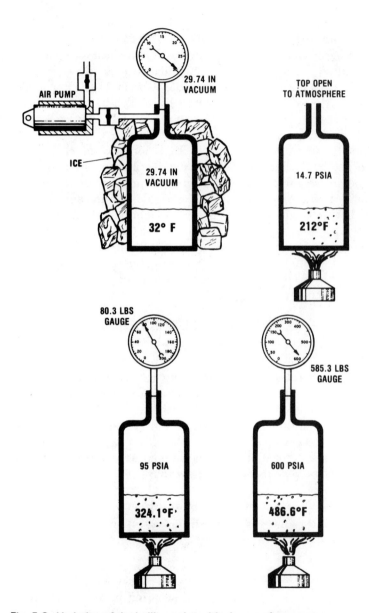

Fig. 5-8. Variation of the boiling point with change of pressure.

POWER PLANT ENGINEERS GUIDE

To answer the question, "What is the weight of 20 ft^3 of steam at 150 psia," go to column 8. The weight of 1 ft^3 of steam at 150 psia is 0.332 pounds. Twenty ft^3 will then weigh 6.64 pounds.

Another example is finding how much more heat is required to generate 26 pounds of steam at 150 psia than at 90 psia. In column 5, we see that the total heat in steam at 150 psia is 1193.4 Btu, while at 90 psia the total heat is 1184.4 Btu. This makes 9 Btu excess heat required per pound (weight) of steam, or a total of 234 excess Btu for 26 pounds.

Finally, suppose you need to find out how much heat is absorbed by the cooling water if a condensing engine exhausts 17 pounds of steam per hour at a terminal pressure of 18 pounds absolute into a 28.5-inch vacuum. Go to column 5 and note that the total heat in steam at 18 psia is 1154.2. Column 4 shows the total heat in the water at a 28.5-inch vacuum is 58.0 Btu. Subtracting the two figures, we see that there are 1096.2 Btu to be absorbed per pound of steam. With 17 pounds of steam, this means 18,635.4 Btu total heat to be absorbed by the cooling water per hour.

SUPERHEATED STEAM

How is superheated steam produced?

Answer: If a closed vessel containing water and steam is heated, the pressure of the steam will gradually rise until all the water has been evaporated. At this point, the further addition of heat will not produce any appreciable increase in pressure but will cause a rise in temperature. In this condition, the steam is said to be *superheated*; so superheated steam is defined as *steam heated to a temperature above that due to its pressure.*

What is the object of superheating steam?

Answer: It reduces, and in extreme cases, prevents condensation, thus giving better economy.

What are the disadvantages in using superheated steam?

Answer: Increased difficulty of securing proper lubrication, higher first-cost and depreciation.

Steam and How a Boiler Makes It

What is the saving due to superheating?

Answer: According to Ripper, the condensation at cut-off is reduced 1% for each 7.5 degrees of superheat. The saving varies with the type of engine, degree of expansion and other factors.

What conditions favor superheat?

Answer: High degree of expansion of steam, slow speed, constant load, and high fuel-cost.

How much superheat should be given to the steam?

Answer: For the maximum saving possible, the degree of superheating should be such that the steam is exhausted in a saturated state (the ideal case).

When is there too much superheat?

Answer: When the superheat is high enough to give a superheated exhaust.

In practice, does this occur often?

Answer: In practice, it is very seldom that the superheating is carried to the extent of giving a superheated exhaust; in fact, the exhaust is usually not even saturated, but wet instead.

The volume of superheated steam is shown in the steam table in Chapter 30, but how has it been calculated?

Answer: By means of Linde's equation, as follows.

$$pv = 0.5962T - p(1 + 0.0014p) [(150,300,000/T^3) - 0.0833]$$

where,

p is expressed in psi,
v is expressed in cubic feet,
T is the absolute temperature F.

You can see that the table will save a lot of calculation.

How can I find the specific heat of superheated steam?

Answer: The table 5 in Chapter 30 lists the mean specific heats from the temperature of saturation to various temperatures at several pressures. It has been calculated by Knoblauch and Jakob from Peabody's Tables.

EVAPORATION

Discuss factors of evaporation.

Answer: It takes more fuel to generate steam at high pressure than at low pressure; accordingly, in the rating of steam boilers, some standard of evaporation must be adopted in order to obtain a true measure of performance. This involves two items: (1) temperature of the feed water and (2) pressure at which the steam is generated.

With respect to the first item, more fuel would obviously be used if the feed water were supplied at 60°F. instead of 150°F., and no comparison of the performance of two boilers working under these conditions could be obtained unless a factor were introduced to allow for feed-water temperature difference. With regard to item (2), more fuel is required to generate steam at high pressure because the external work of vaporization is greater. That is, more work is done in the formation of the steam in making room for itself against a high pressure than against a low pressure.

How is a standard of vaporization obtained?

Answer: By finding the equivalent vaporization "from and at 212°F."

Define "factor of evaporation."

Answer: A factor of evaporation is a quantity which, when multiplied by the amount of steam generated at a given pressure from water at a given temperature, gives the equivalent evaporation from and at 212°F.

How is the factor of evaporation obtained?

Answer: Most easily, from the table in Chapter 30. However, it is equal to the heat in the steam at the pressure generated, minus the heat in the water, the difference divided by the latent heat of

Steam and How a Boiler Makes It

steam at atmospheric pressure. Expressed another way, it is equal to latent heat at generated pressure divided by latent heat at atmospheric pressure. Expressed as a formula, it reads as follows:

$$F = \frac{H - h}{H' - h'}$$

where,

F is the factor of evaporation,
H is the heat above 32°F. in the steam at a given pressure,
h is the heat above 32°F. in the water at a given temperature,
H' is the heat above 32°F. in steam at atmospheric pressure,
h' is the heat above 32°F. in water at atmospheric pressure.

In this formula, the expression of H'-h' may be converted to 970.4 and used as a constant (it is obtained by subtracting 180 from 1150.4—see the evaporation table in Chapter 30).

As an example, suppose you need to find the factor of evaporation for steam at 200 psia when the feed water is delivered to the boiler at a temperature of 150°F. From the steam table, the heat in the steam at 200 psia (H) is 1199.2 Btu. The heat in the feed water at 150°F. *above* 32°F. (*h*) is 150 minus 32, or 118 Btu. Substituting these values in the formula, you get:

$$F = \frac{1199.2 - 118.0}{970.4}$$
$$= \frac{1081.2}{970.4}$$
$$= 1.114$$

The meaning of it is that if the boiler were generating 1000 pounds of steam per hour at 200 psia from feed water at 150°F. It would absorb the same amount of heat from the fuel as when generating 1000 x 1.112, or 1112 pounds of steam from and at 212°F. (steam at atmospheric pressure from feed water at 212°F.). The latter factor of evaporation was taken from the table in Chapter 30.

HOW A BOILER MAKES STEAM

What is the basic principle in steam-making?

Answer: An upset of hydraulic thermal equilibrium, causing circulation; that is, convection currents.

What causes convection currents?

Answer: Basically, a variation in water temperature in different parts of a boiler. Water at high temperature weighs less than water at low temperature; accordingly, low-temperature water sinks to the bottom (Fig. 5-9) of a containing vessel and pushes high-temperature water up to the top. In Fig. 5-10, as heat is applied to the *upflow* or *riser* the water in it expands and becomes less dense. It is then displaced by the colder and heavier water in the *downflow*, causing the water to circulate by convection as indicated by the arrows. Early automobile cooling systems depended solely on this phenomenon before the advent of forced circulation by a water pump. It is still in use, however, in hot-water heating systems.

Is this difference in temperature all that causes convection currents, that is circulation?

Answer: No.

In boiler operation, what additional condition accelerates the circulation?

Answer: The formation of steam bubbles (Fig. 5-11) results in a mixture of steam and water in the upflow side much lighter than the solid column of cooler water in the downflow side.

If a pot is filled with water and placed on an open fire, what happens?

Answer: When it boils, the water rises at the sides and sinks in the center (Fig. 5-9).

Why?

Answer: The water is heated most at the sides of the pot, causing it to expand, become lighter, and rise. It is pushed up by the heavier water in the central part of the vessel.

What happens as the rising, expanded water reaches the top of the vessel?

Answer: The surface is cooled somewhat, which causes the water to contract; becoming denser, it naturally sinks.

Where does the formation of steam take place?

Answer: In the water directly in contact with the pot, especially in the lower part where the temperature of the metal is highest.

Describe the formation of steam.

Answer: A particle of water in contact with the metal is heated until it is changed into steam, which first appears as a small bubble that for a time clings to the metal (Fig. 5-11).

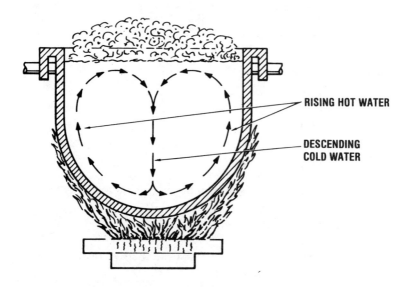

Fig. 5-9. Circulation of water in boilers.

What happens to the bubble?

Answer: Its size gradually increases by the addition of more steam from the surrounding water until it finally disengages itself from the metal.

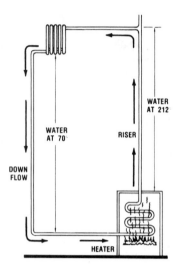

Fig. 5-10. Elementary hot water heating system illustrating thermocirculation.

After disengagement of the bubble of steam, what happens?

Answer: Since it is much lighter than the water, it quickly rises and bursts on reaching the surface, allowing the steam to escape into the atmosphere.

What takes place during the ascent of the bubble and why?

Answer: It expands because of the gradual reduction in pressure due to the decreasing hydraulic head.

What kind of circulation is that just described as distinguished from another kind of circulation?

Answer: Natural; or *undirected* circulation.

What is the other kind of circulation?

Answer: Directed circulation.

Steam and How a Boiler Makes It

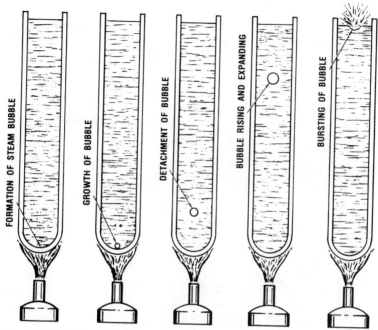

Fig. 5-11. The formation of steam, showing what happens to each bubble.

Why does a pot boil over with undirected circulation?

Answer: Steam bubbles will rise from all points at the bottom in such quantities as to impede the downward flow of water, in which case the pot "boils over."

In what kind of boiler is the "pot" circulation virtually reversed?

Answer: In the vertical, or so-called "upright" shell boiler (Fig. 5-12).

Why?

Answer: The coldest part of the boiler is at its shell.

Describe the circulation.

Answer: In the vertical boiler, the current descends along the shell to the lowest point and rises next to the walls of the furnace; then it continues upward along the tubular heating surface.

105

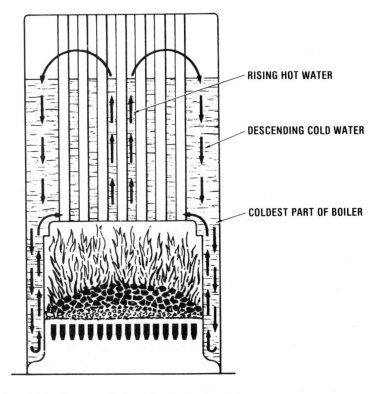

Fig. 5-12. "Reversed" circulation in vertical boiler.

What is the importance of a free circulation in boilers?

Answer: Among other things, it maintains the metal of the boiler at a nearly uniform temperature, which prevents unequal expansion on its various parts, especially in boilers having thick plates. It also facilitates the escape of steam from the heating surface as soon as it is formed—a condition necessary to prevent overheating of the plates, which would occur unless they are maintained in constant contact with the water.

Why?

Answer: Water is a poor conductor and transmits heat principally by convection; that is, by circulation.

Steam and How a Boiler Makes It

In steam-making, is all the heat applied transferred to the water?

Answer: No—a considerable amount of heat generated by the fuel is lost.

What is the spheroidal state?

Answer: The condition of a liquid, as water, when being thrown on the surface of a highly heated metal. It rolls about in spheroidal drops or masses, at a temperature several degrees below ebullition and without actual contact with the heated surface. This phenomenon is due to the repulsive force of heat and the intervention of a cushion of vapor.

CHAPTER 6

Types of Boilers

What is a steam boiler?

Answer: A closed vessel, made of iron or steel, which is partly filled with water when in use. Heat produced in an internal or external furnace is transferred to the water, turning it into steam. A boiler converts the heat produced by the furnace into energy to drive turbines, heat buildings, and many other industrial operations.

BOILER CLASSIFICATIONS

List basic parts of a steam boiler.

Answer: (1) Water space, holding the water from which steam is generated; (2) steam space, the steam storage space above the water level; (3) heating surface, all parts of the boiler with water on one side and fire or hot gases on the other; (4) grate area, where fuel is burned in solid-fuel types; (5) combustion chamber, where combustible gases from fuel are mixed with air and burned; (6) boiler setting, brickwork enclosing the furnace of externally fired boilers (those with internal fireboxes do not use a setting); (7) boiler fittings, all attachments needed to operate the boiler.

How would you describe the water and fire lines?

Answer: The water line is the level of the water in the boiler; the fire line is the highest level subjected to direct heat from the furnace (an imaginary line, usually).

Why is there such a great variety of boiler types?

Answer: It is due to the many different kinds of service for which they are intended, the varied conditions accompanying their use, and the competition among engineers who have sought to produce, at moderate cost, boilers that will be safe, durable, compact, and economical.

Can boilers be grouped into a few general divisions?

Answer: All boilers may be broadly classed with respect to service as (1) stationary, (2) locomotive, and (3) marine. Stationary boilers may be subdivided into those used primarily for heating or those used for generating power. With respect to the form of construction, boilers may be classed as sectional, fire-tube, water-tube, or electrically operated types. Some illustrations of boiler construction and types are found in the following illustrations.

What is a fire-tube boiler?

Answer: A boiler with tubes surrounded by water (Fig. 6-1A) through which the products of combustion pass. Common types are: vertical dry-top boiler, vertical submerged or wet-top boiler, return tubular, and packaged fire-tube boiler.

What is a water tube?

Answer: One with the tubes surrounded by the products of combustion with the water inside the tubes (Fig. 6-1B). A good way to remember the difference is to bear in mind that fire passes

Types of Boilers

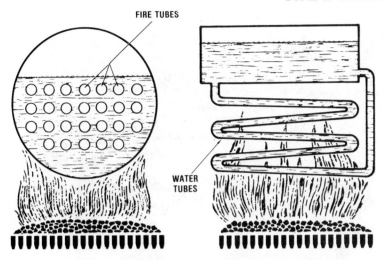

Fig. 6-1. The difference between fire tubes (A) and water tubes (B).

through fire tubes while water passes through water tubes. Common water-tube types are: straight-tube, bent-tube, packaged, field-erected, and utility boilers. Straight- and bent-tube types are older types still found in some installations. Fig. 6-2 is a photo of a modern water-tube boiler.

What is the difference between a water tube and a flue or fire tube?

Answer: A tube is a lap-welded, or seamless cylindrical shell made in small sizes. A flue is a large cylindrical shell; it may be seamless, lap-welded, or riveted.

How is a tube fastened to a sheet?

Answer: By expanding, as in Fig. 6-3.

What is a sheet?

Answer: A flat plate or head having holes for receiving the tubes or flues.

How is a flue fastened to a sheet?

Answer: By flanging and riveting, as in Fig. 6-3.

POWER PLANT ENGINEERS GUIDE

Courtesy Cleaver Brooks

Fig. 6-2. A newer-style, water-tube steam boiler with a capacity of 16,000 pounds per hour.

Why are boilers with flues practically obsolete?
Answer: Because of inadequate heating surface.

What is a single-tube boiler?
Answer: One made up of plain tubes, as in Fig. 6-4.

What is a double-tube boiler?
Answer: One having an auxiliary tube placed inside each main tube to improve circulation, as in Fig. 6-4.

What is a nonsectional boiler?
Answer: One in which all the tubes are in communication with a common header at each end.

What is a sectional or cast-iron boiler?
Answer: One in which the boiler is divided into sections instead

TYPES OF BOILERS

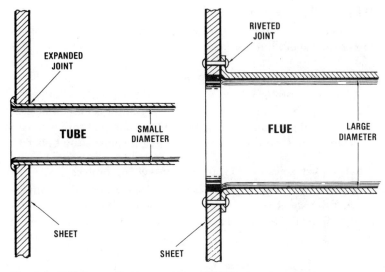

Fig. 6-3. Differences between a tube and a flue.

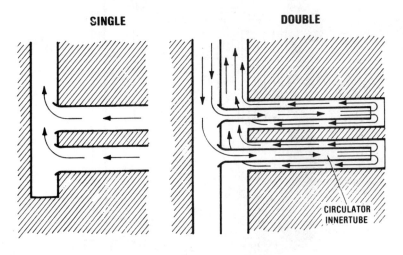

Fig. 6-4. Difference between single and double water tubes.

of tubes (Fig. 6-5). As many sections as needed can be joined or fastened together. Cast-iron sectional boilers are sometimes called "pork chop" boilers because each section resembles a pork chop.

113

Power Plant Engineers Guide

What is a through-tube boiler?

Answer: A vertical shell boiler having tubes extending from the lower tube sheet the full length of the shell, as at left in Fig. 6-6.

What is a vertical tube dry-top boiler?

Answer: A boiler where the water level extends two thirds of the way up the tubes. The top third of the tubes have hot gases inside and steam on the other, resulting in unequal expansion and contraction in the tube sheet. (Same as a through-tube boiler.)

What is a vertical submerged-tube or wet-top boiler?

Answer: A vertical boiler having tubes extending from the lower tube sheet to an upper *submerged* tube sheet, the latter being connected to a head at the top by a "cone," as at right in Fig. 6-6.

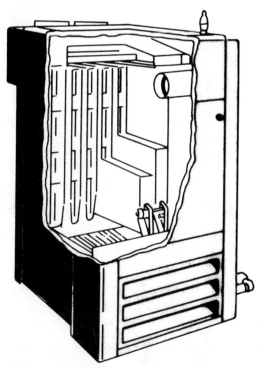

Fig. 6-5. Modern cast-iron sectional boiler.

Types of Boilers

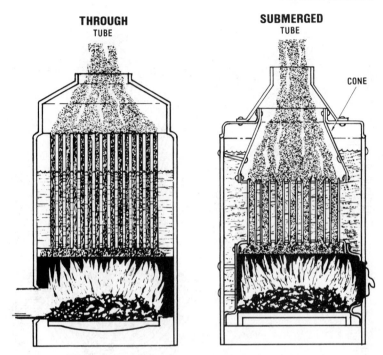

Fig. 6-6. Through (sometimes called flush) tube and submerged tube vertical boilers.

What is a bent-tube boiler?

Answer: A water-tube boiler with banks of bent tubes connected to drums (Figs. 6-7 and 6-8) mounted at different levels (vertically) in the furnace. One of the upper drums accumulates steam, and the lowest is the mud drum, which collects sediment. The mud drum has a blow-down connection.

What is a Sterling boiler?

Answer: A three-cross-drum boiler (Fig. 6-7), with a mud drum suspended by bent tubes from the cross drums. Feed water is applied to the rear upper drum, circulates down the rear tube bank and up the front banks. Steam collects in the center top drum.

What is a Kidwell boiler?

Answer: A modification of the Sterling (Fig. 6-8) with three top

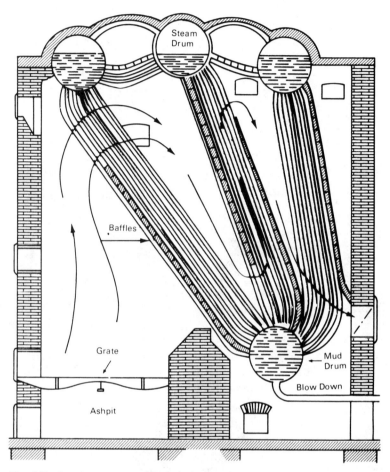

Fig. 6-7. Sterling cross drum bent tube boiler.

drums and a mud drum. Water is said to circulate faster in the Kidwell because of the "ring flow" design, in which water flows from the rear top drum down to the mud drum, then to the intermediate level drum. The Kidwell is supposed to allow freer steam release.

What is a firebox boiler?

Answer: One having the fire within a firebox, although external

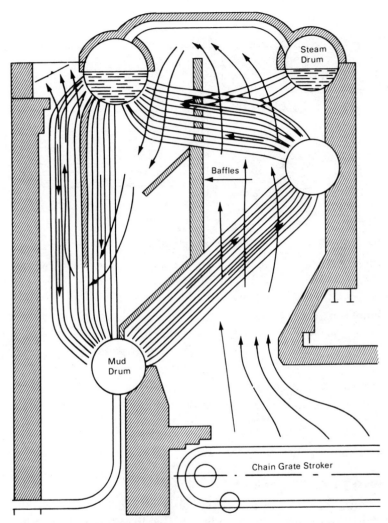

Fig. 6-8. Kidwell ring flow boiler.

to the shell, is rigidly connected to it, as in Fig. 6-9. The firebox is usually of steel plates instead of bricks.

What is a horizontal return tubular (HRT) boiler?

Answer: One so arranged that the products of combustion, after passing along the length of the shell, return in an opposite

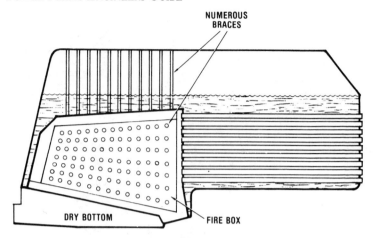

Fig. 6-9. Marine type of locomotive boiler with dry-bottom fire box.

direction through the tubes before passing up the stack. See Fig. 6-10. It is an externally fired boiler.

What is a Scotch boiler?

Answer: A horizontal fire-tube boiler in which the combustion chamber at the end is entirely surrounded by water. See Fig. 6-11.

What is a Clyde boiler?

Answer: A boiler similar to a Scotch boiler, *but* instead of a water space at the back end of the combustion chamber, it has a removable back which is lined with some insulating material such as asbestos or fire tile (Fig. 6-12).

What are Galloway tubes?

Answer: Obsolete transverse tubes placed in a flue and attached to openings in the side of a flue to increase the heating surface, as in Fig. 6-13.

What is the Morrison corrugated boiler?

Answer: A furnace with corrugations around its circumference. The corrugated effect gives the furnace added strength, allowing for greater expansion and contraction.

TYPES OF BOILERS

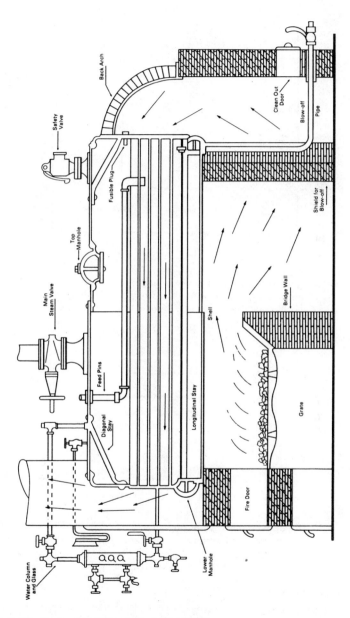

Fig. 6-10. Horizontal return tube boiler (HRT).

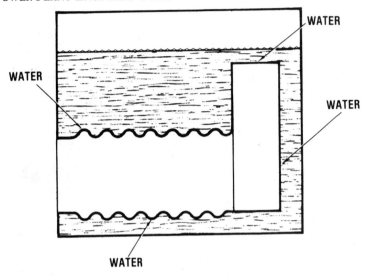

Fig. 6-11. Modern Scotch boiler showing general arrangement. Note that the combustion chamber is entirely surrounded by water except for the opening where it is attached to the furnace.

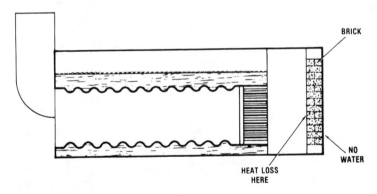

Fig. 6-12. Modified Cycle dry-back boiler with short fire tubes at the rear of the corrugated furnace flue, the object being to provide circulation under the flue.

What is the Adamson ring furnace?

Answer: A furnace made up of cylindrical flanged sections, which are then riveted or welded together to form the required

Types of Boilers

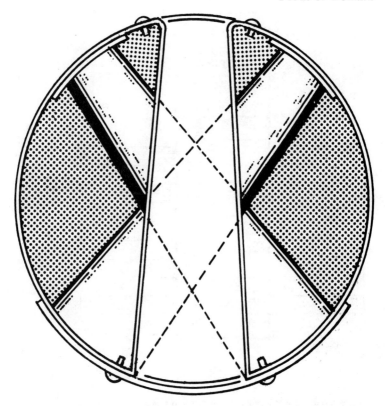

Fig. 6-13. Galloway tubes or flues.

length. With modern welding, this type of fabrication has been replaced with the ring-enforced furnace.

What is the difference between cornish and lancashire boilers?
Answer: They are, respectively, one- and two-flue boilers.

What is a porcupine boiler?
Answer: One having a vertical drum into which are screwed a number of short, horizontal, radial tubes having their outer ends closed and the ends of square section (Fig. 6-14).

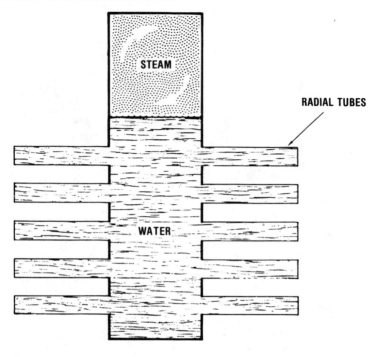

Fig. 6-14. Porcupine boiler.

Describe a locomotive boiler.

Answer: A fire-tube type with a furnace formed by its crown sheet and water legs. In stationary use the boiler has poor circulation, but in mobile use it is an efficient steam generator. See Fig. 6-15.

What is an internally fired boiler?

Answer: One in which the furnace is *within* the shell, being surrounded by water. An example is the familiar vertical shell boiler.

Types of Boilers

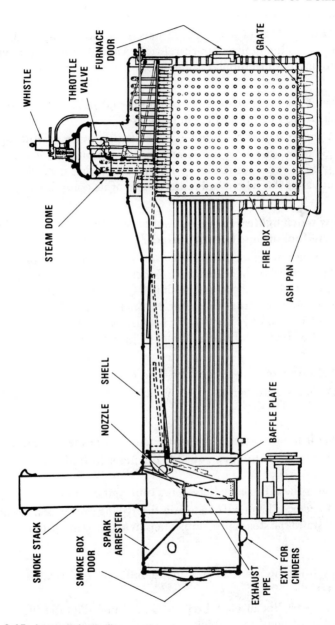

Fig. 6-15. Locomotive boiler.

What is an externally fired boiler?

Answer: One in which the furnace is outside the shell, the furnace walls being usually of firebrick. An example is the familiar horizontal return tubular boiler as shown in Fig. 6-10.

What is the comparison between tubes and pipes with respect to their use as heating surface in boilers?

Answer: Tubes are much lighter than pipes. In case of repair, single tubes may be removed. In the case of pipe boilers, several lengths of pipe are made up into a coil, which necessitates the removal of the whole coil to get at any of its component lengths.

How much lighter is a square foot of tube heating-surface than a square foot of pipe heating-surface?

Answer: It depends upon the size of the tubes and pipes. For instance, the weight of standard, 2-inch (0.095-inch wall thickness) boiler tube is 1.932 pounds per lineal foot; the weight of standard 1-inch pipe (external diameter 1.9 inches) is 3.631 pounds per lineal foot. To have one square foot of external surface, you will need 1.91 feet of tube or 2.01 feet of pipe.

What is the lightest-gauge tube used in standard practice?

Answer: Number 13 BWG, which is a wall thickness of 0.095 inch.

What is the smallest-size tube used, except in special cases?

Answer: That with an outside diameter (od) of 1 inch.

What is the difference between a tube and a pipe?

Answer: The metal of a tube is thin, being proportioned only to withstand the steam pressure, whereas a pipe is made of relatively thick metal.

Why is a tube made of thin metal and a pipe of relatively thick metal?

Answer: Because a tube is *expanded* into the tube sheet but a pipe must be of extra thickness because of the screwed joint.

TYPES OF BOILERS

STEAM GENERATORS

What is the difference between a boiler and a generator?

Answer: A boiler carries a considerable volume of water in proportion to its heating surfaces, and is therefore not very sensitive to sudden changes in the rate of combustion. A generator carries no excess volume of water but converts the water into steam as it traverses the heating surface progressively from one end to the other.

Name two types of generators.

Answer: Semiflash and flash. A semiflash generator is a combination of a shell and flash boiler. It consists of a drum or shell holding a body of water and a coil of pipe forming the heating surface. A flash generator consists of a long length of tubing formed into a coil, water usually entering at the top and being "flashed" into steam at some intermediate point and coming out of the lower layer as superheated steam. The term "boiler" is frequently used in place of "generator."

The steam generator in Fig. 6-16 is a shop-assembled once-through (coil) steam generator. Water is forced through the coils at high velocity, and heated with forced-draft firing. Water recircu-

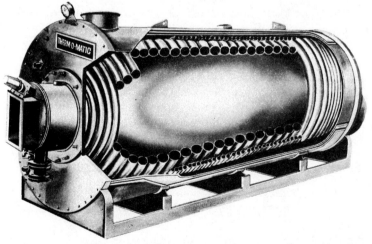

Courtesy Vapor Corp.

Fig. 6-16. Therm-O-Matic steam generator.

125

Courtesy Vaper Corp.

Fig. 6-17. Circulatic steam generator.

lates in the steam generator in Fig. 6-17. It is pumped from the overhead reservoir drum, through the boiler water tubes and back into the reservoir. Steam is taken from the flanged fitting on the front of the reservoir.

WASTE-HEAT RECOVERY BOILERS

What is a waste-heat recovery boiler?

Answer: See Fig. 6-18. A boiler that uses heat left over from industrial processes or other applications to produce steam.

What materials can a solid-waste boiler use?

Answer: Any biomass waste: wood sander dust, sawdust, shav-

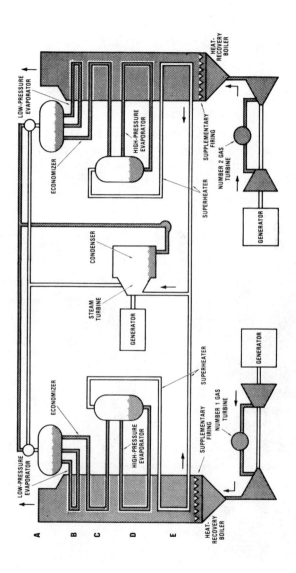

Fig. 6-18. Operating diagram of Southern California Edison's electrical generating station near Daggett, California, using large-scale, heat-recovery boilers.

ings, chips, hogged lumber, veneer trim, or agricultural residue (rice hulls, cotton waste, or nut shells).

What is "cogeneration"?

Answer: The production of electricity or mechanical energy along with the production of steam.

What is topping-cycle cogeneration?

Answer: A method of cogeneration where high-pressure steam drives a turbine and generator to produce electricity. After passing through the turbine, the low-pressure steam is used to operate machinery or for heating.

Courtesy Cam Industries

Fig. 6-19. Packaged electric boiler.

TYPES OF BOILERS

What is bottoming-cycle cogeneration?

Answer: A cogeneration method where steam or heat from an engine or turbine is recovered as a byproduct of electricity production.

What is an economizer?

Answer: A system that recovers heat from boiler flue gases and returns the heat to the boiler feed water.

PACKAGED BOILERS

What is a "packaged" boiler?

Answer: A self-contained system including boiler, firing, or heating equipment, draft fans, feed pump, and automatic controls mounted on a single base. Such units are often shipped completely assembled (except for a few systems which are easily installed on site). Most are oil- or gas-fired; some will use either. The steam generators in Figs. 6-19 and 6-20 are packaged units.

Courtesy Reimers Electra Steam, Inc.

Fig. 6-20. Package electric steam boiler.

129

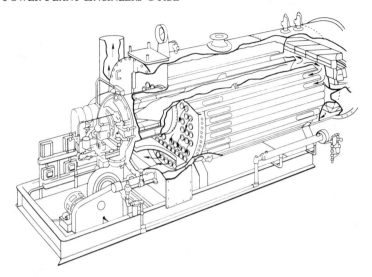

Fig. 6-21. Cutaway drawing showing construction of 4-pass packaged steam generator.

What are 3- and 4-pass steam generators?

Answer: Usually oil- or gas-fired boilers in which combustion gases make three or four passes through the heating area before reaching the stack. See Fig. 6-21.

What types of packaged electrically operated boilers are in use?

Answer: High-voltage electrode and immersion-element boilers and hot-water systems.

What are the differences between electrically operated electrode and immersion types?

Answer: Electrode boilers use the conductive and resistive properties of water to carry electric current to generate steam and hot water. An AC current flows from an electrode of one electrical phase through neutral to another phase, using water as the conductor. Immersion-element boilers have an assembly of fixed-resistance elements that generate heat, which is transferred directly to water.

CHAPTER 7

Shell or Fire-Tube Boiler Construction

What is the distinguishing feature of shell boilers?
Answer: The use of fire tubes instead of water tubes.

What are the common types of fire-tube boilers?
Answer: Vertical and horizontal fire tubes. See the illustrations in Chapter 6.

Why does a fire-tube boiler have numerous small tubes instead of a few large ones?
Answer: More heating surface is added to the boiler.

What are tube sheets?
Answer: The boiler heads at each end or top and bottom of the boiler. The flues or tubes are, respectively, riveted or expanded to the sheets.

Upon what important detail of construction does the strength of a boiler shell depend?

Answer: The longitudinal welded or riveted joint of the shell.

The tube sheets of a fire-tube boiler are what shape?

Answer: Flat. Tube sheets are slightly convex at the edges, because they're rolled over for fastening to the drum. Tube sheets are parallel with each other throughout the part where the tube holes are.

What purpose does the rear arch serve in an HRT boiler?

Answer: It directs the hot combustion gases into the fire tubes for a second pass or returns them to the front end of the boiler. (See Fig. 6-10.)

How is a gas-tight fit maintained between the rear arch and the rear tube sheet?

Answer: With a seam made of noncombustible material, usually asbestos rope.

Two methods have been used to join the edges of boiler plates together. What are they?

Answer: Welding and riveting.

Which is most used today?

Answer: Welding. Riveted boiler shells are still in service, however.

WELDED JOINTS

Are welded joints strong?

Answer: Thanks to advancements in welding techniques and properly made seams, factory welds are every bit as strong or stronger than the plate itself.

Why are welded boiler seams and seams in any pressure vessel carefully examined by X-ray?

Answer: To make sure the weld has properly penetrated the

Shell or Fire-Tube Boiler Construction

metal through the complete length of the weld and that no hidden flaws exist in the base metal or weld itself.

Are welds used to repair boilers?

Answer: Yes, due to improved welding techniques and properly trained welders, welds are acceptable for shell repairs.

What standards are used to assure the use of proper technique and to determine quality?

Answer: The ASME code (American Society of Mechanical Engineers) specifies how welds should be made and provides tests to determine welder capability. The code also contains direction on stress-relieving welds after they're made and on where and how X-rays are to be taken.

How can welds be stress-relieved?

Answer: Rapid, light blows with a ballpeen hammer to the weld area. A weld is best stress-relieved (according to ASME code) by heating the area surrounding the weld to at least 1100-1200°F. (593-649°C.) or higher if the metal doesn't become distorted in so doing. Heat is to be applied slowly and held at the maximum for a period of time equal to one hour per inch (2.54 cm) of thickness. Then it is to be cooled slowly in still air to 600°F. (316°C.) or lower before rapid cooling is allowed to occur.

What is the most efficient welded seam or joint?

Answer: A joint or seam with the weld ground flush with the boiler plate (100% efficient). When the weld isn't ground flush (95% efficient), stress results among the edge of the weld because of the difference in thickness of the metal (plate and weld).

RIVETED JOINTS

How strong are riveted joints?

Answer: The strength of a riveted joint is always *less* than the strength of the plate.

What is the difference between a lap joint and a butt joint?

Answer: The plate ends overlap to form a lap joint and register with each other to form a butt joint, as in Figs. 7-1 and 7-4.

POWER PLANT ENGINEERS GUIDE

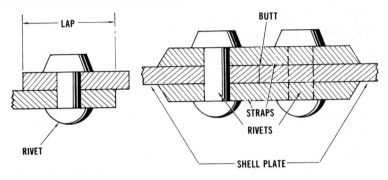

Fig. 7-1. Single-riveted lap joint and single-riveted butt and double-strap joint.

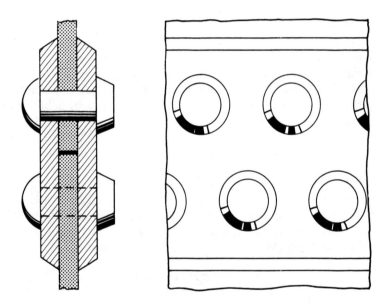

Fig. 7-2. Double-strap, single-riveted butt joint with two cover plates.

What is the objection to a lap joint?

Answer: The plate ends through which the rivets pass are in different planes, so the pull is not direct but tends to twist the plates, frequently causing them to bend.

Shell or Fire-Tube Boiler Construction

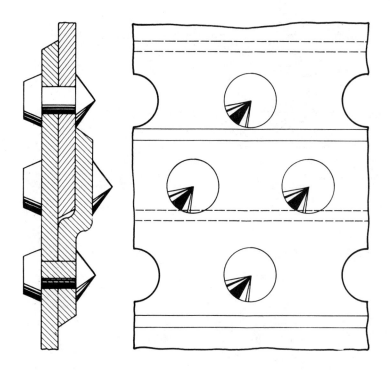

Fig. 7-3. Lap joint with cover plate or offset strap. It is safer than the plain lap joint, but not as good as the butt joint with double strap.

In designing a riveted joint, what is the chief object?

Answer: To so proportion it that it will be equally strong against failure in all its parts.

What is the difference between centers of adjacent rivets in the same row called?

Answer: The pitch. See Fig. 7-5.

What is diagonal pitch?

Answer: Where there are two or more rows of rivets, the distance between the centers of diagonally adjacent rivets is the diagonal pitch.

135

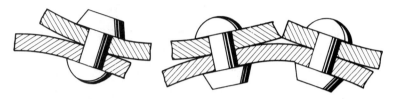

Fig. 7-4. Effect of pull on lap joints and butt joints with one cover plate.

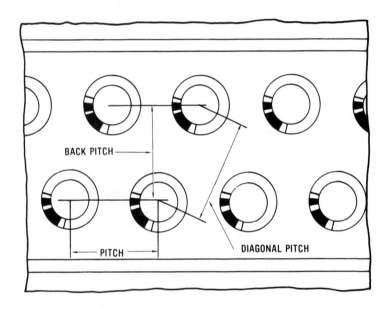

Fig. 7-5. Double-riveted butt joint illustrating pitch, diagonal pirch, and back pitch.

What is a single-strap butt joint?

Answer: A butt joint or seam with only one cover plate.

What is a double-strap butt joint?

Answer: A butt joint or seam with two cover plates, one on each side of the connected plates. The two plates may be equal or different widths. When the widths are different, the widest plate always goes inside the boiler. All rivets go through both cover

Shell or Fire-Tube Boiler Construction

plates when they're equal in width. When they're not equal, the outer row of rivets goes through the wider inner plate only.

How are the rivets in a joint counted?

Answer: By the number of rows of rivets that have to be sheared to allow rupture of the seam. Therefore, all rows are counted in a lap joint, but only those one side of the seam are counted in a butt joint. Single-riveted joints have only one row of rivets resisting rupture (but two rows total); double-riveted have two rows resisting rupture (but a total of four rows); etc.

Which are better, drilled or punched rivet holes?

Answer: Drilled, because they result in a more accurate and stronger joint. When holes are punched, the metal around them is weakened because of the crushing of the metal during punching.

Why should riveted joints be caulked?

Answer: It's almost impossible to obtain a steam-tight joint by just riveting. Riveted joints should be caulked by spreading the edge of the strap between the rivet heads and the lower plate. Seal welding, where a thin layer of metal is applied along the seam, is permissible in circumferential lap joints.

What is back pitch?

Answer: The distance between the center lines of any two adjacent rows of rivets.

What is the difference between single and double shear?

Answer: Single shear occurs in one plane as in a lap joint, and double shear in two planes as in a butt and double-strap joint. See Fig. 7-6.

What is the advantage of double shear?

Answer: It is twice as strong as single shear. In practice, it is taken as 1¾ to allow for imperfection of construction.

What is the efficiency of a riveted joint?

Answer: The ratio which the strength of a unit length of a riveted joint has to the same unit length of the solid plate.

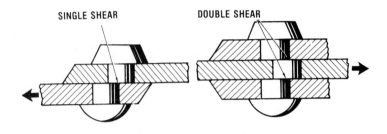

Fig. 7-6. Lap and butt-and-double strap joint, illustrating single and double shear, respectively.

How does the efficiency of a riveted joint increase?

Answer: It increases as the rivet diameter and pitch is increased. There is a practical limit to this, owing to mechanical difficulties encountered.

What basic consideration determines the strength of a riveted joint?

Answer: It depends upon whether the plate or the rivets are the strongest. See Fig. 7-7.

What is a ligament?

Answer: The metal between tube holes.

STAYS

How are flat surfaces, including tube sheets, strengthened and why?

Answer: All flat surfaces must be stiffened or supported by stays or braces. Otherwise, the steam pressure would bulge them outward and tend to make them spherical or cylindrical in shape.

What is the difference between a brace and a stay?

Answer: The chief difference is in size; that is, a brace is a large stay.

Fig. 7-7. Failures of riveted joints. At the top, a fracture between rivets; at the bottom, a split and double-shear between rivets and the edge of the plates. The first is caused by the rivets being too close together; that is, not enough metal in the plate between rivet holes, and the second is due to insufficient metal between rivet holes and edge of the plate.

Into what two classes may all reinforcing members be divided?

Answer: They may be classed as independent or connecting members.

Two types of stays are used in HRT boilers. What are they called?

Answer: Longitudinal and diagonal stays. Stays are further catagorized as screwed, through, radial or crown, and diagonal.

What conditions will effective staying fulfill?

Answer: Stays must be adequate in number and size to carry the total pressure on the area stayed, not counting the strength of the boiler plate itself. Spacing between stays must allow sufficient room for inspection. Water circulation must not be impeded by stays. Each stay should be at a right angle, or as nearly so as possible, to the plate it supports.

What is a stay bolt?

Answer: A metallic pin or rod used to hold objects together and generally having screw threads cut at one end, and sometimes at both, to receive a nut, as in Fig. 7-8.

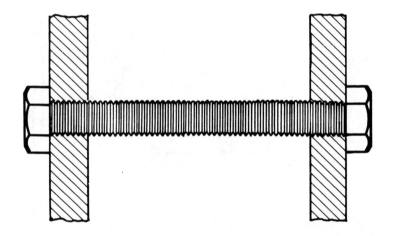

Fig. 7-8. Stay-bolt, consisting of a threaded length of rod with a nut at each end, or a forged head at one end and a nut at the other end.

How are boiler plates tapped so that a threaded bolt or stay will properly screw into both plates without stripping the threads?

Answer: By the use of a long stay-bolt tap which threads both plates in one operation, as in Fig. 7-9. Threads in the two plates must be continuous so there is no tendency for the stay to draw the plates together or force them apart.

Shell or Fire-Tube Boiler Construction

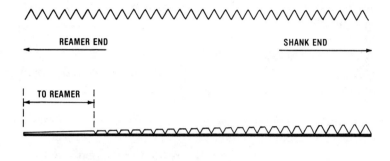

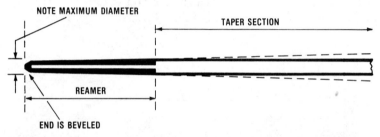

Fig. 7-9. Stay-bolt tap. Both plates are tapped in one operation so that the bolt will thread into both plates without interference.

What thread is used for stay-bolt taps?

Answer: All sizes of stay-bolt taps have 12 threads to the inch, the approved form being the Unified National (formerly known as American National or United States Standard); the V thread was sometimes also used.

What diameter of a screw stay is taken in calculating its strength?

Answer: The *least* diameter. For a continuous thread, the least diameter is at the bottom of the thread, and at the middle section of turned stays. Some manufacturers turn the threads off stays between the plates. Without threads where not needed, stays corrode much less quickly.

If nuts are used on screw stays, must they have a true, flat bearing on the plate?

Answer: Yes. If the stay plates aren't parallel, a tapered washer must be used between the nut and the plate to compensate for the angle. Also, heads of screw stays must be hammered down when first installed.

What are riveted stays?

Answer: Stays in which the threaded ends are riveted instead of having a nut at each end (Figs. 7-10 and 7-11).

Describe a riveted stay as used on the sides of the firebox in vertical and locomotive boilers.

Answer: It consists of a rod threaded at the ends and turned down along the middle section to a diameter slightly less than that of the root of the threads, as in Fig. 7-11.

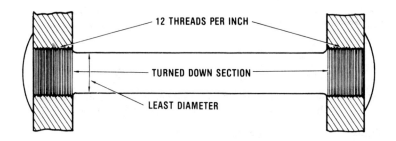

Fig. 7-10. Riveted screw stay or so-called stay bolt. The standard sizes vary from ¾ to 1½ inches in diameter and all have 12 threads per inch.

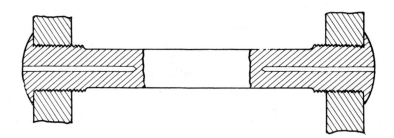

Fig. 7-11. Hollow or drilled riveted screw stay.

Shell or Fire-Tube Boiler Construction

Why slightly less than that of the root of the thread?

Answer: So that in passing the bolt between the sheets, the reduced diameter of the body will clear the threaded hole in the sheet easily.

What is the object of drilling holes through the ends of screw stays, as in Fig. 7-11?

Answer: To indicate, by a steam leak through the drilled holes, a break in the stay and its location. See Fig. 7-12.

Where is the break most likely to occur?

Answer: Near the plate.

What precaution should be taken with drilled stays?

Answer: The holes should not be allowed to become closed by corrosion.

Why do threaded stays sometimes break?

Answer: Since the inner plate is exposed to the intense heat of the furnace, it expands more than the outer one and this sets up a seesaw motion in the stay with changes in temperature. The continued bending of the stays causes their metal to crystallize in time and, eventually, to crack. By drilling the stay as in Fig. 7-12, an extensive crack will be indicated by a leak or a blow through the

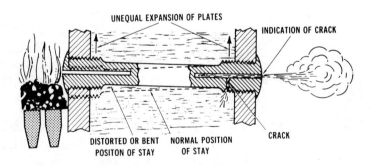

Fig. 7-12. Hollow stays crack.

hole, provided the latter has not become plugged by dirt or corrosion.

What is a socket stay?

Answer: A stay bolt consisting of a rod and socket. See Fig. 7-13.

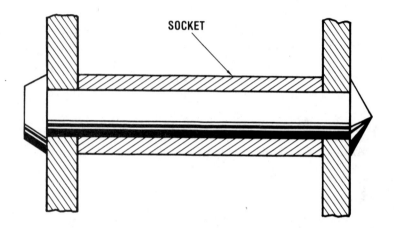

Fig. 7-13. Socket stay or so-called socket bolt. The socket acts as a strut and holds the plates at a proper distance apart.

Describe the placement of a socket stay bolt.

Answer: The socket is placed between the plates to be stayed, the rod threaded through the plates and socket, and the ends riveted.

What is a palm stay?

Answer: A round rod having forged on one end a plate or "palm" and a thread and nut connected at the other end, as in Fig. 7-14.

What is a stay rod or through-stay?

Answer: A stay used to reinforce the parts of the heads not stayed by the tubes. The most common and simple form is a plain rod 1¼ to 2½ inches (3.175 to 6.35 cm) in diameter and threaded at both ends. The rod passes through the steam space and the ends

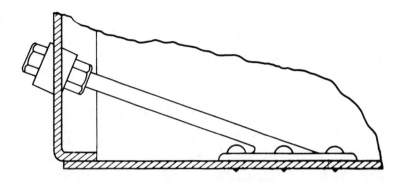

Fig. 7-14. Palm stay, so called because it has a palm-like plate forged at the end. Since the threaded end passes through the head obliquely, two diagonally cut washers are used as connectors between the nuts and plate.

are fastened to the heads but not threaded through as are bolt stays. Length is adjusted in various ways, the simplest being by nuts and washers as in Fig. 7-15. Sometimes the rod is bolted to angle irons which are riveted to the heads, and in this case, turnbuckles are used in place of the nuts for adjusting the length. Often during manufacture through-stays are swelled at one or both ends.

Where are they used?

Answer: Chiefly in marine boilers of the Scotch, Clyde, and HRT types to stay those parts of the heads not stayed by the tubes. They run right through the boiler from head to head.

Why?

Answer: Because these boilers are short and of large diameter, the considerable amount of flat surface in the heads not reinforced by the tubes is conveniently stayed with through-stays without rendering the interior inaccessible.

Most through-stays are designed so that the rear nut is not in contact with heat from the fire. Why is this?

Answer: If the nuts were exposed to the heat of combustion, they would become weak and eventually burn off completely.

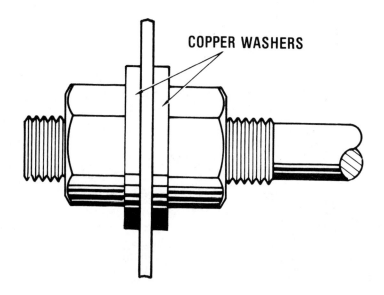

Fig. 7-15. Stay rod or through-stay.

Why are large copper washers used?

Answer: They prevent abrasion of the plates by the nuts and act as packing in securing a tight joint.

How far apart should the stays be spaced and why?

Answer: At least 14 inches (35.6 cm), so that a man can pass between them.

What is a stay tube?

Answer: A thick tube with threads on the ends, one end being larger than the other so that the tube may be slipped through the large hole. See Fig. 7-16. These tubes are practically obsolete because many tests show that the holding power of expanded

SHELL OR FIRE-TUBE BOILER CONSTRUCTION

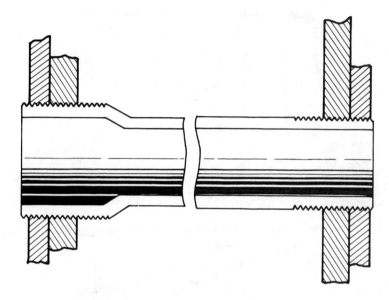

Fig. 7-16. Stay tube ends.

tubes is more than is necessary to support the pressure coming on the plate.

What are gusset stays?

Answer: A type of diagonal stay consisting of a flat piece of plate iron secured to the shell and head by angle irons (Fig. 7-20). Other diagonal stays are shown in Fig. 7-17 through 7-19 and a crow-foot stay in Fig. 7-21.

How is the stress distributed in a gusset stay?

Answer: Because of the character of the stress coming on a gusset stay, it should be proportioned for a larger factor of safety

Fig. 7-17. Diagonal stay bent to form from a flat steel plate.

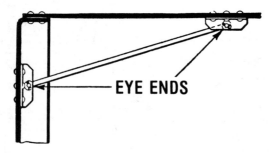

Fig. 7-18. Diagonal stay with eye ends. It is attached to the boiler by angle irons and pins.

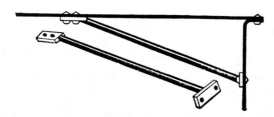

Fig. 7-19. Diagonal stay with forged ends.

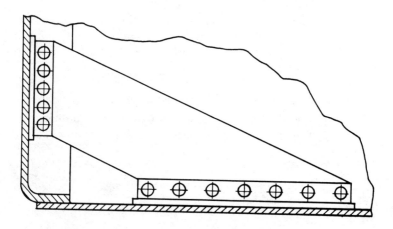

Fig. 7-20. Gusset stay.

SHELL OR FIRE-TUBE BOILER CONSTRUCTION

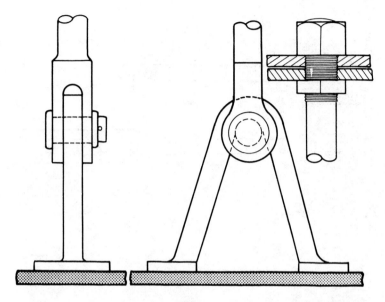

Fig. 7-21. Crow-foot stay, consisting of a rod with forked end, attached by a pin to a V-shaped end with palms or so-called crow foot, the palms of which are riveted to the flat plate to be stayed.

than that for ordinary diagonal stays. The tension is not uniform, but is greater near one edge.

What is a jaw stay?

Answer: A round bar having jaws forged at one end and a flat plate at the other inclined at the proper angle for riveting to the boiler shell, as in Fig. 7-22. The jaw end is attached by a pin to a T iron which is riveted to the head.

What is a steel angle stay?

Answer: Two lengths of steel angle riveted together to form a T-shape piece and then riveted to the plate. See Fig. 7-23.

Where are they used?

Answer: They are used for staying the segment of heads above the tubes on boilers not over 36 inches (91.4 cm) in diameter and limited to 100 pounds pressure.

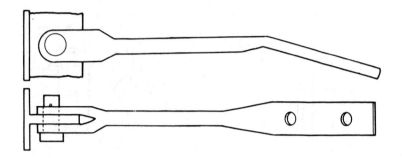

Fig. 7-22. Jaw stay.

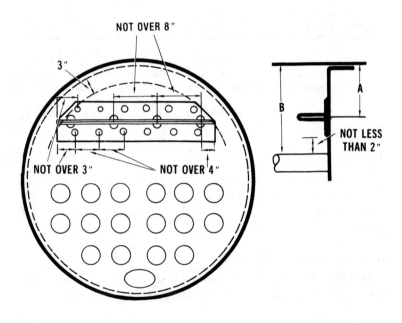

Fig. 7-23. Staying of head in tubular boiler with steel angles.

SHELL OR FIRE-TUBE BOILER CONSTRUCTION

What are crown or radial stays?

Answer: Either screw or through types used to reinforce curved surfaces where the radius of each is different.

What is a crown or roof bar?

Answer: It consists of a solid bar or one made up of two plates welded together at the ends and having a depth of about 4 to 6 inches (10.2 to 15.2 cm) with proper thickness to support the load placed on it. Either bolts or rivets may be used to keep the plates which form the girder from spreading.

What is the application of crown or roof bars?

Answer: They are used for supporting the flat tops of fireboxes and combustion chambers, especially in locomotive and marine boilers (Fig. 7-24).

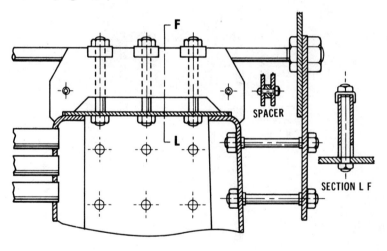

Fig. 7-24. Details of crown bar construction supported by side plates.

BOILER TUBES

How are boiler tubes made?

Answer: They're made of either thin plate, rolled into shape and welded (lap-welded), or they are drawn from solid iron or steel.

How are tubes attached to the tube sheet?

Answer: The end of the tube is expanded into the end of the sheet, with enough of the tube end extending through the sheet to allow it to be beaded over against the outside surface of the tube sheet, as shown in Fig. 7-25. In many cases the bead is seal-welded, especially on older boilers where it may be difficult to seal by rolling and beading. Several other methods are shown in Fig. 7-25.

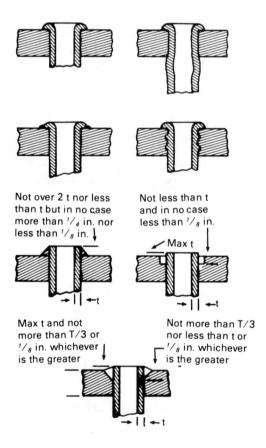

Fig. 7-25. Acceptable methods of fastening fire tubes to tube sheets.

Shell or Fire-Tube Boiler Construction
VERTICAL TUBULAR BOILERS

Name two types of vertical shell boilers.
Answer: The through-tube dry-top and the submerged-tube wet-top.

Describe a through-tube vertical boiler.
Answer: An outer cylindrical shell encloses the water and steam space. Within this shell is a smaller cylinder extending about one-third of the way up which forms the furnace, combustion chamber, and ash pit. The cylindrical furnace is flanged outward at the bottom until it meets the outer shell, dispensing in this way with a lower head. In one side it flanges to the shell to form an opening for the furnace door; the top is flat and into it are expanded a multiplicity of vertical tubes, the upper ends of which are expanded into a similar flat surface at the top of the shell. These flat surfaces are called, respectively, the lower and upper tube sheets. The water level in a dry-top boiler covers two thirds of the tube length. In a wet-top boiler the entire length of the tubes is covered (see Figs. 7-26 and 7-27).

How is the cylindrical furnace fastened to the outer shell?
Answer: By a proper number of stay bolts. See Figs. 7-28 through 7-30.

What would be the tendency in the absence of these stay bolts?
Answer: The cylindrical furnace would tend to collapse.

What is the construction of submerged tubes of vertical boilers?
Answer: The top head of the shell is riveted or welded to a conical submerging chamber, called a "cone," of sufficient depth that the upper tube sheet attached to the lower flange of the cone is submerged below the water level.

What is the objection to this arrangement?
Answer: It complicates the construction and renders the upper tube sheet less accessible.

POWER PLANT ENGINEERS GUIDE

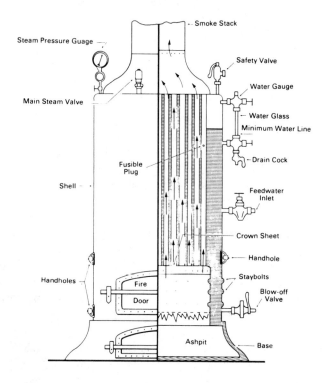

Fig. 7-26. Vertical dry-top tubular boiler.

Where is a fusible plug placed?

Answer: In an outer tube heavier than the others and located at a distance not less than one-third of the length of the tube from the bottom tube sheet. (See Figs. 7-26 and 7-27.)

How is the fusible plug attached to the tube?

Answer: It is screwed into the tube.

Are fusible plugs reliable?

Answer: Yes, if they are inspected regularly and replaced when deterioration is evident. If neglected, they are no more reliable than other neglected safety devices.

Shell or Fire-Tube Boiler Construction

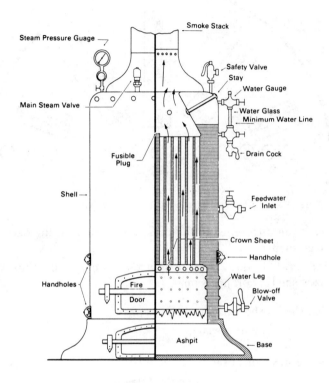

Fig. 7-27. Vertical wet-top tubular boiler.

Why are tube sheets thicker than shell plates?

Answer: They are subject to strains put on them by the expansion and contraction of the tubes. Moreover, they must be thick enough to have sufficient strength to accept the tubes without expanding the holes.

What data is stamped or placed on the nameplate?

Answer: Such items as the name of manufacturer of the steel, its tensile strength, the name of the manufacturer of the boiler, the boiler number, and other data relating to specific boiler in question.

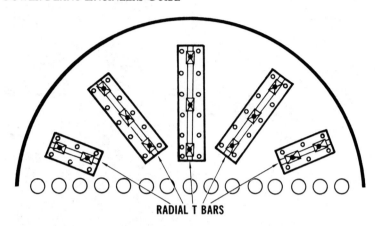

Fig. 7-28. Radial T bars for fastening stays to heads.

What is the difference in the pressure exerted on the outer and furnace shells?

Answer: The outer pressure is in tension and the pressure on the furnace shell is compression.

Which shell is the weaker of the two?

Answer: The inside shell.

Why are stay bolts used?

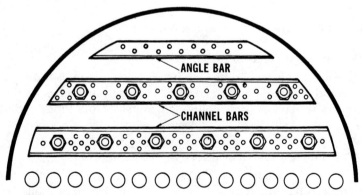

Fig. 7-29. Angle and channel bars for through-stay connections.

Shell or Fire-Tube Boiler Construction

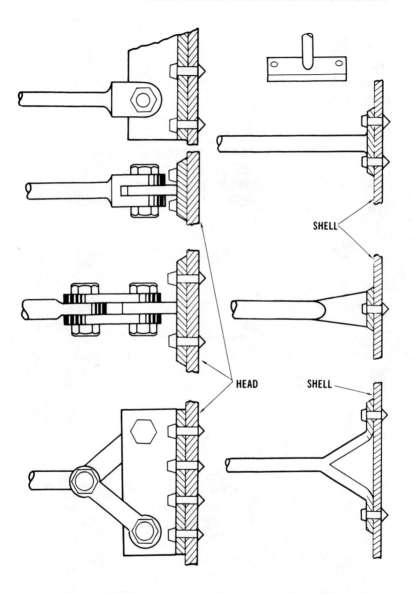

Fig. 7-30. Methods of fastening stays.

Answer: To prevent the bulging or the collapse of the furnace inward.

How about the number of stay bolts provided?
Answer: A large number of small stay bolts is better than a few larger ones.

How is a broken stay bolt detected?
Answer: If of the drilled type, a break is detected by leakage.

How many tube sheets does a vertical boiler have?
Answer: A top tube sheet and a bottom tube sheet.

HORIZONTAL FIRE-TUBE BOILERS

How are horizontal fire-tube boilers fired?
Answer: Either internally or externally.

How is a horizontal return tubular boiler fired?
Answer: An HRT boiler is fired externally by a furnace enclosed in a bricked setting. Combustion gases sweep the lower part of the boiler shell before returning to the front of the boiler through a number of small tubes (see Fig. 6-10) and leave the furnace through the stack. Return tubes are usually larger than 2 inches (5 cm) and smaller than 4 inches (10.2 cm).

How is an HRT boiler mounted?
Answer: Pitched toward the back so the bottom blow-down line will work effectively. Water level is held at least two inches over the top of the tubes at the front (high end) of the boiler.

How are HRT boilers protected against low water?
Answer: By a fusible plug mounted in the tube sheet so it is normally covered by water. If the water level drops below the predetermined safe level, the fusible plug melts and allows steam to escape into the combustion chamber.

Shell or Fire-Tube Boiler Construction

Can HRT boilers carry overloads?

Answer: Yes, due to the large steam space, large overloads can be carried for rather long periods. But HRT boilers are slow steamers.

How does water circulate in an HRT boiler?

Answer: From the hot surfaces upward. Cooler water displaces it, following the curvature of the shell, in a natural circulation pattern.

Do the tube sheets need extra strengthening beyond that given by the tubes?

Answer: Yes, by longitudinal stays placed just below the bottom row of tubes and by diagonal stays between the upper part of the shell and the tube sheet above the top row of tubes (see Fig. 6-10).

How is a Scotch marine boiler fired?

Answer: Internally. Combustion gases make one pass from the furnace, through the fire tubes to the smoke box.

How is a Scotch marine boiler built?

Answer: It has a large cylindrical shell of high strength with the firebox in one end and the smoke box at the stack uptake in the other. The forward tube sheet is located in the smoke box and the rear tube sheet forms the forward wall of the combustion chamber. The fire and stay tubes pass through the water space and support the tube sheets. Fire tubes are beaded over and stay tubes are screwed into and beaded over the tube sheets. The combustion chamber is supported to the shell by short stays, and longitudinal stays support the front and back boiler heads.

What are two furnace types used in Scotch marine boilers?

Answer: Morrison corrugated and the Adamson ring or ring-enforced type. See Chapter 6 for more details.

What is a packaged fire-tube boiler?

Answer: Usually a self-contained steam generator with boiler firing equipment, draft fans, feed pump, and automatic controls, usually shipped completely assembled. Most packaged boilers are oil- or gas-fired (or dual-fired) with an internal 3- or 4-pass furnace.

SHELL BOILER OPENINGS

Why are openings provided leading into the water and steam spaces of shell boilers?

Answer: They are necessary for proper inspection, care, and operation of the boiler.

How are the openings classified?

Answer: They are divided into two classes known as major openings, and minor openings. See Figs. 7-31 and 7-32.

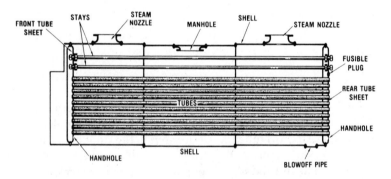

Fig. 7-31. Major and minor openings in a horizontal return tubular boiler.

Name the major openings.

Answer: Manholes and handholes.

What are the two classes of minor opening?

Answer: The steam openings and the water openings.

Name the steam minor openings.

Answer: (1) Main outlet, (2) outlet for safety valve, (3) outlets

SHELL OR FIRE-TUBE BOILER CONSTRUCTION

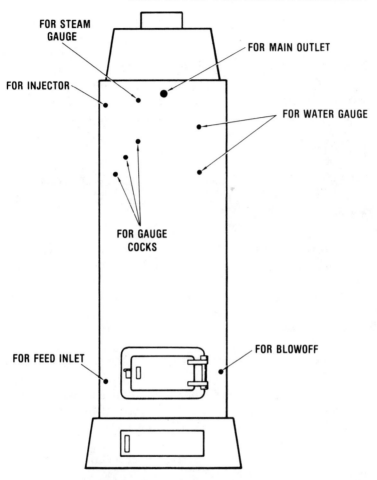

Fig. 7-32. Minor openings in a vertical boiler. In this type boiler, the fusible plug is tapped into one of the fire tubes and is reached through a handhole at the low-water level.

for auxiliary steam, (4) outlet for injector, (5) outlets for gauge cocks, (6) feed-water inlet, and (7) blow-off outlet.

When used for inspection and wash-out purposes, what size should threaded openings be?

Answer: At least 1-inch (2.54 cm), standard tapered pipe

thread, except where other sealing surfaces are provided to prevent leakage.

How are handholes and manholes located?

Answer: Handholes are placed in such a position that accumulations of sediment can be removed and tools can be inserted for cleaning boiler tubes and shell. Manholes are placed so that a person can enter to examine and replace stays, braces, tubes, and pipe connections.

Where is the manhole for a horizontal boiler usually placed?

Answer: In the top of the shell or, for large boilers, in the head above the water line.

What is the usual shape and size of a manhole and how is it placed?

Answer: Elliptical manholes are usually 11 by 15 or 10 by 16 inches (27.9 by 38.1 or 25.4 by 40.6 cm), placed so that the longer measurement is at a right angle to the axis of the boiler shell. Circular manholes are 15 inches (38.1 cm) in diameter at least.

How does removing a section of plate for the manhole affect the strength of the shell?

Answer: The strength of the shell is reduced; therefore, reinforcement must be used, either in the form of flanging over the shell or riveting a collar around the opening.

Why are only one or two bolts sufficient for securing a handhole or manhole cover to the boiler?

Answer: Because the steam pressure does not come on the bolts, but on the boiler plate. The bolts serve merely to hold the cover firmly in place against its gasket when there is no internal pressure on the boiler.

Where is the handhole placed in horizontal tubular boilers?

Answer: In each head below the tubes. In vertical boilers, they are placed opposite the lower tube sheet and at the bottom of the water leg.

Shell or Fire-Tube Boiler Construction

How large are handhole openings?

Answer: A minimum of 2¾ by 3½ inches (7 by 8.9 cm).

How is a tight-joint secured on a handhole or manhole?

Answer: By means of a gasket, which should be painted with a mixture of heavy oil or graphite or prepared commercial lubricant with a large percentage of graphite to keep gaskets from adhering to the cover plate.

Name the three parts of the manhole or handhole assembly.

Answer: The *cover*, the plate that covers the opening; the *yoke*, the parts that bridge across the opening to provide a surface on which to tighten the bolt; the *bolt*, used to draw the cover tightly in place. (See Fig. 7-33.)

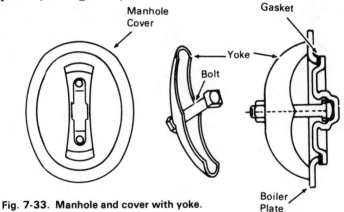

Fig. 7-33. Manhole and cover with yoke.

What is a steam dome?

Answer: A chamber on the top of a boiler.

What is the object of a steam dome?

Answer: To get rid of some of the moisture in the steam.

What is the present status of steam domes?

Answer: They are practically obsolete except on some special boilers.

What size opening is cut in the shell to communicate with the dome?

Answer: Some manufacturers place the upper manhole in the dome, while others make the opening just large enough to pass the steam at the proper velocity.

What is the usual proportion of a steam dome?

Answer: The diameter and height are usually about one-half the diameter of the boiler.

CHAPTER 8

Water-Tube Boiler Construction

What is the main difference between a water-tube boiler and a shell or fire-tube boiler?
Answer: The water is inside the tubes of a water-tube boiler and outside the tubes in a shell boiler, as illustrated in Fig. 8-1.

What are the advantages of a water-tube boiler?
Answer: Good circulation, less prone to explosion since it is made up of small tubes, easier replacement of defective or worn parts, and steam and water drums are not directly in the path of hot gases. Water-tube boilers are rated in Btu of heat absorbed or in the amount of steam generated per hour.

BOILER TYPES AND CONSTRUCTION

How are water-tube boilers classed?
Answer: Straight-tube, bent-tube, packaged, field-erected, and utility.

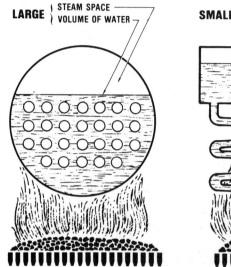 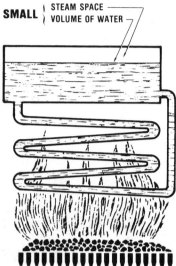

Fig. 8-1. The shell boiler (left) with its large steam and water spaces is less sensitive to sudden load changes than the water-tube boiler at right.

What size are the water tubes in a water-tube boiler?

Answer: Usually under 4 inches (10.16 cm) in diameter.

How can unit steam-generating capacity be increased in a water-tube boiler?

Answer: By making the tubes longer and adding more of them.

In early water-tube boilers, how were the tubes arranged?

Answer: Vertically, extending between the water (mud) drum at the bottom and the steam drum at the top.

Why have vertical water-tube boilers passed from favor?

Answer: Large size made them difficult to transport; poor economy and power-to-space occupied ratio.

Describe the next step in water-tube boiler design.

Answer: The longitudinal drum straight-tube boiler design using box headers (Fig. 8-2). Tubes enter the headers at right

WATER-TUBE BOILER CONSTRUCTION

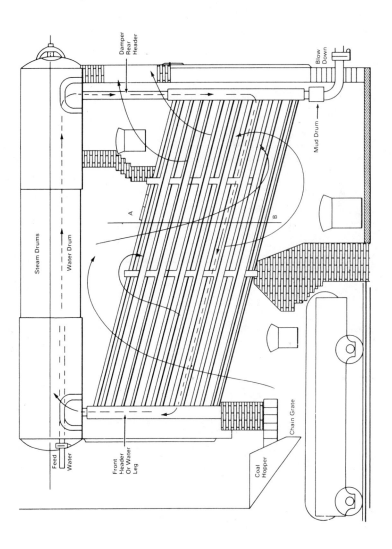

Fig. 8-2. Cross-section drawing of a straight-tube box header boiler.

angles and are expanded into the tube header mounting holes. The section of tubes is inclined at an angle of 15° to the horizontal steam drum. Headers are connected to the steam drum by short lengths of tubing through a forged steel nozzle or are saddle-riveted to the shell. Rear headers are connected by short tubes to a mud drum at the lower end. Each tube can be replaced through individual handholes on each header. (See Fig. 8-20.)

How does water circulate in a longitudinal drum straight-tube boiler?

Answer: Feed water enters the steam drum through the front end and is discharged toward the rear through an internal pipe. The discharge is in the same direction as circulation. Water flows down the rear box header and upward through the water tubes to the front header and then on to the front end of the drum. Steam is released from the drum to the space. Combustion gases flow from the burner or grate over and around the tubes to the flue (Figs. 8-2 and 8-3).

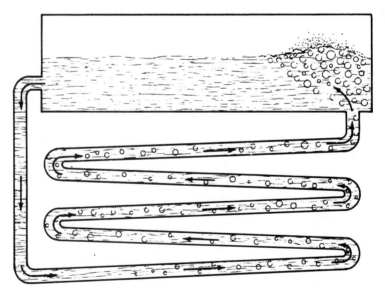

Fig. 8-3. Upflow water circulation.

WATER-TUBE BOILER CONSTRUCTION

What are the methods of grouping the tubes in water-tube boilers?
Answer: Sectional and nonsectional; as shown in Fig. 8-4.

Of what does the tubular heating surface consist?
Answer: Either tubes or wrought pipe. See Fig. 8-4.

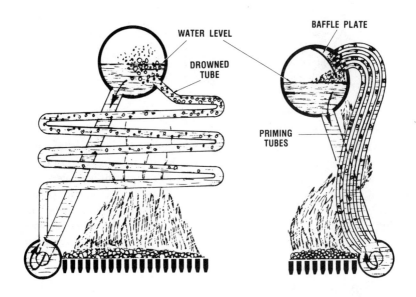

Fig. 8-4. Sectional (left) and nonsectional (right) arrangements of the heating surface.

What are sinuous sectional headers?

Answer: Sectional headers are built up of vertical sections or units. Each unit has one or two vertical rows of tubes, usually staggered so the tubes in one horizontal row are midway between the tubes in the next rows above and below (Fig. 8-5). The number of sections needed can be clamped together to form the necessary boiler width. Headers can be vertical with tubes entering at an angle, or headers can be tilted at an angle with tubes entering at right angles. Each section is separate, so circulation through a header can follow a number of paths.

POWER PLANT ENGINEERS GUIDE

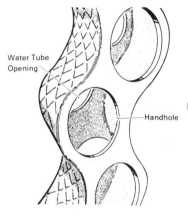

Fig. 8-5. Detail of sinuous sectional header showing construction.

How are sectional headers made?

Answer: Of wrought steel. Each has a handhole opposite each tube end (Fig. 8-5), with the cover held in place by a dog either inside or outside (Fig. 8-20). Two types of handhole covers are used: an outside cover with a machined joint that needs no gasket and an inside which seats against a gasket. Handholes allow cleaning of the tubes and tube replacement. Sectional boilers are suspended from a steel framework, which allows free movement during expansion and contraction.

Describe bent-tube boilers.

Answer: Banks of bent tubes connect into drums or, in a few cases, large headers. Two or three drums is most common, but more are used in some boilers. The upper drum is used for steam release and the lower drum or header is used as a mud drum and has blow-down connections (see Chapter 6). Multidrum designs have several upper drums with all but one immersed. Some designs have what is called a dry drum into which steam is discharged from the steam release drum.

What size tubes are used in bent-tube boilers?

Answer: Three or 3½ inches (7.62 or 8.9 cm) in diameter. Later design package units have smaller tubes, 2 or 2½ inches (5.08 to 6.35 cm).

WATER-TUBE BOILER CONSTRUCTION

How does water usually circulate in a bent-tube boiler?

Answer: Feed water is injected into an upper drum, usually the steam release drum, and distributed across its length to avoid the stresses of adding comparatively cold water. In some cases, water is fed into the mud drum, then it circulates up through the tubes.

What is a packaged boiler? A field-erected boiler?

Answer: A packaged boiler is shipped as one unit, ready to install. A field-erected boiler is shipped to the site in sections for assembly there. Packaged boilers are limited in size due to shipping restrictions. Most packaged-boiler designs have a large number of small diameter tubes. Packaged boilers are considered hard to clean.

What is the difference between a box header boiler and a sinuous sectional boiler?

Answer: A box header has one large section with tubes arranged in parallel rows, one above the other. Sinuous headers are made up of narrow sections with tubes staggered. Generally, bent-tube boilers have replaced straight-tube header types because bent tubes are more efficient in heat and space utilization.

How would you describe a miniature packaged boiler?

Answer: An extremely compact water-tube, or in some cases tubeless, high-pressure boiler designed to generate a comparatively large quantity of steam in a relatively small space.

What is a modular boiler?

Answer: A field-directed boiler with the cost benefits of a packaged boiler because the modules are factory-assembled, necessitating only minimal field erection. Utility and larger power plants use this method of construction.

What minor openings are provided for the ordinary horizontal tubular boiler?

Answer: (1) Two main steam outlets, one for the main steam supply and the other for auxiliaries, (2) an independent outlet for the injector only, (3) two outlets for water column connections, (4) opening for fusible plug, (5) inlet for feed water, (6) outlet for scum cock, (7) outlet for the blow-off cock. See Chapter 12.

How are the fire-door openings constructed on vertical boilers?

Answer: It is necessary to provide an opening through both furnace plate and outer shell for firing. The simplest is "ring construction" in which the plates are riveted against the ring and caulked to give a tight joint.

Describe the door construction.

Answer: The door is made of cast iron and should be sufficiently large to permit the convenient handling of a shovel and slice bar. See Fig. 8-6.

What is the usual size of the door?

Answer: From 12 by 10 inches (30.48 by 25.4 cm) to 16 by 20 inches (40.64 by 50.8 cm).

How is the door protected thermally on the inside?

Answer: By a lining plate of cast iron which should be perforated so that air entering the damper of the door will be divided into fine streams to cool the plate.

What is liable to happen to the lining plate?

Answer: It is likely to warp or crack from the intense heat.

WATER TUBES

How are boiler tubes listed and described?

Answer: By the *outside* diameter. See Fig. 8-7.

How are wrought pipes listed?

Answer: By the nominal *inside* diameter. Thus, a 1-inch wrought pipe is nearly 1¼ inches (3.2 cm) outside diameter, whereas a 1-inch boiler tube is exactly 1 inch (2.54 cm).

WATER-TUBE BOILER CONSTRUCTION

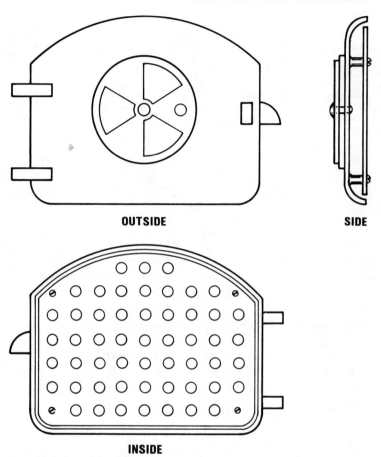

Fig. 8-6. Furnace door detail.

How are seamless tubes manufactured?

Answer: From solid billets by passing the white-hot billet through a piercing mill.

What is the holding power of the boiler tubes?

Answer: Experiments by Yarrow & Co., on steel tubes 2 to 2½ inches (5 to 6.4 cm) in diameter expanded into tube sheets, gave results ranging from 7,000 to 41,715 pounds (3150 kg to 18,772 kg). Most tests give 20,000 to 30,000 pounds (9000 to 13,500 kg).

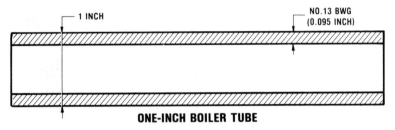

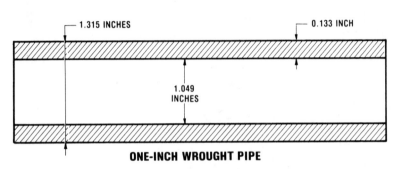

Fig. 8-7. The difference between a tube and a pipe.

How are boiler tubes fastened to the tube sheets?

Answer: By expanding the metal of the tube against the tube plate with a tube expander, and then beading over the ends with a beading tool.

What shapes are given to the tubes or pipes?

Answer: They may be straight, curved, coiled, closed, or porcupine. See Fig. 8-8.

How are the tubes arranged?

Answer: They may be connected (1) all in series, (2) all in parallel, (3) sections in series, (4) sections in parallel, or (5) sections in series-parallel. See Figs. 8-9 to 8-11.

What positions are given to the tubes?

Answer: They may be horizontal, inclined, or vertical. See Fig. 8-12.

Water-Tube Boiler Construction

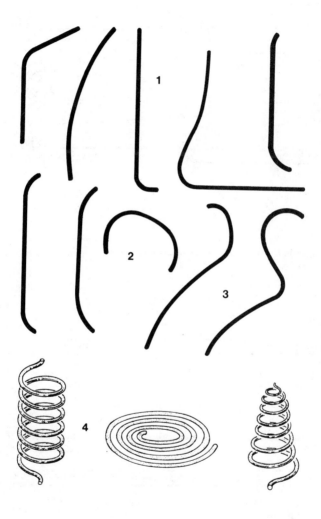

Fig. 8-8. Various forms of curved or bent tube as used in curved-tube boilers. They may be classed as: **1**, single curve; **2**, double curve; **3**, triple curve, etc.; **4**, circular form as helix, flat and cone-shaped spirals, etc.

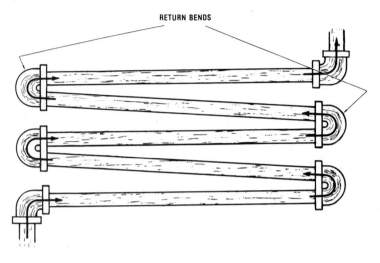

Fig. 8-9. Series connection showing arrangement of pipes joined by return bends.

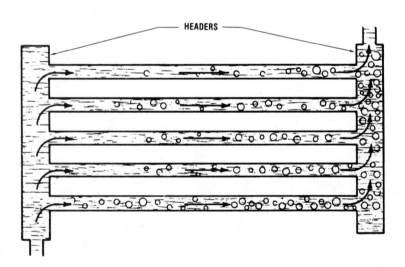

Fig. 8-10. Parallel connection, showing light arrangement of tubes expanded into two headers.

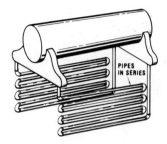

Fig. 8-11. Series light parallel connection, showing-arrangement of pipes tapped into headers; here, only two coils are shown.

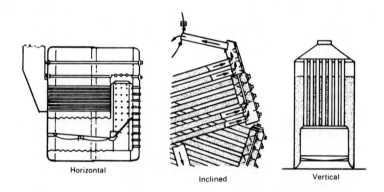

Horizontal Inclined Vertical

Fig. 8-12. Inclination of the heating surface showing tubes in horizontal (left), inclined (center), and vertical (right) positions.

How are tubes classed with respect to circulation?

Answer: As (1) upflow, (2) downflow, (3) over discharge (priming tube), (4) under discharge (drowned tube), or (5) directed flow (double tube). See Figs. 8-3 and 8-13 to 8-15.

WATER-TUBE BOILER PARTS

What are the essential parts of any water-tube boiler no matter how complex its construction may be?

Answer: The parts necessary are (1) steam-water drum, (2) sometimes separate water drum, (3) downflow tubes, (4) lower header, (5) upflow tubes, (6) sometimes upper header, (7) feed-

Power Plant Engineers Guide

water heater, (8) sometimes superheater, (9) grate, (10) ash pan. See Fig. 8-16.

How are these elements assembled?

Answer: Into one unit by means of suitable fittings and connections. The assembly is then placed in an insulated casing.

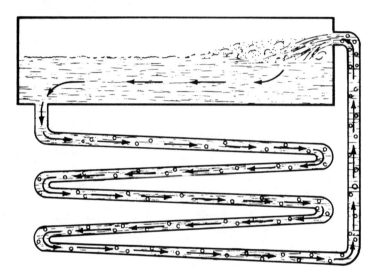

Fig. 8-13. Downflow circulation, as in the porcupine boiler.

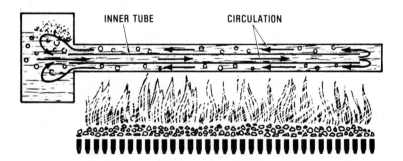

Fig. 8-14. Directed-flow circulation, as in the porcupine boiler.

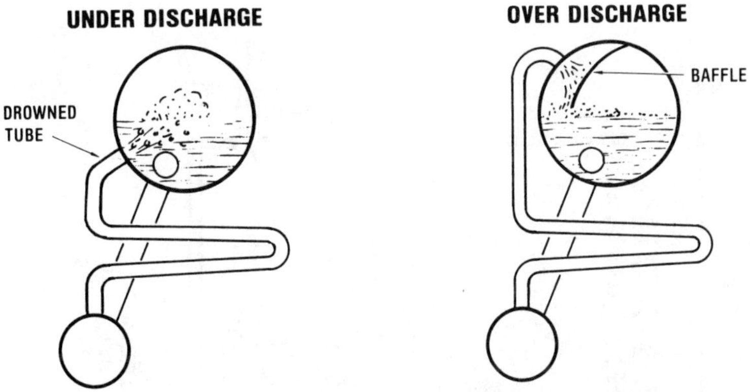

WATER-TUBE BOILER CONSTRUCTION

Fig. 8-15. Principles of under discharge (drowned tube) and over discharge (priming tube). In the latter method a baffle plate is necessary to protect the outlet from spray, especially in the absence of a dry-pipe.

SECTIONAL AND NONSECTIONAL CONSTRUCTION

What is the construction of a nonsectional boiler?

Answer: It consists essentially of a mass of tubes expanded *in parallel* into two headers which connect at the ends of a combined steam-and-water drum, as in Figs. 8-17, 8-18, and 8-19.

What is the advantage of the parallel arrangement of tubes?

Answer: All the tubes are accessible for cleaning.

What is the construction of a sectional boiler?

Answer: The heating surface is divided into a number of sections, each section consisting of a few tubes in parallel expanded into small headers, or a few pipes joined in series by return bends and connected to a header or direct to the drum. Details of construction are shown in Figs. 8-5, 8-20, and 8-21.

What is the advantage of the parallel arrangement?

Answer: The tubes are accessible for cleaning (especially

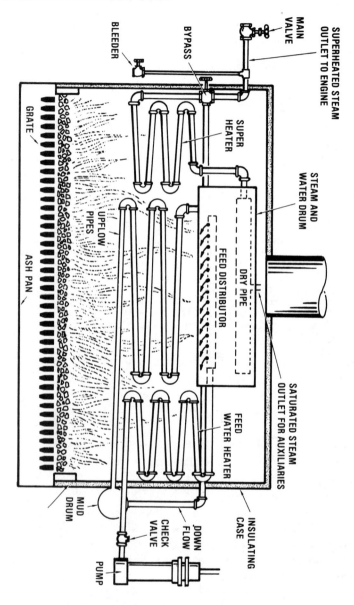

Fig. 8-16. Elementary water-tube boiler, showing essential parts.

WATER-TUBE BOILER CONSTRUCTION

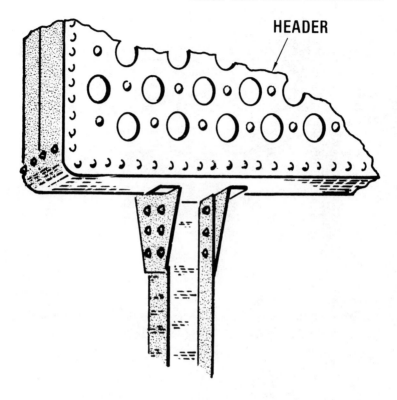

Fig. 8-17. Detail of header for nonsectional arrangement of the tubular heating surface.

important when operated with impure feed water) and are also easy to replace.

What should be noted about the series arrangement?

Answer: Since the tubes are not accessible for cleaning, the series arrangement precludes the use of impure feed water.

What advantages have sectional boilers over nonsectional boilers?

Answer: The sectional boiler can be more easily transported over difficult routes than the nonsectional type because it can be

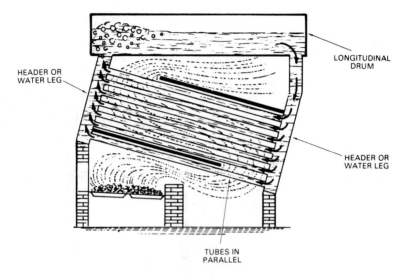

Fig. 8-18. Elementary nonsectional boiler with longitudinal drum, two headers or water legs, and mass of tubes in parallel.

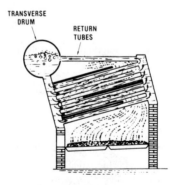

Fig. 8-19. Elementary nonsectional boiler with transverse drum and return tubes. Since only one header is connected to the drum, the return tubes complete the path for circulation.

WATER-TUBE BOILER CONSTRUCTION

Fig. 8-20. Sectional headers showing replacement of a tube which can be accomplished without waiting for the boiler to cool down.

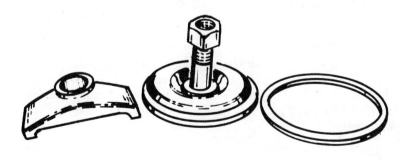

Fig. 8-21. Details of sectional-header handhole plate. The oval shape of the plate permits removal.

knocked down into a number of comparatively light units. The sectional construction avoids the use of stay bolts, and a tube or pipe failure is easily repaired by removing the section. However, in the case of expanded tubes in a nonsectional boiler, the headers are arranged with openings opposite the tube ends so that any tube may be removed.

ADVANTAGES OF VARIOUS TUBE TYPES

Name several advantages of small water tubes over large water tubes.

Answer: A boiler with smaller pipes produces steam faster, reaches operating pressure quicker. Therefore, it can better adjust to varying loads. Small-tube boilers are more compact (generate more steam per hour for the floor space occupied) and more efficient (greater turbulence in the furnace due to the smaller furnace space.)

Name several disadvantages of small-tube boilers.

Answer: A greater tendency for scale buildup and overheating. Since each row of tubes has a different bend, a larger stock of replacement spares is needed. Tube replacement often requires disturbing others to get to a defective tube.

What are the advantages of a straight-tube or header-type boiler?

Answer: Less expensive to build and maintain because it contains more standard parts including identical tubes in any given section. Each tube is easily accessible for cleaning and replacement. Rigid construction adjusts more readily to expansion and contraction.

What are the advantages of bent-tube boilers?

Answer: Natural circulation is better because the tubes are almost vertical; thus, water absorbs heat more efficiently. Bent-tube boilers can be operated at higher pressures and can be forced.

Water-Tube Boiler Construction

Compare the disadvantages of straight- and bent-tube boilers.

Answer: Straight-tube: more expensive initially because they are field-erected, take up more floor space, and are slow steamers. Bent-tube: difficult tube replacement, harder to clean and maintain, feed-water quality is more critical, spare tube stock is more expensive because a larger number of sizes has to be carried.

Why can a bent-tube boiler be forced harder from a cold start?

Answer: Because it has smaller diameter tubes and more of them, which exposes a greater heating surface to direct radiation. Water circulates more readily, too.

What is meant by forcing? Is it a good practice?

Answer: When a boiler is fired hard to bring it up to operating temperature and pressure quickly. It is not a good practice because of the unequal stresses resulting from the sudden change in temperature.

What results are obtained by the use of bent tubes?

Answer: The results obtained are (1) provision for expansion and contraction, (2) longer tube length, (3) flexible disposition of the heating surface, (4) in large boilers, one manhole to be removed instead of individual-tube handhole plates for cleaning, (5) ease of making repairs, depending on the design, (6) curved tubes designed for over discharge give a large space above the grate, thus improving the combustion efficiency.

Under what condition is provision for expansion and contraction especially important?

Answer: For operation under forced draft.

What is the advantage of longer tubes?

Answer: It reduces the number of expanded joints.

What is the advantage of flexible disposition of the heating surface?

Answer: In special cases the heating surface may be suitably located without mechanical difficulties to give good circulation.

What is the point with respect to larger boilers having one manhole?

Answer: The drum is of such size that a man can enter through the manhole and gain access to any of the tubes for cleaning.

What should be noted with respect to straight and curved tubes?

Answer: A straight tube is easier to clean than a curved tube.

How does ease of repair depend upon design?

Answer: In some boilers, any tube may be removed without disturbing the others, whereas in other tube arrangements, it is necessary to start at the beginning of the row and remove all tubes up to the one damaged.

What is the advantage of curved tubes designed for over discharge?

Answer: This arrangement gives a large space above the grate which improves combustion efficiency.

What is the object of tubes arranged in series-parallel?

Answer: This arrangement is found in some large, multidrum boilers. It lends itself to very large boilers, the unit virtually comprising several boiler units combined into one.

What are some shortcomings of direct-draft and baffled-draft designs?

Answer: See Fig. 8-22. With direct draft, there is some short-circuiting of the hot gases which renders portions of the heating surface correspondingly less effective. On the other hand, baffles become coated with ashes or other debris over a period of time and require frequent cleaning; this is especially true with the horizontal type of baffle shown in Fig. 8-22.

MISCELLANEOUS BOILER TYPES

What is a pipe boiler?

Answer: One made of wrought pipe and malleable fittings (Fig. 8-23).

Water-Tube Boiler Construction

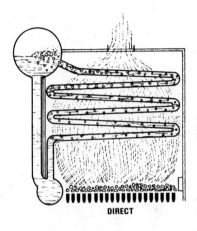

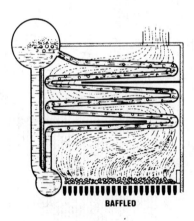

Fig. 8-22. Direct-draft and baffled-draft combustion chambers.

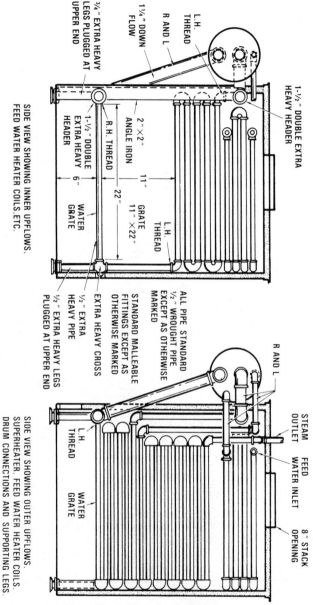

Fig. 8-23. Small, water-grate boiler pipe boiler.

WATER-TUBE BOILER CONSTRUCTION

What are they usually called?
Answer: Water-tube boilers.

What kind of pipe is used?
Answer: The wrought pipe used is listed according to the nominal diameter rather than the actual diameter. There is considerable difference, especially in the smaller sizes.

What weight of pipe is used?
Answer: Standard or, in some cases, extra heavy.

Describe a typical pipe boiler.

Answer: It is built up in sections, each section being composed of a few lengths of pipe connected in series by return bends. The lower end of each section is connected by a right-and-left, long nipple to a bottom header or side-pipe, and the upper end by a short right-and-left nipple to the drum. The left-hand thread connection is in the side-pipe and drum.

What are the features of pipe boilers?
Answer: The material of which they are constructed is cheap and easily obtained in case of repairs. They can be shipped knocked down, facilitating transportation over difficult routes, and are easily assembled by any pipefitter of ordinary intelligence. High steam-pressure may be safely carried.

In the selection of a pipe boiler, what are the points to be noted?
Answer: (1) Accessibility for repairs, especially the location of the right-and-left connections which have to be reached to remove sections, (2) special fittings—these are preferably avoided in design, especially for boilers used in remote places because of the delay in sending to the factory for new parts in case of repairs, (3) provision for cleaning, (4) construction of casing, (5) mud drum and blow-off, (6) lifting ring for connection to hoist tackle in installing.

Power Plant Engineers Guide

Describe the water-grate boiler shown in Fig. 8-23.

Answer: This is a small, sectional, series-pipe boiler designed to furnish steam for experimental purposes. Its chief features include easy and low-cost construction, a water grate, and a furnace enclosed on three sides by the water-heating surface. It can be made up entirely of pipe and fittings, although a lighter drum may be made from a large tube with heads turned out of boiler plate and properly stayed. The sections are made up of ½-inch pipe and return bends. Right-and-left elbows at the lower ends are connected to the upper heads by right-and-left nipples. There are ten upflow sections, eight inner sections, two side sections, and two superheater sections (one on each side). Upflows total 26.6 square feet, feed-water heater 13.3 square feet, superheater 7.1 square feet, total heating surface 47 square feet, grate area 1.92 square feet, and ratio 1 : 25.4. Total length of ½-inch pipe is 212 feet. The grate is made from extra heavy, ½-inch pipe spaced 1⅜ inches between centers, and the case indicated by dotted lines is made from thin sheet-iron lined with asbestos board.

What is a water grate?

Answer: It consists of a series of pipes connected close together in parallel to a header at one end and to the upflow elements at the other. This construction avoids the sagging and burning-out com-

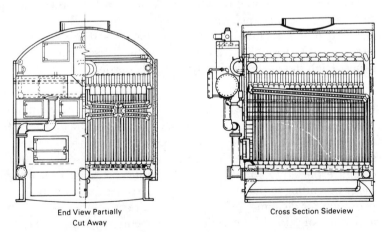

End View Partially Cut Away   Cross Section Sideview

Fig. 8-24. Almy Class E water-tube pipe boiler.

WATER-TUBE BOILER CONSTRUCTION

mon to ordinary grates, especially when forced. In early times, water grates were tried out by James Gurney, Frank Graham, and others. See Fig. 8-25.

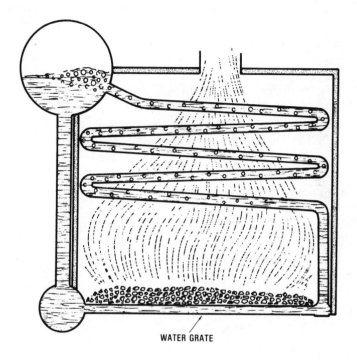

Fig. 8-25. Elementary water grate.

Describe the pipe boiler shown in Fig. 8-24.

Answer: It consists of a large number of pipes screwed into upper and lower manifolds and is composed of side and fore-and-aft sections, a feed-water heater, and steam dome. The top manifold extends across the front and along the sides of the boiler, while the bottom manifold extends along the sides and across the back, below the grates. Between these manifolds are the tubes, which form the heating surface. A drum extends along the front of the boiler, forming a water reservoir connected at the bottom with the lower manifold and at the top to a steam dome where the steam rises or separates from the entrained water. The feed enters the feed-water heater at the top and then passes to the bottom of the horizontal reservoir. It then flows through the large tubes into the lower manifold, passes to the tubes, becomes heated, enters the top manifold as a mixture of steam and water, and flows to the separator. The water from the separator falls to the bottom of the reservoir and continues in the circulation path until evaporated.

What is the difference between upflow and downflow boilers?

Answer: According to the way in which the water passages are arranged, the circulation may be directed upward or downward (Figs. 8-3 and 8-13). Most boilers work on the upflow principle.

In what special-type boiler is the downflow principle used?

Answer: In flash boilers.

What is a flash boiler?

Answer: A boiler consisting of a series of coils of steel tubing. Water is supplied by a pump which delivers the water to the top coil, from which it circulates through the other coils, becoming heated in its descent and issuing from the lower coil as highly superheated steam.

CHAPTER 9

Some Big-Boiler Features

The title of the chapter is appropriate as the size of some boilers has reached such proportions that some of them occupy the space of a building and are over 100 feet high with a capacity of 4,000,000 pounds of steam per hour or more. Many big boilers are field-erected from modular units that are factory-assembled. Modular construction requires only a minimum of assembly in the field, thus offering some of the cost benefits of packaged units.

CLASSIFICATIONS

What are the two general types of these large boilers?
 Answer: Straight tube and curved, or bent tube.

How are the curved-tube boilers classed and why?

Answer: As "drum type" because all tubes are connected directly to drums (no headers).

In some designs, how have these characteristics been combined?

Answer: To provide for furnace wall cooling.

What are the points relating to large, straight-tube boilers?

Answer: They have vertical or slightly inclined headers into which the tubes which constitute the heat-absorbing surface are connected.

What are the two general types of headers?

Answer: Box and sectional.

Describe box headers.

Answer: They consist essentially of two large, flat plates welded or riveted to plates at the edges and secured by stay bolts or tubes at intervals as required to support the plate surfaces. One of these plates is drilled for tube holes and the other for handholes opposite the tube holes.

Describe sectional headers.

Answer: Boilers of this type (Fig. 9-1) have the front and rear headers divided into "sections" (usually one tube wide except for marine boilers) made sinuous to provide for staggering the tubes in the tube bank. See Figs. 9-2 to 9-6.

How are marine sectional-header boilers constructed?

Answer: They usually have single tubes in the lower two or three rows and clusters of small tubes above. As the headers are narrow, they do not require additional stays.

In the case of both the box- and sectional-header boilers of the cross-drum type, how is the drum located and connected?

Answer: It is located above the rear (low) header or headers and is connected to both front and rear headers by steam and water circulators.

BIG BOILERS

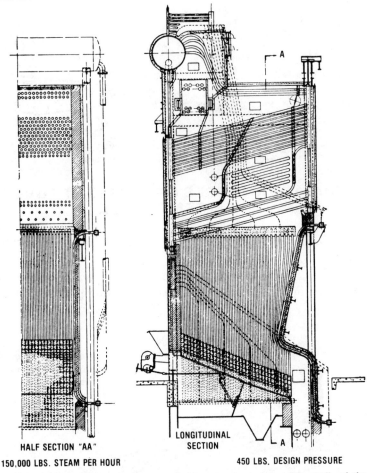

Fig. 9-1. Sectional-header, straight-tube steam generator. The bulk of the heating surface consists of straight tubes connected to the headers.

Describe a modified arrangement.

Answer: Box-header boilers and a few sectional-header boilers have been made with longitudinal drums which are connected to the headers by circulating tubes or (in the case of the box-header boilers) by extending the headers to throats which are connected directly to the drum sheet.

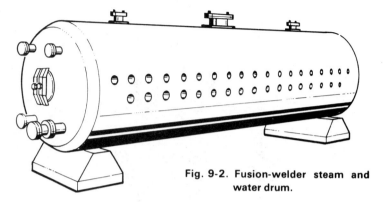

Fig. 9-2. Fusion-welder steam and water drum.

MAIN ADVANTAGES OF STRAIGHT TUBES OVER BENT TUBES

Describe in detail the advantage of straight-tube over bent-tube designs.

Answer: The straight-tube boilers have an advantage over the bent-tube boilers in that all tube-bank tubes are plain, identical,

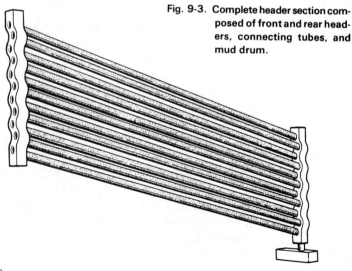

Fig. 9-3. Complete header section composed of front and rear headers, connecting tubes, and mud drum.

SOME BIG-BOILER FEATURES

and can be withdrawn directly through the handholes without disturbing any other tubes in the bank. A tube cleaner will pass through these tubes readily, and to some extent, vision examination is possible. Straight-tube boilers usually require less headroom for the same heating surface.

Fig. 9-4. Vertical and inclined baffle assemblies.

How about the circulation?

Answer: The circulation is not as positive and active as in the bent-tube type, and some trouble has been experienced in large, high-pressure boilers of this type due to inadequate or reversed circulation at high steaming rates, especially in high tube banks.

Are straight-tube boilers preferred?

Answer: They are preferred for some applications, but in general, the trend has been toward the use of bent-tube boilers, fewer and larger units, and higher pressures for a given steam demand. The high head pressure and relatively uncongested circulation make bent-tube boilers more suitable for high evaporation rates.

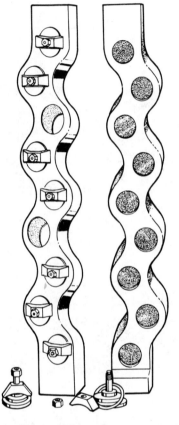

Fig. 9-5. Handhole-side and tube-hole-side views of a typical, forged-steel tube header.

State what size tubes are used in the standard large sectional-header boilers and give the spacing.

Answer: They have 4-inch (10.1 cm) tubes spaced horizontally on 7-inch (17.8 cm) centers, although one well-known boiler uses 3½-inch (19 cm) tubes.

What about baffling?

Answer: Sectional-header boilers are usually partly cross-baffled. (Also see Chapter 8.) Box-header boilers may be either cross-baffled or horizontally baffled; if cross-baffled, they have a somewhat wider tube spacing to permit the use of rotating soot blowers. Horizontally baffled boilers have stay tubes in the head-

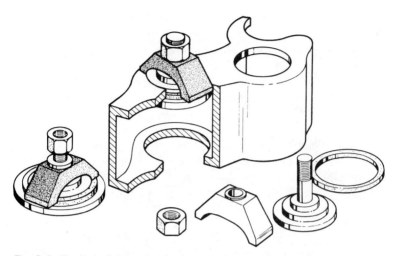

Fig. 9-6. Section of through-tube header, showing handhole closure assembly opposite a tube hole.

ers to permit placing the baffle tile, and hollow stay bolts to accommodate the soot blowers, which blow parallel with the tubes. As the rate of heat transfer from the gas to the tube depends, among other things, on the film thickness or resistance at the boundary layer, a high gas-velocity and turbulent flow are desirable. This means that for the same gas velocity, crossflow increases the heat-transfer rate. For a given tube size, tube spacing is limited by ligament strength and the necessity of providing access for cleaning. With a longitudinally baffled boiler, the exit-gas temperature rises rapidly as the rating increases, so these boilers are not economical for sustained high ratings. They are particularly suitable for heating or processing installations where low headroom is imposed by building limitations, because it is possible to get a large amount of surface in relatively small volume due to the close tube-spacing in the direction of gas flow.

Compare straight- and bent-tube boilers with respect to cost of construction.

Answer: The conventional design of straight-tube boilers (particularly sectional header) is usually heavier and more expensive than bent-tube boilers, for most sizes.

Give construction details of bent-tube boilers with respect to drums.

Answer: Customarily, all bent-tube boilers are arranged with one lower, or mud, drum and from one to three upper drums. There are a few exceptions, as in the case of double-set boilers and boilers having integral economizer sections. The tubes are always arranged in rows in both directions, and at least in the case of the front tube bank, these rows are so spaced (at right angles to the direction of gas flow) as to provide lanes which permit withdrawal and insertion of tubes anywhere in the bank without the destruction of other tubes.

Describe a few details of some well-known, low-pressure boilers.

Answer: Typical are the four-drum and low-head, three-drum boilers made with 3¼-inch (8.3 cm) tubes spaced alternately on 5¼-inch (13.3 cm) and 6¾-inch (17.2 cm) centers. The other three-drum boilers have 3-inch (7.6 cm) tubes spaced on 5¾- and 6¼-inch (14.6 and 15.9 cm) centers or 5⅞-and 6⅛-inch (14.9 and 15.6 cm) centers.

What should be noted about the use of 2-inch (5 cm) tube banks?

Answer: They are desirable from the standpoint of heat-absorption because more heat-absorbing area can be placed in a given space and they can be cross-baffled to give more turbulent flow. However, the spacing is such that many of these 2-inch tubes cannot be withdrawn without the destruction of others. The gas temperatures in the rear bank are relatively low and there is little danger of overheating tubes.

In the case of failure of these 2-inch (5 cm) tubes, how can they be cut off?

Answer: The tube holes may be plugged. The present practice in treating feed water usually prevents formations which might cause serious trouble.

Some Big-Boiler Features

WATER-COOLED FURNACES

In the development of the water-cooled furnace, the first step was the installation of a water screen in the furnace incident to the development of pulverized-fuel firing. The next step was the placing of riser tubes from the bottom screen header against the inside surface of the rear furnace wall. The apparent advantage of this arrangement led naturally to the extension of water-cooling to side- and front-wall surfaces. See Figs. 9-7 and 9-8.

In recent years, water-cooled furnaces have become standard practice in connection with nearly all medium-sized and larger boiler installations.

The tubular absorbing surface is built with fin tubes, plain tubes, and bifurcated tubes, or with various combinations of these tubes, spaced to meet the operating requirements.

With pulverized-fuel-fired installations, it is customary to cool the bottom of the furnace as well as the walls. Dry-bottom furnaces may have plain tubes located above the floor to form a

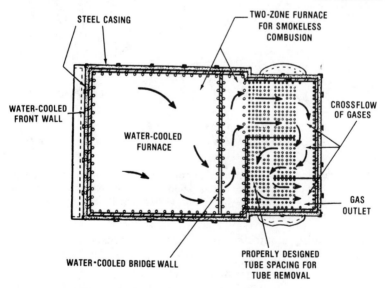

Fig. 9-7. View of integral furnace boiler, showing path of gases.

201

screen or, in some forms of hopper-bottom furnaces, the water screen is omitted and the furnace-wall tubes are continued downward to fan the surface of the inclined walls of the hoppers. In either case, the ash is cooled and the lower part of the furnace is protected from excessive heat.

In slagging-bottom furnaces, water-circulating tubes cool the furnace floor on which the slag bed rests. The slag spout is cooled by water supplied from an outside source.

With stoker-fired installations, those portions of the lower wall surfaces which are subjected to direct contact with the fuel or ash are protected against abrasion by heavy-gauge, finned tubes. Corner openings are sealed by silicon carbide blocks which are bolted to the tubes.

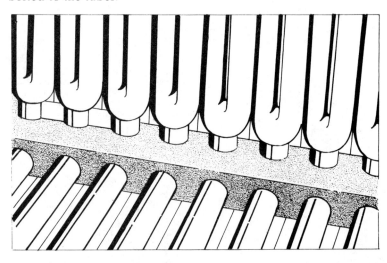

Fig. 9-8. Water-cooled furnace wall constructed with bifurcated tubes to reduce the number of rolled joints and handholes in headers. This all-metal wall is sealed against leakage and is constructed to minimize heat-loss.

How are tubes connected to headers and drums?

Answer: By rolling. An exception is to be found in high-pressure, forced-circulation units where stub tubes are welded to the drum in the shop and the furnace tubes welded to these in the field.

Describe the process.

Answer: In this process, the metal of the tube is forced into firm contact with the metal of the tube seat by the use of an expanding device consisting of a series of rolls and a tapered mandrel. In high-pressure boilers, the tube seats are grooved to obtain a stronger joint. The metal of the tube end is deformed and hardened by cold-working with the expander and is held firmly by the tube seat, which is not correspondingly deformed.

How is it determined when the tube has been sufficiently rolled?

Answer: By measuring the elongation of the tube end with a dial gauge fastened to the tube and in contact with the drum or header.

FORCED-CIRCULATION BOILERS

How is natural circulation affected as the working pressure increases? How remedied?

Answer: The difference between the density of water in the downflow and the water/steam mixture in the upflow sections becomes less. For very high pressure, it becomes necessary to place the drum at considerable height in order to provide the required hydraulic head to assure adequate natural circulation.

What was done because of this condition?

Answer: It led to the development of forced circulation systems. With forced circulation, small tubes are employed; a drum may or may not be required, depending upon the design. Flexibility in the arrangement of heating surfaces is possible, and positive circulation is attainable.

How are large, water-tube boilers usually classed?

Answer: With respect to the number of drums and whether the tubes are straight or bent.

CHAPTER 10

Strength Of Boiler Materials

It is essential that anyone engaged in the design, construction, erection, or operation of a steam boiler should be familiar with the materials that enter into its construction and understand some of their characteristics.

METALS USED

Alloy cast irons: Irons that owe their properties chiefly to the presence of an element other than carbon.

Alloy steels: Steels that owe their properties chiefly to the presence of an element other than carbon.

Basic pig iron: Pig iron containing so little silicon and sulfur that it is suited for easy conversion into steel by the basic open-hearth process. This process is restricted to pig iron containing not more than 1% silicon.

Bessemer pig iron: Iron that contains so little phosphorus and sulfur that it can be used for conversion into steel by the original or acid Bessemer process. This process is restricted to pig iron containing not more than 0.1% phosphorus.

Bessemer steel: Steel made by the Bessemer process, irrespective of its carbon content.

Blister steel: Steel made by carburizing wrought iron by heating it while in contact with carbonaceous matter.

Cast iron: Iron containing so much carbon or its equivalent that it is not malleable at any temperature. It is recommended that the line between cast iron and steel be drawn at 2.2% carbon content.

Cast steel: The same as crucible steel. "Cast steel" is an obsolete term that should be replaced by *crucible steel* or *tool steel*.

Charcoal hearth cast iron: Cast iron from which silicon and usually phosphorus have been removed in the charcoal hearth; it still contains so much carbon as to be distinctly cast iron.

Converted steel: The same as blister steel.

Crucible steel: Steel made by the crucible process, irrespective of its carbon content.

Gray pig iron and gray cast iron: Pig iron and cast iron in the fracture of which the iron itself is nearly or quite concealed by graphite, so that the fracture has the color of graphite.

Malleable castings: Castings made from iron which, when first made, are in the condition of cast iron and are made malleable by subsequent treatment without fusion.

Malleable iron: The same as wrought iron.

Malleable pig iron: An American trade name for the pig iron suitable for conversion into malleable castings through the process of melting, treatment while molten, casting in a brittle state, and then making malleable without remelting.

Open-hearth steel: Steel made by the open-hearth process, irrespective of its carbon content.

Pig iron: Cast iron that has been cast into "pigs," or crude castings, direct from the blast furnace.

Puddled iron: Wrought iron made by the puddling process.

Puddled steel: Steel made by the puddling process, and necessarily slag-bearing.

Refined cast iron: Cast iron that has had most of its silicon

Strength of Boiler Materials

removed in the refinery furnace but that still contains so much carbon as to be distinctly cast iron.

Shear steel: Steel, usually in the form of bars, made from blister steel by shearing it into short lengths, piling, and welding the bars by rolling or hammering them at welding heat. If this process of shearing is repeated, the product is called *double-shear* steel.

Steel: Iron that is malleable at least in some range of temperatures; and in addition, it is cast into an initially malleable mass, is capable of hardening greatly by sudden cooling, or is both so cast and so capable of hardening.

Steel castings: Unforged and unrolled castings made of Bessemer, open-hearth, crucible, or any other steel.

Washed metal: Cast iron from which most of the silicon and phosphorus have been removed by the Bell-Krupp process and which still contains enough carbon to be cast iron.

Weld iron: An obsolete term for wrought iron.

White pig iron and white cast iron: Pig iron and cast iron in which little or no graphite is visible in a fracture; the fracture is silvery and white.

Wrought iron: Slag-bearing, malleable iron that does not harden materially when suddenly cooled.

CHARACTERISTICS OF BOILER METALS

The following definitions of terms are used to define the properties of materials involved in boiler construction. These definitions should be noted and understood.

Brittle: Breaking easily and suddenly with a comparatively smooth fracture; not tough or tenacious. This property usually increases with *hardness*. The hardest and most highly tempered steel is the most brittle. White iron is more brittle than gray, and chilled iron more brittle than any other. The brittleness of castings and malleable work is reduced by *annealing*.

Cold short: The name given to metal that cannot be hammered, rolled, or bent when cold without cracking at the edges. Such a metal can be worked or bent when heated, but not below the temperature assigned to dull red for that metal.

Cold shut: A foundry term applied to cooled metal that does not properly unite at the point of meeting after passing around the two sides of a mold.

Ductile: Easily drawn out, flexible, pliable. Material such as iron is ductile when it can be extended by pulling.

Elasticity: The property that allows materials to resist deformation and which causes them to return to the original form after the deforming force is removed. The *elastic limit* or *yield point* is the smallest force that will permanently deform the material (exceed its elastic property).

Fusible: Capable of being melted or liquefied by the action of heat.

Hardness: The quality of being not easily penetrated or separated into parts; the quality of not yielding to pressure by machining, chipping, and filing.

Homogeneous: Of the same kind or nature throughout. As applied to boiler plates, it means even-grained. In steel plates, there are no layers of fibers and the metal is as strong in one direction as in another.

Hot short: More or less brittle when heated, such as hot short iron.

Malleability: A malleable material can be bent, drawn, or lengthened by beating with a hammer without breaking or fracturing. A malleable iron or steel can be forged by a blacksmith. Flattening tests are used to measure malleability.

Melting point: The temperature at which a solid becomes a liquid. All metals are liquid above their respective melting points and probably all turn into gas or vapor at an even more elevated temperature. Their melting points range from minus 39°F. (the melting point of mercury) to well above 3000°F.

Plasticity: The capacity allowing a material to be molded or altered in shape, to retain a shape formed under pressure after the molding pressure is removed. By applying force, a plastic material is easily formed or molded into a different shape, then holds the new form or shape after the applied force is removed. Steel and iron are plastic when hot. Softer metals, lead and lead-base babbit, for example, are plastic when cold.

Resilience: The quality of elasticity; the property of springing back or recoiling upon removal of a pressure, as with a spring.

Specific gravity: The weight of a given substance relative to an equal *bulk* of some other substance which is taken as a standard of comparison. Water is the standard for liquids and solids and air the standard for gases. If a certain mass is weighed first in air and again in water, and the weight in air divided by the *loss of weight* in water, the resulting figure will be the specific gravity of the mass. For example, a piece of cast iron weighing 10 pounds in air is reweighed with the scale pan submerged in a bucket of water and the scale reads 8.6 pounds. This represents a loss in weight of 1.4 pounds. Dividing 10 by 1.4 produces 7.14, which is the specific gravity of cast iron.

Stiffness: The resistance to bending. Stiff materials are usually brittle and hard.

Strength: Power to resist force; solidity or toughness; the characteristic enabling a material to endure the application of force without breaking or yielding.

Tenacity: The attraction which the molecules of a material have for each other, giving them the power to resist tearing apart; the strength with which any material opposes rupture; its tensile strength.

Tough: Having the quality of flexibility without brittleness; capable of resisting great strain; able to sustain hard usage; not easily separated or cut. Material such as iron is said to be tough when it can be bent first in one direction and then in the other without fracturing. The greater the angle it bends through, coupled with the number of times it bends, the tougher it is.

Weldable: Suitability for welding into a designed structure. The nearer the properties of the welded material to the unwelded material, the more weldable it is. Plate and bar steel can be welded by heating and either pressing or hammering the parts together; gray and alloy cast iron cannot be welded in this manner, however. Most of the metals used in boiler construction can be welded by oxyacetylene or arc processes. Fig. 10-1A shows an example of oxyacetylene welding. The material to be welded has been bevel-ground at the joint and the torch flame will bring the rod as well as the material to a molten state. Fig. 10-1B shows a coated rod about to become part of a joint when an electric arc brings the materials to a molten state. Fig. 10-1C is a cross-section of a welded joint, showing penetration of the weld and a slight excess of weld material.

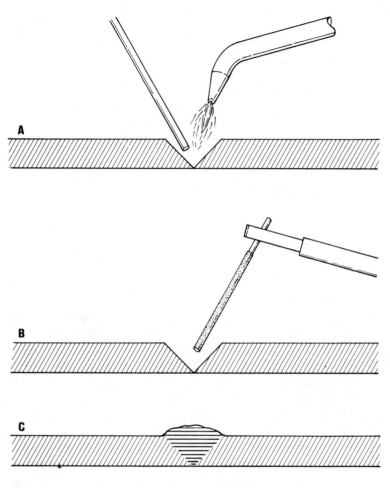

Fig. 10-1. Gas welding (A), arc welding (B), and (C) a cross section of a finished weld.

METAL TESTING

Metals are tested in fairly standardized ways to determine their characteristics as to tensile strength, compression, transverse stress, shear strength, torsional stress, hardness, bending, and

Strength of Boiler Materials

homogeneity. Samples of standardized dimension are taken from each batch of material, tested, and the results compared to the desired specification. Material which fails to meet specifications is rejected.

Bending Stress

This is produced by a force acting upon some member of a structure and tending to deform it by bending or flexing. The effect of the force causes bending strain on the fibers or molecules of the material of which the part is composed. Bending tests measure a material's resistance to bending. Flange testing is very similar. An example of pure bending stress is the deflection of a lever produced by pulling on the end of it while performing work.

Bending Tests

The cold-bend test consists essentially of bending the cold sample through 180° flat on itself, if the sample is 1-inch thick or less; if over 1-inch thick, it is bent into a U-shape around a pin equal in diameter to the thickness of the sample. The quench-bend test involves heating the sample to a light cherry red (not less than 1200°F.), quenching it at once in water between 80° and 90°F., and then bending it through 180° flat on itself. To pass either test, the specimen must not crack on the outside of the bent portion. Rivets used in boiler construction and repair are cold-bend tested in the same manner. In addition, the rivet head must flatten while hot to a diameter 2½ times the diameter of the shank without cracking at the edges.

Compression

Compression involves pressing or pushing the particles of a member closer together as is done, for example, by the steam pressure in a boiler acting upon the fire tubes. The principle of compression testing is shown in Fig. 10-2, where the sample to be tested is placed between a movable ram and a fixed base. Two reference marks a standard distance apart are placed on the sample, and the compression load represented by the movable weight is gradually increased. Changes in the distance between the refer-

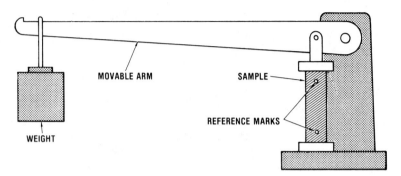

Fig. 10-2. Elementary device for compression testing.

ence marks and the distance between the longitudinal edges of the sample are observed according to the force applied and compared to specifications.

Deformation

This represents a change of shape or a disfigurement, such as the elongation of a sample of iron or steel under test for tensile strength. In general, all tests are applied to samples until the latter are permanently deformed or, in some cases, destroyed.

Etch Test

A chemical agent is applied to a sample of the material to "eat into" the surface to allow examination of the grain of the metal.

Factor of Safety

The ratio between the breaking load and what is selected as the safe working load is known as the factor of safety, or safety factor. For example, if the breaking load of a bolt is 60,000 psi and the working load is 6000 psi, the safety factor is 60,000 ÷ 6000, or 10.

Force

A force is that which changes or tends to change the state of a body at rest or that which modifies or tends to modify the course

Strength of Boiler Materials

of a body in motion. A force always implies the existence of a simultaneous, equal, and opposite force called a *reaction*.

Hardness Test

There are two methods of testing for hardness. The first is by pressing a hardened steel ball into the sample under a fixed pressure and noting the diameter of the indentation. The second is by letting a weight fall from a given height onto the sample and noting the height of the rebound. In these tests, the hardest material will have the smallest indentation and cause the highest rebound.

Hydrostatic Test

A test designed to test the quality of the workmanship of the boiler manufacturer. It is also used after each boiler repair.

Load

Load is the total pressure acting upon a surface. For example, if an engine piston has an area of 200 square inches and is subjected to steam pressure of 150 pounds per square inch (psi), the load or total pressure acting on the piston surface is 200×150, or 30,000 pounds.

Member

A part of a structure, such as a brace, rivet, or tube, that is subject to stresses is called a member.

Modulus of Elasticity

Also called the *coefficient* of elasticity, this represents the load per unit or per section, divided by the elongation or contraction per unit of length. Within the elastic limit, when the deformation is proportional to the stress, the modulus of elasticity is constant; beyond the elastic limit, however, it decreases rapidly. There is a well-defined elastic limit to wrought iron and steel, and the modulus within that limit is nearly constant.

Modulus of Rupture

This is a value obtained by experiment upon a rectangular bar sample supported at the ends and loaded at the middle. Its numerical value comes through the use of the following formula:

$$R = \frac{3Pl}{2bd^2}$$

where,

P is the breaking load in pounds,
l is the length of the bar in inches,
b is the width of the bar in inches,
d is the thickness of the bar in inches.

Machinery used to perform the experiment is similar to that used in the transverse test (Fig. 10-3).

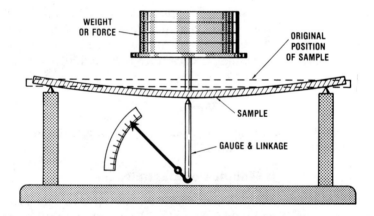

Fig. 10-3. Elements of transverse testing. Deflective force is increased at the sample's midpoint until the breaking load is reached.

Permanent Set

When a metallic piece is stressed beyond its elastic limit, permanent deformation occurs. This alteration in form is known as *permanent set* and results from the piece being stretched, crushed, bent, or twisted, according to the nature of the stress.

Resilience

This is the property of springing back or recoiling upon removal of a pressure, as with a spring. Without special qualifications, the term is understood to mean the work given out by a piece (strained similarly to a spring) after being strained to the extreme limit within which it may be repeatedly strained without rupture or permanent set.

Shear

A member is said to be *in shear* when acted upon by external forces so as to cause adjacent sections of the member to slip past each other. The principles of shear testing are shown in Fig. 10-4. The sample is placed in a holder and measured stress is applied until the sample is sheared. The device shears the metal in a single plane for single-shear testing and in two planes for double-shear testing.

Strain

According to Wood, strain is the name given to the kind of alteration produced by stress. This distinction between strain and stress is not always observed in technical writings because there is much confusion among writers as to these terms.

Stress

Stress is an internal action or internal force set up between the adjacent molecules of a body when acted upon by external forces. It is the force or combination of forces which produces strain.

Tensile Strength

This is the cohesive power by which a material resists an attempt to pull it apart in the direction of its fibers (if any), but it bears no relation to the capacity of a material to resist compression. Tensile strength is tested by a device operating on the principle of that shown in Fig. 10-5. Two reference marks are made a standard distance apart on a standard specimen of the material to be tested. One end of the sample is secured to the fixed end of the machine and the other end secured to the movable head, thus placing the

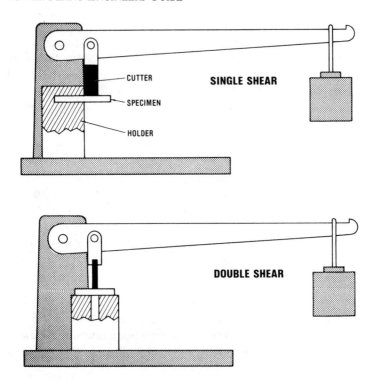

Fig. 10-4. Shear testing, illustrating single-shear and double-shear.

sample in tension. Increasing stress is applied (in this case, by sliding the weight to the left) and the distance between the reference marks is measured at each increase until the sample ruptures. Besides elongation, observations are also made as to its contraction of area for various loads, its elastic limit, and the breaking load.

Tension

Tension is the stress or force causing or tending to cause the stretching or extension of a member. When thus stressed, the member is said to be *in tension*.

Strength of Boiler Materials

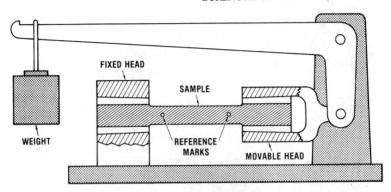

Fig. 10-5. Working principle of a device used for testing tensile strength.

Torsion

Torsion, or torque, is a twisting force expressed as the product of force times distance, such as in inch-pounds, foot-pounds, or newton-meters. Torsional testing is used to determine the torsional elastic limit and ultimate torsional strength of a material. A machine operating on the principle of that shown in Fig. 10-6 holds one end of the sample material immovable. An arm of a given length, securely attached to the opposite end of the sample, applies a twisting force of increasing intensity so that an element on the surface of the sample becomes distorted from a straight line into a spiral. The amount of twist depends on the intensity of the torsional force and the resisting force of the metal. It can be read in degrees by means of a deflection indicator attached to the arm-end of the device, and is then expressed as degrees per inch. For example, if the indicator shows 20° and the sample is 20 inches long, the deflection in twist is stated as 20° ÷ 20 inches, or 1° per inch. If the force applied to the arm is 100 pounds and the arm is 30 inches long, the torsional stress is expressed as 100 pounds × 30 inches, or 3000 inch-pounds.

Transverse Test

In this test, the sample is placed on two supports and an increasing load is applied at its midpoint until the sample breaks (Fig. 10-3). The amount of deflection is noted for each increase in load, as is the total load present at the time of rupture.

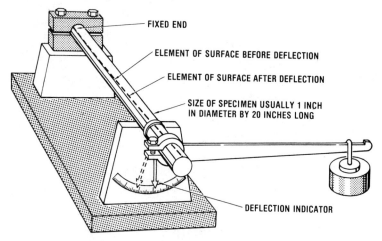

Fig. 10-6. Principle of torsion-testing apparatus.

Ultimate Strength

This is the maximum unit stress developed at any time before rupture occurs.

Yield Point

The yield point is the point at which the stresses and strains become equal and deformation or permanent set occurs. It is the point at which the stresses equal the elasticity of a test piece.

X-Ray Test

The use of X-rays to examine the internal structure of a boiler for flaws. It is used extensively to test welds.

CHAPTER 11

Boiler Calculations

SHELL PRESSURE AND STRESS

What must be considered in determining the strength of a boiler shell?

Answer: It is necessary to consider steam pressure, diameter of shell, tensile strength of the steel plate, thickness of shell, efficiency of joint or ligament, and the safety factor.

How much of the shell is considered in making the calculation and why?

Answer: A section of the shell *one inch long* is taken and its diameter is expressed in inches because the steam pressure, as indicated by the steam gauge, means the pressure acting upon each square inch.

How is the thickness of the shell expressed?

Answer: As a fraction of an inch.

Give an example illustrating *total pressure* to which the metal of the shell is subjected.

Answer: Consider a one-inch length of a shell 10 inches in diameter and suppose the lower half to be filled with concrete and the upper half subjected to a steam pressure of 50 psi as shown in Fig. 11-1. Since the shell is 10 inches in diameter and 1 inch long, the area of the concrete surface exposed to the steam pressure is:

$$10 \times 1 = 10 \text{ sq. in.}$$

and as there is 50 pounds steam pressure acting on each square inch, the total pressure on the concrete is:

$$50 \times 10 = 500 \text{ lbs.}$$

One half of this is carried by the metal of the shell at each side, making the load on the metal 250 pounds.

What would be the *stress* in the shell if it were 1 inch thick?

Answer: Its sectional area would be $1 \times 1 = 1$ sq. in., so the stress in the shell would be 250 psi.

What would be the stress in the shell if it were only ¼ inch thick?

Answer: The sectional area would be $¼ \times 1 = ¼$ sq. in. and the total pressure or load of 250 lbs. would be carried by only ¼ sq. in. of metal. The stress would then be increased 4 times; that is, the metal would be subjected to a stress of

$$250 \div \frac{1}{4} = 250 \times \frac{4}{1} = 1000 \text{ psi}$$

State the method of determining the stress in a shell in the form of a rule.

Answer: Multiply steam pressure in psi by the radius of the shell (in inches) and divide by the thickness of the shell expressed as a fraction of an inch.

What important factor remains to be considered?

Answer: The welded or riveted joint.

Boiler Calculations

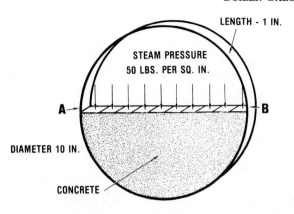

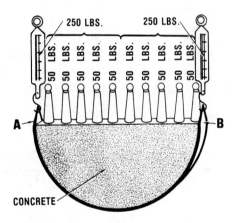

Fig. 11-1. Section of boiler shell, illustrating total pressure to which the metal of the shell is subjected.

Why?

Answer: Because the strength of a riveted joint is always *less* than the strength of the plate.

What is the ratio of the strength of the joint to the strength of the plate called?

Answer: The efficiency of the joint.

Why is a riveted joint weaker than the solid plate?

Answer: Because part of the metal plate is cut away for holes for the rivets. For example, see Fig. 11-2. If the plate is ¼-inch thick and the total pressure on the full plate section is 250 psi, the stress at point P would be 250 ÷ ¼, or 1000 pounds. If the efficiency of the joint is 50%, the area of section J will be one-half of P, or ⅛-inch, making stress along the joint 250 ÷ ⅛, or 2000 pounds. The same result can be obtained more quickly by dividing the stress on the solid plate by the efficiency of the joint, i.e., 1000 ÷ 0.5 = 2000 pounds.

What are some rules covering the figuring of pressure and stress in the shell?

Answer: In order to find the total pressure, or load, to be carried by the shell, multiply the gauge steam pressure in psi by the radius of the shell in inches. In order to find the stress coming on the shell, divide the total pressure just determined by the area of the solid plate per inch of length of longitudinal section and by the efficiency of the joint. Expressed as a formula, this becomes:

$$\text{Stress in Shell} = \frac{\text{Steam Pressure (psig)} \times \text{Radius of Shell (inches)}}{\text{Solid Plate Thickness (inches)} \times \text{Efficiency of Joint}}$$

WORKING AND BURSTING PRESSURE

Upon what does the bursting pressure depend?

Answer: Upon tensile strength of the shell, thickness of the shell, radius of the shell, and efficiency of the joint. If the internal pressure acting on the shell is enough to stress the metal to its tensile strength, the shell will rupture at the weakest section of the joint. For example, see Fig. 11-3. Suppose the thickness of the solid plate is ¼ inch, the diameter of the shell 10 inches, the efficiency of the joint 50%, and the tensile strength 60,000 psi. What would the bursting steam-pressure be? The equivalent thickness of metal for

Boiler Calculations

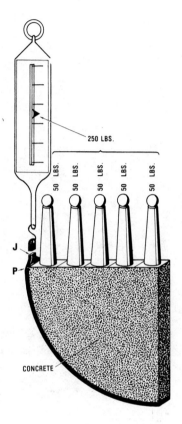

Fig. 11-2. Half-section of shell, illustrating efficiency of the joint.

50% efficiency would be 50% of ¼, or ⅛ inch; since tensile strength is 60,000 psi, only ⅛ of this force, or 7500 pounds, would be the corresponding force necessary to rupture the joint, *acting on the half-section.* In actuality, this pressure is distributed over an area of 5 square inches, making the equivalent steam pressure *per square inch* only 1500 pounds.

What is the rule for determining bursting pressure?

Answer: Multiply the thickness of the shell in inches by the efficiency of the joint and by the tensile strength of the metal. Divide the product by the radius of the shell and the result will be the bursting pressure in psi.

POWER PLANT ENGINEERS GUIDE

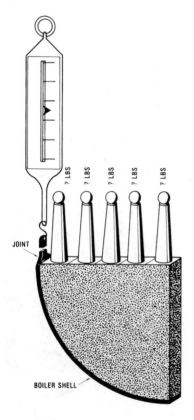

Fig. 11-3. Half-section of shell, illustrating the bursting pressure. The concrete indicates a uniform distribution of pressure due to the weights.

What is the factor of safety?

Answer: The ratio of the bursting pressure to the working pressure.

What is meant by "working pressure"?

Answer: It is the maximum safe pressure to which a boiler is subjected.

What determines the safe working pressure?

Answer: Since it is the maximum pressure to be carried on a boiler consistent with the factor of safety employed in the design, it can be determined by dividing the bursting pressure by the safety factor. For example, if the bursting pressure of a certain boiler is 500 pounds and it is desirable to have a safety factor of 5, the working pressure would be 500 ÷ 5, or 100 psi.

Upon what does the working pressure depend?

Answer: Tensile strength, thickness of shell, radius of shell, efficiency of the joint, and the safety factor. For example, suppose a given boiler has a tensile strength of 60,000 pounds, has plates ⅜-inch thick, is 50 inches in diameter, has joint efficiency of 87%, and a safety factor of 5. What would the maximum allowable working pressure be? (The traditional way to calculate working pressure follows.) A tensile strength of 60,000 pounds corresponds to a stress of 22,500 pounds in a ⅜-inch plate per inch length of section (60,000 × ⅜), and for a safety factor of 5, the maximum load allowable on the solid metal of the shell would be 22,500 ÷ 5, or 4500 pounds. Because the efficiency of the joint, however, is 87%, this figure must be reduced to 87% of its value, which then becomes 3915 pounds—*not* pounds-per-square-inch, but maximum allowable force distributed over the radius of the shell. Since the boiler diameter is 50 inches, a one-inch length of the shell would have a radius of 25 inches and an area of 25 square inches; the maximum allowable working pressure (Fig. 11-4) is 3915 ÷ 25, or only 156.5 psi. Expressed as a formula, the problem is as follows:

$$\text{Working Pressure} = \frac{T \times t \times E}{R \times F}$$

where,

  T is the ultimate tensile strength stamped on the plates in psi;
  t is the minimum thickness of the shell plates in their weakest course, in inches;
  E is the efficiency of the longitudinal joint or of ligaments between the tube holes, whichever is least;
  R is the inside radius of the shell;
  F is the safety factor, or the ratio of ultimate strength to allowable stress.

Is there any other way to calculate maximum working pressures?

Answer: Since many different grades of steel are used in boil-

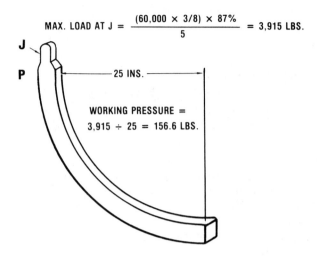

Fig. 11-4. Half-section of shell, illustrating a method of determining the working pressure.

ers, and due to the fact that working temperatures vary widely, the practice of using a number as a factor of safety has been replaced in some cases by a maximum allowable stress factor related to the grade of steel and operating temperature. These values have been determined by the ASME and appear on table P-7 of the ASME Power Boiler Code. The Maximum Allowable Stress table in Chapter 30 lists some of them. The formula used to calculate maximum allowable working pressure is:

$$P = \frac{SEt}{R + 0.6t}$$

where,

P = maximum allowable working pressure in psi,
S = maximum allowable unit working stress in psi determined from the above table,
E = efficiency of the longitudinal joint or of ligaments between tube openings,

t = minimum thickness of steel plates in the weakest course in inches,

R = inside radius of the weakest course of the shell or drum in inches.

SHELL THICKNESS

After figuring the size of a boiler for a given capacity, what is usually the first problem to solve?

Answer: Determining the necessary thickness of the shell for proper safety.

Upon what does the thickness of the shell depend?

Answer: The working steam pressure, radius of the shell, efficiency of the joint, tensile strength of the plate, and the safety factor. For example, suppose you have a 50-inch boiler suitable for 125 pounds working pressure; it is to be made from plates having 60,000 pounds tensile strength, will have joint efficiency of 82%, and is to have a safety factor of 5. What shell thickness (Fig. 11-5) is required? Because the diameter is 50 inches, the radius must be 25 inches; multiplying this figure by the working pressure (125 psi) gives us 3125 pounds, or the total pressure to be carried by the shell. Since the safety factor is 5, the shell must be able to withstand 5 times this load, or 15,625 pounds. With 100% joint efficiency and 60,000 pounds tensile strength, the shell thickness would be 15,625 ÷ 60,000, or 0.26 inches, but we know that the joint efficiency is only 82%. Dividing 0.26 by 0.82 gives us the correct shell thickness of 0.317, or about 5/16 inch.

RIVETED JOINTS

In review, name two general types of riveted joint.

Answer: Lap joints, butt, and single- and double-strap joints.

Name the various types of lap joint.

Answer: Single-riveted, double-riveted, and cover plate.

POWER PLANT ENGINEERS GUIDE

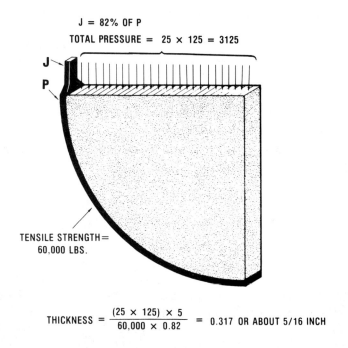

Fig. 11-5. Half-section of shell, illustrating a method of determining the working pressure.

Name the various types of butt and double-strap joints.

Answer: Double-riveted, triple-riveted, quadruple-riveted, and quintuple-riveted.

What is the difference between a lap and a butt joint?

Answer: The plate end overlaps to form a lap joint and register with each other to form a butt joint. A butt joint is reinforced by one or two straps to form tie pieces in riveting.

What is the objection to a lap joint?

Answer: The plate ends through which the rivets pass are in different planes. The pull is not direct, but tends to twist the plates, frequently causing them to bend and sometimes result in an explosion.

Boiler Calculations

In designing a riveted joint, what is the chief object?

Answer: To so proportion it that it will be equally strong against failure by all possible ways of breaking.

What should be noted about the shearing strength of the rivets of lap and butt joints?

Answer: In the case of a lap joint the rivet will shear at only one section, whereas in the case of the butt and double-strap joint, shearing of the rivet is resisted by two sections.

What is the efficiency of a riveted joint?

Answer: The ratio which the strength of a unit length of a riveted joint has to be the same unit length of the solid plate.

In calculating the strength of any form of riveted joint, upon what is the calculation based?

Answer: A unit length or *element* of the joint is taken because the strength of the entire seam is the same as the strength of this element.

How long is the element?

Answer: The length of the element considered depends upon the arrangement of the rivets and is equal to the greatest pitch.

Why should no further length of the seam be considered?

Answer: Because the entire seam is composed of similar elements having the same symmetrical arrangements of rivets and plate material.

In calculating any riveted joint, what are the three things to be considered?

Answer: The strength of the plate, the strength of the rivets, and the efficiency of the joint. The Efficiency of Riveted Joints table in Chapter 30 lists the efficiencies of various types of riveted joints.

How are riveted joints calculated?

Answer: The method can be illustrated with the single lap joint as an example, which is the simplest joint; in Fig. 11-6, ABCD is the

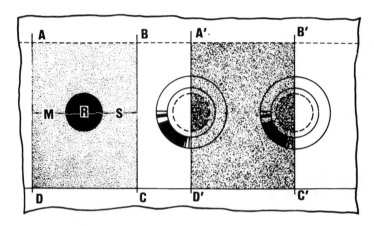

Fig. 11-6. Single-riveted lap joint, illustrating the element of the seam to be considered in determining the strength of the joint. The shaded portion, ABCD, is the element; its equivalent A'B'C'D' may also be considered as an element.

element of the joint to be considered. Since R has been cut out of the plate to accommodate the rivet, the solid metal left to resist the pull is M plus S (the *pitch* is equal to M + R + S). From this and from the preceding information, we can express the strength of the plate by the following formula:

$$\text{Strength of Plate} = t(P - d)T$$

where,

- t is the thickness of the plate,
- P is the pitch,
- d is the diameter of the rivet,
- T is the tensile strength.

The strength of the rivet is equal to its section area times shearing strength, according to this formula:

$$\text{Strength of Rivet} = 0.7854 d^2 \times S$$

where,

- d is the diameter of the rivet,
- S is the shearing strength.

Boiler Calculations

Now, if the tensile strength of the plate is, say, 60,000 pounds but the strength of the rivet is 40,000 pounds, the shearing strength of the rivet will then be only ⅔ the tensile strength of the plate; expressed by formula

$$S = \tfrac{2}{3} T$$

If we substitute this value for S in the preceding formula, we get:

$$\text{Strength of Rivet} = 0.78154 d^2 \times \tfrac{2}{3} T$$
$$= 0.524 d^2 T$$

Obviously, for equal strength of plate and rivet, the values obtained in the formulas for strength of plate and strength of rivet must be equal; that is, $t(P - d)T$ in the first formula must equal $0.524 d^2 T$ as in the last formula. To simplify, $t(P - d) = 0.524 d^2$.

To determine the efficiency of the joint, the strength of the *solid* plate must be considered, and this is equal to the product of its thickness times its pitch times its tensile strength, or tPT.

Since efficiency of the joint is equal to the strength of the joint divided by the strength of the solid plate, and since the strength of the plate at the joint is equal to the strength of the rivet, we can use information from the previous equations to set up this formula for efficiency:

$$\text{Efficiency} = \frac{t(P - d)T}{tPT}$$

$$= \frac{0.524 d^2 T}{tPT}$$

or, reduced to lowest terms:

$$\text{Efficiency} = \frac{P - d}{P}$$

$$= \frac{0.524 d^2}{tP}$$

One item not considered in these calculations is the resistance of the plate to shearing between the edge of the plate and the rivet. In practice, such failure is guarded against by placing the rivet hole so that its center rests 1½ times its diameter from the edge of the plate.

Although a single lap joint was chosen for illustration here, the foregoing calculations show enough of the principles involved that there should be no real difficulty with more complicated joints if they are thoroughly understood. Other calculations follow; they differ slightly and pertain to several types of joint construction.

In the following five examples, you will find the following common data and letter designations, not all of which will appear in each example. Other designations and data will be given in each example as they apply to that particular calculation:

- TS is the tensile strength stamped on the plate, expressed in pounds-per-square-inch;
- t is the thickness of the plate, in inches;
- b is the thickness of the butt strap, in inches;
- P is the pitch of the rivets, in inches, on the row with greatest pitch;
- d is the diameter of the rivet after driving, in inches, or diameter of rivet hole;
- a is the cross-sectional area of the rivet after driving, in square inches;
- s is the shearing strength of a rivet in single shear, in pounds-per-square-inch;
- S is the shearing strength of a rivet in double shear, in pounds-per-square-inch;
- c is the crushing strength of mild steel, in pounds-per-square-inch;
- n is the number of rivets in single shear in a unit length of joint;
- N is the number of rivets in double shear in a unit length of joint.

Lap Joint—single-riveted, either longitudinal or circumferential (see Fig. 11-7). Additional data for this joint calculation is as follows:

BOILER CALCULATIONS

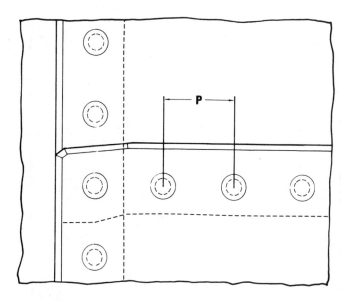

Fig. 11-7. Example of lap joint, longitudinally and circumferentially single-riveted.

A is the strength of the solid plate and is determined by $P \times t \times TS$;

B is the strength of the plate between rivet holes and is determined by $(P - d) t \times TS$;

C is the shearing strength of one rivet in single shear and is determined by $s \times a$;

D is the crushing strength of the plate in front of one rivet and is determined by $d \times t \times c$.

To perform the calculation, divide B, C, or D, whichever is least, by A and the quotient will be the efficiency of a single-riveted lap joint as shown in Fig. 11-7.

Lap Joint—double-riveted, either longitudinal or circumferential (see Fig. 11-8). Additional data for this joint calculation is as follows:

A is the strength of the solid plate ($p \times t \times TS$);

233

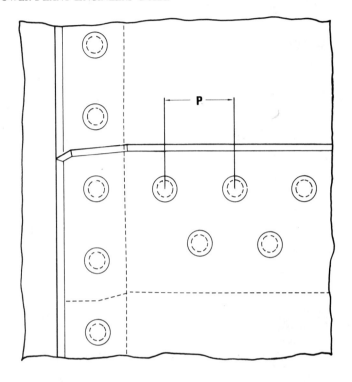

Fig. 11-8. Example of lap joint, longitudinally and circumferentially double-riveted.

B is the strength of the plate between rivet holes ($[P - d]t \times TS$);

C is the shearing strength of *two* rivets in single shear ($n \times s \times a$);

D is the crushing strength of the plate in front of *two* rivets ($n \times d \times t \times c$).

To perform the calculation, divide B, C, or D, whichever is least, by A and the quotient will be the efficiency of a double-riveted lap joint, as in Fig. 11-8.

Butt and Double-Strap Joint—double-riveted (see Fig. 11-9). Additional data for this joint calculation is as follows:

BOILER CALCULATIONS

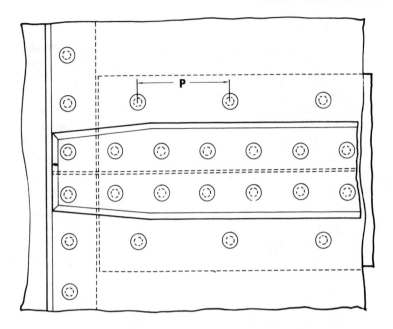

Fig. 11-9. Example of a butt and double strap joint, double-riveted.

A is the strength of the solid plate ($P \times t \times TS$);
B is the strength of the plate between rivet holes in the outer row ($[P - d]t \times TS$);
C is the shearing strength of two rivets in double shear, plus the shearing strength of one rivet in single shear and is determined by $(N \times S \times a) + (n \times s \times a)$;
D is the strength of the plate between rivet holes in the second row, plus the shearing strength of one rivet in single shear in the outer row and is determined by $(P - 2d)t \times TS + (n \times s \times a)$;
E is the strength of the plate between rivet holes in the second row, plus the crushing strength of the butt strap in front of one rivet in the outer row and is determined by $(P - 2d)t \times TS + (d \times b \times c)$;
F is the crushing strength of the plate in front of two rivets, plus the crushing strength of the butt strap in front of one

235

rivet and is determined by $(N \times d \times t \times c) + (n \times d \times b \times c)$;

G is the crushing strength of the plate in front of two rivets, plus the shearing strength of one rivet in single shear and is determined by $(N \times d \times t \times c) + (n \times s \times a)$;

H is the strength of the butt straps between the rivet holes in the inner row and is determined by $(P - 2d)2b \times TS$. This method of failure is not possible for normal butt-strap thicknesses and the computation need be made only for very old boilers in which thin butt straps have been used. For this reason, this method of failure will not be considered in other joints.

To perform the calculation, divide B, C, D, E, F, G, or H, whichever is the least, by A and the quotient will be the efficiency of a butt and double-strap joint, double-riveted, and in Fig. 11-9. One rivet is in single shear and two are in double shear in a unit length of joint.

Butt and Double-Strap Joint—triple-riveted (Fig. 11-10). Additional data for this joint calculation is as follows:

A is the strength of the solid plate $(P \times t \times TS)$;

B is the strength of the plate between rivet holes in the outer row ($[P - d]t \times TS$);

C is the shearing strength of *four* rivets in double shear, plus the shearing strength of one rivet in single shear, as $(N \times S \times a) + (n \times s \times a)$;

D is the strength of the plate between rivet holes in the second row, plus the shearing strength of one rivet in single shear in the outer row and is determined by $(P - 2d)t \times TS + (n \times s \times a)$;

E is the strength of the plate between rivet holes in the second row, plus the crushing strength of the butt strap in front of the rivet in the outer row and is determined by $(P - 2d)t \times TS + (d \times b \times c)$;

F is the crushing strength of the plate in front of *four* rivets, plus the crushing strength of the butt strap in front of one rivet, as $(N \times d \times t \times c) + (n \times d \times b \times c)$;

G is the crushing strength of the plate in front of *four* rivets,

BOILER CALCULATIONS

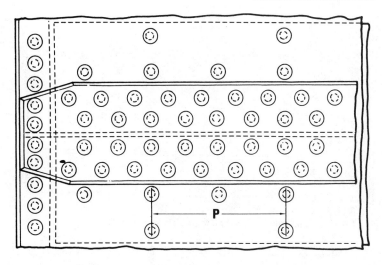

Fig. 11-10. Example of a butt and double-strap joint, triple-riveted.

plus the shearing strength of one rivet in single shear, as
$(N \times d \times t \times c) + (n \times s \times a)$.

To perform the calculation, divide B, C, D, E, F, or G, whichever is least, by A and the quotient will be the efficiency of a butt and double-strap joint, triple-riveted, as in Fig. 11-10. One rivet is in single shear and four are in double shear in each unit length of joint.

Butt and Double-Strap Joint—quadruple-riveted (see Fig. 11-11). Additional data for this joint calculation is as follows:

A is the strength of the solid plate ($P \times t \times TS$);
B is the strength of the plate between rivet holes in the outer row ($[P - d] t \times TS$);
C is the shearing strength of *eight* rivets in double shear, plus the shearing strength of *three* rivets in single shear, as $(N \times S \times a) + (n \times s \times a)$;
D is the strength of the plate between rivet holes in the second row, plus the shearing strength of one rivet in single shear in the outer row ($[P - 2d]t \times TS + [1 \times s \times a]$);
E is the strength of the plate between rivet holes in the third

237

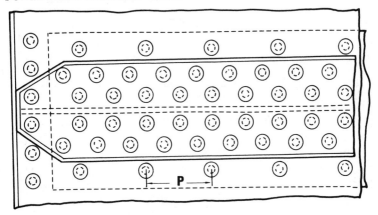

Fig. 11-11. Illustration of a butt and double-strap joint, quadruple-riveted.

row, plus the shearing strength of two rivets in the second row in single shear and one rivet in single shear in the outer row and is determined by $(P - 4d)t \times TS + (n \times s \times a)$;

F is the strength of the plate between rivet holes in the second row, plus the crushing strength of the butt strap in front of one rivet in the outer row and is determined by $(P - 2d)t \times TS + (d \times b \times c)$;

G is the strength of the plate between rivet holes in the third row, plus the crushing strength of the butt strap in front of two rivets in the second row and one rivet in the outer row. It is determined by $(P - 4d)t \times TS + (n \times d \times b \times c)$;

H is the crushing strength of the plate in front of *eight* rivets, plus the crushing strength of the butt strap in front of *three* rivets, as $(N \times d \times t \times c) + (n \times d \times b \times c)$;

I is the crushing strength of the plate in front of *eight* rivets, plus the shearing strength of two rivets in the second row and one rivet in the outer row, in single shear, as $(N \times d \times t \times c) + (n \times s \times a)$.

To perform the calculation, divide B, C, D, E, F, G, H, or I, whichever is least, by A and the quotient will be the efficiency of a butt and double-strap joint, quadruple-riveted, as in Fig. 11-11.

There are three rivets in single shear and eight rivets in double shear in each unit length of joint.

WELDED CONSTRUCTION

What is meant by "welded" construction?

Answer: An arc-welding process is used instead of the previously described lap, and butt and strap riveted joint.

What are the advantages of welding?

Answer: There are several. The metal is not weakened by the single, double, or triple rows of holes needed in the riveted system; less material is needed for a welded butt joint than would be used to form safe riveted laps or butt strap joints; less labor time is needed because extensive layout, drilling, and setting of rivets is eliminated; and X-ray testing eliminates leaking possibilities due to stresses incurred during the shipping/transfer process.

Doesn't the welding operation create crystallization and hard spots?

Answer: An annealing operation ensures that the welded part is the same as the adjacent material.

Can welded and riveted boilers of equal material thickness and capacity withstand equal pressures?

Answer: Yes. Both are tested with pressures in excess of their rated capacity.

LIGAMENTS

What is a ligament?

Answer: The section of metal not cut away between two adjacent tube holes.

Where is the section or ligament?

Answer: It lies in a plane passing through the axis of the holes;

that is, the line joining the centers of the holes. When a head is drilled for tubes, a good deal of the metal is cut away, so the efficiency of the metal between the tube holes or the *ligament* must be considered.

Name two arrangements of the tube holes.

Answer: Holes drilled in line *parallel* to the axis of the shell, and holes drilled in a line *diagonal* with the axis of the shell.

How are the holes in each row spaced for Case 1?

Answer: The pitch may be equal or unequal.

For equal spacing, as in Fig. 11-12, how is the efficiency of the ligament determined?

Answer: By means of the formula $(p - d) \div p$, in which p equals the pitch of the tube holes in inches and d equals the diameter of the tube holes in inches. Converting fractional measurements to decimals will make the calculations much faster and simpler, and will provide the final answer directly in percentage.

How is ligament efficiency determined when there is unequal pitch, as in Figs. 11-13 and 11-14?

Answer: By means of the formula $(p - nd) \div p$, in which p is the unit length of the ligament in inches, n is the number of tube holes found in that unit length, and d is the diameter of the tube holes in inches.

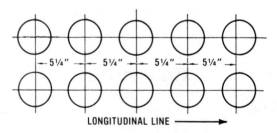

Fig. 11-12. Example of tube spacing, with the pitch of the holes equal in every row, illustrating the efficiency of the ligament.

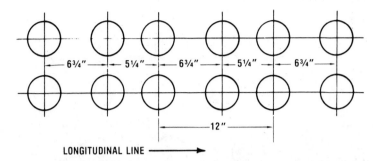

Fig. 11-13. Example of tube spacing with the pitch of the holes unequal in every second row, illustrating the efficiency of the ligament.

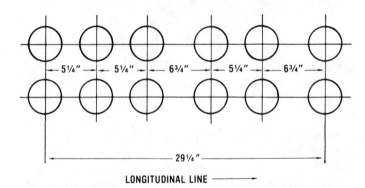

Fig. 11-14. Example of tube spacing with the pitch of the holes varying in every second and third row, illustrating the efficiency of the ligament.

Do either of these formulas apply in the case of a ligament having tube holes located in a line *diagonal* with the axis of the shell? See Fig. 11-15.

Answer: Only one of them, and then only partially. Two formulas are used and the efficiency is equal to the least of the two results obtained:

(Formula 1) Efficiency of Ligament = $\dfrac{0.95(p_1 - d)}{p_1}$

(Formula 2) Efficiency of Ligament = $\dfrac{p - d}{p}$

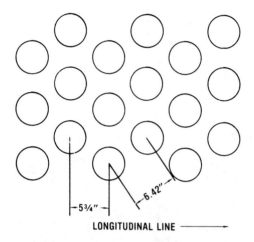

Fig. 11-15. Example of tube spacing with the tube holes on diagonal lines, illustrating the efficiency of the ligament.

where,

p_1 is the diagonal pitch of the tube holes, in inches;

d is the diameter of the tube holes, in inches;

p is the longitudinal pitch of the tube holes, or the distance between the centers of tubes in a longitudinal row of tubes, in inches;

0.95 is a constant which applies only as long as $p_1 \div d$ is 1.5 or more.

For example, the diagonal pitch of the tube holes in Fig. 11-15 is 6.42 inches, the diameter of the holes $4\frac{1}{32}$ inches (4.031), and the longitudinal pitch of the holes 11.5 inches. Substituting these values in Formulas 1 and 2, we get:

(Formula 1) $\qquad \dfrac{0.95(6.42 - 4.031)}{6.42} = 0.353$

(Formula 2) $\qquad \dfrac{11.5 - 4.031}{11.5} = 0.649$

Since the least of the two results is 0.353, this will be the efficiency of the ligament.

REINFORCEMENT OF FLAT SURFACES

How are flat surfaces usually reinforced?
Answer: By stays or braces.

Into what two classes may all reinforcing members be divided?
Answer: They may be classed as *independent* and *connecting* fastenings.

What is a stay bolt?
Answer: By definition, a metal pin or rod used to hold objects together and generally having screw threads cut at one end (sometimes at both ends) to receive a nut.

How are boiler plates tapped so that a threaded bolt or stay will properly screw into both plates without stripping the threads?
Answer: By the use of a long stay-bolt tap which threads both plates in one operation.

What thread is used for stay-bolt taps?
Answer: All sizes of stay-bolt taps have 12 threads to the inch.

What diameter of a screwed stay is taken in calculating its strength?
Answer: The *least* diameter.

What is a stay rod?
Answer: A through-stay especially adapted to short boilers of large diameter.

What is the maximum stress allowed on stay bolts?
Answer: 7500 pounds per square inch.

What is the area supported by one stay bolt?

Answer: It is the area enclosed by the two pitches less the area of one stay-bolt hole.

If the diameter of a bolt or rod is increased how will its stress-bearing capability change?

Answer: It will increase in direct proportion to the cross-sectional area of the bar. Stress in pounds per square inch on a bolt or bar equals the pressure in pounds per square inch divided by the cross-sectional area in square inches of the bolt or bar. Stated differently, the cross-sectional area in square inches equals the total load in pounds divided by the stress in pounds per square inch. These relationships can be abbreviated as:

$$S = \frac{P}{A}$$

and

$$A = \frac{P}{S}$$

How would you calculate the stress on four 1-inch American Standard bolts used to support a load of 7 tons? What is the stress per square inch?

Answer: Look up the diameter of the bolt at the bottom of the threads in the American Standard Bolt Sizes table in Chapter 30 (0.8376 for a 1-inch bolt). Square the diameter of the bolt and multiply the product by 0.7854, which will give you the cross-sectional area of one bolt. For four bolts multiply the area of one bolt by 4, which produces the combined load-bearing cross-section of 4 bolts. Divide the load bearing area into the load:

$$\text{Area of 1 bolt} = 0.8376 \times 0.8376 = 0.7016 \times 0.7854$$
$$= 0.5510 \times 4$$
$$\text{Area of 4 bolts} = 2.2041 \text{ sq. in}$$

BOILER CALCULATIONS

Stress per sq. in. on
the bolts = 14,000 ÷ 2.2041
= 6351.8 lbs.

How would you calculate the diameter of 4 bolts needed to carry a load of 7 tons?

Answer: Calculate total load in pounds (7 × 2000 = 14,000). Divide by the number of bolts (14,000 ÷ 4 = 3500 lbs.). Assuming a stress per square inch of 6000 pounds, calculate the area needed in each bolt (3500 ÷ 6000 = 0.5833). Determine the bolt diameter at the bottom of the threads from the area; divide the area by 0.7854 and find the square root of the quotient (0.5833 ÷ 0.7854 = 0.7427; the square root of 0.7427 is 0.8618). Look up in American Standard Bolt Sizes table in Chapter 30 the diameter nearest the size you calculated. In this case, the diameter of a 1-inch bolt is 0.8376, a bit lower than 0.8618, so the next larger size (1⅛ inch) would be needed.

Can a boiler be constructed without using tubes or through-braces to stay the heads?

Answer: Yes, but the circumferential joints must have sufficient strength to resist any method of failure from the total longitudinal force acting upon the joint, and must do this with a safety factor of 5. When 50% or more of the total force is relieved by the effect of tubes or through-stays, the strength of the circumferential joints can be less, but must be no less than 70% of what it would be without the tubes or stays. Total longitudinal force is calculated with this formula:

$$F = \tfrac{1}{4} \pi D^2 P$$

where,

- F   is the total longitudinal force, in pounds;
- D   is the diameter in inches of the circular area acted upon by the pressure;
- P   is the pressure, in psi;
- π is 3.1416.

Areas of Heads to be Stayed

Where flat heads are used, it is necessary to provide stays or braces for that part which is unsupported by the tubes, and finding the area in square inches of the segment of the head to be braced is a problem usually presented during the examination for an engineer's license. See Figs. 11-16 and 11-17.

The area to be stayed is enclosed by lines drawn 2 inches from the tubes and at a distance d from the shell. The value of d in inches can be equal to the outer radius of the flange, not exceeding 8 times the thickness of the head, or it can be found by the formula $5T/\sqrt{P}$ (where T is thickness of the head in sixteenths of an inch and P is the maximum allowable working pressure in psi), whichever is larger.

Once d has been determined, the net area to be stayed in a segment of a head (in square inches) may be calculated by this formula:

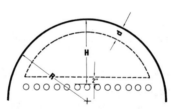

Fig. 11-16. Upper segment of head to be stayed.

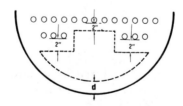

Fig. 11-17. Lower segment of head to be stayed.

$$\text{Area} = \frac{4(H-d-2)^2}{3}\sqrt{\frac{2(R-d)}{(H-d-2)} - 0.608}$$

H is the distance from the tubes to the shell, in inches;
d is the larger of the two values previously calculated;
R is the radius of the boiler head, in inches.

Should the value of d equal 3 inches, the area to be stayed in a segment may be determined in square inches by the following:

$$\text{Area} = \frac{4(H-5)^2}{3}\sqrt{\frac{2(R-3)}{H-5}}$$

Should there be a manhole opening in the portion of the head below the tubes in an HRT boiler (lower segment), and the opening has a flange formed from solid plate and turned inward to a depth of at least three times the thickness of the head (measured from the outside), the area to be stayed in Fig. 11-17 may be reduced by 100 square inches. The surface around the manhole must be supported by through-stays with nuts inside and outside at the front head.

Diagonal and Gusset Stays

Multiply the area of a direct stay required to support the surface by the diagonal length of the stay; divide this product by the length of a line running perpendicular from the supported surface to the center of the palm of the diagonal stay (Fig. 11-18). Expressed as a formula, this becomes:

$$A = \frac{aL}{l}$$

where,

- A is the sectional area of diagonal stay, in square inches;
- a is the sectional area of direct stay, in square inches;
- L is the length of diagonal stay, as in Fig. 11-18;
- l is the length of a line running perpendicular from the boiler head to the center of the palm of the diagonal stay, as in Fig. 11-18.

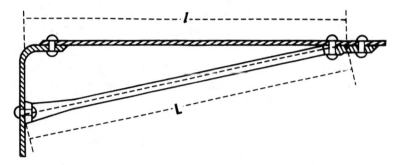

Fig. 11-18. Measurements for determining stresses in diagonal stays.

Stay Tubes

When stay tubes are used in multitubular boilers to give support to the tube plates, the sectional area of the tubes is determined in square inches by the following calculation:

$$\text{Total Section of Stay Tubes} = \frac{(A-a)\,P}{TS}$$

where,

- A is the area of that portion of the tube plate containing the tubes, in square inches;
- a is the aggregate area of the holes in the tube plate, in square inches;
- P is the maximum allowable working pressure, in psi;
- TS is the tensile strength (not to exceed 7000 pounds per square inch).

Circular Flues

There are two formulas for determining the maximum allowable working pressure for seamless or welded flues having a maximum diameter of 18 inches and a minimum diameter greater than 5 inches. Formula 1 is used when the thickness of the wall is not greater than 0.023 times the diameter and Formula 2 is used for thicker walls:

(Formula 1) $$P = \frac{10,000,000 t^3}{D^3}$$

(Formula 2) $$P = \frac{17,300 t}{D} - 275$$

where,

- P is the maximum allowable working pressure, in psi;
- D is the outside diameter of the flue, in inches;
- t is the thickness of the wall of the flue, in inches.

These formulas can also be applied to riveted flues of the size specified provided that the sections are not over 3 feet in length and provided that the efficiency of the joint is greater than PD/20,000t.

Adamson Flues

When plain, horizontal flues are made in sections not less than 18 inches long and 5/16-inch in thickness, they must be flanged with a radius measured on the fire side of not less than three times the thickness of the plate, and the flat portion of the flange outside of the radius must be at least three times the diameter of the rivet holes. The distance from the edge of the rivet holes to the edge of the flange must not be less than the diameter of the rivet hole, and the diameter of the rivets before driving needs to be at least ¼ inch larger than the thickness of the plate.

Another requirement is that the depth of the Adamson ring between the flanges be not less than three times the diameter of the rivet holes, with the ring substantially riveted to the flanges. The fire edge of the ring must terminate at or near the point of tangency to the curve of the flange, and the thickness of the ring must not be less than ½ inch.

Maximum allowable working pressure in psi is determined by this formula:

$$P = \frac{57.6}{D}(18.75\,T - 1.03L)$$

where,

P is the maximum allowable working pressure, in psi;
D is the outside diameter of the furnace, in inches;
L is the length of the furnace section, in inches;
T is the thickness of the plate, in sixteenths of an inch.

BOILER OUTPUT

What is one boiler horsepower?

Answer: The evaporation of 30 pounds of water from an initial temperature of 100°F. to steam at 70 pounds gauge pressure, which is equivalent to 34.5 pounds of water evaporated per hour from a feed-water temperature of 212° into dry steam at the same temperature.

What is the abbreviation for the expression: "from a feed-water temperature of 212° into dry steam at the same temperature."

Answer: It is abbreviated to *"from and at 212°."*

If a boiler evaporates 600 pounds of steam at 105 pounds pressure per hour from feed water at 100°F, what is the equivalent evaporation "from and at 212°F."

Answer: The factor of evaporation (see Chapter 5) is 1.158. Equivalent evaporation equals 600 × 1.158, or 694.8 pounds "from and at 212°F."

If the rate of evaporation is given at 8 pounds of steam per pound of coal, how much coal is required to evaporate 600 pounds of steam under the conditions just given?

Answer: Total coal per hour equals total evaporation per hour times the factor of evaporation divided by the rate of evaporation. Substituting:

$$\text{Total coal per hour} = \frac{600 \times 1.158}{8} = 86.85 \text{ lbs.}$$

BOILER CALCULATIONS

How much heat is needed to convert 1 pound of water at 32°F. into steam at 170 pounds per square inch absolute?

Answer: Refer to the steam table in Chapter 30. Find 170° in the absolute pressure column and read the heat in the steam (H) as 1195.4 Btu.

A boiler is operating at 145 psi and evaporating 5 pounds of water per hour per square foot of heating surface. Feed-water temperature is 90°F. How much heat is needed per square foot to produce steam?

Answer: In the steam table, at 145 psig or 160 psia, 1194.5 less 90-32, 1194.5-58 or 1136.5. Btu are needed to produce 1 pound of steam. Therefore, 5 times 1136.5 or 5682.5 Btu are used to heat 5 pounds of steam. This relationship can be expressed:

$$H = h - (t - 32)$$

where,

- H = total heat needed to convert 1 pound of feed water into steam at any pressure or temperature,
- h = total heat needed to convert 1 pound of water at 32°F. into steam at any pressure (listed on the steam table),
- t = temperature of feed water.

If all the heat contained in one Btu could be converted into work, how much would be produced?

Answer: 778 foot-pounds, which is the mechanical equivalent of heat used to determine thermal efficiency.

What is the thermal efficiency of a boiler operating at 155 psig with feed water at 150°F. if 6 pounds of steam is generated from each pound of coal burned, assuming the coal has a heat value of 10,000 Btu per hour?

Answer: Heat in steam divided by heat in fuel equals thermal efficiency. Therefore, this boiler's efficiency would be 6 times the Btu per pound of steam (1195.4 on the steam table) minus the temperature above 32°F. (150 - 32 = 118) times 100. This total is divided by 10,000, the Btu is per pound of coal. So:

251

$$\frac{6 \times [1195.4 - (150 - 32)] \times 100}{10{,}000} = 64.6\%$$

If 400 pounds of fuel with a heat value of 8000 Btu per pound is burned per hour to produce an engine output of 175 horsepower, what is the overall thermal efficiency of the plant?

Answer: In this case, thermal efficiency is calculated as the engine output of foot pounds per hour divided by mechanical equivalent of heat in foot pounds in fuel burned per hour. In this example,

$$\frac{175 \times 33{,}000 \times 60}{400 \times 8000 \times 778} \times 100 = 13.9\%$$

What is the horsepower of a boiler with a working pressure of 145 psig, a feed-water temperature of 105°, and a rate of evaporation of 2500 pounds?

Answer: From the steam table in Chapter 30, the factor of evaporation at this pressure and temperature is 1.159, so $2500 \times 1.159 = 2897.5$ lbs. To convert to boiler horsepower, divide by $34.5 = 83.99$ or 84 hp.

Some boilers are rated in output at Btu per hour or in units of evaporation. How would the boiler in the above question rate at 1000 Btu per hour?

Answer: Horsepower (84) times the amount of evaporation (34.5) times the heat needed to change water into steam at atmospheric pressure times (970.4 Btu) = 2812 Btu.

FURNACE

What determines the size of grate needed in a furnace?

Answer: It depends upon the rate of combustion and amount of solid fuel to be burned per hour.

What is an ordinary rate of combustion in stationary practice?

Answer: About 12 pounds of coal per square foot of grate per hour.

Boiler Calculations

If the total coal burned per hour is 86.85 pounds and the combustion rate is 12 pounds what is the required area of grate?

Answer: Area of grate = 86.85 ÷ 12 = 7.24 sq. ft. = 7.24 × 144 = 1042.6 sq. ins.

How do you find the diameter of the grate after finding the area?

Answer: Calculate it, or look it up in the properties of circles table in Chapter 30. Take the nearest value *larger* than that corresponding to the given area.

Would a boiler ordinarily be built with such a dimension for a grate and why?

Answer: No. Standard dimensions such as 36 inches (rather than 36½ inches) are used where the difference is not too great.

Why not use the exact dimension as calculated?

Answer: A boiler is far from being "a hard-boiled egg" like a gas engine. In other words, the range of output of a boiler is very great, whereas a gas engine has one maximum rating above which further forcing is not practical owing to its inherent defects.

For a vertical boiler, how is the diameter determined to accommodate a grate of given diameter?

Answer: Space must be allowed for (1) thickness of metal of furnace sheet, and (2) thickness of water leg. These allowances are multiplied by 2, since allowance must be made for both sides of the furnace. Expressed as a formula:

Inside diameter of shell = diameter of grate + 2 × (thickness furnace sheet × width of water leg).

For a 36-inch grate, 1½-inch water leg, and furnace wall ½-inch thick, the inside diameter of the shell = 36 + 2 (½ + 1½) = 36 + 1 + 3 = 40 inches.

If a boiler has a grate area of 7.07 square feet and the ratio of heating surface divided by the grate area is 25, how much heating surface must be provided?

Answer: 7.07 × 25 = 176.8, say 177 sq. ft.

Power Plant Engineers Guide

HEATING SURFACE

Of what does heating surface consist, considering a vertical tubular boiler?

Answer: (1) Furnace heating surface, and (2) tubular heating surface.

Suppose the furnace is 36 inches in diameter and 18 inches high. What is its heating surface?

Answer: Reducing dimensions to feet: 36 inches = 36 ÷ 12 = 3 feet; 18 inches = 18 ÷ 12 = 1½ feet. From which the area of the furnace heating surface = (3.1416 × 3) × 1½ = 9.43 × 1½ = 14.2, say 14 square feet.

Considering the same boiler having 177 square feet of heating surface and deducting the furnace heating surface, how much tubular heating surface must be provided?

Answer: 177 − 14 = 163 sq. ft.

What governs the diameter and length of tubes?

Answer: The type of boiler and service requirements.

For an ordinary vertical shell boiler having a 36-inch lower tube sheet which will easily accommodate 91 two-inch tubes, how would you find the length of tubes giving 163 square feet of tubular heating surface?

Answer: In the table of "Properties of Boiler Tubes" in Chapter 30, the length of two-inch tube required per square foot of inside surface is 2.11 feet. Hence, the total length of tubes = 163 × 2.11 = 343.9, say 344 feet. For 91 tubes length of each tube = 344 ÷ 91 = 3.78 feet or 3.78 × 12 = 45.4 inches.

Would such a tube dimension as 45.4 inches be used in boiler construction?

Answer: No.

Why?

Answer: The dimension of tube length would be such that the

Boiler Calculations

tubes could be cut to that length without wastage. That is, stock lengths would be selected by the designer.

What length would be used in place of 45 inches?

Answer: Say 42 inches or 48 inches, increasing or decreasing the number of tubes respectively to keep the tubular heating surface the same.

Suppose you didn't have a table of "Properties of Boiler Tubes" available. How would you figure the tubular heating surface?

Answer: Multiply number of tubes by the inside circumference by the length and divide by 144. Expressed as a formula:

Tubular area in sq. ft. = (Number of tubes × inside circumference × length) ÷ 144.

What is the internal heating surface of 91 two-inch tubes 48 inches long?

Answer: From the table, the inside diameter of a 2 inch tube = 1.81 inches. Applying the rule,

$$\text{area} = \frac{91 \times (1.81 \times 3.1416) \times 48}{144} = 172.5 \text{ square feet}$$

The following examples will illustrate the great convenience of the table.

Example—How many square feet of inside heating surface in 40 four-inch tubes, each 16 feet long? From the table, the length of a 4-inch tube per square foot of inside surface is 1.024 feet. Hence,

$$\text{total heating surface} = \frac{40 \times 16}{1.024} = 625 \text{ square feet}$$

Example—How many feet of one-inch tube are required for 140 square feet of inside heating surface, and what is the weight? From the table the length of one-inch tube per square foot of inside surface is 4.479 feet. Hence, the amount of one-inch tube required

Power Plant Engineers Guide

= 140 × 4.479 = 627 feet. From the table, one-inch tube weighs 0.9 pounds per lineal foot. Hence, the weight of 627 feet of one-inch tube = 627 × 0.9 = 564 pounds.

What comprises the heating surface of a horizontal return tubular boiler?

Answer: The shell surface, the tubular surface, and net area of tube sheets.

What part of the sheet surface is taken and how much?

Answer: That part exposed to the hot gases in combustion; in amount about 200° of circumference.

How is the heating surface of an HRT boiler calculated?

Answer: Take the dimensions in inches:

Shell: Multiply length by the fraction of circumference in contact with hot gases.

Tubes: Multiply the sum of the inside circumferences of all of the tubes by their common length.

Tube sheets: Multiply twice the diameter squared by 0.7854 and subtract twice the sum of the internal areas of all the tubes (similarly obtained).

Total heating surface: Divide the sum of the several areas just obtained by 144.

What should be noted in finding the heating surface of the tubes?

Answer: The sizes of boiler tubes as given are the outside diameters, whereas in calculating the heating surface of horizontal boilers the inside diameter is taken to determine the surface exposed to the fire and hot gases.

Example—What is the heating surface of a horizontal return tubular boiler 54 inches in diameter, 16 feet long and having 40 tubes each 4 inches in diameter?

Shell: Area = $54 \times 3.1416 \times \dfrac{200}{360} \times (16 \times 12) = 18{,}096$ square inches

Boiler Calculations

where,

3.1416 (π) is the number which multiplied by the diameter of a circle gives the circumference,

200 ÷ 360 is the fractional part of the shell in contact with the hot gases,

(16 × 12) is the length of the shell in inches.

Tubes: Area = 40 × (3.732 × 3.1416) × (16×12) = 90,040 square inches

where,

40 is the number of tubes,
3.732 is the inside diameter of a standard 4-inch tube,
(3.732 × 3.1416) is the inside circumference,
(16 × 12) is the length of each tube in inches.

Tube Sheets: Gross area = 2 × 54^2 × 0.7854 = 4580 square inches.

where,

2 is the number of sheets,
54 is the diameter of the boiler in inches,
0.7854 is that fraction which, multiplied by the diameter squared, gives the gross area of one sheet.

Gross area of sheets is the area not including that cut away by holes for tubes.

Twice area of tubes = 2 × 40 × 3.732^2 × 0.7854 = 875 square inches.

where,

2 is the number of sheets,
40 is the number of tubes per sheet,
3.732 is the inside diameter of one tube, in inches,
0.7854 is the multiplying fraction.

This area figure refers to cross-sectional area. The inside diameter of the tube is used because the thickness of each tube is considered as forming a part of the tube sheet, as it is regarded as sheet heating surface.

Net area of sheets = ⅔ × 4580 − 875 = 2178 square inches

where,

⅔ is the amount of sheet area ordinarily exposed to hot gases,
4580 is the gross area of the sheets,
875 is twice area of the tubes.

The factor ⅔ is an approximation; to be exact, the exposed portion of the sheet area must be measured.

Total Heating Surface:
  Total area = 18096 + 90040 + 2178 = 100,314 square inches

where,

18096 is area of shell in square inches,
90040 is area of tubes in square inches,
2178 is net area of sheets in square inches.

To find area in square feet, divide the figure by 144.

How do you find the inside diameter of a boiler tube?

Answer: From a table giving "Properties of Standard Boiler Tubes." These diameters have a nasty decimal fraction and it is better to look up cross-sectional areas in the table than to calculate them as the example just given will indicate.

How do you calculate the tubular heating surface of a water-tube boiler?

Answer: By basing the calculations on the outside area of the tubes instead of the inside area.

What should be noted about tube and pipe diameters?

Answer: A tube diameter is given as the actual outside diameter whereas a pipe diameter is given as the *nominal* diameter which is considerably different from the listed diameter. For instance a wrought pipe listed as ½-inch pipe has an outside diameter of 0.84 and an actual inside diameter of 0.622.

How many linear feet of ¾-inch wrought pipe is required for 163 feet of heating surface?

Answer: From the table of "Properties of Wrought Pipe," Chapter 30, a length of ¾-inch pipe per square foot of external surface = 3.637 feet. Hence, length of ¾-inch pipe for 163 square feet of heating surface = 163 × 3.637 = 592.8, say 593 feet.

How much heating surface is ordinarily allowed per boiler horsepower?

Answer: 12 square feet.

How would you find the length of a 2-inch boiler tube per square foot of inside heating surface?

Answer: A standard 2-inch boiler tube (from the table in Chapter 30) is No. 13 BWG whose decimal equivalent is 0.095 inches. Twice thickness = 2 × 0.095 = 0.19. Inside diameter = 2 − 0.19 = 1.81 inches. Circumference = 1.81 × 3.1416 = 5.686 inches. Length per square foot of inside heating surface =

$$\frac{144 \div 5.686}{12} = 2.11 \text{ square feet.}$$

CHAPTER 12

Boiler Fixtures, Fittings and Attachments

Various names such as fixtures, attachments, fittings, trimmings, and mountings have been given to the numerous devices fastened to the boiler which are necessary for its proper operation. The term "fixtures" relates rather to the grate, ash pit, doors, dampers, funnel, and smoke hood than to fittings such as steam gauge cocks. All fittings attached to a boiler must conform to designs approved by the ASME boiler code. Any pipe connection must be made to nozzles that are either welded or riveted to the drum. See Fig. 12-1.

What valves are necessary on a boiler?
　Answer: Safety or relief valve, stop valve, check valve, blow-off valve (cock). Some are shown in Fig. 12-2.

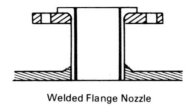
Welded Flange Nozzle

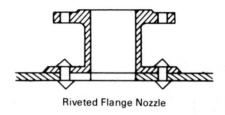
Riveted Flange Nozzle

Threaded Riveted Nozzle

Threaded Welded Nozzle

Fig. 12-1. Acceptable boiler pipe connections.

SAFETY VALVES

What is considered by many to be the most important valve on a boiler and why?

Answer: The safety valve, because upon its proper operation

Boiler Fixtures, Fittings and Attachments

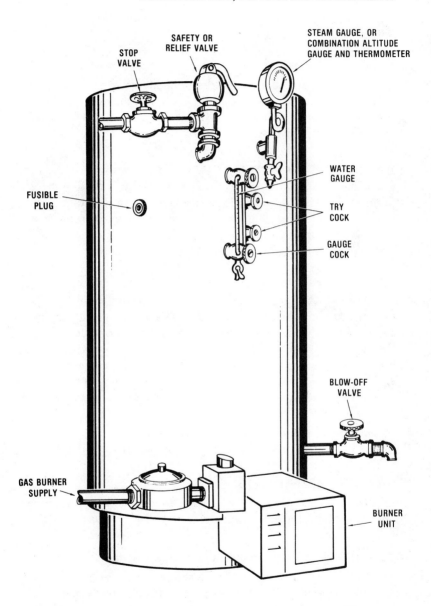

Fig. 12-2. Vertical boiler and fittings.

depends the safety of those in charge of the boiler. (More on safety valves in Chapter 29.)

What is a safety valve?

Answer: An automatic loaded or weighted valve which opens at a predetermined pressure and releases steam from the boiler, thus keeping the steam pressure from rising above that for which the valve is set. See Fig. 12-3.

What care should a safety valve receive?

Answer: It should be kept clean and should be raised by hand periodically. Most engineers in plants with pressures up to 250 psig test the safety valve when the boiler is put back into service after cleaning and overhaul and on a monthly basis thereafter. On low-pressure plants, they should be checked weekly.

Why should it be raised so often?

Answer: So that it cannot stick in its seat through the accumulation of dirt and scale. However, excessive opening of the safety valve tends to weaken the spring. Also, frequent opening causes the valve seat and disc to become wire-drawn or to steam out. Sometimes during a test a valve will fail to close properly due to scale or other foreign material, in which case it may be necessary to allow steam to blow off (escape) through the valve to clear it.

What types of safety valves are in general use?

Answer: The lever safety valve and the spring safety valve.

What is the application of each type?

Answer: The lever-type is generally used in stationary practice; the spring-type for marine plants.

How should a safety valve be attached to a boiler?

Answer: It should be attached to a separate outlet at the highest point in the steam space. On boilers having only one main outlet, it may be attached to a tee on main steam pipe as close to the boiler as possible without any valve between it and the boiler.

BOILER FIXTURES, FITTINGS AND ATTACHMENTS

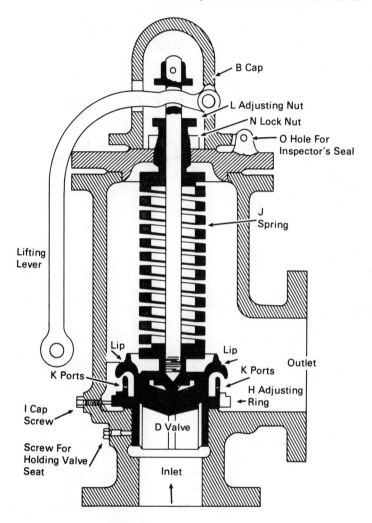

Fig. 12-3. Typical boiler safety valve.

May an escape pipe to discharge steam released by the safety valve be used?

Answer: Yes. Its cross-sectional area must not be any smaller than the total area of the valve outlet or than the total of the areas of the outlets discharging into it. The discharge from the escape pipe must not be near platforms, gangways, or other places where persons normally work or pass by. It must be as straight as possible and properly drained.

STOP AND CHECK VALVES

What must be connected to every steam outlet of a boiler?
Answer: A stop valve. When the outlet pipe is over 2 inches (5.08 cm) in diameter, the stop valve must be an outside-screw-and-yoke type, where the screw part of the stem is outside the valve body.

What actually opens a stop valve?
Answer: Steam pressure. The stem simply holds the disc firmly against the seat when the valve is closed; the stem has no lifting connection.

How did the gate valve get its name?
Answer: The discs or wedges in a gate valve operate much as does a gate being open or closed.

What is a stop valve?
Answer: A *nonreturn* valve having a hand wheel and screw stem which acts only to close the valve. A counterbalance spring tends to keep the valve open. See Fig. 12-4.

What is the mistaken idea about a stop valve?
Answer: It is erroneously applied to all hand-control valves.

What is the object of a stop or nonreturn valve?
Answer: The valve is designed to close automatically if any pressure part fails in a boiler to which it is attached. By preventing back flow from the steam header to the point of failure, it automatically isolates a defective boiler from any other boiler supplying the same header.

Boiler Fixtures, Fittings and Attachments

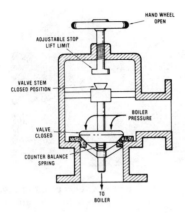

Fig. 12-4. Stop or nonreturn valve. A form of check valve that can be opened or closed by hand control when the pressure in the boiler is greater than that in the line, but cannot be opened when the pressure within the boiler is less than that in the line. The counterbalance spring slightly overbalances the weight of the valve and tends to hold the valve open, thus preventing movement of the valve with every slight fluctuation of pressure.

What is a check valve?

Answer: A form of nonreturn valve used to control the flow of water as in pump operation. See Figs. 12-5 to 12-8.

Fig. 12-5. Swing or hinged-type check valve. The design gives a valve opening area equal to that of the connecting pipes. The valve disc is attached by a nut to the carrier, which is pivoted. The two side plugs serve as bearings for the pivot pin. Should the movement of the pin cause the plugs to wear, they can be easily renewed at small expense. To prevent the disc locknut jarring loose, a hole is drilled through both the locknut and threaded end of the disc, through which a wire is inserted. To regrind, unscrew the bonnet and place some powdered glass or sand and soap or oil on the seat; also unscrew the plug opposite the disc, which permits inserting a screwdriver in the slot of the disc.

Fig. 12-6. Disc check valve. The check-valve disc has integrally cast wing guides, which snugly engage within the guide; auxiliary guides are provided below the disc. To regrind, remove the bonnet, lift out the guide, place a little fine sand or ground glass and water on the disc face; replace some in the body and apply a screwdriver to the slot in the disc stem. Rotate back and forth until a good bearing is obtained, then carefully wipe off the ground glass or sand, replace the valve guide, and screw on the bonnet.

Fig. 12-7. Ball check valve. This form of check consists of three parts: seating casting, ball, and bonnet. It meets the requirements for users of this type of check valve, but is not desirable for sizes above 3 inches because of the high cost and weight of the ball.

Name the different types of check valve.

Answer: Disc, ball, swinging, adjustable.

What should be placed between a check valve and boiler, and why?

Answer: A globe or preferably a gate valve as in Figs. 12-9, 12-10, and 12-11 to permit cleaning or repairs to the check valve.

BOILER FIXTURES, FITTINGS AND ATTACHMENTS

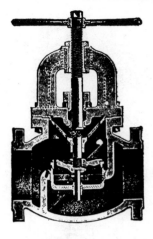
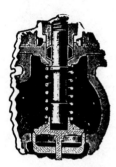

Fig. 12-8. Adjustable-lift check valve showing (left) the valve without the spring bearing on the disc, and (right) the valve with the spring valve with the spring bearing on the disc.

Fig. 12-9. Globe valve. A commonly used type of valve that takes its name from the globe-shaped casting forming the body of the valve. It should be noted that whereas the entire assemblage of parts here shown is ordinarily called a valve, the term "valve," strictly speaking and in accordance with the definition, means the disc at the end of the valve spindle.

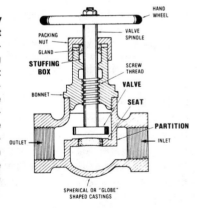

BLOW-OFF AND BLOW-DOWN VALVES

What is a blow-off valve?

Answer: A valve of special construction used to provide the means for discharging mud, scale, and other impurities from the

269

POWER PLANT ENGINEERS GUIDE

Fig. 12-10. Right way and wrong way to connect a globe-type valve in a pipe line, showing the disastrous result of attempting to repack a wrongly connected valve. In the illustrations the partition and disc are shown in dotted lines from which the proper position of the valve is clearly seen.

boiler water and for rapidly lowering the boiler water level if it's too high. The blow-off connection is used also to drain the boiler; therefore, it must be connected at the lowest part of the shell, drum, or water leg. If the blow-off pipe must run through the combustion chamber, it must be protected from the direct sweep of the flames and hot gases. Where the pipe passes through the setting, a sleeve must be built into the wall so the pipe can slide freely during expansion and contraction.

Why is an ordinary valve not suitable?

Answer: Because when the valve is open small pieces of scale and other foreign matter are hurled against the seat with great force and grind the surface of the seat and valve away, causing the valve to leak.

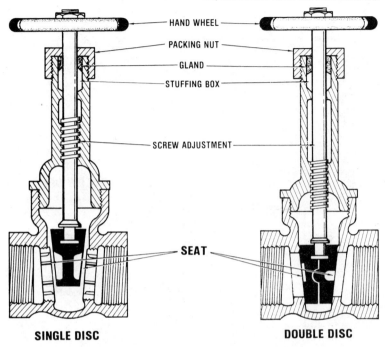

Fig. 12-11. Single-disc and double-disc forms of gate valve. Each is operated by raising or lowering the disc. When closed, the two faces of the disc are tightly pressed (by wedge action) against the seats, thus effecting a double seal. It should be noted that there are two seats for both the single and double disc types.

How is this avoided in a blow-off valve?

Answer: The valve is so constructed that the valve and seat, when open, are out of the path of the escaping water and impurities. There are no internal corners or pockets where sediment can collect. See Fig. 12-11.

How should a blow-off valve be connected to a boiler?

Answer: A stationary boiler must have two blow-off valves, either two slow-opening valves or one slow-opening and one quick opening, as shown in Fig. 12-12.

271

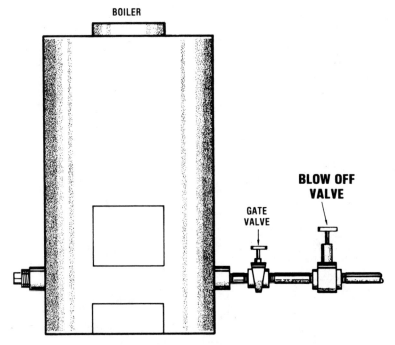

Fig. 12-12. Approved method of connecting a blow-off valve. Suppose something lodged under the blow-off valve and you couldn't shut it, then you would have another chance by closing the gate valve—that is why it is put there.

Why?

Answer: To ensure a tight outlet and to provide additional means of shutting off the connection in case anything happens to the blow-off valve.

How should a blow-off connection be made on a horizontal return tubular boiler?

Answer: The boiler shell is tapped at the rear end for the blow-off pipe. The latter should preferably be run straight down to below the floor level of the combustion chamber and then out, the pipes in the combustion chamber being protected from the heat by some insulating material such as tile or brick.

Boiler Fixtures, Fittings and Attachments

Where should the blow-down connection be made to a packaged water-tube boiler?

Answer: At either the front or rear of the boiler, depending on the direction it is pitched. The connection must be made at the bottom of the lower drum or mud leg at the lowest point.

Is it important to have the gauge glass located so the engineer can see it during blow-down?

Answer: Yes, so the operator can see how much water is being blown down. If the gauge glass can't be seen at the valve position, the operator will have to leave the valve to look at the glass.

Are blow-off and blow-down, as discussed here, the same as continuous blow-down?

Answer: No. Continuous blow-down is used to control the chloride content of boiler water.

What are the minimum and maximum sizes of blow-off piping and fittings?

Answer: Minimum is 1 inch (2.54 cm) and maximum is 2½ inches (6.35 cm). One exception: for boilers with 100 square feet or less of heating surface the minimum is ¾ inch (1.91 cm).

What is the difference between a Y blow-off valve and a globe valve?

Answer: In a Y valve the flow is straight through. See Fig. 12-13.

What is a slow-opening valve?

Answer: A valve that takes at least five 360° turns of the handwheel to open it.

What is a quick-opening valve?

Answer: A valve that can be opened in less than five complete turns of the handwheel.

What is the difference between a quick-opening and slow-opening valve?

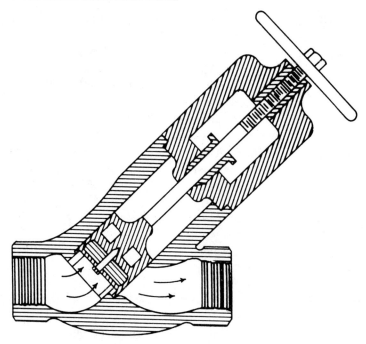

Fig. 12-13. Y-type blow-off valve.

Answer: In a quick-opening valve a disc slides across the seat when the handle is moved, instead of a disc that is slowly lifted from the seat.

Why is a blow-off tank used?

Answer: To avoid discharging live steam into a sewer or the atmosphere.

How does a blow-off tank work?

Answer: A blow-off tank has a fixed water level inside during normal operation. Steam and water from the boiler flow into the tank under the water level or seal during blow-off. See Fig. 12-14. The steam condenser and the water is cooled. Sometimes cold water is added to the tank to make sure boiler discharge water is adequately cooled. When the water in the tank reaches the over-

BOILER FIXTURES, FITTINGS AND ATTACHMENTS

flow point, water flows into the sewer. Blow-off tanks are vented to the atmosphere to prevent pressure buildup and an anti-siphon device in the vent line keeps the tank from draining dry.

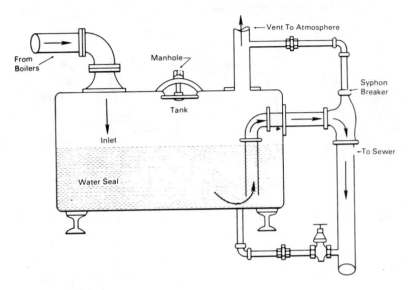

Fig. 12-14. Horizontal blow-off tank.

How is the continuous blow-down connection made?

Answer: To an internal pipe in the boiler just below the water line. Usually this pipe has a series of holes drilled along the length of the pipe (the combined areas of the holes must equal the cross-sectional area of the pipe). Boiler water escapes through this pipe, controlled by a metering or needle valve. The setting of this regulator gives the engineer control of the chloride concentration in the boiler.

COCKS

What is a cock?

Answer: A device for regulating the flow of fluids through a pipe; a typical straightway cock is shown in Figs. 12-15 and 12-16.

Power Plant Engineers Guide

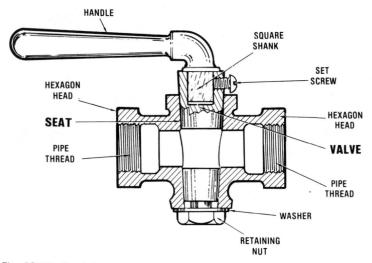

Fig. 12-15. Straightway cock showing ends tapped for connection in the run of a pipe line and detachable handle.

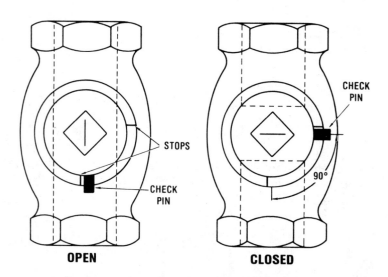

Fig. 12-16. Straightway cock with stops and check pin in open and closed positions. This control device is especially desirable on three-way and waste cocks.

Boiler Fixtures, Fittings and Attachments

How is it usually constructed?

Answer: It usually consists of a tapered conical plug having a hole or port in it, and working in a shell of iron or brass bored out to receive the plug and provided with passages to connect into pipes at either end.

How is a cock operated?

Answer: Rotation of the plug (90°) controls the passage of fluids by bringing the opening in the plug opposite those in the shell or away from them. Fig. 12-17 shows *how not* to turn a cock.

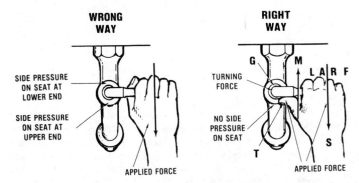

Fig. 12-17. Wrong and right way to open a ground cock. Grasping the handle and simply pulling it toward you (left) brings considerable pressure against the seat and tends to warp or distort the seat, causing leakage. The handle should be turned as at right, pushing with the thumb (T), and pulling with the other fingers (L,A,R,F), producing forces M and S. Force M prevents side pressure due to S coming on the valve seat, giving a resultant turning force (G) around the valve axis.

How many cocks are ordinarily attached to a boiler?

Answer: One or more blow-off cocks and three gauge cocks.

What is a gauge cock?

Answer: A device for determining the water level in the boiler.

How is the water level determined by means of gauge cocks?

Answer: Each cock is opened slightly and the presence of water or steam escaping from the cock tested by its appearance, sound, and feel to the hand. The right and wrong methods are shown in Fig. 12-18.

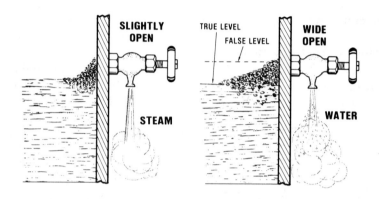

Fig. 12-18. Right and wrong way of testing water with gauge cocks. When the cock is only slightly opened, the water level is not materially raised by the outrushing steam; but is opened wide, the reduction of pressure inside causes a considerable disturbance of the water level near the cock, resulting in a false level as shown. Ths precaution should be remembered, especially when using the lower cock, because if opened wide the water is lifted surprisingly high. Hence, unless the lowest cock is at a liberal height above the crown sheet, it may, when opened wide, indicate water, though the true level may be dangerously low.

Why is the cock only slightly opened in testing?

Answer: Because a full opening tends to raise the water level at the gauge, thus giving a false indication.

What are the two principal types of gauge cock, classed with respect to the means employed for closing?

Answer: Compression and pressure.

What is a Mississippi cock?

Answer: One in which steam pressure keeps the cock closed and a push button is provided to open the cock.

BOILER FIXTURES, FITTINGS AND ATTACHMENTS

What other name is given to gauge cocks?
Answer: "Pet" cocks.

WATER GAUGE

What is a water gauge?
Answer: A device used to indicate the height of water within a boiler. Typical construction is shown in Fig. 12-19.

How is it constructed?
Answer: It consists of a strong glass tube, long enough to cover the safe range of water level, having its ends connected to the boiler interior by fittings. Since both ends of the tube are in communication with the boiler, the water level in the tube will be *approximately* the same as that in the boiler.

Why approximately?
Answer: It registers a false level lower than the level in the boiler because of the difference in temperature of the water in the tube and in the boiler.

What is a water column?
Answer: A boiler fixture consisting of a cylindrical piece to which are attached the water gauge cocks, thus combining the two into one unit. The top and bottom have outlets which connect it with the boiler below and above the water level, as shown in Fig. 12-19.

Does the ASME code allow the use of valves between the water column and boiler drum?
Answer: Yes, gate valves of the outside screw-and-yoke type. Provision should be made to lock them.

What fittings should be used for these connections?
Answer: Tees with plugs, not elbows, to allow cleaning of the pipes connecting the water column to the boiler.

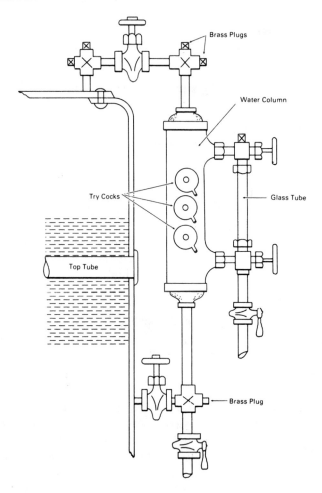

Fig. 12-19. Water gauge and gauge glass.

Why?

Answer: So that by removing the plugs, the connecting pipes may be cleaned of any foreign matter.

What attention should be given to the water-gauge glass when the boiler is down for repair or cleaning?

BOILER FIXTURES, FITTINGS AND ATTACHMENTS

Answer: It should be checked for worn and leaky ends, and gaskets should be replaced when they are hard and brittle.

Why is it important that the water column and glass gauge be thoroughly blown down at regular intervals on low-pressure boilers?

Answer: Under normal pressure, thorough blow-down may not occur.

What causes the gauge to close automatically when the glass breaks?

Answer: In normal operation, the gauge inlet valve is held open by equal pressure on each side of the valve ball. If the glass breaks, the pressure balance is upset and the ball drops, closing the valve.

What is another name for gauge cock?

Answer: Tricocks.

On a manually operated gauge, how is correct water level checked with gauge cocks?

Answer: The top gauge blows steam when opened; the middle water and steam; and the bottom blows water.

Are automatic controls used?

Answer: Yes, there are electrical and mechanical alarms. Both alert the operator of water-level discrepancies. The electrically operated gauge can be used to start the feed-water pump at low water or cut off fuel to the furnace.

FEED-WATER CONNECTION

What is an injector?

Answer: A device for forcing feed-water water into a boiler against steam pressure by means of a steam jet.

Where is the feed-water connection?

Power Plant Engineers Guide

Answer: At the coolest part of the boiler, through a special bushing or flange. On an HRT boiler the feed inlet is usually on top of the shell. On a packaged water-tube boiler, feed water enters either the front or rear upper drum heads. See Fig. 12-20.

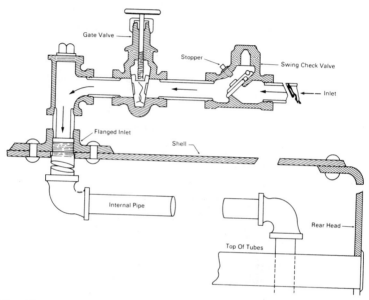

Fig. 12-20. Feed-water connection used on HRT boilers.

How is feed water mixed with hot water in the boiler?

Answer: A pipe inside the boiler tempers and distributes the fresh water beneath the boiler water line. Very often the inboard end of the internal pipe is plugged and water discharges through holes drilled in the sides of the pipe. The total diameter of the holes must at least equal the diameter of the pipe. Usually, such discharge pipes have heavy angle-iron baffles welded onto them to help disperse the water before reaching any hot metal surface.

Why is a check valve used in the feed-water line in addition to a stop valve?

Answer: To keep pressure in the boiler from pushing water back into the water line. Check valves must be installed properly

BOILER FIXTURES, FITTINGS AND ATTACHMENTS

or they will allow water to flow the *wrong* way. The stop valve must be in the feed line between the check valve and the boiler.

What size boilers require more than one water feed line?
Answer: One with more than 500 square feet of heating surface.

STEAM GAUGE

What is a steam gauge?
Answer: A device indicating *gauge* pressure as distinguished from *absolute* pressure. The most popular type is the Bourdon; it is made in two types, single-tube and double-tube.

How does a steam gauge work?
Answer: A steam gauge (according to type) works on one of two principles: (1) the expansion of a corrugated diaphragm when pressure is applied, and (2) the tendency of a curved tube to assume a straight position when under pressure. The two types are shown in Fig. 12-21 and operation of the curved tube gauge in Fig. 12-22.

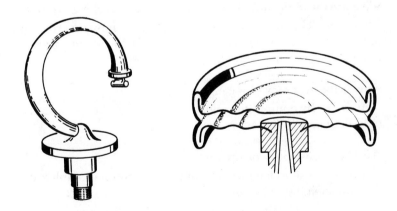

Fig. 12-21. Diaphragm and bent tube as used in the two classes of steam gauge.

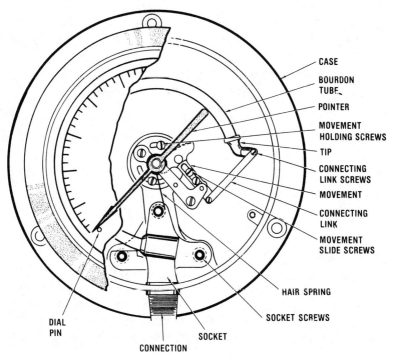

Fig. 12-22. Typical C-type Bourdon gauge showing names of parts.

What are the indications that a steam gauge is working properly?

Answer: The index or pointer (Fig. 12-23) moves easily with every change of pressure in the boiler. If the steam drops to atmospheric, the hand should go back to zero.

How is the accuracy of a gauge tested?

Answer: By comparing it with a test gauge of known accuracy or with dead-weight tester.

What is a goose neck or siphon?

Answer: A short length or coil of pipe having one complete turn and to which the steam gauge is attached. See Fig. 12-24.

What's the idea of curving the goose neck or siphon?

Answer: It traps condensate, gradually fills, and the cool water

Fig. 12-23. Steam pressure gauge.

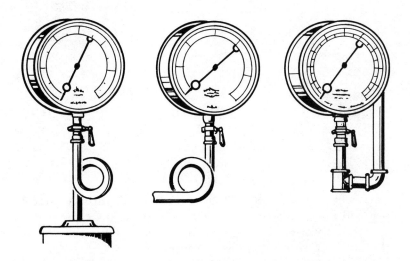

Fig. 12-24. Various forms of connection for a steam gauge. The pocket formed by the connection becomes filled with condensation, which protects the spring from the heat of the steam.

prevents live steam from touching the corrugated diaphragm tube. However, it should be filled with water when the gauge is installed.

How should a siphon be connected, and why?

Answer: It should first be filled with water to protect the gauge mechanism from the hot live steam.

FUSIBLE PLUGS

What is a fusible plug?

Answer: A safety device which protects a boiler from damage due to low water. It consists of an alloy of 93.3% pure tin, 0.5% copper, 0.1% lead and not over 0.7% impurities, which fills a hole in a brass plug, as shown in Fig. 12-25.

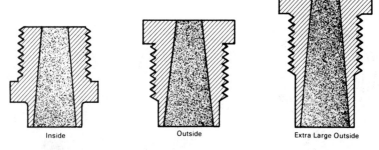

Fig. 12-25. Three sizes of fusible plugs.

Where is a fusible plug located?

Answer: It is screwed into the plate above the fire line so that one end is exposed to fire or hot gases while the other end is covered by water. The large end of the plug is always inserted in the tapered hole so it is under boiler pressure.

What should be noted about fusible plugs as a safety device?

Answer: They should not be used in boilers operating at pres-

BOILER FIXTURES, FITTINGS AND ATTACHMENTS

sures beyond 250 lbs. The temperature at this pressure equals or is above the melting point of tin.

What precaution should be taken with fusible plugs?

Answer: They should be renewed at least once every year and examined and cleaned every time boilers are washed out. Fusible plugs are not used as extensively now as in the past, although many older boilers (and some new ones) use them.

STEAM TRAPS

What is a steam trap?

Answer: An automatic device which allows the passage of water but prevents the passage of steam. (Fig. 12-26.)

What is a trap used for and where should it be located?

Answer: To drain pipes of condensate. Traps should be located on the ends of the main header, on the end of each branch line, on each heat exchanger where the steam gives up its heat.

In construction how is the device shaped so as to introduce centrifugal force?

Answer: By introducing a partition or equivalent means to suddenly change the direction of the fast flowing steam, usually through 180°.

What may be said of steam traps?

Answer: According to Crane, "A trap is a trap and it is unfortunate that it is impossible to get along without them." According to the author, "They are frequently too much annoyance and breeding devices for profanity." This goes also for injectors when not properly installed.

Are there many types of traps on the market?

Answer: There is an undue multiplicity of traps on the market (each one better than the others according to the maker).

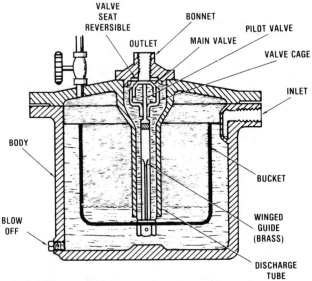

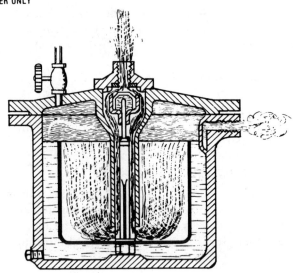

Fig. 12-26. Bucket pilot valve steam trap before discharge (left) and during discharge (right). In operation, condensation enters the trap at

Boiler Fixtures, Fittings and Attachments

the inlet against the baffle plate and overflows into the bucket, which, when full, drops and opens the pilot valve. This relieves the pressure under the main valve, and the pressure on top of this valve under the seat then forces it to the bottom of the cage. The condensation is then discharged from the bucket down to the low water line which keeps the end of the discharge tube always water-sealed. The bucket then rises and the valves close.

Name the most popular trap in use today.

Answer: The most popular type of trap in general use today is the nonreturn type. The return type is found on some systems, located beside the boiler and a little above it. Condensate from the return type is discharged directly into the boiler. The nonreturn type sends condensate to the condensate system.

SEPARATORS

What is a separator?

Answer: A device designed to remove as much moisture as possible from steam after it leaves the boiler. See Figs. 12-27 and 12-28.

Where is it placed and why?

Answer: Close to the engine or steam load so as to avoid any further condensation between separator and load.

What is the outstanding basic force which causes the separation of moisture?

Answer: Centrifugal force.

Any other force?

Answer: Sometimes gravity as a secondary force.

How does the centrifugal force act?

Answer: Change in direction of the steam flowing at 6000 to 8000 feet per minute creates very great centrifugal force which, acting on the heavy globules of condensate, hurls them out of the path of the steam.

What is usually provided on a separator and why?

Answer: A reservoir, having a glass gauge and drain cocks, to catch the condensate. The collected condensate is usually automatically removed by a trap.

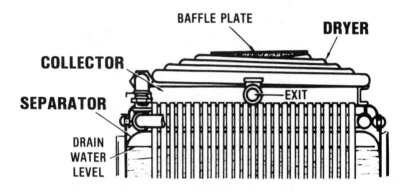

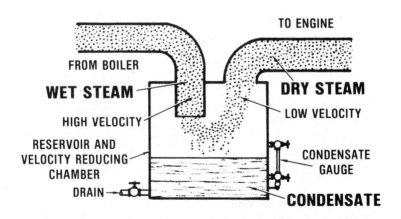

Fig. 12-27. Elementary steam separator illustrating principles of operation.

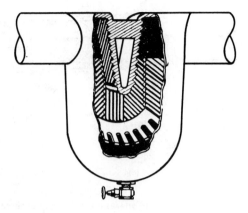

Fig. 12-28. Typical separator showing construction. The baffle plate, which serves to change the direction of the steam flow, is not set at right angles to the entering steam current, but is set at an angle so that when the steam is impinged against it, the particles of water rebound at an opposite angle. This sets up a rotating motion in the steam, bringing the latter in contact with the inside walls of the separator. These walls are heavily corrugated, as is also the surface of the baffle plate, and all corrugations are designed so as to carry the drainage out of and away from the course of the steam. Any moisture not caught by the upper baffle plate and by the inner walls is subject to further separation process by means of additional baffle plates located in the well or receiver portion of the separator. One of these plates is shown in the illustration. The separator is adapted for steam flow in either direction.

COLLECTORS

What is a collector, or so-called dry pipe?

Answer: A pipe placed inside a boiler at a high point and having small perforations extending its length so as to take off steam at a multiplicity of points and thus avoid turbulence which comes from taking off steam at only one point (Fig. 12-29).

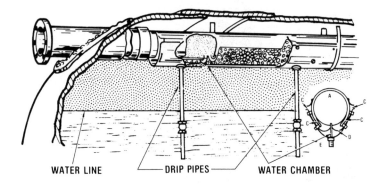

Fig. 12-29. Dry pipe or collector. In operation, the steam enters through two narrow slots and in passing down through the thin passages is brought in contact with the perforated plates or lining, through which moisture is conducted to the water chamber. Steam from the passages then enters the main pipe. Water in the water chamber is conducted to the water in the boiler by drip pipes, which are provided with check valves as shown.

What is the matter with calling a collector a separator?

Answer: It's distinctly a misnomer, because the function of a collector is to collect steam from a multiplicity of points in small amounts from each point, thus bringing into action disengagement over practically the whole disengaging surface.

STEAM LOOP

What is a steam loop?

Answer: An ingenious "thermal pump" consisting of an arrangement of piping wherein condensate is returned to the boiler.

Why call it a thermal pump?

Answer: Because its basic principle of operation is the expenditure of heat to cause condensation and pressure difference.

What are the four essential parts of a steam loop?

Answer: (1) Riser, (2) goose neck, (3) condenser, (4) drop leg. See Fig. 12-30.

Boiler Fixtures, Fittings and Attachments

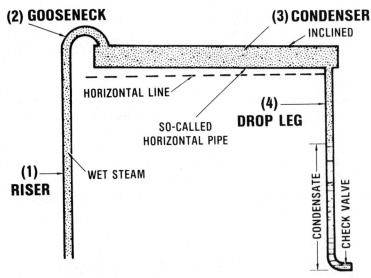

Fig. 12-30. Essential parts of a steam loop: (1) riser; (2) goose neck, or nonreturn device; (3) condenser, commonly and erroneously called the horizontal pipe; (4) drop leg, or balancing pipe. A check valve is placed at the end of the drop leg to prevent surging or fluctuating of the water level, thus rendering the operation stable. There are two conditions necessary for proper operation of a loop: (1) sufficient length of the drop leg to balance the pressure reduction due to weight and friction of the mixture in the riser, (2) sufficient condenser cooling surface to condense at a rate that will give the proper flow in the riser.

What determines the capacity and strength of the system?

Answer: The proportions of the four parts.

How does a steam loop work?

Answer: The *riser* does not contain a solid body of water, but a mixture of water and steam. The steam part of this mixture is readily condensed by means of the *condenser* at the top, usually and erroneously called the "horizontal pipe." This condensation reduces the pressure in the system which causes an upward flow of the mixture in the riser; that is, the riser is constantly supplying steam, conveying large quantities of water in the form of a fine spray to take the place of the steam condensed in the condenser.

As soon as the water mixed with steam passes the *goose neck*, it cannot return to the riser; hence the contents of the pipes constantly work from the riser toward the boiler, the condenser being slightly inclined toward the *drop leg* so as to readily drain the condensate into the drop leg.

The condensate will accumulate in the drop leg to a height such that its weight will balance the weight of the mixture in the riser.

FURNACE

Define the term "furnace."
Answer: The part of the boiler designed for burning the fuel.

Describe the grate assembly for a horizontal return tubular boiler.
Answer: It consists of (1) the grate, (2) front support or dead plate, (3) rear support or bridge wall, (4) main combustion chamber, (5) supplementary combustion chamber. An oil- or gas-fired boiler, of course, has no need for a furnace grate.

How is the grate constructed?
Answer: It is of cast iron made up in sections each containing numerous so-called grate bars.

What duties are performed by the grate?
Answer: It serves the purpose of holding the fuel while it burns and of admitting sufficient air so that it can burn.

Why are there numerous types of grate bars?
Answer: Because of the great variety of fuels which require different conditions for best combustion.

How much air space is provided?
Answer: From 30% to 50% of the grate area.

Why are grate bars tapered to give a wider opening between the bars at the lower end?
Answer: For free flow of air and cooling.

Boiler Fixtures, Fittings and Attachments

How wide are the air spaces?

Answer: They vary from ⅜ to 1 inch wide.

How long should grate bars be made?

Answer: Not longer than three feet.

What precaution should be taken in installing grate bars?

Answer: Ample space should be provided for expansion.

Why do some grate bars have a shallow groove running along the top?

Answer: To fill with ashes which tends to prevent clinker.

Mention some smoke prevention arrangements used.

Answer: (1) Large combustion chamber, (2) supplementary combustion chamber, (3) baffle plates, (4) brick checker work, (5) supplementary air supply, (6) Dutch oven, (7) down draft furnace.

What is the advantage of preheating the air?

Answer: The affinity of heated air for carbon being much greater than that of cold air, it raises the intensity of combustion.

CHAPTER 13

Boiler Feed Pumps

Some power plants use a pump to assure a steady supply of feed water ("new" water from the mains or "recycled" steam) to the boiler. The size of the pump needed is determined, of course, by the amount of water to be pumped during a given period of time. As you would expect, feed-water pumps range widely in configuration, depending on plant size and age of the plant. To adequately prepare you for any situation, all types are covered, including generally obsolete types, on the chance that a few may still be in service.

PUMP TYPES

Either two small feed-water pumps may be installed, using one at high speed and using the other as a spare, or one large pump may be installed and run at slow speed, using an injector in case the pump breaks down. In the first instance, if the growth of the plant requires increased boiler capacity, both pumps must be run at the same time. Hence, there is no reserve pumping capacity in case of breakdown. With the large pump and injector, there will be considerable margin to meet the increasing load without using both pump and injector at the same time. If a line shaft is available, a power pump with gear or belt transmission is desirable, using an injector for emergency feed. While the main engine is idle, there should always be two devices (as two pumps or a pump and an injector) for feeding the boiler to guard against interruption in case of breakdown of one.

What is a pump and how does it move a fluid or gas from one point to another?

Answer: A pump is a mechanical device that moves liquids or gases from one point to another by increasing the pressure on the liquid or gas to be moved. To draw a liquid, a pump reduces the pressure in a suction line, which allows atmospheric pressure to push the liquid to the pump.

What is a feed pump?

Answer: The pump which supplies a boiler with "feed" water.

What is the *net* feed water?

Answer: The quantity of water necessary to supply a stated evaporation in a given interval of time.

What is the *gross* feed water?

Answer: The *net* feed water plus the quantity provided from the mains for that blown out.

What two general classes of feed pump are used?

Answer: Reciprocating and centrifugal.

Name two well-known types of reciprocating pumps.

Answer: Simplex and duplex.

Boiler Feed Pumps

Name two kinds of reciprocating pumps classed in general with respect to the method of drive.

Answer: (1) Engine-driven and (2) independent.

What does the term "engine-driven" mean?

Answer: It means that the pump is driven directly by the main engine, by steam produced by the boiler.

What is an independent feed pump?

Answer: One having its own power unit attached.

What are the advantages of the engine-driven pump?

Answer: It has the advantages of the superior economy of the engine and variable speed proportional to the demand for steam by the main engine.

What is the advantage of an independent pump?

Answer: It can feed the boiler when the main engine is not running. If the pump is powered by an electric motor, control is much simplified.

Explain the difference between a single- and a double-acting pump.

Answer: A single-acting pump discharges in one direction and draws in the opposite direction. When each piston or plunger stroke draws on one side and discharges on the other it is double-acting.

What is the difference between a piston and plunger pump?

Answer: A piston pump piston moves in a liner or bored cylinder. In a plunger type, the plunger moves through a packing gland, which is either inside or outside the fluid chamber.

Why are pressure gauges installed on the discharge and suction lines of a pump?

Answer: To make sure pressure is normal in both lines. Abnormal readings usually indicate some blockage or obstruction in the line.

What are the various forms of power transmission for an engine-driven pump?

Answer: (1) Direct connection to crosshead, (2) walking beam, (3) reduction gears, (4) eccentric, (5) Scotch yoke, (6) direct drive. See Fig. 13-1. Figs. 13-2 and 13-3 show several ways of improving the efficiency of a direct-connected pump.

THE SIMPLEX SYSTEM

The word *simplex* is used here to distinguish the single-cylinder, direct-acting pump from the two-cylinder duplex, or twin type. Henry R. Worthington in 1840 invented the simplex, or direct-acting reciprocating pump. In this pump, the essential feature is that the movement of the steam piston (which is directly connected to the water piston or plunger) is automatically operated by the movement of the steam piston. See Fig. 13-4.

What is the difficulty in obtaining automatic action at the steam end?

Answer: The necessity of providing a special valve gear, somewhat complicated, is because of the absence of a rotating part which prevents the use of an eccentric.

Simplex Pump Valve Gears

In most cases, the necessary movements of the valve gear are obtained from three sources:

1. The movement of the piston,
2. The movement of the piston rod,
3. The steam pressure.

A valve gear thus operated usually consists of:

1. A main valve, which admits and exhausts steam from the cylinder.

BOILER FEED PUMPS

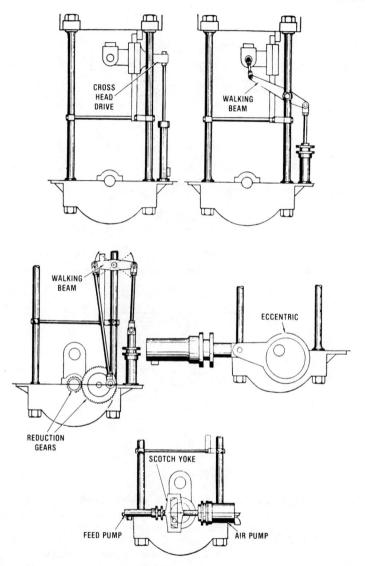

Fig. 13-1. Various methods of driving engine-driven feed pumps. Upper left, direct crosshead drive; upper right, walking-beam crosshead drive; center left, walking-beam reduction gear; center right, eccentric drive; bottom, Scotch yoke drive.

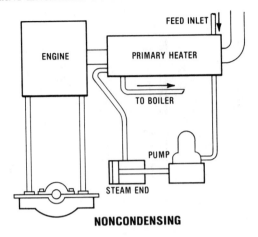

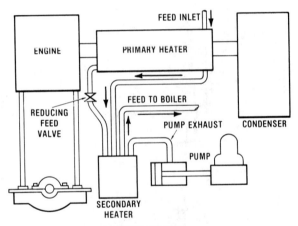

Fig. 13-2. Feed-water heater method of improving the efficiency of a direct-connected pump. In the upper illustration, the exhaust for the pump is piped to the main or primary heater and a large part of its heat is recovered in heating the feed water. When a condenser is used, as in the bottom illustration, a small or secondary heater should be provided, into which the pump exhausts. Any steam that is not condensed in the secondary heater passes to the primary heater. In such cases a reducing valve is placed between the two heaters so as to maintain a predetermined pressure in the secondary heater. In operation, as the feed water passes through

the primary heater it is heated to temperatures ranging from 110° to 130°F., more or less depending upon the vacuum maintained in the condenser. The water thus heated now passes through the secondary heater where additional heat is imparted to it from the exhaust of the pump and other auxiliaries. Its final temperature is within a few degrees of that of the exhaust. By means of the adjustable reducing valve, any pressure desired may be maintained in the secondary heater, thus varying the back pressure on the pump and its working temperature range to some value as may be found by test to give the best economic effect. Of course, increasing the back pressure involves increasing the size of the pump cylinder to do the work.

2. An auxiliary piston, connected to the main valve and moving in a cylinder formed on the valve chest.
3. An auxiliary valve, controlling the steam distribution to the auxiliary piston cylinder and operated with suitable gear by the main piston or piston rod.

How does the gear just mentioned work?

Answer: As the main piston approaches the end of the stroke, it moves the auxiliary valve.

What results from this movement?

Answer: It causes steam to be admitted to one end of the auxiliary piston and exhausted from the other, resulting in a movement of the auxiliary piston.

What happens when the auxiliary piston moves?

Answer: The movement of the auxiliary piston moves the main valve. The steam distributed to the main cylinder, thus affected, reverses the motion of the main piston, and the return stroke takes place, completing the cycle.

What detail varies mostly in pumps of different makes?

Answer: The auxiliary valve and the method by which it is operated. With respect to these features the majority of pumps may be divided into two classes as these having a separate auxiliary valve or an auxiliary valve and auxiliary piston combined.

Power Plant Engineers Guide

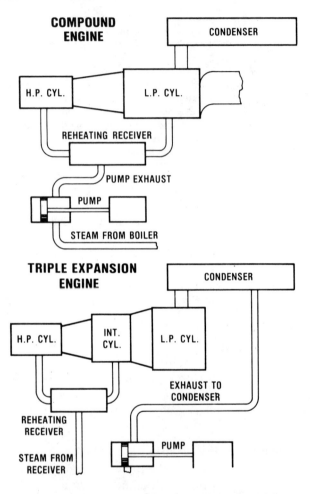

Fig. 13-3. Compounding method of improving the efficiency of a direct-connected pump. Top, arrangement with compound engine; bottom, arrangement with triple expansion engine. At the top, the pump receives its steam from the boiler, and exhausts into a reheating receiver where more or less of the condensate it contains is reevaporated. The exhaust then does useful work in the low-pressure cylinder. At the bottom, steam is taken from the receiver and exhausted into the condenser. If the pump is suitable for high pressures, a more economical arrangement would be to take steam from the boiler and exhaust into the receiver, thus

Boiler Feed Pumps

obtaining the advantage of expansion in the intermediate and low-pressure cylinders.

Simplex Gears with Separate Auxiliary Valve

In pumps of this type the auxiliary valves usually have stems or tappets which project into the cylinder at the ends and are moved by contact with the main piston as it nears the end of the stroke.

An example of this class is shown in Fig. 13-5. Each auxiliary valve "X" has a short stem which projects into the cylinder.

What happens when the piston strikes one of the auxiliary valves?

Answer: The valve is driven back and opens exhaust passage E, from the corresponding end of the auxiliary piston F, which immediately shifts under pressure of live steam on the opposite side of the auxiliary piston head.

What is the object of the little hole in each end of the auxiliary piston?

Answer: When both auxiliary valves are closed the steam passing through these holes leaves the auxiliary piston entirely surrounded by live steam and accordingly in perfect balance endwise.

How long does this balance last?

Answer: Until the main piston strikes the stem in the opposite cylinder head, at which time the valve-moving operations are repeated in the opposite direction.

With what does the space back of the auxiliary valves communicate?

Answer: The steam chest. The connecting passages are shown in dotted lines. The valve is therefore closed by steam pressure as soon as the piston moves back from the stem.

What should be noted about the piston?

Answer: It closes the exhaust passage before the end of the stroke.

POWER PLANT ENGINEERS GUIDE

What is the reason for this?

Answer: To trap the steam so as to form a cushion between the piston and the cylinder head.

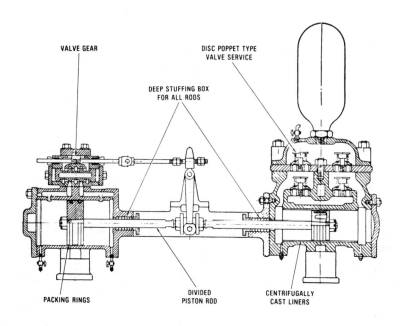

Fig. 13-4. Simplex horizontal pump showing general construction.

How is the piston started on the return stroke?

Answer: Sufficient steam is admitted through a little passage cut in the cylinder wall to start the piston.

How does the auxiliary valve shift the main valve?

Answer: In the direction of the piston travel at the end of the stroke. That is, opposite to that of a common slide valve. This valve therefore has two cavities, each of which alternately puts the cylinder in communication with the steam chest and the central exhaust port. In Fig. 13-5 see lever L, by means of which the auxiliary piston may be reversed by hand when expedient.

Boiler Feed Pumps

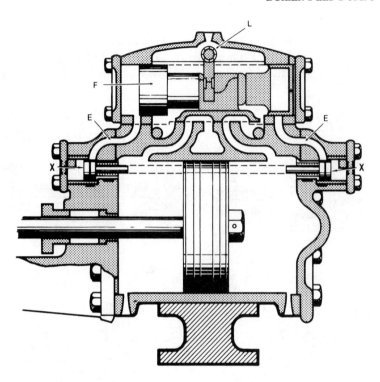

Fig. 13-5. Simplex pump of the separate auxiliary valve type. In operation the piston as it nears the end of each stroke strikes the stem and lifts the valve (x) off its seat. This allows the exhaust steam behind the piston valve to escape. The live steam pushes the piston toward the exhausted end, carrying the main slide valve along with it.

Simplex Gears with Auxiliary Valve and Auxiliary Piston Combined

With this arrangement an initial rotary motion is given the auxiliary piston by the external gear causing it to uncover ports which give the proper steam distribution for its linear movement. An example of this class is shown in Fig. 13-6. The main valve is operated by a positive mechanical connection between it and the main piston rod, and also by the action of the steam on the valve pistons.

Describe the construction.

Answer: The steam end consists of the cylinder M, valve A, and valve pistons B and B. These pistons are connected with sufficient space between them for the valve A to cover the steam ports F and F.

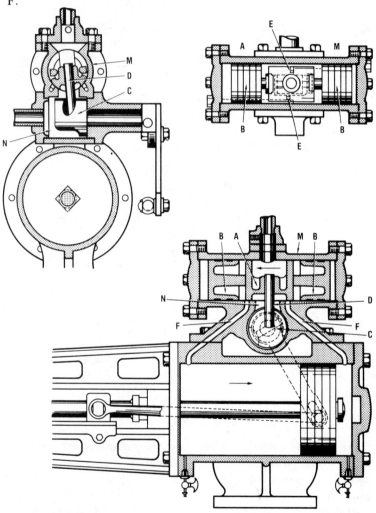

Fig. 13-6. Simplex pump of the combined auxiliary valve and auxiliary piston type as described in the text.

How is the valve operated?

Answer: By the steel cam C, acting on a steel pin D, which passes through the valve into the exhaust port N, in which the cam is located.

What is provided in addition to this positive motion?

Answer: Steam is alternately admitted to and exhausted from the ends of the valve piston through the ports E and E, which moves the pistons B and B.

Describe the operation of the pump.

Answer: Assume the pump to be at rest with valve A covering the main steam ports F and F. In this position, the cam C holds the main steam valve by means of the valve pin D, so that ports E and E admit steam to one end of the valve piston; at the same time it connects the other end with the exhaust port. The steam acting on the valve pistons moves both of them and opens the main ports F and F, admitting steam to one end of the steam cylinder and opening the other end to the exhaust.

What happens if the valve is in any other position than the one described?

Answer: The main ports F and F will be opened for the admission and exhaust of steam. It is accordingly clear that the pump will start from any point of the stroke.

What happens on the admission of steam to the cylinder?

Answer: The main port F, the main piston, cam, and valve will move in the direction indicated by the arrows in Fig. 13-6 (bottom).

What does the first movement of the cam do?

Answer: It oscillates the valve preparatory to bringing it into proper position for the opening of one of the auxiliary steam ports (E) to live steam and the other to exhaust. Also, it closes the valve mechanically just before the main piston reaches the end of its stroke. This causes a slight compression and fully opens one of the ports (E) to steam and the other to exhaust. By the admission of steam to one end, the other being open to the exhaust, the valve pistons move the valve to allow the admission and exhaust of steam from the cylinder for the return stroke.

Simplex Gears, Piston Steam-Valve Type

The single, direct-acting pump, steam-valve mechanism uses piston steam valves instead of the conventional slide valve. Large, unbalanced slide valves may be sluggish, resulting in uneven and noisy pump operation and in rapid wear on the steam valve and on the cylinder face on which the valve operates. This gear is designed not only for high-pressure, high-temperature steam, but also for service where the pump must operate without steam cylinder lubrication, as in marine installations.

In the valve operating mechanism shown in Fig. 13-7 the reciprocating motion of the piston rod is transmitted to the pilot valve 11, through the piston rod spool 34, the lever 39, the lost-motion block 47, adjustable valve-rod link 491, with tappet collars, and the steam-valve rod, 54. The small pilot valve is of the D-slide type.

The slide-type pilot valve is used in this gear, as such a valve is more easily tightened originally and reconditioned when worn than is a piston valve of the very small size which would be required in these pumps. The pilot valve is inverted and operated on a separate valve plate, with both the plate and the valve readily removable without disturbing any other parts. Steam and exhaust ports drilled in the pilot-valve plate register with ports drilled in the steam chest (18) which lead to the ends of the piston valve chamber in the chest and to the exhaust opening.

Reciprocation of the pilot valve alternately admits steam to and permits exhaust from the chamber in which the piston valve operates. This piston-valve chamber in the chest is bored and honed for the ground piston valve. The piston valve (12) reciprocates, admitting steam alternately from the chest to the two ends of the main steam cylinder and from the cylinder to exhaust. This alternate admission of steam to the ends of the main steam cylinder with synchronized exhaust from the opposite end results in the required reciprocating movement of the steam piston (7) in the steam cylinder (1). The steam valve is of the balanced-piston type.

The provision for externally changing the length of travel of the pilot valve permits easy adjustment of the valve gear to care for wide variations in speed and in relative pressures in the steam and liquid ends of the pump.

The steam chest is constructed with five main ports. Live steam

BOILER FEED PUMPS

enters through the two outside ports. The two intermediate ports connect to the ends of the cylinders and are combined steam inlet and exhaust ports. The central port is the exhaust outlet.

In Fig. 13-7, the steam inlet to the chest is at the left and the exhaust at the right. The chest is symmetrical and can be reversed if required to provide steam at the right and exhaust at the left.

In the piston valve (12) three ports are provided, the two outer ports for steam inlet and the center for exhaust. The valve is hollow and the steam ports intercommunicate. There is consequently a flow of steam from two sources when the ports are uncovered. There is but one main port to each end of the main steam cylinder. Two communicating openings are provided between each of these ports and the cylinder bore.

The small opening at the end of the cylinder has two functions. Through it, a small volume of steam is admitted to the cylinder at the beginning of the stroke, while the piston covers the inner and larger ports, preventing admission of steam at that point. By retarding the admission of steam at this beginning of the stroke, the moving parts of the pump and water column are gradually actuated up to the point where the main steam port is uncovered by the steam piston, thus ensuring that the pump will operate without shock or water hammer.

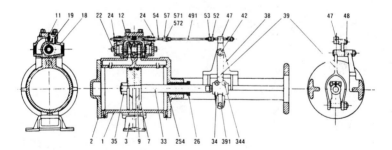

Fig. 13-7. Simplex pump gears, piston steam-valve type.

When the piston advances so that the inner, larger steam port is uncovered, steam enters the cylinder through both openings. Near the end of the stroke, the larger inner port through which the spent steam has been passing to exhaust is closed by the piston. The steam remaining in that end of the cylinder is, therefore, trapped

311

and compressed, providing cushion to absorb the inertia of the moving parts and of the water column, giving smooth, quiet deceleration. As the trapped steam then passes out through the starting port, the piston continues its movement slowly until reversal occurs.

The stroke of the pump may be lengthened by an increase in the distance between the nuts on the outside adjustable valve link. This increase in the lost motion retards the movement of the pilot valve. Conversely, a decrease in the amount of the lost motion will result in a corresponding shortening of the stroke of the pump. In general, it is desirable to so adjust the lost motion as to give the longest possible stroke attainable, without permitting the piston to hit the heads, as a long stroke results in the most satisfactory operation and minimum steam consumption.

THE DUPLEX SYSTEM

By definition, a duplex direct-acting pump is a combination of two pumps arranged side by side and so connected that the piston rod of one pump, in making its stroke, acts through a simple mechanism to move the valve which admits steam to the cylinder of the other. After it finishes its stroke, it waits for its own steam valve to be acted upon by the movement of the piston of the other side before it can make its own return stroke.

Name a desirable characteristic of the duplex pump.

Answer: There is no dead point at any stage of the stroke.

Why?

Answer: Because one or the other of the steam ports is always open.

What is the comparison between duplex valves and simplex valves?

Answer: The valves of a duplex pump are mechanically operated and not the steam-thrown valves necessary with the simplex pumps.

BOILER FEED PUMPS

Why does a duplex pump have five ports for each cylinder?

Answer: In addition to the three ports necessary for operation, two other ports are required to provide cushioning. It will be seen from Fig. 13-8 that the valve seat has five ports, giving separate steam and exhaust passages and a central exhaust cavity as shown. The passages Q and K nearest the ends are steam passages, and the inner passages O and R are for exhaust. These inner passages are covered or closed by the piston just before the end of the stroke whereby a portion of the exhaust steam is compressed and made to act as a cushion between the piston and cylinder head, thus preventing the piston from striking the cylinder heads when operating at high speed.

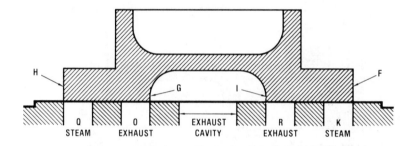

Fig. 13-8. Duplex pump steam valve and valve seat. H and F are the steam edges of the valve and G and I, the exhaust edges. Q and K are the steam ports and O and R the exhaust ports. The exhaust cavity is at the center. Note there are five ports in all.

Describe the steam distribution with the piston approaching the end of the stroke.

Answer: In the position shown in Fig. 13-9, the valve covers four ports. Steam is here admitted through the passage at the right end, causing the piston to move as indicated by the arrow. During this the steam port at the other end is closed and the exhaust port at that end is open, but the exhaust passage is closed by the piston, which traps some of the steam and cushions the piston.

How is the degree of cushioning regulated in some pumps?

Answer: By cushion valves.

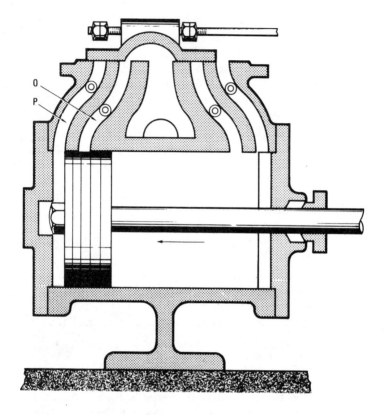

Fig. 13-9. Steam end of duplex pump showing cushion relief passages.

Where are they located?

Answer: These valves are placed in a passage leading from the steam port to the exhaust port at each end as indicated in Fig. 13-9.

Explain their operation.

Answer: When a cushion valve is partly open, some of the steam compressed in the clearance space and its steam port escapes into the exhaust port, thus reducing the cushioning effect.

What effect has this on the piston movement?

Answer: The piston moves closer to the cylinder head before it

BOILER FEED PUMPS

stops. The need for cushioning increases with the speed of the pump. Where the speed is increased it is necessary to increase the cushioning effect by partly closing the cushion valves. These valves provide a simple means of obtaining a full stroke of the piston without danger to the pump, whether its speed is fast or slow.

How are the valves proportioned?

Answer: They have no outside lap nor inside lap.

What is "lap"?

Answer: It is that portion of the valve face which overlaps the ports when the valve is in its central or neutral position, as in Fig. 13-10.

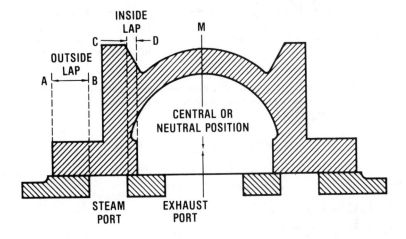

Fig. 13-10. Expansive working valve for flywheel pump. In this class of pump the dynamic inertia stored up in the flywheel during the first part of the stroke (up to cut-off) keeps the pump going during the expansion period where the pressure in the steam cylinder is not high enough to overcome the resistance at the water end.

What is the difference between outside lap and inside lap?

Answer: Outside lap (AB) is lap referred to the steam port and inside lap (CD) is lap referred to the exhaust port, as in Fig. 13-10.

Why do some valves have no lap?

Answer: Because a direct-acting pump must take steam the whole length of the stroke and have no compression except that needed for cushioning. See Fig. 13-11. In neutral positions the valves just cover the steam ports leading to opposite ends of the cylinders. Note the relation as shown.

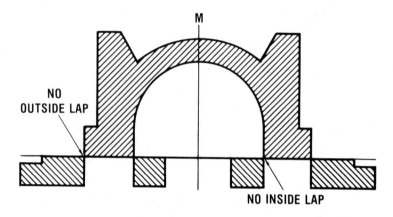

Fig. 13-11. Nonexpansive working valve for a direct-connected pump. The absence of lap allows the cylinder to take steam the full length of the stroke.

The Duplex Pump Valve Gear

The mechanism or gear which operates the valves in each cylinder of a duplex pump consists of a:

1. Crosshead
2. Long rocker arm
3. Rocker shaft
4. Short rocker arm
5. Connecting link
6. Valve stem

On the pump there are two sets of these parts—one set for each cylinder. However, in order to clearly show the assembly, one set only is shown in Fig. 13-12. Referring to the illustration, the source of motion for operating the valve of the left steam cylinder is

Boiler Feed Pumps

obtained from the piston rod of the cylinder on the other side. As shown there is a cross head (1), attached to the rod in which works a long rocker arm (2). The reciprocating motion of the rod imparts an oscillating motion to the long rocker arm. This rocker arm is attached to a rocker shaft (3). The rocker shaft works in a long bearing which is not shown in order to make working parts visible. At the other end of the rocker shaft is attached a short rocker arm (4), which in operation rocks or oscillates in unison with the long rocker arm. The oscillating motion of the short rocker arm imparts a reciprocating motion to the valve stem (6), through the connecting link (5).

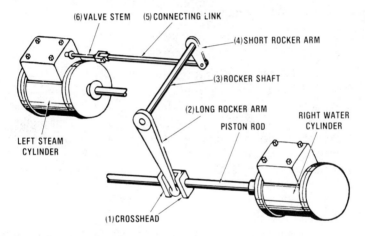

Fig. 13-12. Duplex valve gear shown for one cylinder of the pump.

Name one detail not shown in Fig. 13-12.

Answer: Means for introducing "lost motion" in the operation of the gear.

Duplex Valve Gear "Lost Motion"

There is *always* a lost motion between the slide valve and the valve rod operating it so that the valve does not move until the piston on the one side which operates it has traveled some distance. This affords a short pause in the flow of water at the end of a stroke and gives the water valve an opportunity to seat quietly before the reverse stroke takes place. The piston on the other side, having renewed its movement, tends to lessen pressure and flow fluctuations.

How long is the lost-motion pause in general practice?

Answer: From about one-quarter to possibly one-half the whole stroke of the piston, depending upon the amount of lost motion in the valve gear.

Is the lost motion adjustable?

Answer: On some pumps, yes; on others, no. The lost motion should be adjustable on all pumps.

"Lost-Motion" Mechanism

On some pumps, this detail is inside the valve chest where it can't be reached without stopping, cooling the pump and taking off the valve chest head. On other designs, it is placed outside as a part of the connecting link (see [6], Fig. 13-12) where it ought to be—easily accessible. With this arrangement, the lost motion may be adjusted while the pump is in operation. In fact, the pump *should* be in operation to properly "tune up" the lost motion.

The lost motion may be either
 1. Fixed, or
 2. Adjustable,
and the mechanism may also be classed with respect to location as
 1. Inside,
 2. Outside.

Fig. 13-13 shows an inside *fixed* or nonadjustable mechanism. As shown, it consists of a block threaded on the valve rod, the block being a loose fit between two lugs on the steam valve and the amount of space between the block and the lugs constituting the lost motion. This arrangement might go on very small or cheap pumps, but in a first class design such makeshift would not be tolerated.

According to Raabe: "It is seldom the case that the amount of lost motion has to be altered and unless the operator is thoroughly familiar with the details and design of the pump he should not

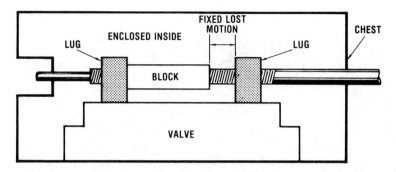

Fig. 13-13. Inside fixed lost-motion mechanism as made for small or cheap pumps.

undertake such alterations, as the designer knows best the requirements."

Complete Duplex Valve Gear

The complete valve gear, that is, the two separate gears (one for each pump), is shown in Fig. 13-15. It will be seen that although there are two "sets" of gears, the parts making up the mechanism are the same for both sides excepting long rocker arms C and L, which are of different lengths.

It will also be noted that short rocker arm G (which is moved by long rocker arm C) points down, while short rocker arm J (moved by long rocker arm L) points up. This arrangement of the short rocker arms is necessary in order that one of the slide valves which they move (both valves being alike and of the plain "D" type) shall admit steam at the opposite end of the steam cylinder so as to reverse the piston stroke. With regard to the long and short rocker arms, while they are of different lengths the two moving together have the same ratio of lengths as the other two and as each rocker arm has the same length of travel, the short rocker arms also move the same distance. On account of the reversal of the short rocker arms (one pointing down and the other up) the difference in long rocker arms is a mechanical necessity in order to bring the valve rods to a common elevation.

On simplex pumps, when is a "D" and "B" valve used?

Answer: The "D" valve is used on gears in which the valve

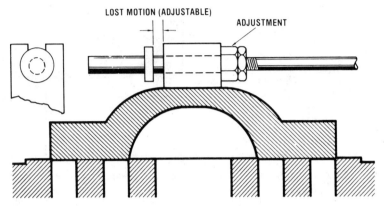

Fig. 13-14. Inside adjustable lost-motion mechanism shown in detail.

follows the piston movement; the "B" valve is used on gears in which the valve moves in reverse direction to the piston movement.

What names are usually objectionably used for short rockers and for long rockers?

Answer: Cranks for short rockers and levers for long rockers.

Duplex "D and B" Type Valve Gear

In order to avoid the separate passage at each end of the cylinder, a type of gear has been devised which employs a "D" valve on one cylinder and a "B" valve on the other cylinder. The difference between these valves is shown in Fig. 13-16. Using a D valve and a B valve permits a unique arrangement of steam passages. The prime reason for this design as just stated is to get rid of the separate passage at each end of the cylinder used for cushioning.

What is the difference between a port and a passage?

Answer: A port is the entrance at the valve seat to either a steam passage leading to the cylinder, or the exhaust passage leading to the exhaust pipe. A port has only two dimensions and a passage three dimensions; in other words, a port has *area* and a passage, *volume*.

BOILER FEED PUMPS

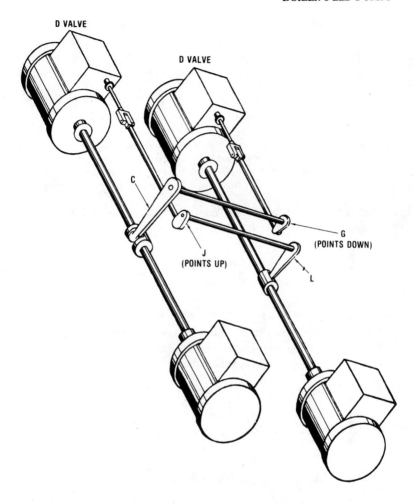

Fig. 13-15. Duplex horizontal pump showing valve gear for D valves.

Describe the D-B valve gear.

Answer: The gear is shown in Figs. 13-17 to 13-21. In Fig. 13-21, one and five are D valve steam passages. Two and six are B valve steam passages. Three and four are exhaust passages. Seven and eight are valve stem nuts.

321

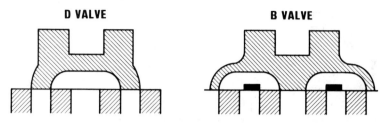

Fig. 13-16. Difference between D and B valves. The reason for these two types of valves is explained in the text. B valves are used on pumps having a separate auxiliary valve in which the piston contacts with stems or tappets which project into the cylinder at the ends.

What is the feature of the rockers 9 and 10 in Fig.13-21?
Answer: They are duplicates and therefore interchangeable.

What is the object of the small drilled holes?
Answer: They are for starting.

Describe the starting passage and its action.
Answer: This is a small drilled passage which admits steam to the cylinder when the piston covers the main passage as the piston passes this passage when cushioning.

How is a pump rated?
Answer: Diameter of steam cylinder by diameter of water cylinder or plunger diameter by the length of the stroke; for example, $10 \times 5 \times 10$.

What is displacement and how is it calculated?
Answer: Displacement is the volume moved through by the piston during one stroke. The displacement of the water end on a $10 \times 5 \times 10$ pump would be: $5 \times 5 \times 0.7854 \times 10 = 196.35$ cubic inches, less the volume of the piston rod.

How much pressure could a $10 \times 5 \times 10$ pump produce with 160 pounds of steam pressure?
Answer: Since the relationship between circles equals the squares of diameters, $10^2 \div 5^2 = 100 \div 25 = 4$. With 160 pounds of

Boiler Feed Pumps

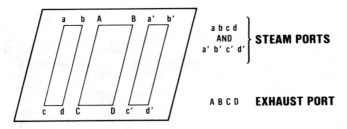

Fig. 13-17. Valve seat illustrating ports.

steam pressure on the 10-inch piston, the pump would generate 640 psi on the 5-inch piston.

How much water will this pump discharge per hour if it makes 15 double 8-inch strokes per minute?

Answer:

$$\text{Area} = \frac{22d^2}{28} \times \frac{22}{28} \times \frac{5}{1} \times \frac{5}{1} \times \frac{1}{144} = \text{sq. ft.}$$

$$\text{Volume} = \text{area} \times \text{length} = \frac{22}{28} \times \frac{5}{1} \times \frac{5}{1} \times \frac{1}{144} \times \frac{8}{12} = \text{cu. ft.}$$

Volume per minute = volume of one stroke × number of strokes =

$$\frac{22}{28} \times \frac{5}{1} \times \frac{5}{1} \times \frac{1}{144} \times \frac{8}{12} \times 2 \times 15$$

Volume per hour = volume per minute × 60 =

$$\frac{22}{28} \times \frac{5}{1} \times \frac{5}{1} \times \frac{1}{144} \times \frac{8}{12} \times \frac{2}{1} \times \frac{15}{1} \times \frac{60}{1}$$

Number of gallons = cu. ft. × 7.48 gals. per cu. ft.

$$= \frac{22}{28} \times \frac{5}{1} \times \frac{5}{1} \times \frac{1}{144} \times \frac{8}{12} \times \frac{2}{1} \times \frac{15}{1} \times \frac{60}{1} \times \frac{7.48}{1} =$$

1224 gal. per hour

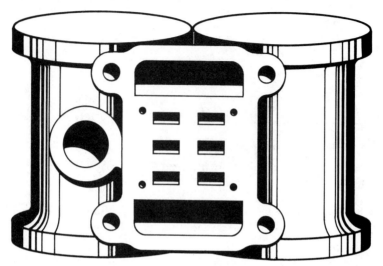

Fig. 13-18. Top view of the steam cylinders of a duplex D-B pump showing the ports. Note only three ports to each cylinder. The two outside ports of each cylinder are the steam ports and the center port the exhaust port.

Fig. 13-19. Bottom of one B and one D valve forming a part of the D-B system.

CENTRIFUGAL PUMPS

How does a centrifugal pump operate?

Answer: An impeller is rotated at high speed within a casing. See Figs. 13-22 and 13-23. The centrifugal force generated by the

BOILER FEED PUMPS

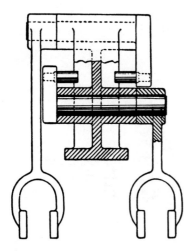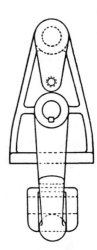

Fig. 13-20. Views of ordinary duplex rockers and standard.

impeller is transmitted to the liquid in the casing, thus pushing it through the discharge opening.

Does a centrifugal pump offer any advantages for boiler feed-water supply?

Answer: Yes, a centrifugal pump is simple and reliable. It has only one moving part and no valves. By staging centrifugal pumps high pressures can be developed, which makes them ideal for high-temperature, high-pressure boiler feed operations.

How about turbine driven centrifugal pumps as boiler feeders?

Answer: They are very extensively used, especially in large plants like electrical power plants.

Why?

Answer: Because they are continuous in action, put no severe strain on the piping, if the feed valves were all shut off accidentally, no dangerous pressure would result. There is also less attention required and a consequent saving in maintenance.

What type centrifugal pump is used?

Answer: Multistage, from two to five stage or more depending upon the boiler pressure.

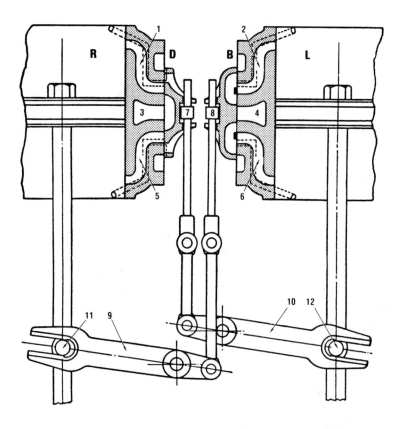

Fig. 13-21. Duplex D and B, valve gear, the object of which is to avoid separate exhaust passages for cushioning and the steam waste caused by an unduly large clearance.

What is the steam consumption of the turbine-driven pump?

Answer: About the same as the ordinary slide valve engine, or less, depending upon the quality of the steam and other conditions.

BOILER FEED PUMPS

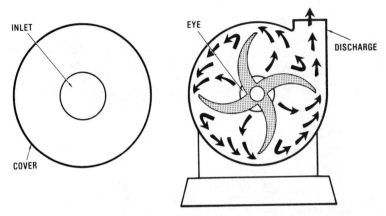

Fig. 13-22. Cover and section of centrifugal pump commonly called a **volute** pump due to the shape of the casing. The illustrations show the inlet, eye, and discharge.

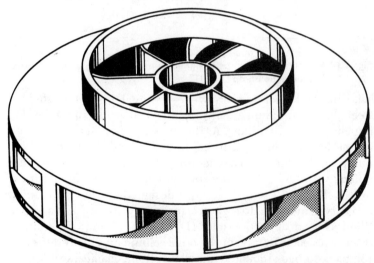

Fig. 13-23. Enclosed double-inlet impeller. The impeller is cast in one piece of bronze except in cases of pumping liquids requiring special metals, such as chrome, monel, nickel or other suitable alloys.

What type of pumps are used where boiling water must be moved?

Answer: Pumps designed for this environment, such as those shown in Fig. 13-24. They feature bronze-fitted construction,

POWER PLANT ENGINEERS GUIDE

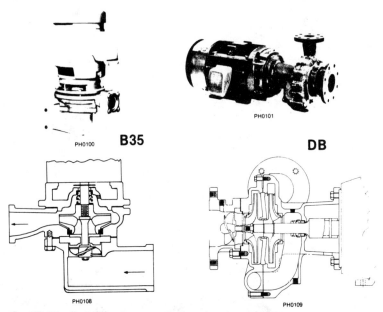

Fig. 13-24. Centrifugal pumps designed to handle boiling water. Courtesy Domestic Pump Div., ITT.

hand-finished enclosed bronze centrifugal impellers, stainless steel shafts, premium quality mechanical seals, and axial-flow impellers as the first stage. As the cutaway views show, the pump on the left has one centrifugal impeller, and the high-pressure pump on the right has two, in series. Both are designed to pump boiling water with 24 inches submergence. The pump illustrated in Fig. 13-25 is a two-stage type designed for low to medium capacity in medium- to high-pressure applications. Parts are identified as vertical suction inlet (1), case wear rings (2), impeller discharges each 180° from the other, eliminating shaft deflection (3), back-to-back mounted impellers (4), mechanical seal (5), and close-coupled construction (6).

MAINTENANCE

No piece of machinery can be expected to continue in satisfactory operation unless it receives proper and periodic attention to

BOILER FEED PUMPS

correct any faults or derangements that may arise. This comes under the heading of ordinary servicing rather than repairs. See Figs. 13-26 to 13-34 for helpful illustrations.

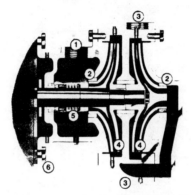

Courtesy Industrial Steam Div., Kewanee Boiler Corp.

Fig. 13-25. Two-stage centrifugal pump.

Reciprocating Pump Troubles

What causes a pump to fail to discharge?

Answer: (1) Pump not properly primed, (2) lift too high for temperature of water, (3) inlet valves stuck, (4) valve seats in bad condition.

What should be the maximum lift for water at ordinary well temperature?

Answer: Not over 22 feet.

How can hot water be handled?

Answer: To handle hot water, *negative* lift is necessary, in an amount depending upon the temperature of the water.

What is *negative* lift?

Answer: Head on inlet. That is, the water level of the supply is at a higher elevation than the inlet valve.

329

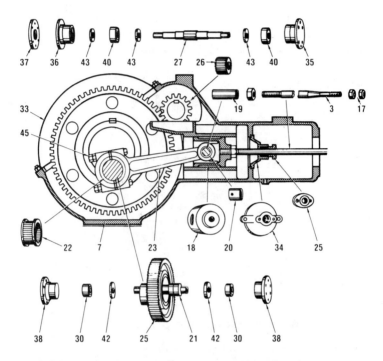

Fig. 13-26. Crank end of horizontal duplex double-acting piston pump.

What causes a pump to pound and vibrate?

Answer: (1) Air in the liquid, (2) leaky inlet line, (3) leaky packing on inlet valve chamber cover, (4) leaky packing around plungers or piston rods, (5) lift too high, (6) excessive speed, (7) vacuum chamber filled with air, (8) air chamber filled with water.

Parts List for Illustrations

1. Pump cylinder
2. Pump cylinder removable liner
3. Piston rod
4. Pump cylinder head
5. Valve plate
6. Valve chest cover
7. Frame
8. Pump piston head
9. Pump piston follower

10. Fibrous packing
11. Inner packing ring (metal)
12. Outer packing ring (metal)
13. Piston rod gland
14. Valve seat
15. Valve spring
16. Valve
17. Piston rod nut
18. Crosshead
19. Crosshead pin
20. Crosshead brass
21. Crankshaft
22. Crank brass
23. Connecting rod
24. False head
25. Crank gear
26. Pump pinion gear
27. Pinion shaft
28. Valve stem
29. Valve guard
30. Crankshaft bearing
31. Liner securing stud
32. Liner securing stud nut
33. Frame cover
34. Piston rod oil flange
35. Pinion shaft bearing housing
36. Pinion shaft bearing housing (pulley side)
37. Pinion shaft bearing housing cover
38. Crankshaft bearing housing
39. Pump cylinder foot
40. Pinion shaft bearing
41. Pinion shaft oil seal
42. Crank thrust ring
43. Pinion shaft thrust ring
44. Pinion bearing outer race lock ring
45. Connecting rod cap bolts
46. Cradle head gasket
47. Valve pot cover
48. Valve stem and washer
49. Valve wing guide
50. Valve stem lock washer
51. Valve stem locknut

What causes short-stroking?

Answer: (1) Pump not properly lubricated, (2) piston rings strike shoulders worn at ends of cylinder, (3) packing too tight, (4) excessive cushioning in steam cylinder, (5) excessive head, (6) steam valves not properly set.

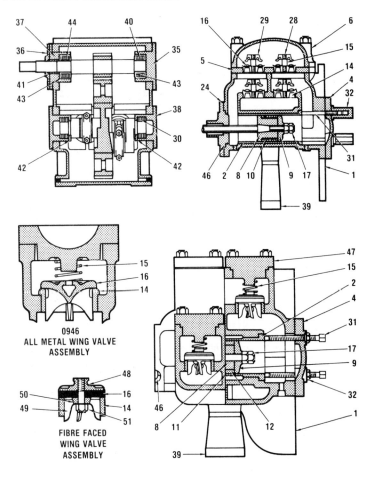

Fig. 13-27. Water end of horizontal duplex double-acting piston pump.

What makes pistons jump at beginning of the stroke?

Answer: Air trapped somewhere in the water end of the pump.

Under what conditions is this most likely to occur?

Answer: On pumps operating with high lift or from allowing the pump to race when starting.

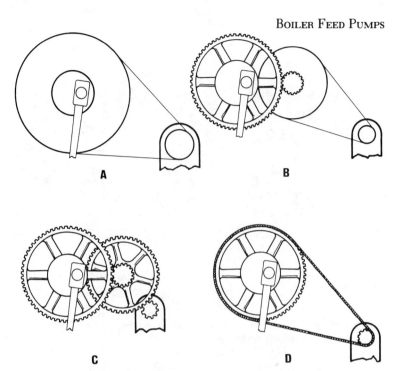

Fig. 13-28. Various forms of drive for power pumps. A, belt single reduction; B, combined belt and toother gear double reduction; C, toother gear, double reduction; D, chain-single reduction.

How should the pump be started?

Answer: Operate slowly for a while with the air vents open.

What is the effect of excessive lift?

Answer: It prevents the pump cylinder filling with water before the discharge stroke begins.

What causes a piston to strike the cylinder head?

Answer: Improper adjustment of cushion valve or improper adjustment of lost motion.

Give directions for servicing the valve gear when the pump refuses to start.

Answer: Disconnect the auxiliary valve stem from the oper-

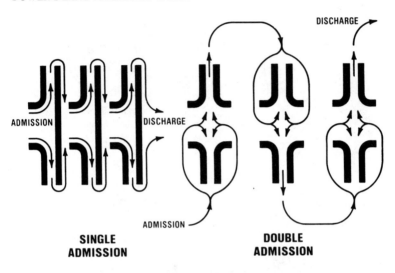

Fig. 13-29. Multistage impeller assemblies showing path of flow from admission to discharge.

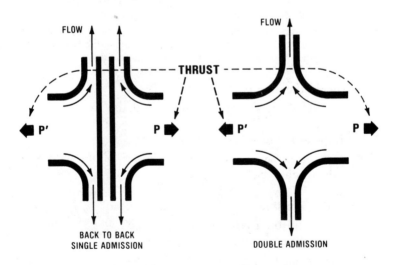

Fig. 13-30. Natural balancing (left) single-stage with two single-admission, back-to-back-impellers and (right) single-stage with one double-admission impeller.

Boiler Feed Pumps

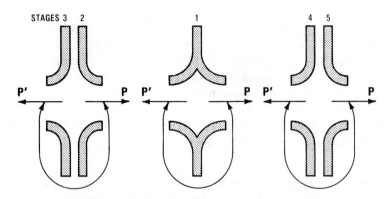

Fig. 13-31. Assembly of one double-admission and two pairs of single-admission, opposing, back-to-back impellers.

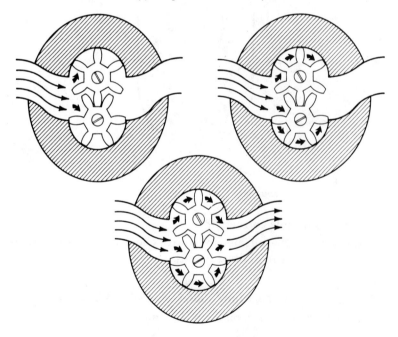

Fig. 13-32. Operation of external spur gear pump. Top left, liquid entering through the admission opening; top right, liquid being carried between the gear teeth toward the discharge side; bottom, liquid being forced out of the pump.

ating gear without disarranging adjustment of tappet collars. Open exhaust, inlet, and discharge valves and then crack throttle. Work auxiliary valve by hand; it should work freely.

Should the pump still refuse to start, secure it. Remove valve chest cover and examine the main valve to see if it has over-ridden or stuck. If the pump cannot now be started a complete overhaul of working parts of the steam end is necessary to stop steam leakage either in steam piston or valves, which is the most probable cause of the pump not starting.

What does jerky operation in starting indicate?

Answer: Failure of the water supply to follow the water piston.

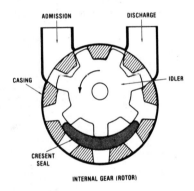

Fig. 13-33. Internal spur gear pump.

How is this trouble corrected?

Answer: See that all inlet line stop or check valves are open and that the line is clean of obstructions.

When a feed pump is vapor bound what should be done?

Answer: Turn a hose on the water end.

If a pump races without increasing its output, what is the cause?

Answer: A leaky plunger, leaky, broken, or stuck water valve, or air leakage.

BOILER FEED PUMPS

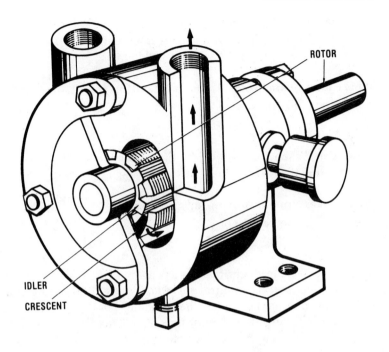

Fig. 13-34. Gear-type rotary pump. Cutaway view showing casing, rotor gears, and crescent in part.

What should be done in such case?

Answer: Stop the pump as soon as practical in order to ascertain and correct the trouble. Should the pump, after it has been running properly, suddenly lose pressure on one stroke, look for a broken valve at once.

How would you adjust the cushion valve?

Answer: Run the pump without water load and open the cushion valve until the piston strikes the head. If the piston strikes too hard, adjust the lost motion; do the same if the piston does not strike. Now place the pump under load and close the cushion valve until the piston stops striking the cylinder head.

What should be done if the pump has no cushion valves?
Answer: Adjust lost motion so there is just enough to permit the pump to make a full stroke without striking.

What causes steam to blow through the stuffing box at stroke end?
Answer: Shoulders.

Why?
Answer: When the piston hits these shoulders, they lift the piston rod in the stuffing box packing, allowing steam to escape.

What makes a pump stop on "dead center"?
Answer: Slide valve worn.

Explain.
Answer: Wear reduces the lap at each end so that when the valve is in the central or neutral position, the edge of the valve and edge of the port are no longer "line and line" but the wear gives a slight port opening at each end, thus equalizing the pressure on both sides of the piston, causing it to stop.

Centrifugal Pump Troubles

In most cases, what causes faulty operation?
Answer: Faults on the inlet line.

What causes failure to discharge?
Answer: Improper priming.

Explain.
Answer: If the priming valve is completely opened, allowing water to rush into the pump, it will trap air in the top of the casing, which will cause a reduction in discharge.

How should the pump be primed?
Answer: Allow the priming water to run slowly and open the petcock on top of the pump.

BOILER FEED PUMPS

How can an accumulation of air in the top of the pump casing be avoided?

Answer: Tap in a small pipe with valve from the top of the casing and operate with the valve "cracked" allowing no more leakage than necessary.

Mention a common cause of reduced capacity.

Answer: Air leaking through the packing of a pump operating with lift.

What is a good remedy to prevent this?

Answer: Install a lantern ring. In some cases a grease cup substitutes for the pressure fluid connection to the lantern ring. The lantern ring is kept filled with grease which works out between the shaft and packing to form an airtight seal.

What attention should be given to inlet screens?

Answer: They should be kept clean.

With proper priming, why in some cases does a pump not deliver full capacity?

Answer: Too much lift.

What sometimes causes a great reduction in capacity?

Answer: Impeller put on backwards by mistake after repairs.

What causes a great reduction in both capacity and head?

Answer: Reverse rotation.

What is the effect of worn sealing rings and why?

Answer: When sealing rings between impeller and casing become worn, the capacity is decreased because some of the discharge flows back into the inlet.

What causes noisy operation?

Answer: (1) Air, (2) cavitation when operating on heads too low for the design.

Mention a cause of vibration.

Answer: Misalignment which is magnified at high temperatures.

What is the cause of packing trouble?

Answer: Too much tightening adjustment.

How can trouble by sand lodging in the lantern ring be avoided?

Answer: Take sealing water from a clean supply and dispose of any solids that might get into the lantern ring by installing a drain pipe with a valve connecting with the stuffing box below the ring.

Rotary Pump Troubles

What should be noted about the discharge of a rotary pump?

Answer: Its discharge is positive; accordingly, it will develop excessive pressure if the discharge line valve is closed. On this account, a safety relief should be provided and set to the proper relieving pressure.

What about troubles in general?

Answer: Many of them are similar to troubles with centrifugal pumps.

What constitutes a major trouble?

Answer: Wear of the rotating parts and casing.

What causes vibration?

Answer: Excessive wear and misalignment.

What precaution should be taken with the inlet and discharge pipe lines?

Answer: Support these lines so that they do not put a strain on the pump casing.

BOILER FEED PUMPS

Why?

Answer: Such a condition tends to cause internal rubbing.

What reduces capacity?

Answer: Air traps or air leaks.

What is necessary on the inlet line and why?

Answer: A strainer to prevent solid materials entering the pump which would cause damage.

What precaution should be taken with the packing?

Answer: The packing should not be too tight or dry in order to prevent scoring of the shaft.

What other trouble is encountered with tight packing?

Answer: It wastes power, especially on pumps having four stuffing boxes.

VARIABLE STROKE POWER PUMPS

The variable stroke pump (duplex and triplex) forms a very desirable unit especially with automatic control for boiler feed when operated at constant pump speed. The pump delivery can be controlled either manually or automatically to suit varying requirements and the power consumption is in almost direct proportion to the rate of output.

Variable capacity demands are flexibly afforded without excessive capacity, intermittent start and stop operation or of inlet valve control all of which reduce the economy of the system.

In general it might be said that where there is sufficient boiler capacity that forcing is not necessary on peak load, the rate of feed should be proportional to the steam demand.

In plants where a constant and variable feed pump is desirable, the Aldrich-Groff variable capacity triplex boiler feed pump (called by the manufacturer the "Powr-Savr" pump) serves as an illustration. See Figs. 13-35 to 13-37.

The construction of the pump, which is of the triplex or three-cylinder type, features a three throw crankshaft with a crank sequence of 120 degrees; the fluid end is located on top of the pump frame, allowing the discharge valves to be positioned directly above the plungers, with inlet valves at the side of the cylinder.

This arrangement affords a streamlined flow of liquid into and out of the pumping chamber. The upright design prevents entrance of any air or lubricating oil into the pumping chamber through stuffing boxes, which is an important feature particularly for boiler feed service and when handling light liquids.

What is the characteristic of the 120° sequence throw crankshaft arrangement?

Answer: In this arrangement, the strokes overlap one another, producing almost a continuous and almost uniform rate of discharge.

How are the plungers packed?

Answer: The plungers are outside-packed permitting adjustment of glands during operation.

What is the principal advantage of the Aldrich pump?

Answer: Being of the variable stroke type, the length of plunger stroke and corresponding delivery can be adjusted, either manually or automatically from zero to full stroke, that is to say, from no delivery to full hydraulic capacity.

Describe the mechanism of the Aldrich variable stroke pump.

Answer: Fig. 13-35 is a sectional view through one of three identical cylinders; it shows the simple mechanical principles incorporated in the design of these units.

The lower end of each pump plunger is connected to a crosshead and through a wrist pin to a link. The link is connected to a bronze guide block which slides in a curved track on the stroke transformer. The "stroke transformer" is supported on two end trunnions to permit tilting. Tilting to vary the length of stroke is accomplished automatically by means of the adjustment cylinder (or a manually operated screw mechanism).

Power Pumps — Variable Stroke

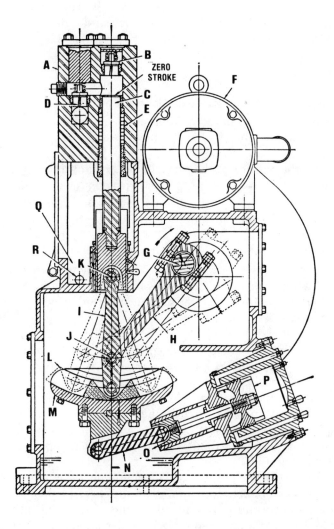

Fig. 13-35. Position of curved link for zero stroke. A, liquid end; B, discharge valves; C, plungers; D, inlet valves; E, stuffing boxes; Q, slump; R, drain; G, crank pins; J, connecting rod pins; K, wrist pins; L, sliding blocks; M, curved link; N, connecting link; P, servo piston; O, crosshead.

In operation, a pendular motion is imparted to the link by the connecting rod operated by the crankshaft. The motion of the plunger is variable from zero to maximum stroke, depending upon the angular positioning of the stroke transformer. The delivery of the pump is thus infinitely controllable from zero to the rated maximum output.

What in fact is the so-called stroke transformer?

Answer: It is a curved adaptation of the old Hackworth link which was the pioneer of radial valve gears.

What is the position of the curved link for zero stroke?

Answer: The position is shown in Fig. 13-35 in which the arc of the F curved link (M) being such that the connecting link (N) swings as a pendulum about the crosshead (O) pin as a center.

What is the position of the curved link for full stroke?

Answer: This is shown in Fig. 13-36 in which the lower end of the F connecting link (N) receives its reciprocating motion from the vertical component of the motion imparted to the guide block by the curved link (M).

Automatic control of pump delivery may be accomplished either by air or water control—pneumatically or hydraulically. Automatic control has been developed for such applications as boiler feeding, process charging, oil burner supply, and other similar uses where control of delivery must be accurately maintained and at any one point over the whole range from zero to full capacity. The method of operating a pump equipped with this system may be understood by referring to its diagrammatic equivalent in Fig. 13-37.

Assume that the pump has been operating at its full stroke and capacity as a boiler feed pump with the curved link or so-called stroke transformer M, positioned as indicated by solid lines in Fig. 13-37, by a hydraulic servo-piston P, under automatic control from a conventional spring-balanced diaphragm regulator 1. This regulator 1 may be automatically controlled in turn via the compressed air or hydraulic line 2, connected to a remotely located water level regulator of conventional design, to a conventional differential

Boiler Feed Pumps

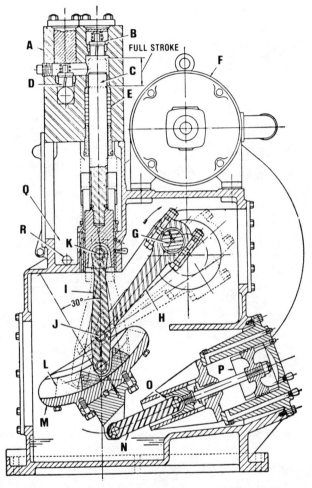

Fig. 13-36. Position of curved link for full stroke. The reference letters correspond to those of Fig. 13-35.

pressure regulating means, to other desired means such as are responsive to rate of steam flow, or to a rate of fuel supply, etc., depending upon the particular conditions of the installation.

Upon reduction in boiler load and feed-water requirements, the pressure in pipe 2 is reduced by the remotely located primary regulator which permits the spring of regulator 1 to move the link 3

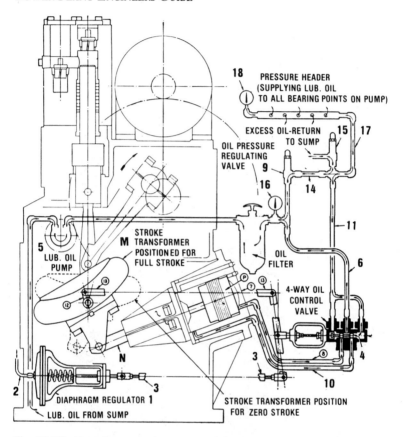

Fig. 13-37. Type-A automatic stroke control system suitable for boiler feeding, oil burner, and similar applications.

to the left for a certain definite distance, depending upon the reduction in load. This moves piston 4 of the four-way oil control valve to the left of its central position and causes oil, supplied by the lubricating oil pump 5 via supply pipe 6, to enter the outer end of the double acting cylinder 7 via pipe 8 and at a pressure of 120 psi (8.4 kg/cm^2). Gauge 16 indicates this pressure, which is maintained in pipe 6 by the regulating valve 9. Simultaneously, oil is discharged from the other end of cylinder 7 via pipes 10 and 11, against a back pressure of about 20 psi (1.4 kg/cm^2). This permits servo piston P to move to the left which it continues to do until the follow-up lever 12 on the curved link or so-called stroke transformer and link 13 return control valve position 4 to its central position as shown. In this position of piston 4 both ends of the hydraulic cylinder 7 are sealed against escape of oil and the entrapped oil serves to positively position the piston P and curved link or stroke transformer M at some definite point between its zero-stroke and full-stroke position. The paths of the oil travel through the four-way control valve during the foregoing described movement of the curved link M are indicated by the arrows on the pipes connected thereto.

Operation of the pump continues with the curved link so positioned until there is a subsequent change in boiler output to require either more or less feed water which will be automatically taken care of by respectively increasing or decreasing the pump stroke and delivery as before.

It is to be noted that there is no interruption or alteration in the pressure supply of lubricating oil to the pump bearings during operation of the hydraulic servo-piston P. For during such operation the amount of oil bypassed from the ordinary bearing supply pipe 14, and supplied to either end of the cylinder 7 at 120 psi (8.4 kg/cm^2) via pipe 6, is substantially equalled by the amount of oil displaced from the opposite end and supplied to the bearings via pipe 11.

Relief valve 15 is set to maintain a constant back pressure of about 20 psi (1.4 kg/cm^2) on the oil discharge from cylinder 7 and on the supply to the pump bearings via pipe 17, which pressure is indicated by gauge 18.

CHAPTER 14

Feed-Water Heaters and Economizers

What is a feed-water heater?
Answer: An apparatus for raising the temperature of boiler feed water by abstracting some of the heat from exhaust steam or from the hot gases of combustion.

TYPES OF HEATERS

How are these distinguished by name?
Answer: They are called *primary* feed-water heaters and *secondary* feed-water heaters. The source of heat is exhaust steam or the hot gases of combustion, respectively.

What name is generally given to a secondary feed-water heater?
Answer: Economizer.

Give a further classification of primary feed-water heaters.

Answer: They may be further classed:

1. With respect to provision for getting rid of the air (see Fig. 14-1) as:
   a. Deaerating,
   b. Nondeaerating.
2. With respect to hook up and other features, as:
   a. Open or semi-closed system.
   b. Single stage,
   c. Multistage,
   d. Steam tube,
   e. Metering.

What is the object of heating the feed water?

Answer: To save some of the heat going to waste in the exhaust steam and, in some cases, to purify the feed water; that is, to eliminate scale-forming substances by precipitation. Also, feed water entering the boiler at a high temperature puts less strain on the shell than would cold water.

Why do you say "to save *some* of the heat . . . "?

Answer: It is impossible to save all of the heat.

Why?

Answer: In any practical construction there is a temperature difference, that is, the exhaust steam leaves the heater at a higher temperature than the final temperature of the water.

How would you determine the saving by heating the feed water?

Answer: The following example will illustrate the method.

Example—If the temperature of the feed water is 60° F. (15.6° C.) before entering the feed-water heater, and the heater raises its temperature to 210° F., what is the saving with boiler pressure at 99.3 psig (6.95 kg/cm^2)?

Feed-Water Heaters and Economizers

Total heat in 1 pound (0.454 kg.) of steam
at 99.3 psig (6.95 kg/cm^2) .... 1188.7 Btu

Total heat in 1 pound (0.454 kg.) of feed water
entering the boiler at 60°F. (15.6°C.) .... <u>28.08</u>

Heat required to form 1 pound (0.454 kg.)
of steam..................................... 1160.62 BTU

Total heat in water leaving the feed-water heater at 210°F. (98.9°C.) = 177.5 Btu (approx.).

Heat saved by heating the feed water from 60° F. (15.6 C.) to 210° F. (98.9°C.) = 177.5 − 28.08 = 149.42 Btu

Percentage of saving = 149.42 ÷ 1160.62 = .121

That is 12.1% which is equal to 1% for each 149.42 ÷ 12.8 = 12.3 degrees that the temperature is raised by the heater.

Is a saving of fuel realized where the feed water is heated by an injector or live steam purifier and why?

Answer: No, because both devices draw from the boiler the heat which is imparted to the feed water.

Name two general types of feed-water heaters.

Answer: Open or through and closed.

OPEN HEATERS

What is an open heater?

Answer: One in which the exhaust steam and feed water are brought into intimate actual contact (see Figs. 14-1 and 14-2) by spraying the water through the steam. Both heated water and condensate go to the boiler.

What are the advantages of the open heater?

Answer: It is more efficient than the closed heater, but the principal advantage is the separation of scale-forming substances from the feed water by precipitation.

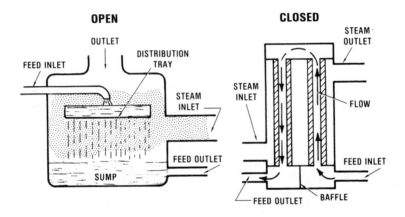

Fig. 14-1. Essential features of open and closed feed-water heaters.

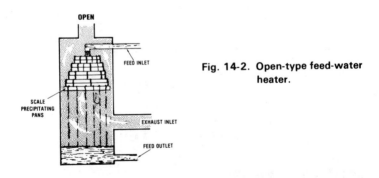

Fig. 14-2. Open-type feed-water heater.

What is the arrangement for separating the scale-forming substances called?

FEED-WATER HEATERS AND ECONOMIZERS

Answer: A purifier, that is, the purifier is a part of the open heater.

What is the disadvantage of the open heater?

Answer: Since the condensate goes into the boiler along with the feed water, the latter is contaminated with oil in the exhaust steam.

What provision is made to get rid of the oil?

Answer: Some form of oil extractor must be used.

Does an open heater free feed water from all of the impurities?
Answer: No.

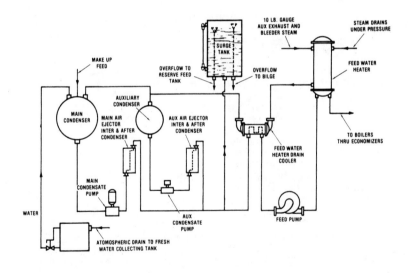

Fig. 14-3. Open or semi-closed feed-water system.

Where is the open heater located with respect to the feed pump?
Answer: Always on the inlet side. (See Fig. 14-3.)

Why?

Answer: Because it is a low-pressure heater, and is placed at a sufficient elevation to prevent the formation of steam at the inlet to the pump.

Explain.

Answer: The heater is located higher than the pump so as to operate under negative lift. Otherwise, it would be subject to partial vacuum and the hot water would flash into steam.

How high should the heater be placed?

Answer: The elevation above the pump depends upon the temperature of the water. Usually 12 ft. (3.7 m.) *negative* lift is sufficient for open heaters.

What happens if the feed pump becomes steam-bound?

Answer: It will race and so may be damaged.

How about the steam supply to open heaters?

Answer: It is seldom over 3 to 5 psig (0.21 to 0.35 kg/cm^2).

What provision is made?

Answer: The heater shell is usually vented to atmosphere through a small line.

What is the maximum temperature obtainable?

Answer: Slightly over 212°F. (100° C.)

What safety device is provided?

Answer: The shell is protected by an atmospheric relief valve set at the maximum pressure—usually 15 psi (1.05 kg/cm^2).

Feed-Water Heaters and Economizers
CLOSED HEATERS

What is a closed heater?

Answer: One in which the exhaust steam and feed water are *separated* by a metal surface through which the heat must pass. Figs. 14-4 through 14-7 show an elementary heater and several system diagrams.

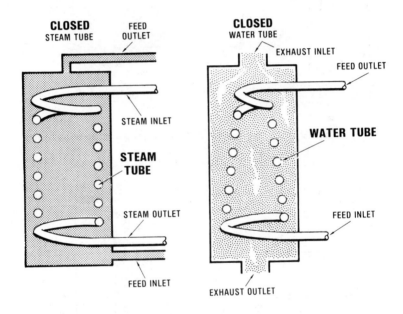

Fig. 14-4. Closed-type feed-water heaters. Elementary diagrams showing three types of feed-water heaters.

What is the advantage of this arrangement?

Answer: No oil can enter the boiler.

What is the disadvantage?

Answer: The heater is not as efficient as the open heater.

Power Plant Engineers Guide

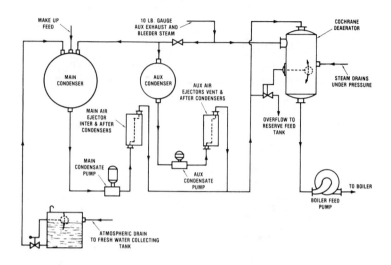

Fig. 14-5. Single-stage feed system with direct-contact deaerating heater.

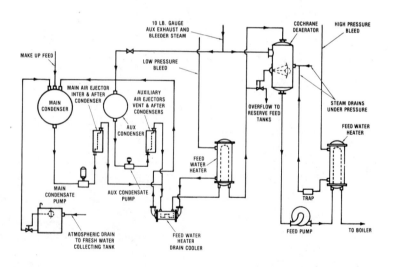

Fig. 14-6. Multistage feed system with direct-contact deaerating heater.

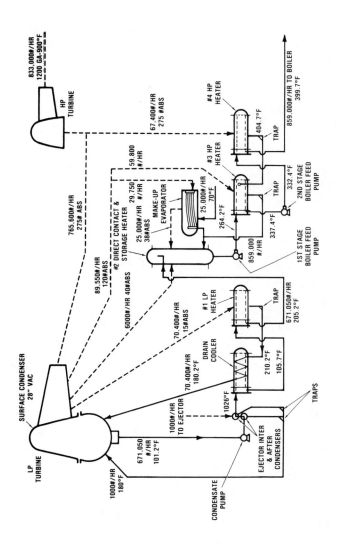

Fig. 14-7. Typical regenerative feed-heating diagram. Three closed-type heaters, a direct-contact heater, a drain cooler, and an evaporator result in a calculated reduction of heat consumption of approximately 12.6%.

Power Plant Engineers Guide

Why?

Answer: Because the steam does not come into actual contact with the water. Heat from the steam has to be transmitted through the metal forming the heating surface.

What precaution should be taken in the case of closed heater?

Answer: A closed heater should not be used unless the feed water is reasonably free from scale-forming substances.

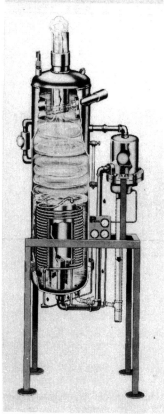

Courtesy Pennsylvania Separator Corp.

Fig. 14-8. Blow-down enters this heater in a spinning manner, releasing flash steam in the center area. Reusable steam is recycled and is fed from the top connection to the deaerator. The remainder of the blow-down heat is used to heat feed water flowing through the copper coil in the bottom of the unit.

Feed-Water Heaters and Economizers

Why?

Answer: The metal forming the heating surface is subject to deposits of oil on the steam side and scale on the water side, which, being good insulators, decrease the efficiency of the heater.

Name two types of closed heater.

Answer: The water-tube heater and the steam-tube heater (see Fig. 14-4).

Describe construction of the water-tube heater.

Answer: A typical water-tube heater consists of two or more coils (depending on size) connected in *parallel* and placed within a

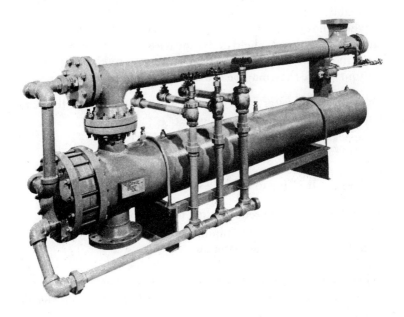

Fig. 14-9. This packaged blow down system is designed to serve two primary functions: automatically control the surface blowdown to maintain the desired level of total dissolved solids (TDS) in the boiler, and recover the heat from the high temperature blowdown and transfer it to the incoming cold feed water.
Courtesy Sentry Equipment Corp.

casing which is provided with an inlet and outlet for the exhaust. The ends of the coils are spaced so that ample area is provided for the flange of the exhaust without undue back pressure. Figs. 14-8 and 14-9 show blow-down recovery heaters.

Describe the construction of the steam-tube heater.

Answer: A typical heater has a shell of heavy steel plate which contains the water to be heated and the steam tubes. Each tube is U shape, both ends being expanded into a tube head bolted to the lower part of the shell. The head has a partition so that steam enters one end of each tube and leaves at the other end.

What is the application of this heater?

Answer: It is used to a considerable extent in plants where the inequalities of service compel an irregular feeding of the boilers, or where the exhaust steam supply is intermittent. In such cases the large volume of water held in storage is essential to absorb the heat from the exhaust steam. Moreover impurities tend to settle to the bottom where they may be drawn off from time to time.

Does the feed water from a heater enter the boiler at the same temperature as on leaving the heater?

Answer: No, there is a loss in temperature.

Why?

Answer: It is due to transmission loss.

How can this loss be reduced to a minimum?

Answer: By having the piping short, direct, and well covered with the best insulating material.

Where should the feed water enter a closed heater?

Answer: At the top where the exhaust steam is leaving the heater.

Why?

Answer: Because if located otherwise the cold water coming in contact with the high-temperature steam would cause the tubes to become brittle or crystallized.

FEED-WATER HEATERS AND ECONOMIZERS

Of what material are the coils made?
Answer: Brass or copper.

How would you test a heater for a leaky coil?
Answer: Stop the engine, partially close the valve on the discharge near the boiler and start the pump. Open the drip; a leaking tube or tubes will be indicated by water running out of the drip.

How would you know a heater was leaking with the plant in operation?
Answer: A leak would be indicated by the pump running faster and an extra amount of spray coming out of the exhaust.

What should be installed at the bottom of a closed heater?
Answer: A trap.

Should condensate from the trap be returned to the boiler and why?
Answer: No, because it contains too much oil.

How do you clean scale from the tubes?
Answer: With sal soda. Use about 10 lbs (4.5 kg.) per week for each 75 hp of boiler.

How is it applied?
Answer: Feed it into the inlet of the pump.

What provision in the piping should be made so that a heater may be repaired?
Answer: A bypass for the exhaust.

The corrosion problem is completely answered by a deaerating unit which when operated at a pressure above atmospheric, delivers oxygen-free water and has the additional advantage of supplanting the conventional feed tank and one of the feed-water heaters.

The open surge tank and steam blanketed or semi-closed systems very frequently do not give satisfactory protection against corrosion. The oxygen content of the feed water passing through these systems is generally in the neighborhood of 0.1 to 0.4 cc per liter. The closed system, if maintained absolutely tight to prevent air infiltration, would theoretically give better results; however, the oxygen content of water in such systems has been found to be as high as 0.2 cc per liter. The feed water with this oxygen content is not good enough for the modern boiler since considerable corrosion will occur as long as there is any trace of oxygen in the water.

Not only do closed feed systems fail to give the highest degree of freedom from oxygen but they are also slow and inefficient in removing the oxygen from the feed system when starting up after shut down. Furthermore, in such systems, where the main condenser is used for removal of oxygen from the water, efficiency in removal will be less in proportion as the temperature of the water leaving the hot well is below the exact steam temperature corresponding to the vacuum. Whenever this temperature is lower than the vacuum temperature, complete removal of oxygen is impossible according to the law of solubility.

ECONOMIZERS

What is an economizer?

Answer: A second-stage or *secondary* feed-water heater which utilizes the hot gases *leaving* the boiler, instead of exhaust steam as in a *primary* feed-water heater. See Fig. 14-10.

What distinction should be made between the terms "feed-water heater" and "economizer"?

Answer: They are entirely different heating devices. The word feed-water heater is always applied to the *first stage* heater using exhaust steam as the heating medium, whereas the economizer uses the hot gases of combustion as the heating medium.

Where is an economizer of the independent type located?

Answer: Between the boiler setting and the chimney.

Feed-Water Heaters and Economizers

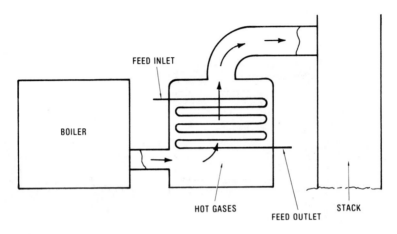

Fig. 14-10. Economizer construction and installation.

What is the usual construction of an economizer?

Answer: The usual construction consists of a series of steel tubes (coils or in parallel) through which the water flows, being surrounded by a casing to guide the hot gases over the heating surface. See Fig. 14-11.

How does an economizer operate?

Answer: Feed water is forced through the tubes while the gases circulate around them.

What was the typical construction formerly used?

Answer: It consisted of a series of cast-iron tubes made up in sections, connected at the ends, and placed in a brick chamber through which the gases pass from the boiler to the chimney or fan. Feed water was forced through the tubes while the hot gases circulate around them.

What device was provided to keep the tubes clean?

Answer: For removing soot from the tubes, scrapers were provided, and so arranged as to move slowly up and down each tube while driven by an engine.

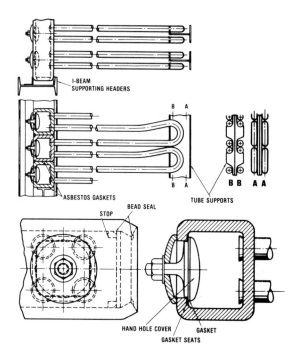

Fig. 14-11. Header-type economizer construction. The economizer consists of groups of rectangular, seamless steel headers connected by seamless steel tubes. Each tube has one U-bend. Separate return bends are also used. Two or two and one-half-inch OD tubes are used. One end of each tube is expanded into the upper tube hole of the header and the other end is expanded into the lower hole of the header immediately above. The tubes are spaced parallel and with a lane of sufficient width between alternate rows of tubes to permit tube removal. The tubes from the top header are normally connected to the number three drum. Handhole covers, each of which serves four tube holes and is held in place by a single integral forged stud and nut, are drop forgings. Gasket seats in headers are carefully machined. Armored gaskets are used. Headers are installed in groups of the desirable number with enlarged U-tubes connecting the groups. This construction permits intermediate header supports, permits free expansion of each group, and ready access to the economizer. Headers are

Feed-Water Heaters and Economizers

supported on structural members with separate supports for each group of headers. Headers are held in place at each end by a channel or other structural members. This construction permits free expansion vertically and horizontally. The U-bends are supported by H-columns, one flange of which is specially notched to hold the upper loop of the U-bend. The entire economizer is enclosed in a paneled steel, insulated casing properly constructed and reinforced to assure gas-tightness at all ratings. The ends are equipped with gas-tight, hinged doors to give quick and easy access to the economizer.

Why were the tubes made of cast iron?

Answer: Because cast iron resists corrosion longer than steel when exposed to gases of combustion at comparatively low temperatures.

What was necessary with very-high-pressure boilers when cast-iron tubes were used?

Answer: A secondary feed pump was used to relieve the tube from excessive pressure.

What provision should be made to cope with the condition under which economizers work?

Answer: Since soot is a very efficient heat insulator, any deposit on the tubes will considerably reduce the efficiency of the heating surface. Hence, some device should be provided for continuously or periodically removing the soot from the heating surface of the tubes.

What is the prevailing practice as to tubes and why?

Answer: Steel tubes are used, having been made practical by advances in oxygen control; thus, cast-iron pipes are not so common as formerly.

What degree of heat is available for heating feed water by economizers?

Answer: The temperature of the hot gases escaping up the chimney ranges from 500° to 700° F. (260 to 371° C.) which represents a considerable margin for the second-stage heating of the feed water.

What limits the amount of heating surface that can be installed in an economizer?

Answer: It is limited by the final temperature at the exit of the economizer. See Fig. 14-12.

Mention a late practice in boiler construction.

Answer: Some manufacturers are building their boilers with an integral bank of feed-water heating tubes inside the boiler casing in the gas-pass of the boiler. This is generally called a feed-water heater because it is placed within the boiler casing. Strictly speaking, the term "economizer" means a feed-water heating device that not only employs the hot gases as the heating medium, but is located outside of the boiler casing; that is, it is not a part of the boiler.

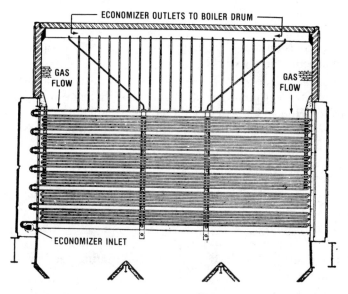

Fig. 14-12. Return-bend, loop-tube-type economizer. Economizers such as this one have been built in sizes having 357 to over 31,200 square feet of heating surface and for working pressures up to 1400 pounds per square inch.

Feed-Water Heaters and Economizers

Upon what does the economy of an economizer depend?

Answer: The economic result from using an economizer when the draft is sufficient depends upon the normal temperature of the flue gases escaping from the boiler and of the feed water supplied to the economizer.

How is the gain calculated?

Answer: By the formula.

$$\text{Gain in percent} = \frac{100(T - t)}{H - t}$$

where,

T is the number of heat units (Btu) in one pound (.453 kg) of feed water above 32°F. (0°C.) after heating,

t is the number of heat units (Btu) in one pound (.453 kg) of feed water above 32°F. (0°C.) before heating,

H is the number of heat units (Btu) in one pound (.453 kg) of steam at boiler pressure above 32°F. (0°C.).

For example, if the temperature of the feed water is raised from 190° F. (87.8°C.) to 310°F. (154°C.) by an economizer and the guage steam pressure is 150 pounds (67.9 kg), what is the gain? The steam tables show that T is 280 Btu, t is 161 Btu, and H is 1195 Btu (approximately). Therefore,

$$\text{Gain} = \frac{100(280 - 161)}{1195 - 161}$$

$$= \frac{100 \times 119}{1034}$$

$$= 11.5\%$$

CHAPTER 15

Feed-Water Regulators, Injectors, Deaerators, and Ejectors

Slow and uniform feeding of boilers by regulators causes a saving of fuel in the feed-water heater or economizer, because the regular speed of the water through the latter allows the feed water to be always heated to a higher degree than is possible when the flow varies, as is always the case with hand feeding.

When feed-water pressure is provided by reciprocating, steam-driven feed pumps, it is necessary to regulate the speed of these pumps, rather than to throttle feed-water flow with a differential pressure valve. For this service, the pump governor should be installed in the steam line to the feed-water pump. In this position, it maintains a constant excess feed-water pressure by regulating pump speed.

A properly designed boiler will not prime or permit its water to be carried over into the steam pipes if the water level is kept at the middle gauge. Most cases of boiler priming and water in the engine cylinders are due to the carelessness of the attendant in allowing too much water to the boiler. Other cases of priming are due to bad water or undue forcing. Keep the water to the middle gauge rather than to the top gauge.

What is a feed-water regulator?
Answer: An automatic device for controlling the flow of feed water into a boiler.

What is the ideal condition in the production of steam?
Answer: A constant flow of feed water in quantity corresponding to the amount of steam being produced.

Does a feed-water regulator meet this requirement?
Answer: No, only approximately.

How is the output of a boiler affected?
Answer: Inversely by the rate at which the feed water is supplied.

What happens when the feed is intermittent?
Answer: The generation of steam *decreases* when the feed valve is open and *increases* when the valve is closed.

Name an example where intermittent feed is used to advantage.
Answer: Boiler feed for steam locomotives. When going down grade where the power is shut off, the feed is turned on. This prevents the safety valve blowing and fills the boiler to a high level, so that on up-grade, where maximum power is required, the feed may be shut off while the excess water is being evaporated.

Name the various types of feed-water regulators.
Answer: They may be classified as:

1. Displacement which contains a buoyancy float or a specific gravity body,
2. Evaporation,
3. Expansion.

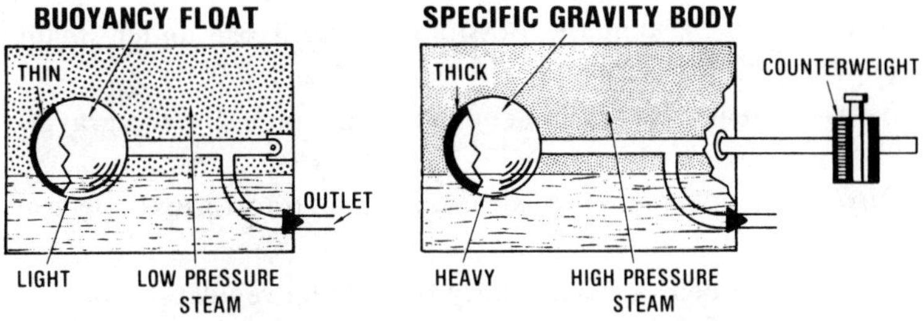

Fig. 15.1. Elementary diagrams showing the difference between a buoyancy float and a specific gravity body, and conditions for which they are suitable.

DISPLACEMENT REGULATORS

Describe a displacement regulator.

Answer: The simplest form consists of a hollow metal buoyancy float placed in the water column, which moves up and down with changes in the water level and operates the feed supply valve. See Fig. 15-1.

By what two methods is this motion transmitted to the feed supply valve?

Answer: By (1) direct lever connection or (2) through an auxiliary valve and diaphragm.

What is a specific gravity body as applied to a regulator?

Answer: The term "specific gravity body" means a hollow cylindrical body of such substantial construction that it is too heavy to float, but of course weighs less when submerged in water than when in the steam space.

What is used to offset the excess weight of the specific body and why?

Answer: A counterweight is used so that the combination of the two is equivalent to a light, or buoyancy, float.

What is the reason for such construction?

Answer: To prevent the possibility of collapse under steam pressure.

What is the real effect of the counterweight on a specific gravity body?

Answer: It gets rid of the extra weight and transforms a specific gravity body into a float.

How does a specific-gravity-body feed-water regulator work?

Answer: As shown in Fig. 15-2. When the water level in the column is below the opening of the special nipple, steam will enter the chamber of the regulator and the water in it will be displaced, falling through the pipe at its bottom to the level of the water in the water column. The displacement body in the chamber will then fall by gravity to the bottom of the chamber and the counterweight and lever will rise, holding the actuating valve against its top seat the exhaust valve will be open to the air. There will then be no pressure on the piston of the controlling valve, which will be wide open as the boiler takes water. When the boiler fills up to the opening of the special nipple, it will be sealed by the rising water and steam will be cut off from entering the chamber. The steam which was in it will be condensed, slightly reducing the pressure therein so that the water from the boiler will fill it to the top. The displacement body now weighs less than it did when the chamber contained steam only, by the weight of water which it displaces. The counterweight is now heavy enough to overbalance the displacement body and goes down, while the displacement body goes up. As the outside lever goes down the actuating valves go down, opening the steam connection and shutting the exhaust. This admits the steam pressure to the piston chamber of controlling valve, forces the piston and valve down, and shuts off the feed water at once. No more water can enter the boiler until the water level falls to the opening of the special nipple, when steam is admitted to the top of the chamber, the water in it falls to the old level, all the operations are reversed and the controlling valve opens again. These operations are repeated as the water gets above or below the desired point and the variation does not exceed one-half inch. When the counterweight is up, it indicates the boiler is feeding, and when down, that it is not.

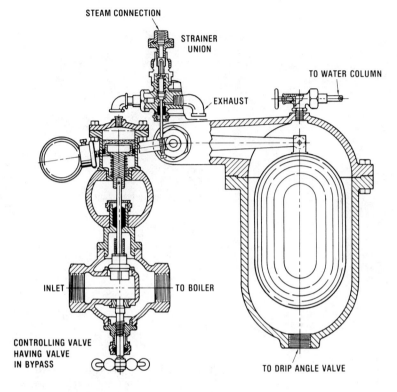

Fig. 15-2. Specific gravity body feed-water regulator.

EVAPORATION REGULATORS

What is an evaporation regulator?

Answer: A type of regulator whose operation depends upon the evaporative effect of a metal tube containing steam and water to actuate a diaphragm valve and thus regulate feed water (Fig. 15-3).

Describe its operation.

Answer: As the water level drops in the boiler, it drops a corresponding amount in the steam tube and allows the consequently entering steam to heat the water in the generator. As the generator

373

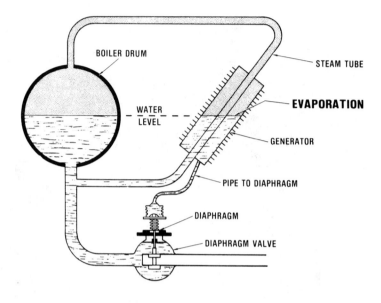

Fig. 15-3. Evaporation-type regulator.

water becomes hotter, it expands and brings pressure to bear on the diaphragm valve, which opens and allows feed water to enter the boiler. Rising water levels in the boiler and steam tube cool the generator, relax the pressure on the diaphragm valve, and stop the flow of feed water when it reaches the regulated level. Fig. 15-4 shows construction of a diaphragm valve in which hydraulic pressure in the bellows is opposed by at least 600 pounds of adjustable spring pressure. The spring pressure may be changed by adjusting the nuts under the lower spring plate in order to compensate for various types of boiler operating characteristics or for small changes in the desired average boiler water level.

EXPANSION REGULATORS

What is an expansion regulator?

Answer: The basic principle of this regulator is the expansion

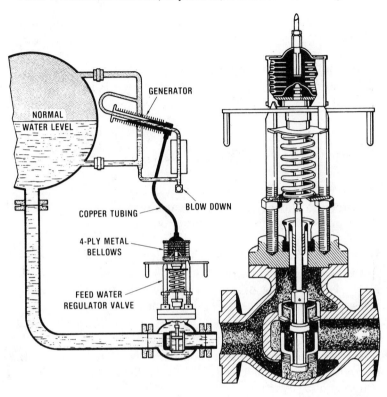

Fig. 15-4. Evaporation-type feed-water regulator assembly for 250- and 400-pound standards. This type is also called thermo-hydraulic control. At right is a diaphragm valve.

and contraction of a metal tube subjected to variable temperature (Figs. 15-5 and 15-6).

How about the movement thus produced?

Answer: It is multiplied by a system of levers which operate the feed control valve, being so arranged that the expansion of the tube opens the valve and contraction closes it.

What is essential for proper operation?

Answer: The feed pump must be provided with a governor giving such control as to maintain an excess pressure in the feed

Power Plant Engineers Guide

line so as to meet any demand of the feed-water regulator control valve.

Name another expansion-type regulator.

Answer: The air-operated control illustrated in Fig. 15-7.

INJECTORS

What is an injector?

Answer: A device for forcing water into a boiler against the boiler pressure by means of a steam jet.

What is the principle upon which an injector works?

Answer: An injector forces water into a boiler because the kinetic energy of a jet of steam is much greater than that of a jet of water escaping from the boiler under the same conditions.

What are the essential parts of an injector?

Answer: The simplest form consists of (1) a steam nozzle, (2) combining tube, (3) delivery nozzle with check valve, (4) inlet for water, and (5) overflow.

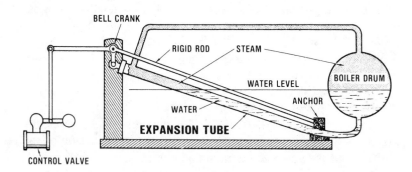

Fig. 15-5. Expansion-type regulator. The expansion tube and rod are fixed (anchored) at one end and the tube is free to move due to expansion or contraction with temperature changes.

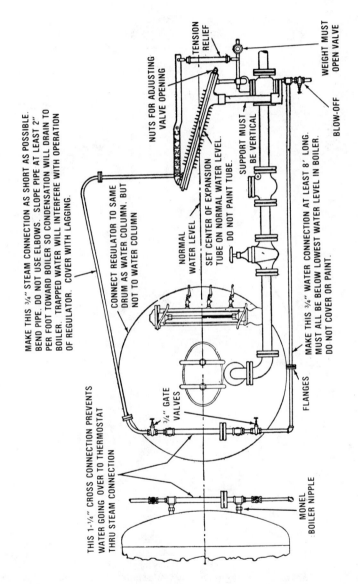

Fig. 15-6. Expansion-type regulator installation.

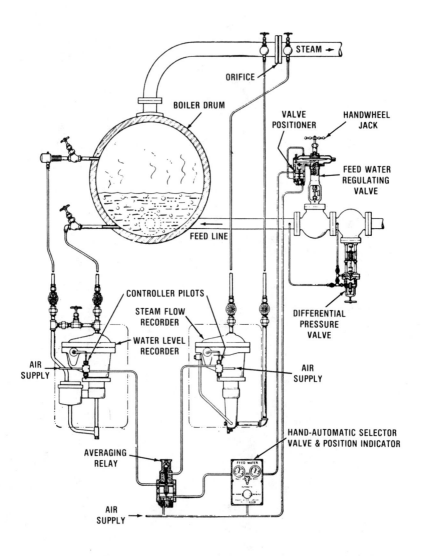

Fig. 15-7. Hookup and elements of a two-element, air-operated feed-water control.

Feed-Water Regulators, Injectors, Deaerators, and Ejectors

How does an injector work?

Answer: See Fig. 15-8. Steam from the boiler, entering the steam nozzle, passes through it, through the space between the steam nozzle and combining tube, and then out through the overflow. This produces a vacuum which draws the water in through the water inlet. The incoming feed water condenses the steam in traversing the combining tube, and the water jet thus formed is driven at first out through the overflow. As the velocity of the water jet increases, sufficient momentum is obtained to overcome the boiler pressure, with the result that the water enters the delivery tube and passes the main check valve into the boiler. See Fig. 15-9.

How are the different types of injectors classed?

Answer: (1) Nonlifting, (2) lifting, (3) positive, (4) automatic, (5) adjustable-nozzle, (6) exhaust.

What is a double-tube injector?

Answer: Virtually two injectors in one, the first acting to lift the water and the second to force it into the boiler. See Fig. 15-10.

What are the two classes of lifting injector?

Answer: (1) Diverging-tube, (2) double-tube.

What is the difference in construction between nonlifting and lifting injectors?

Answer: In order that an injector may be able to lift its water supply, the pressure at the end of the steam nozzle must be less than that of the atmosphere. This result is secured by using a diverging nozzle, Fig. 15-11. Also, see Fig. 15-12. In operation, a momentary period elapses between the opening of the injector steam valve and the establishment of the jet of water to the boiler. During this interval, the steam and water must not be allowed to back up in the inlet line, so a series of exhaust openings are provided at intervals along the jet passages. At first, the steam and water exhaust from all these openings or "spills"; then, the establishment of the jet of water begins at the top and proceeds downward. It is therefore possible for the exhaust from lower spills to be drawn again into the upper spills. Under such conditions, the upper zone of the injector may be established and have ceased to "spill" thus inducing an inward draft of upper spills. Now this upper zone was established because the steam and water quantities were rightly proportioned for that instant and that zone. However, if the lower zone (which is not yet established) is allowed to throw its surplus back up and into the upper zone, the upper zone will again be disturbed and unbalanced. Therefore, the ring valve is used to prevent such an occurrence. When water is taken from a pressure source, a special, large steam-jet is employed. Hot water is not favorable for low starting. Standard temperature for local tests is 74°F.

What should be noted about the nonlifting injector?
Answer: It's almost obsolete.

What other method is used to lift the water supply and adapt the injector to high lifts?
Answer: By providing two sets of nozzles and tubes as in the so-called double-tube type.

What is a positive injector?
Answer: One with a hand-operated overflow valve.

What is the object of the hard-operated overflow valve?
Answer: This permits operation at high pressure by stopping the drizzle from the overflow.

What is an automatic injector?
Answer: One that is self-starting after its operation has been stopped by the interruption of its water supply.

What is the automatic feature and how does it work?
Answer: It comprises two check valves, which when seated progressively close the combining and the overflow chambers to the atmosphere.

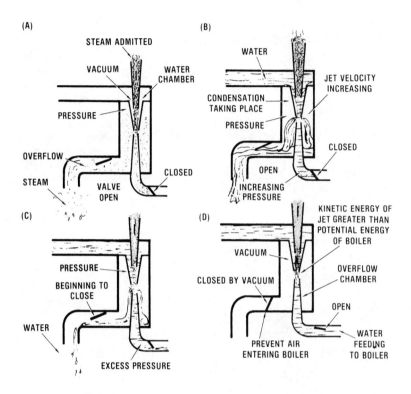

Fig. 15-8. Starting cycle of the injector. (A) Steam flows through the overflow check valve and passes out through the overflow outlet; the action of the steam jet creates a vacuum in the water chamber; (B) vacuum created in the water chamber draws in the water, which condenses the steam in the combining tube, and the jet of steam issuing from the overflow becomes a jet of water rapidly increasing in velocity and building up pressure against the boiler check; (C) the continued increase in velocity of the jet causes pressure against the boiler check valve to exceed boiler pressure and the latter begins to open; as it does a part of the jet water enters the boiler and part flows out through overflow; (D) velocity of the jet has become so great that all resistance is overcome; the check valve is forced wide open and all the water entering the boiler. The action of the jet now creates a vacuum in the overflow chamber, which causes the overflow valve to close, thus shutting out the air that would otherwise be forced into the boiler.

POWER PLANT ENGINEERS GUIDE

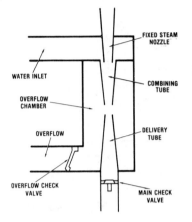

Fig. 15-9. Rudimentary fixed-nozzle, single-tube injector. The assembly of a steam nozzle, combining, and delivery tubes is called a **single-tube** injector, as distinquished from a **double** injector, which has two sets of nozzles and tubes.

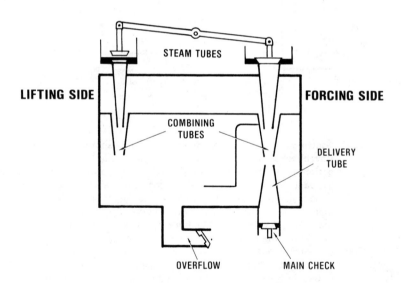

Fig. 15-10. Rudimentary fixed-nozzle, double-tube injector. In this type the lifting tube lifts the water to the injector, and the forcing tube forces the water into the boiler. This type is adapted to high lifts.

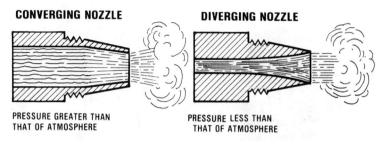

Fig. 15-11. Lifting and nonlifting nozzles. In double-tube injections, the lifting nozzle is made diverging and the forcing nozzle converging.

Describe its restarting action.

Answer: In restarting, steam blowing through the combining tube ejects the air, causing the combining-chamber check to close and establish a vacuum which draws up the water. The jet of water escapes through the overflow until sufficient pressure is obtained to lift the boiler check. At this instant, the vacuum created in the overflow chamber causes the overflow check to close, thus preventing air being carried by the jet into the boiler.

What is the advantage of a double-tube injector?

Answer: It is desirable for installations where the injector must lift the water a considerable distance.

How should an injector be constructed to adapt it to work satisfactorily with a wide range of steam pressures?

Answer: It should be provided with an adjustable nozzle so that the water passage between the steam nozzle and the combining tube can be varied in size automatically, or by hand.

Injector Selection

In considering the selection of the type or model required, it is best to select one which will surely take care of any condition ever to be met in service. If an injector is required to work at its limit of capacity, steam pressure, lift, or temperature of feed water, it will usually give trouble after any appreciable wear occurs.

Note the following points on selection:

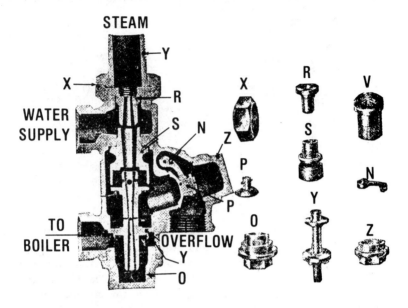

Fig. 15-12. Single-tube, fixed-nozzle, lifting automatic injector. The parts are: R, steam jet; S, inlet jet; T, ring valve; Y, delivery jet; O, plug; V, tail pipe; X, coupling nut; N, overflow hinges; P, overflow valve; Z, overflow cap. Just as soon as a vacuum draft is developed at upper the spills, ring I is drawn upward to its seat on jet S, forming an auxiliary chamber around the upper zone for its protection. This ring valve, T, is to the upper zone just what the outer overflow valve P is to the injector as a whole. Without ring T, the injector could not be operated on lower steam pressures.

1. Determine the lowest and highest steam pressure which will be carried at any time.
2. If water is to be lifted, determine the vertical distance of lift.
3. If water is from an overhead tank or from a city water main, the inlet should be throttled to reduce the pressure enough for the injector to handle it.
4. Determine the temperature of the water supply.

By referring to the working ranges shown in manufacturer's tables for each type or model, the correct type can be selected. In many instances, it will be found that more than one type or model

Feed-Water Regulators, Injectors, Deaerators, and Ejectors

will meet the requirements. The choice, then, is usually based upon individual preference for one particular type over another.

How do you select the proper size when full information is not known?

Answer: A fire-tube boiler will evaporate 30 pounds of water per hour per horsepower at 100% rating. To find maximum capacity of the boiler, multiply 30 pounds by the number of horsepower. If the boiler is to be operated at other than 100% rating, multiply its capacity at 100% rating by the operating rating. To allow for unfavorable or emergency operating conditions, add 10% to maximum capacity.

For a water-tube boiler, the exact same procedure is followed but the maximum capacity will now be obtained in pounds of water per hour. Convert this to gallons by dividing by 8.3. After maximum capacity is determined, the proper injector can be found in the manufacturer's capacity tables. The selection will be governed by operating steam pressure and individual preference as to type.

For example, what size and type of injector would be used to feed a 75-horsepower, fire-tube boiler at 125% rating, operating at 135 pounds?

$$30 \times 75 = 2250 \text{ pounds of water at } 100\% \text{ rating}$$

$$2250 \times 1.25 = 2812.5 \text{ pounds of water at } 125\% \text{ rating}$$

$$2812.5 + 281.25 = 3093.75 \text{ pounds of water, maximum capacity}$$

$$3093.75 \div 8.3 = 373 \text{ gallons per hour (1411.8 liters)}$$

As a second example, what size and type injector would be used on a 400-horsepower, water-tube boiler operating at 250 pounds at 200% rating?

$30 \times 400 = 12{,}000$ pounds of water at 100% rating

$2 \times 12{,}000 = 24{,}000$ pounds of water at 200% rating

$24{,}000 + 2400 = 26{,}400$ pounds of water, maximum capacity

$26{,}400 \div 8.3 = 3180$ gallons per hour (12,035 liters)

Refer to the manufacturer's tables in each case and select the corresponding size.

Automatic Injector Operation

Starting.—See Fig. 15-13. Fully open water-supply globe valve H and then fully open steam-supply globe valve D. If water comes from the overflow, throttle water valve H (*not* the steam valve) until this discharge stops. When water supply comes from city mains or is above the injector (as at K), open valve H only one-half turn and then open steam valve D fully. When two valves H and M are used, valve M should be opened only one-half turn, or a little more, since it is designed to reduce the water pressure to valve H. Valve H is always the one used to regulate the water supply to the injector. Steam valve D must always be opened fully—a lot of trouble comes from opening it only halfway.

A petcock, or a tee with a globe valve, placed anywhere between the check valve F and the injector will help very materially in starting on low pressure; leave open until injector is started, then close. Such an arrangement is also convenient for draining the discharge pipe in the water.

Stopping.—To stop the injector, close steam valve D. Water valve H need not be closed unless the injector is used as a nonlifter or if lift is considerable.

Injector Maintenance

Testing for Leaks.—Plug the end of the water supply pipe, if possible, and fit a piece of wood into the overflow cap (Z, Fig. 15-12) so that when cap is screwed down it will hold valve P in place. Then, turn on the steam and it will locate the leak.

Feed-Water Regulators, Injectors, Deaerators, and Ejectors

Removing Lime in the Jet.—Soak in a solution of one part muriatic acid and seven parts water as long and as often as necessary to keep parts free from lime.

Regrinding the Overflow Valve.—Should this valve become coated with lime or become rough, take off cap Z (Fig. 15-12). Using a screwdriver, turn valve P back and forth to grind it to the seat. A little emery dust on the underside of the valve will be of great advantage.

Cleaning.—To clean the injector, unscrew bottom plug O (Fig. 15-12) and the removable jet Y, which rests on it, will follow the plug out. Turn on not less than 40 pounds of steam and all dirt will be blown out. Examine all passages and holes to see that no dirt or scale has lodged in them. Replace the jet by setting it in the plug (which acts as a guide) and screw the assembly tightly into place. Be careful not to bruise any jets; use no wrenches on the body of the injector.

Repairing.—You can put in new parts yourself, as all jets come out without a tool except for the inlet. This can be removed with a piece of gas pipe slotted at one end.

Order parts from the dealer or manufacturer, stating the injector number and the part name or number.

Connecting Injectors

What should be done before connecting an injector and why?

Answer: Blow out the steam pipe thoroughly so as to remove any dirt, rust, or scale that may have accumulated in the pipes. Otherwise, the injector will become stopped up by this dirt when starting. This is important!

Tail Pipes or Nipples.—Don't use a pipe wrench to screw these into fittings. There is an iron lug in the tool box that fits into the square hole in the end of a tail pipe; use a wrench on this.

Steam.—Take the steam from the highest point possible on the boiler, and never connect with any steam used for other purposes. If the steam pipe happens to be a size smaller than the injector connections, it will work all right if it is not too long.

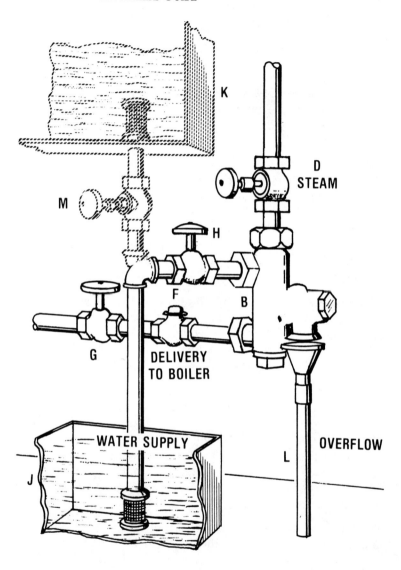

Fig. 15-13. Single-tube, fixed-nozzle lifting automatic injector with connections.

Valves and Supply Piping.—Every valve shown in Fig. 15-13 must be used and valve H put near the injector. Use globe valves, not straightways, for valves H and D, and repack the stem on valve H (nine out of ten leak). Fit valve H so that it shuts against the supply; that is, so the water comes up under the seat.

If the water-supply pressure is above 25 psi (1.75 kg/cm^2), make the supply pipe and valve H one size smaller than the injector couplings; that is, ¾-inch pipe and valve feeding a 1-inch injector. If it is not convenient to do this, use two valves (H and M) in the supply pipe. Never use pipe smaller than the injector connections when the water supply is taken from a lift.

If more than ten feet of water-supply pipe is used on a lift, the pipe and the valve should be a size larger than the injector connections. If there is a very long pull in addition to a very long lift, a pipe and valve two sizes larger should be used, and this pipe must be tight. When forcing through a heater, install a check valve between the heater and injector.

The strainer furnished is a special mesh and it—or one equally as fine—must be used. On a lift of over ten feet, a foot valve on the lower end of the water-supply pipe is a great advantage in starting the injector and results in a great saving of steam.

Discharge.—This pipe should be the same size as the injector connections and may be larger if desired. Place check valve F (Fig. 15-13) at least two feet from the injector, if possible, and be sure the valve lifts freely.

Overflow.—A *short* piece of pipe may be screwed into the overflow to carry off waste. A long piece interferes with vacuum and must not be used on a long lift or draw. Place a large piece of pipe under the overflow, if there is a lift or if a waste pipe cannot be arranged, so that the discharge end is in sight.

Possible Causes of Injector Failures

If the injector is new, it is probable that any trouble is due to faulty installation or conditions. Every injector is tested through its entire range under actual working conditions at the factory.

It is well to check carefully to see that the proper type and size has been installed. Next, check installation and operating instructions given in the catalog. If after checking these points, the trouble is not found, the following suggestions should be checked:

Failure of Injector to Lift the Water:
1. Too low steam pressure for the lift,
2. Water pipe or strainer clogged,
3. Water supply too hot,
4. Wet steam,
5. Combining tube clogged,
6. Overflow valve stuck,
7. Overflow pipe too small or end restricted,
8. Loose lining in water hose,
9. End of water pipe not submerged,
10. A leak in the water pipe,
11. Foreign matter, dirt, scale, iron cuttings, or red lead drawn in or blown in through the steam or water pipe clogging tubes and nozzles and interfering with the proper operation of the valves. This is a common cause of trouble.

Injector Lifts Water but Does Not Force It into Boiler:
1. A leak in the water pipe,
2. Wet steam,
3. Insufficient steam pressure,
4. Steam taken from pipe which furnishes steam for other purposes,
5. Combining tube clogged,
6. Overflow valve leaking or not seating properly,
7. Delivery pipe clogged,
8. Check valve in delivery pipe not lifting sufficiently or not lifting at all,
9. An obstruction between injector and check valve, or between check valve and boiler. Examine the delivery pipe where it enters the boiler. In many cases it will be found choked with lime or sediment which almost closes the opening. Often the coils of the feed-water heater through which the water is being forced are badly clogged.

Feed-Water Regulators, Injectors, Deaerators, and Ejectors

Steam and Hot Water Come Out of the Overflow Together:

1. Steam pressure too high for the lift,
2. Water supply too hot,
3. Water pipe clogged.

Injection Starts, but Breaks Off:

1. Water supply not correctly regulated,
2. Leak in water pipe admitting air to the injector (a common cause of trouble),
3. Dirt, sediment, lime, or some other form of obstruction in the delivery tube. This is a common cause of trouble. An obstruction between injector and check valve or between check valve and the boiler.

Bear in mind that for good results, the steam should be dry, all pipes and valves full size, and the water pipe absolutely air and water tight. If the machine is old, trouble is probably due to certain parts being worn. Repair parts can be bought and inserted by the engineer in charge, or the injector may be sent to the factory for a general overhauling. Unless you are sure that you have facilities to do your own repair work, return the injector to the factory.

DEAERATION

For what was the word "deaeration" coined?

Answer: To describe a particular stage in boiler feed-water conditioning.

Explain deaeration specifically.

Answer: It is the process of separating and getting rid of air, particularly oxygen, in feed water.

Why should these gases be removed?

Answer: Because air, oxygen especially, causes corrosion in boilers (Fig. 15-14).

How does the oxygen and air get into the water?

Answer: Both oxygen and air are soluble in water.

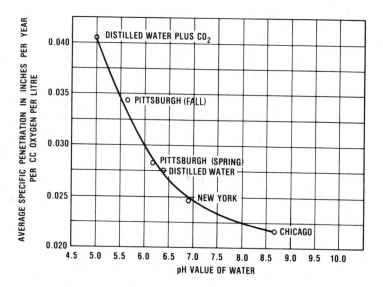

Fig. 15-14. Relation of pH value (alkalinity) to corrosiveness on basis of equal dissolved oxygen content.

Upon what does the solubility depend?

Answer: It depends upon temperature and pressure (Fig. 15-15).

Explain the relation.

Answer: Maximum solubility is at low temperatures and high pressures and minimum solubility is at high temperatures and decreased pressures.

Describe the essentials of deaeration.

Answer: The process is to produce a condition which will separate entrained air and air in solution from the water, and get rid of the air before it can get back into the water again.

How can the air be separated from the water?

Answer: By heating the water and flashing it into a region of pressure with a saturated temperature less than that to which the water has been heated.

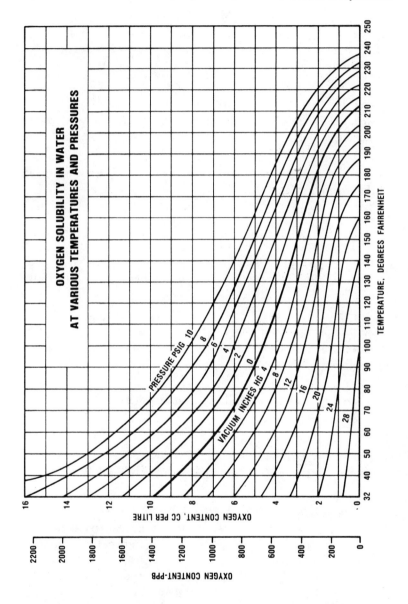

Fig. 15-15. Solubility of oxygen in water exposed to ordinary atmosphere at given pressures.

By what other method can this be accomplished?

Answer: By heating in a direct-contact heater which is proportioned to heat the water to the saturated temperature at the pressure being maintained.

What is the next step?

Answer: To provide an oxygen-free steam "atmosphere" above the water.

What will happen if the bubbles of oxygen, on separating, are left in this steam atmosphere?

Answer: If left in this "atmosphere" over the water surface, they will be reabsorbed due to their solubility.

How are they removed?

Answer: By passing oxygen-free steam over the surface of the water.

Describe the process in more detail.

Answer: The oxygen is concentrated at some point where it can be removed.

Is this removal steam vented to the atmosphere and wasted?

Answer: No.

How is waste avoided?

Answer: It is used as the medium for heating the water.

How does the steam flow in the deaerating apparatus?

Answer: In countercurrent-flow units, this steam passes into the heating section from the deaerating section where the concentration of gases is increased by condensation of the condensable steam. Then, finally, a small percentage is used as a carrier for the gases and is taken into the vent chamber where the concentration is completed by the final condensation of the condensable portion. The gases are then expelled to the atmosphere.

Feed-Water Regulators, Injectors, Deaerators, and Ejectors

Describe the flow in the parallel-flow type.

Answer: In this type, the steam passes directly from the deaerating section of the unit to the vent condenser, since it has previously passed through the heating section.

Describe the original flash-type deaerator.

Answer: It had separate heating and deaerating shells and the water from the heating chamber was passed into the flash chamber which maintained a lower pressure. Explosive boiling liberated the gases and formed steam to carry the gases to the vent condenser.

What is the application of this type?

Answer: It has a definite application where low-temperature, deaerated water is needed and where this temperature difference is not detrimental to the heat cycle.

Describe the marine-type deaerator used aboard ships.

Answer: This type of equipment uses a battery of spray nozzles for spraying the water into the steam "atmosphere" in the heating section, after which the water is collected and put through a deaeration nozzle where the steam and water are intimately mixed and broken up.

How is the steam brought into the unit?

Answer: Through the deaerator nozzle, then into the heating section, and finally, a small percentage is taken into the vent condenser.

What is the feature of this equipment?

Answer: It is developed to operate in spite of the roll and pitch of a vessel (which eliminates consideration of a tray-type).

What should be noted about deaerators?

Answer: A deficiency of steam will defeat the purpose of a deaerator in heating and gas removal.

Mention one item of importance.

Answer: The problem of the inlet valve is one which causes more trouble than any other single item involved in a deaerator. Careful consideration of quality and pressure should be made before actual installation.

Describe in detail the deaerating cycle employed in a well-known deaerator for marine power plants.

Answer: The feed water enters and passes through the vent condenser to the preheater from which it is discharged with practically uniform velocity at all loads into an atmosphere of steam. This initial spraying heats the water almost to the steam temperature and at the same time removes much of the dissolved oxygen.

The preheated and partially deaerated water is collected and distributed uniformly over the atomizer where the incoming steam breaks the water down into a fine mist, thereby exposing a large aggregate liquid surface. The flow of steam agitates and scrubs the water surfaces and carries away the liberated oxygen and other gases. The deaerated water falls into the storage compartment and the steam and liberated gases are delivered to the preheater where nearly all of the steam is condensed in the initial heating of the water. Some of the steam, however, continually flows to the vent condenser carrying with it the liberated gases, which may then be discharged to the atmosphere.

How should the deaerator be located?

Answer: Ordinarily, the design can be elevated sufficiently to give the head required on the inlet of the boiler feed pump. When the available head room is too limited to permit this, a low-head booster pump is used.

Describe the preheater.

Answer: It is of the direct-contact type, using spray nozzles of nonferrous material designed for operation under low water pressures. They spray the water outward and upward in contact with the steam, which is supplied by the atomizer at practically constant differential pressure, in a conical jet.

FEED-WATER REGULATORS, INJECTORS, DEAERATORS, AND EJECTORS

Describe the atomizer

Answer: It consists of a bronze valve, ordinarily in the form of an inverted cap, seating against the end of the downwardly projecting steam pipe and held to its seat by external spring loading. The incoming steam discharges through the atomizer at practically constant differential pressure, in a conical jet.

With a differential steam pressure of ½ pound per square inch ($35g/cm^2$), the velocity of the steam jet exceeds 300 feet (9144cm) per second. Its kinetic energy is employed to atomize the preheated water, creating a fog or mist, thus exposing a maximum water surface for liberation and separation of gas bubbles. The steam and liberated gases flow upward to the preheater and the deaerated water then falls to the storage space.

STEAM-JET AIR EJECTORS

What is a steam-jet air ejector?

Answer: A device for air removal from closed vessels, in which the operating pressure is less than atmospheric is and whose basic principle of operation is similar to the feed-water injector.

How does the ejector work?

Answer: It operates by means of the entrainment action produced by a jet of high-velocity steam.

What are the essential parts?

Answer: A vacuum chamber in which there is an expansion steam nozzle connecting with a venturi-shaped tail piece.

How does the device work?

Answer: Steam at high velocity passing through the expansion nozzle entrains any adjacent gases and carries them through to the diffuser (tail piece). Part of the momentum of the steam is transferred to the incoming air or vapors with a resultant velocity high enough to discharge the mixture at the designated discharge pressure.

How is the mixing and compression of the gases accomplished?

Answer: This takes place in the diffuser (also called a compression tube or tail piece) consisting essentially of a converging section, followed by a minimum section (called the diffuser throat) and then by the diverging section. In the latter, velocity energy is converted back into the pressure energy required to discharge the steam-air (or gas) mixture against the pressure existing at the outlet of the element.

Ejector Applications and Classifications

What are the applications of ejectors?

Answer: Steam-jet ejectors are generally used to remove air, gases or vapors from systems or vessels where the operating pressure required is less than atmospheric pressure (14.7 psi or 1.03 kg/cm^2 absolute at sea level). Occasionally, however, such ejectors are used to handle steam or gases at inlet pressures above atmospheric and to discharge them at somewhat higher exhaust pressures.

What is a thermocompressor or booster ejector?

Answer: One designed to handle steam instead of other gases.

How are ejectors classed?

Answer: With respect to the number of stages, as one-, two-, or three-stage, and so forth.

What are the applications of the single-stage arrangement?

Answer: The single-stage ejector, in which the gas is compressed in one stage, is suitable for vacuums up to 26.5 inches (673.1 mm) of mercury.

Why is the single-stage not used for higher vacuums?

Answer: While single-stage ejectors can be built for higher vacuums, the steam consumption becomes excessive as compared with multistage units.

Feed-Water Regulators, Injectors, Deaerators, and Ejectors

What is the application of the two-stage ejector?
Answer: They are suitable for vacuums of 26.5 to 29.3 inches (673.1 to 744.2 mm).

Of what does the two-stage ejector consist?
Answer: Two ejectors connected in series.

What is used with a two-stage ejector?
Answer: An inter-condenser is placed between stages.

Describe a three-stage ejector.
Answer: It consists of three ejectors connected in series with inter-condensers.

What is its application?
Answer: For vacuums from 29.3 to 29.9 inches (744.2 to 759.46 mm).

Ejector Operation

Starting Single-Stage Ejectors:

1. Open discharge valves.
2. If an after-condenser is provided, turn on water-supply valve.
3. Open the steam valves, admitting steam from a supply line at the proper pressure.
4. Open the air inlet valves.

As soon as the full steam pressure is supplied in the steam chamber, the ejector will start operating. The vacuum will be gradually increased as the air is removed from the system and after a short interval of time the normal operating vacuum will be obtained.

Shutting-Down Procedure:

1. Closes the inlet valves.
2. Closes the steam valves.
3. Closes the water-supply valve.
4. Closes the discharge valves.

Starting Two-Stage Ejectors (Single-Element):

1. Open the air exhaust valve at the discharge of the secondary element or after condenser.
2. Open the air inlet valve at the inlet of the primary element.
3. Open the valve in the inter-condenser drain loop.
4. Open the drain valve in the after-condenser drain line (if the inter-condenser and after-condenser are of the surface type).
5. Start water circulating through the inter- and after-condensers.
6. See that both steam valves are closed.
7. Open the main steam-supply valve, admitting steam at the proper pressure from the steam lines.
8. Open the steam valve to secondary element.

As soon as the full steam pressure is supplied to the steam nozzles, the ejector will start operating. An interval of time should then be allowed to build up the vacuum to approximately 16 to 20 inches (406.4 to 508.0 mm). Then open the steam valve to the primary element. After another short interval of time, the normal operating vacuum will be obtained.

If the ejector is provided with a jet inter-condenser, the water supply should be turned on before the steam is admitted to the primary element. The water quantity should be regulated to provide sufficient cooling water, but care should be exercised to limit the flow to prevent flooding the jet inter-condenser.

The procedure just outlined is for starting up ejectors on a complete system, bringing up the vacuum from atmospheric pressure. If a system should be provided with two steam-jet air ejectors and it is desired to put a second ejector into service while the other ejector is in operation, all operations *except* Nos. 2 and 3 should be performed in their regular order. After full vacuum is established at the inlet of the primary element on the ejector that is to be placed in service, operations 3 and 2 should be carried out *in this order*.

Shutting-Down Procedure:

1. Close the primary air-inlet valve.
2. Close the steam valve to the primary.
3. Close the steam valve to the secondary.

Feed-Water Regulators, Injectors, Deaerators, and Ejectors

4. Close the main steam-supply valve.
5. Close the condensing water-supply valve to the inter- and after-condensers.

Starting Two-Stage Ejectors (Twin or Triple-Element with Isolating Valves):

1. Open the valve in the air-discharge line from the after-condenser.
2. Open the secondary-element discharge valves.
3. Open the secondary-element inlet valves.
4. Open the primary-element discharge valves.
5. Open the primary-element inlet valves.
6. Open the valve in the inter-condenser drain-loop line.
7. Open the valve in the after-condenser drain line.
8. Start circulation of water through the surface-type inter- and after-condenser.
9. See that the steam valves to all elements are tightly closed.
10. Open the main steam-supply valve, admitting steam at the proper pressure to the ejector.
11. Open the steam valves to the secondary elements.

As soon as the full steam pressure is supplied in the steam chambers, the pump will start functioning. An interval of time should be allowed to build up the vacuum to approximately 16 or 20 inches (406.4 to 508.0 mm). Then open the steam valves to the primary elements. After another interval of time, the normal operating vacuum will be obtained.

It is customary to use all sets of elements for establishing full vacuum on the system in the shortest period of time. If the air leakage to the system is at a minimum, then the desired vacuum can be maintained by operating with only one set of elements (one primary and one secondary).

When a condenser is furnished with a priming ejector, it is started up first and allowed to operate until the condenser vacuum reaches 25 to 26 inches (635.0 to 660.4 mm). The ejector is then shut down by closing air-inlet and steam-supply valves in the order named.

Where an ejector is equipped with a raw-water auxiliary cooling section, the water valve should be opened when the ejector is started up. The same instruction applies to recirculation of condensate. Unless the steam load is very light, the raw-water or recirculation valve can be closed after full vacuum is reached.

Shutting-Down Procedure:

1. Close the inlet valve of the element to be shut down.
2. Close the steam valve to the same element.
3. Close the discharge valve of the same element.

When shutting down one set of elements only, it is recommended that the primary element be shut off first and then the corresponding secondary element. In order to maintain the normal vacuum it is necessary to have one primary and the secondary element in service together. It is not possible to obtain high vacuum with two secondaries only without any primaries, nor with two primaries alone and no secondaries. If it is desired to put any additional elements into service, the shutting-down procedure just outlined should be reversed.

To shut down a complete ejector:

1. Close the primary inlet valves.
2. Shut off the main steam supply.
3. Shut off the water supply to the inter- and after-condenser.

Operation of the three-stage ejector is similar to that of the two-stage unit with the additional operation of the isolating and steam valves of its primary stage.

Ejector Troubles

Name six possible causes for faulty operation of an air ejector.

Answer: (1) Insufficient cooling water, (2) steam nozzles plugged with scale, (3) water flooding the inter-condenser due to faulty drainage, (4) low steam pressure, (5) high back pressure at discharge of ejector, (6) loss of water seal in inter-condenser drain loop.

How do you check for insufficient supply of cooling water?

Answer: By observing the temperature of the water entering and leaving the air ejector. If the temperature rise in the ejector is not excessive, the cooling water supply is adequate and the trouble is elsewhere.

What do you do if a scale deposit forms in the throats of the steam nozzles?

Answer: It should be removed with drills of the same diameter as those with which the nozzles were originally drilled.

How do you check for flooding of the inter-condenser?

Answer: By feeling the temperature of the inter-condenser shell.

What causes low-pressure steam?

Answer: Clogging of the steam strainers or orifice plates with pipe scale or sediment, improper operation of the regulating valve, or low boiler pressure.

What causes high back pressure at the discharge of the ejector and what do you do?

Answer: This sometimes occurs where the pump discharges into a common exhaust system with other equipment. If this happens, it will be necessary to provide an independent discharge from the ejector to the atmosphere.

How about loss of water seal in the drain loop?

Answer: This takes place occasionally in installations where the vacuum in the system is subject to sudden fluctuations.

What should be placed on the inter-condenser drain loop to show whether the valve is properly sealed when the ejector is in normal service?

Answer: A gauge glass.

Where is this glass located and what is the indication?

Answer: As near the bottom of the drain loop as possible; if the water is visible anywhere in the glass, the loop is properly sealed.

Feed-Water Regulators, Injectors, Deaerators, and Ejectors

How do you reestablish the seal in the drain loop?

Answer: Close the valve provided for this purpose in the drain-loop line—usually located near the condenser.

Describe how the valve should be operated.

Answer: The valve must be closed for the short period of time required to form sufficient condensate to fill the loop. After the water again shows at the top of the gauge glass, the valve should be opened very gradually.

What will happen if the valve is opened too quickly?

Answer: The difference in pressure will cause surging of the water and again unseal the loop. In certain cases, some drain loops have a tendency to be unstable due to fluctuations in condenser vacuum. In these instances, it is customary to operate with the valve in the drain-loop line partly throttled, opening it just enough to pass the condensate at all times.

What are the indications that the drain loop has become unsealed?

Answer: This is indicated if no water is visible or if it surges violently in the glass.

What is the cause of this?

Answer: When this happens, some of the air, which has been removed from the main condenser by the primary element, is recirculated and flows back through the drain loop to the main condenser, thereby reducing the vacuum.

CHAPTER 16

Feed-Water Testing and Treatment

If all feed waters were pure, that is, if they consisted of nothing but H_2O (2 atoms of hydrogen to one atom of oxygen) water treatment would not be necessary. Or, if all waters carried the same impurities, water treatment could be reduced to a nearly uniform prescription. However, absolutely pure water never occurs in nature, and impurities vary.

The nearest approach to pure water is rain water, but even that contains dissolved oxygen, carbon dioxide, and other chemicals picked up from the air, making it corrosive. Nothing is said here about the purity of water of condensation.

Upon what does impurity in natural waters largely depend?

Answer: On the source.

How are feed waters classed?

Answer: As ground waters (wells and springs) and surface waters (rivers and lakes).

How about ground waters?

Answer: They almost always pick up dissolved hardness as they seep through rock strata containing lime and magnesium, but they are usually clear and free of suspended matter.

How about moving surface waters (rivers)?

Answer: They contain organic matter such as leaf mold and, often, insoluble suspended matter such as sand and silt. They are frequently polluted by industrial waste and sewage.

How about the effect of stream velocity?

Answer: This causes characteristics to change rapidly, depending upon the amount of change in stream velocity. Also a factor is the amount of rainfall and on what part of the watershed it occurs.

What is the characteristic of still surface waters?

Answer: Still waters (lakes) settle out suspended solids.

What is the general requirement of surface waters and why?

Answer: Filtration to remove suspended solids.

What else may be necessary?

Answer: Surface waters may also need *softening*, or removal of dissolved solids.

How about ground waters?

Answer: They require softening but rarely filtering.

Why?

Answer: Because of the natural filtering effect of passage through rock or sand.

Of what does ground-water hardness usually consist and why?

Answer: Usually of calcium and magnesium bicarbonate, called *carbonate hardness*, because it is easily convertible to carbonate form.

FEED-WATER TESTING AND TREATMENT
SCALE AND MUD

What is scale?

Answer: Incrustation within a vessel, caused by mineral substances precipitated from the water.

Describe the formation of scale.

Answer: Ordinary scale is a result of the chemical effects of heat and concentration. Water containing dissolved and suspended solids flows into a boiler continuously. As generated steam is practically pure, it leaves behind in the boiler the dissolved and suspended solids from the water. These build up, forming scale, which becomes attached to the metal of the boiler unless it is removed.

Of what does scale consist?

Answer: In most cases, scale consists of insoluble compounds of calcium and magnesium. Sometimes it is cemented into a hard mass by silica, or under severe conditions scale may form which consists wholly or in part of complex silicates or iron oxide.

What are the effects of scale?

Answer: It restricts water flow and reduces heat-transfer rates, thus altering the system's balance.

How can scale affect heat transfer?

Answer: Scale has a low heat-transfer coefficient. Metals generally are better conductors of heat than nonmetals. Copper has a heat-transfer rating of 221, while lime scale (calcium carbonate) has a heat-transfer coefficient of 0.4; silica scale approximates 0.3 and calcium sulfate, only 0.12. Obviously, if lime scale has a heat-transfer rate of more than 550 times *less* than copper, it can be classified as an insulator.

What is the scale "vicious cycle"?

Answer: As the scale film starts to build on the surfaces of tubing, the rate of water flow and heat transfer falls off, thereby causing an acceleration in the formation of additional scale, and the cycle is established.

Methods of Avoiding Scale Formation

The best way to handle bad feed water is to treat the feed water before it enters the boilers and pass it through a feed-water heater-purifier. Then, see that the water is fed in so that it will first pass to the mud drum, or to the rear of the boiler, where the sediment will be deposited and can be blown out.

What is the object sought in this method?

Answer: The point is to get the scale-forming material deposited as soft mud instead of hard scale and to arrange the circulation so that this mud will be thrown down as far as possible from the hottest part of the boiler in order that it will not bake into the plates.

Scale Treatment

What is frequently used with soda?

Answer: Lime—when lime and soda are added in proper proportions to water containing dissolved impurities, insoluble compounds are precipitated as sludge, leaving sodium compounds as the principal impurities. The lime-soda treatment is explained in detail in other sections. See Cold and Hot Lime-Soda Treatments.

In a badly scaled boiler, how much soda ash would you use to prevent formation of more scale in cases where this reagent is suitable?

Answer: A small quantity of soda ash will act with good results on large quantities of feed water.

What happens if too much soda ash is used?

Answer: It tends to cause priming.

What is the best method of feeding?

Answer: Connect the feed pump or injector to a soda tank so that a supply of soda solution can be drawn at regular intervals. If an injector is used, be sure to attach a strainer to the inlet pipe.

FEED-WATER TESTING AND TREATMENT

HARDNESS OF FEED WATER

There are two kinds of hardness of feed water:
1. Temporary hardness,
2. Permanent hardness.

To what is temporary hardness of feed water due?
Answer: To the presence of bicarbonates of calcium and magnesium.

Why is temporary hardness so-called?
Answer: Because boiling drives off carbon dioxide and precipitates carbonates which, under proper control, do not form scale.

What other name is given to temporary hardness?
Answer: Carbonate hardness.

What other name is given to permanent hardness?
Answer: Noncarbonate hardness.

What causes noncarbonate or permanent hardness?
Answer: It is caused by the sulfates, chlorides, and nitrates of calcium and magnesium.

How do these compounds behave?
Answer: They are not precipitated by heat, but form a hard scale in boilers when water is evaporated.

SATURATION POINT

This is sometimes referred to as solubility. By definition, it is the amount of a solid that can be dissolved in a liquid, as in water under certain conditions. For instance, in the generation of steam, under certain conditions of temperature and pressure, the steam will absorb the maximum amount of moisture and is then said to be *saturated*. If more heat is added, it becomes *superheated*; if it is cooled (pressure remaining constant) *condensation* takes place;

that is, the *dew point* has been reached. Precipitation takes place in the form of condensate or scale.

PRECIPITATION

According to Feller, "In the simple case of boiling or evaporation, the solid comes out of solution because its degree of solubility has been exceeded. When several substances are dissolved in the same solution, however, the situation is more complex. Suppose we have a solid in solution at a given temperature and we add another solid which, by itself, would also be completely soluble. The two dissolved solids may react chemically and one of the products of the reaction may be less soluble than either of the original substances. This product would then precipitate or come out of solution as a result of chemical change."

CHEMICAL TERMS RELATING TO FEED WATER

A few chemical terms are given here to indicate the direction in which the advanced engineer must push his energies in view of the increasing importance laid upon a knowledge of the chemical composition of feed water. The reader will find considerable related matter under BOILER OPERATION (Chapter 19), which should be studied along with this section.

Element: An undecomposed substance.

Reagent: Any chemical used to investigate the qualities of some other chemical. For instance, hydrochloric acid is a reagent for finding carbonic acid in limestone, which, when treated, will give up its free carbonic acid gas.

Oxide: Any element, such as lime or magnesium, combined with oxygen. The rust from iron (formed by oxidation) is the oxide of iron.

Carbonate: An element, such as iron or sodium, which forms a union with carbonic acid. Carbonic acid is found principally combined with lime and magnesium—called carbonate of lime and carbonate of magnesium—and is very destructive to boilers.

Acid: A liquid which contains both hydrogen and oxygen combined with some element such as chloride or sulfur. It turns litmus paper red.

Alkali: The opposite of acid. Principal alkalies are potash, soda, and ammonia. These, when combined with carbonic acid, form carbonates.

Amine: A class of basic compounds derived from ammonia.

Ammonia Nitrogen: Nitrogen combined in the form of ammonia.

Sal Soda: Carbonate of soda.

Chloride: An element combined with hydrochloric acid. A good example is common salt, being sodium united with the element chlorine. All water contains traces of chlorides, but they are not particularly dangerous to boilers.

Chlorine: A solution of sodium chloride used as a disinfecting agent in water purification.

Fluoride: A compound of fluorine; a salt of hydrofluoric acid.

Hydrazine: A corrosive reducing liquid of a weaker base than ammonia used in making salts.

Hydrogen Peroxide: A corrosive compound used as an oxidizing agent.

Lime: A white, alkaline, earthy powder obtained from the native carbonates of lime. The chemical name of lime is *calcium*.

Sulfate: A substance formed by the action of sulfuric acid upon an element such as sodium or magnesium. Sulfates are dangerous to boilers.

Silica: The gritty part of sand and the basis of all fibrous vegetable matter. A familiar example is the ash which shows in packing which has been burnt by the heat in steam. Silica is present in all boiler scale.

Insoluble Matter: In water analysis, this term is silica. Silica is one of the least dangerous of all feed-water impurities.

Magnesium: A fine, light, white powder, having neither taste nor smell. It is almost insoluble in boiling, but less so in cold, water. Exists in two states in feed water: (1) oxide, and (2) a carbonate. When in the latter form and free from iron, it tends to give the yellow coloring matter to scale.

Soda: A grayish-white solid, fusing at a red heat, volatile with difficulty, and having an intense affinity for water, with which it combines with great evolution of heat. The only reagent which is available for distinguishing its salts from those of the other alkalies is a solution of antimonate of potash, which gives a white precipitate in diluted solutions.

Sodium: The metallic base of soda. It is silver-white with a high luster, and crystallizes in cubes. It is of the consistency of wax at ordinary temperatures, completely liquid at 194°F. (90°C.), and volatilizes at a bright-red heat.

Soda Ash: A trade name for sodium carbonate.

Salt: The chloride of sodium, the base of salt springs, sea water, and salt-water lakes.

CHARACTERISTICS OF CHEMICAL SUBSTANCES

The aforementioned chemical substances can be classified with two distinct classes: (1) Incrusting, and (2) noncrusting.

What of the incrusting salts?

Answer: Carbonate of magnesium is the most objectionable and any feed water containing a dozen grains per gallon of magnesium is very injurious to boilers, causing corrosion and pitting. Carbonate of lime, while not as bad as the magnesium carbonate, has a very destructive action on a boiler. Twenty grains per gallon (3.78 liters) is considered bad feed water.

What is the effect of all oxides of iron and aluminum and sulfate of lime?

Answer: They are incrusting. *[looks like glass]*

Caustic Embrittlement: The tendency of caustic sodium hydroxide to concentrate in drum seams, under rivets, or at rolled tube joints, injuring the boiler metal.

Carryover: Entrained moisture and associated solids passing from a boiler with the steam.

Coagulation: Collecting of large particles of suspended matter in a finely divided or colloidal state.

Incrustation: A coating-over, the coating being commonly known as scale.

FEED-WATER TESTING AND TREATMENT

Corrosion: Destruction of the surface of a metal by oxidation.
Pitting: Minute holes eaten into the metal by oxidation.
Acid Radical: A chemical unit. In a molecule of sulfuric acid, for instance, the group SO_4 is called an *acid radical*. All the acids in inorganic water chemistry consist of H (hydrogen) combined with one of these acid radicals.

SIMPLE TESTS

Take a large (or tall) glass vessel and fill it with the water to be tested. Add a few drops of ammonia to the water until the water is distinctly alkaline. Next add a little phosphate of soda. See Figs. 16-1 to 16-4.

What is the action of the phosphate of soda?
Answer: It changes the lime or magnesium into phosphates which are deposited in the bottom of the glass.

What does this indicate?
Answer: The amount of matter thus collected gives a crude idea of the relative quantity of sediment and scale-making material in the water.

How do you test for acid?
Answer: Test with blue litmus paper. If the water turns the litmus paper red, before boiling, acid is present.

How is carbonic acid detected?
Answer: If the blue color can be restored by heating, the water contains carbonic acid.

Where can you buy litmus paper?
Answer: In any drugstore.

How do you test for sulfurous water?
Answer: If the water has a foul odor and gives a black precipitate with acetate of lead, it is sulfurous.

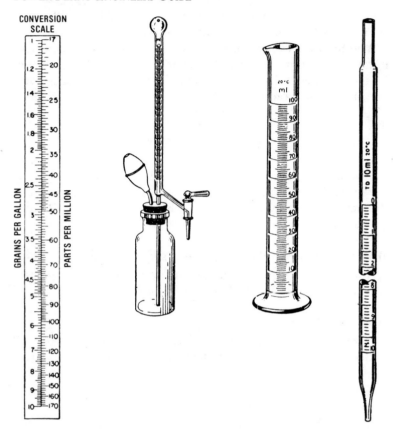

Fig. 16-1. Handy tools that help the plant operator measure samples and reagents for accurate results.

Describe the test for determining the amount of impurities.

Answer: Dissolve common white or other pure soap in a glass of water and then stir into the glasses of water to be tested a few teaspoonfuls of the solution. The precipitate will show the comparative amount of the scale-making material contained in the feed water.

How do you determine the proportion of soda required?

Answer: (1) Add ¹⁄₁₆ ounce (1.7g) of the soda to a gallon (3.7 l) of the feed water and boil. (2) When the precipitate formed by the boiling has settled to the bottom of the kettle, pour off the clear water, and (3) add ½ dram (0.88g) of soda.

Now if the water remains clear, the soda which was first put in has removed the lime; but, if it becomes muddy, the second addition of soda is necessary. By this method, a sufficiently accurate estimate of the quantity of soda required to eliminate the impurities of the feed water can be made and the due proportion added to the feed water.

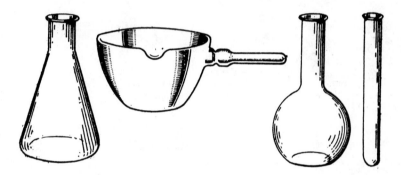

Fig. 16-2. Test vessels. Keep clean and store in a dust-free locker. Wash well between tests.

How do you test for carbonate of lime?

Answer: Into half a tumbler of the feed water put a small amount of ammonia and ammonium oxalate. Heat to the boiling point. If carbonate of lime is present, a precipitate will be formed.

How do you test for sulfate of lime?

Answer: Add a few drops of hydrochloric acid and a small quantity of a solution of barium chloride to ¾ of a tumbler of the feed water and heat the mixture slowly. If a white precipitate forms that will not dissolve on adding a little nitric acid, sulfate of lime is present.

How do you test for organic matter?

Answer: Put a few drops of sulfuric acid into a tumbler of water, add a sufficient quantity of pink-colored solution of potas-

Power Plant Engineers Guide

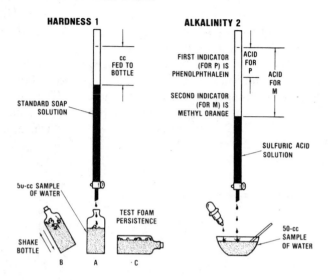
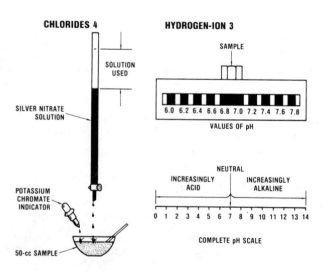

Fig. 16-3. Routine control tests that check plant operation and spot faulty treatment. These tests can easily be carried out by the shift engineer with relatively simple equipment and reagents.

Feed-Water Testing and Treatment

PRESS TO SNAP CHEMet TIP

WATER SAMPLE

Fig. 16-4. Water test kit and device, called a CHEMet, used to analyze oxygen content. The tip of the tester is inserted into a flowing stream of water. When the tip is snapped a vacuum pulls a water sample inside the test chamber where it mixes with the premeasured reagent. The solution forms a reddish-violet color proportional to the dissolved oxygen in the water. The exact amount is determined by comparing the sample with liquid standards. Courtesy CHEMetrics, Inc.

sium permanganate to make the entire mixture a faint rose color. If, after standing 3 to 5 hours, no change in color occurs, no organic matter is present.

How do you test for matter in mechanical suspension?

Answer: By allowing a glass of feed water to stand 8 to 10 hours, mechanically suspended matter (if any) will settle to the bottom and the amount of sediment may be noted.

How do you determine the correct amount of commercial scale-remover necessary for a given system?

Answer: You must first calculate the total volume of water in the system; otherwise, too small an amount would be ineffective and too large an amount would waste money.

How is this done?

Answer: By determining the amount of water contained in the sump and/or holding tank, plus what is in the pipes and condensers.

For example, if the sump or tank has a rectangular shape, measure the length and width of it and the depth of the water which will just cover the pump suction line. Multiply length, times width, times depth (all in feet), times 7.5, which will give the volume in gallons (this figure can be multiplied by 3.785 to obtain the volume in liters).

Problem: The sump is 5 feet long, 4 feet wide, and contains 6 inches of water, which just covers the pump suction line.

$5 \times 4 \times 0.5 \times 7.5 = 75$ gallons (283.8 liters) approximately

For a tank or sump having a cylindrical shape, measure the diameter of the tank and the depth of water. The diameter squared, times water depth, times 6 will give the water volume in gallons.

Problem: The sump is 3 feet in diameter and has a water depth of 3 feet.

$3 \times 3 \times 3 \times 6 = 162$ gallons (613.1 liters) approximately

Follow these steps to find the amount of water in the pipes:
1. Estimate the total footage of pipe from tower to condenser and return (for example, 200 feet).
2. Divide this figure by 50 (200 ÷ 50 = 4).
3. Divide sump and/or holding tank water-volume by 10 (75 ÷ 10 = 7.5).
4. Multiply the result of Step 2 by the result of Step 3 to obtain the total volume of water in the pipes (4 × 7.5 = 30 gallons, or 113.5 liters, approximately).

Add the result of Step 4 to the sump volume in order to get the total system contents (30 + 75 = 105 gallons, or 397.4 liters, approximately). The volume of water in a single condenser can be discounted.

The amount of scale-remover necessary for this system can then be determined from the manufacturer's ratio chart. According to Virginia Chemical, Inc., additional allowances need to be made for cooling units with double condensers.

Test kits designed to analyze water for impurities are available from a number of manufacturers. They're listed in directories, including the Yellow Pages of your local telephone directory.

FEED-WATER TREATMENT

The following methods have been employed successfully for treating feed water in power plants (Fig. 16-5):

1. Evaporators,
2. Hot-process softeners with external or internal phosphate treatment,
3. Cold-process softeners using sodium zeolite or hydrogen zeolite.

Evaporators

The use of evaporators is all right in some instances where the percentage of makeup is small (1 to 4%). For larger percentages of makeup, the evaporator is more expensive than other types of treatment (Fig. 16-6).

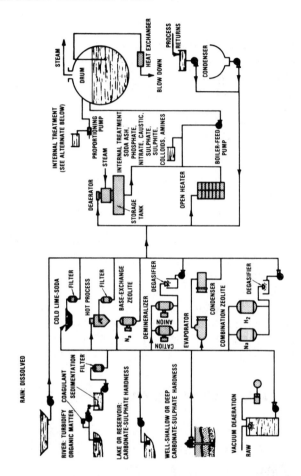

Fig. 16-5. Possible combination of treatment functions. Diagram shows various functions such as filtration, softening, silica removal, deaeration, and internal treatment that may be involved in conditioning water for boiler use. Combination of steps for any given job depends on the nature of the water.

Modern power plant cycles allow losses of condensate which may be due to soot-blower operation, safety-valve blowing, pump-leakage at stuffing boxes, and continuous blow-down.

An evaporator operated at concentrate rating produces a vapor of extremely high purity. Mineral substances dissolved in the raw

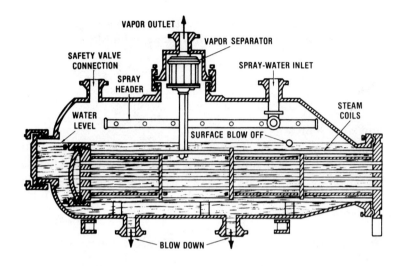

Fig. 16-6. Evaporator. It makes up water by boiling contaminated water. The steam thus generated is condensed for feed water.

water, being nonvolatile in nature, remain in the liquid concentrate and form scale and sludge.

Cold Lime-Soda Treatment

This process involves the addition of the correct proportion of lime (calcium) and soda ash to cold raw water. Settling is allowed to take place for three hours or more and a sludge is precipitated which is removed by filtration. Usually, an excess of chemicals is used. Apparatus required includes a mixing chamber and sedimentation tank with sludge take-off. The softened water is filtered before use. When the chemical reactions take place at "room temperature," the process is termed *cold*.

How does the lime react?

Answer: The lime reacts with (1) dissolved CO_2 and calcium bicarbonate to precipitate calcium carbonate, (2) magnesium bicarbonate to form magnesium and calcium carbonate, (3) magnesium carbonate to form magnesium hydroxide and calcium carbonate, (4) magnesium sulfate to form magnesium hydroxide, and (5) magnesium chloride to form magnesium hydroxide and calcium chloride.

What is soda ash?
Answer: A trade name for sodium carbonate.

How does soda ash react?
Answer: Soda ash reacts with (6) calcium sulfate to form calcium carbonate and sodium sulfate, and (7) calcium chloride (result of reaction 5) to form calcium carbonate and sodium chloride. Note that both lime and soda ash are needed to remove magnesium sulfate and chloride.

What are various remedies for preventing scale, corrosion, carryover, and embrittlement?
Answer: Scale prevention (Fig. 16-7A) requires reducing calcium and magnesium to low limits so they do not produce hard deposits on the heating surfaces and cause overheating. Corrosion protection (Fig. 16-7B) requires maintenance of correct pH and elimination of oxygen, carbon dioxide, and other corrosive gases dissolved in the feed water (Fig. 16-8). Carryover eliminator (Fig. 16-7C) requires maintenance of low alkalinities, dissolved solids, and sludge concentration; also, no soapy matter (an antifoam agent is used). Embrittlement inhibition (Fig. 16-7D) requires low-hydroxide alkalinity, freedom from leaks at stressed metal (Fig. 16-9) and from silica at high pressure.

What is the remedy for the presence of fine particles of calcium carbonate that do not settle?
Answer: The remedy is an overfeed of lime and addition of a coagulant (Fig. 16-10). Excess lime can be removed by introducing carbon dioxide gas, but care should be exercised to prevent formation of bicarbonates if the water is used for boiler feed.

How about tank capacity and why?
Answer: Tank capacity should be sufficient to hold the water-

Feed-Water Testing and Treatment

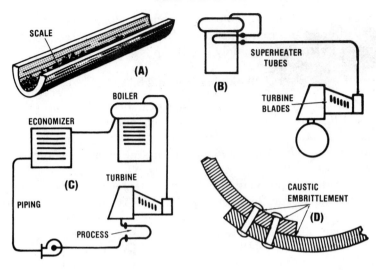

Fig. 16-7. Points at which scale, corrosion, carryover, and embrittlement occur.

chemical mixture from 4 to 8 hours while the sludge settles, because reactions are slow and there is a tendency for precipitate to form in filters and piping. (See Figs. 16-10 and 16-11.)

Is complete softening obtained with the cold process?

Answer: The clarified water is never completely softened because of the long time necessary to complete the reactions and because calcium carbonate and magnesium hydroxide are slightly soluble in water.

Hot Lime-Soda Treatment

The hot lime-soda process employs the same chemicals as in the cold process. Advantage includes a chemical reaction 8 times as fast at the boiling point as the cold process; also, smaller settling tanks are used. Carried out at temperatures above 212°F., hot-process softening employs live or exhaust steam as the heat source. Designs vary, but their softening principles are all similar. Chemical reactions in the lime-soda unit are the same as for the cold process, except they are almost instantaneous and produce softer water. Although available for softening alone, many designs incorporate deaerating sections for both makeup and condensate, or for treatment water only if makeup is large.

Removing Iron and Silica

The hot-process lime-soda softener can be used to remove iron and silica from water. On water of high, noncarbonate hardness, there is no decrease in solids content of the effluent and on water containing only a small amount of natural alkalinity, an increase in solids may result.

The greatest problem in lime-soda softening is maintaining uniform lime dosage. Soda ash is quite readily dissolved in water but

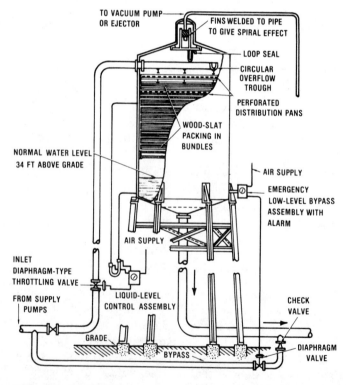

Fig. 16-8. Vapor take-off and controls for a large vacuum deaerator designed to treat cold water. Baffles prevent carryover.

Feed-Water Testing and Treatment

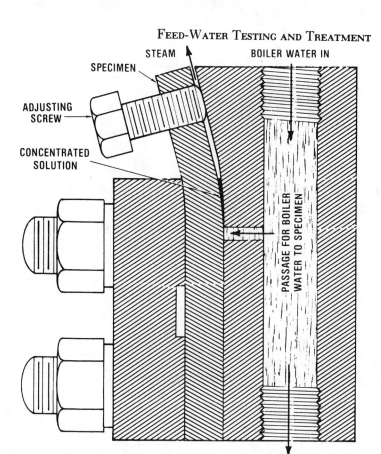

Fig. 16-9. Detector connects to boiler to check water for any embrittlement tendency.

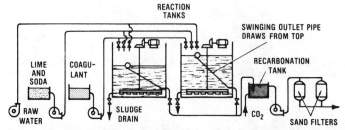

Fig. 16-10. Cold lime-soda softener. Intermittent (batch process) cold lime-soda treatment requires 4 to 8 hours retention time in reaction tanks and recarbonation to prevent after-deposits.

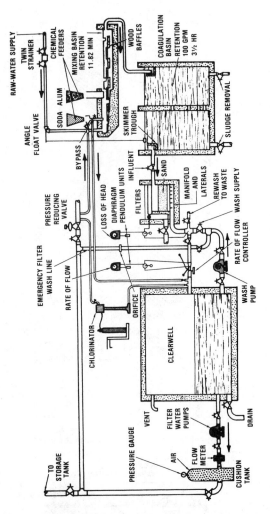

Fig. 16-11. Clarification plant with mixing basin, settling tank gravity sand filters, and storage tank uses alum as a coagulant and soda ash to adjust the pH of the effluent.

lime is only slightly soluble and must be fed as a slurry. In this state of suspension, the lime settles out unless the mixer does a good job.

The hot lime-soda process can be improved by circulating sludge, especially for silica removal, or by introducing water from

FEED-WATER TESTING AND TREATMENT

the boiler continuous blow-down to the softener (Fig. 16-12). This uses some boiler water alkalinity to reduce the amount of lime and soda ash needed and at the same time reduces the boiler-water suspended solids content. Boiler-water suspended solids act as seeding points for quick precipitation in the softener.

Other advantages offered, although all of them may not be attained at one time, are (1) reduction in hardness of the effluent, (2) decrease in sensitivity of chemical control, (3) improvement in silica removal and deaeration, (4) reduction in boiler alkalinity, (5) reduction in boiler internal treatment, and (6) reduced blow-down requirements.

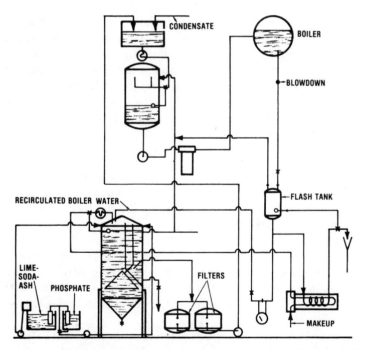

Fig. 16-12. Recirculation sludge from the softener or water from the boiler reduced the amount of chemicals and helps to form floc.

Phosphate Treatment

Phosphates are used either as direct treatment or as adjuncts in the cold and hot process. They can be fed directly to the boiler feed pump.

Owing to the tendency to precipitate, the phosphates are more commonly fed by a pump or hydraulic differential around the feed-water regulator to the main boiler drum. Phosphates have other properties that make them desirable.

Hard water can be economically treated by the combination of lime, soda, and phosphate with resulting soft water, and soft water (not over 50 ppm hardness) is more economically treated with phosphate alone. Sodium sulfate is also commonly injected at the deaerating heater or boiler feed-pump inlet to assist in removing the last traces of oxygen.

Because of the inherent residual hardness in the lime-soda effluent, a supplementary treatment is needed to make the water fit for boiler use. The common method is to introduce sodium phosphate directly to the boiler drum, add a separate phosphate compartment to the lime-soda softener, or install a separate tank between the primary softener and filters. Any of the various sodium phosphates may be used, depending upon the characteristics of the water to be treated. Holding an excess of 5 to 10 ppm to maintain a pH of about 9.7 assures a complete precipitation of calcium carbonate as calcium phosphate.

Definition of "pH"

Exactly defined, pH is the logarithm of the reciprocal of the hydrogen ion concentration. It is a logarithmic function and solutions having pH of 6.0, 5.0 and 4.0 are 10, 100 and 1000 times as acid as one having a pH of 7.0.

A simpler explanation is that the pH value is a number between 0 and 14, which denotes the degree of acidity or alkalinity. The pH scale resembles a thermometer or hydrometer scale—just as a thermometer measures intensity of heat and a hydrometer the density (strength) of a solution, so the pH scale indicates intensity of acidity or alkalinity.

The numbers 0 to 14 express pH values: the midpoint 7.0 represents neutral and a solution having a pH of 7.0 is neither acid nor alkaline.

Numbers below 7.0 denote acidity and those above denote alkalinity.

Zeolite Treatment

The term "zeolite" is applied to a certain class of minerals that have the property of *base exchange*. Zeolite originally applied to natural inorganic sands that entered into a reversible chemical reaction with hardness ions in water. Natural zeolite (greensand) composed of hydrated silicates of iron, aluminum and sodium, is mined from the earth and processed until the granules are separated from any impurities.

When the zeolite is treated with dilute brine (salt water), water containing calcium and magnesium can be made almost completely soft by passing it through a bed of the material.

It is important to know when the absorbing capacity of the zeolite is nearly exhausted. A sample of the softened water is taken and subjected to a saponification test, which indicates directly the relative hardness of the water. If this exceeds 1 to 1.5 grains per gallon (approximately 25 ppm), regeneration should be started.

What is the first step?

Answer: Remove any dirt that may have collected on top of the zeolite bed during normal downflow.

How is the removal accomplished?

Answer: This is done by backwashing. A predetermined amount of solution of common salt is then injected.

Where is the salt solution admitted?

Answer: At the top of the vessel, where it passes down over the zeolite bed evenly.

How does the salt react?

Answer: The salt reacts with the zeolite, removing the calcium and magnesium in the form of soluble chlorides and restoring it to its original active condition.

Power Plant Engineers Guide

What is the last step?

Answer: The last step is to rinse out excess salt and chlorides of calcium and magnesium by flow of water admitted at the bottom of the tank at low velocity, the rinsed material being discarded.

Sodium Zeolite Process

When hard water is brought into contact with sodium zeolite, the zeolite exchanges its sodium base for the calcium and magnesium in the water (Fig. 16-13). A zeolite water softener in which water is treated consists of a shell holding a bed of active zeolite supported by layers of graded gravel lying over a water distribution and collection system. As the water passes downward through the sodium zeolite, the calcium and magnesium in the water are exchanged for the sodium of the zeolite. When most of the available sodium has been exchanged, so that the first trace of hardness

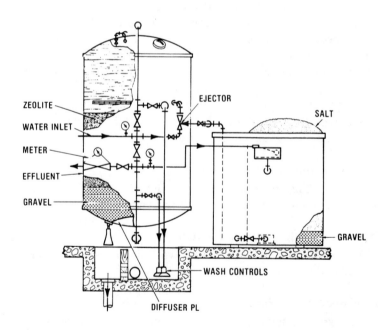

Fig. 16-13. Sodium zeolite softener with individual valves.

appears in the effluent, the zeolite is backwashed and then regenerated or revivified by passing a solution of common salt through it. Sea water, if available, may be employed for regeneration. The excess brine is rinsed out and the unit is restored to service. A meter is provided to guide the operator in regenerating after a definite number of gallons of water has been softened, or to actuate automatic regeneration.

Hydrogen Zeolite Process

This process neutralizes acid by adding caustic soda (Fig. 16-14). Hydrogen zeolite is the only known chemical method of removing sodium bicarbonate from water. Many well waters in various parts of the country contain large amounts of sodium bicarbonate. Hydrogen zeolite transforms such waters into waters of low dissolved solids, free from objectionably high alkalinity. It is possible also to use a mixed bed of hydrogen and sodium zeolite in one unit and regenerate it by subnormal amounts of acid or acid plus sodium chloride to avoid the necessity for neutralization. Sodium zeolite removes hardness but substitutes sodium for it and

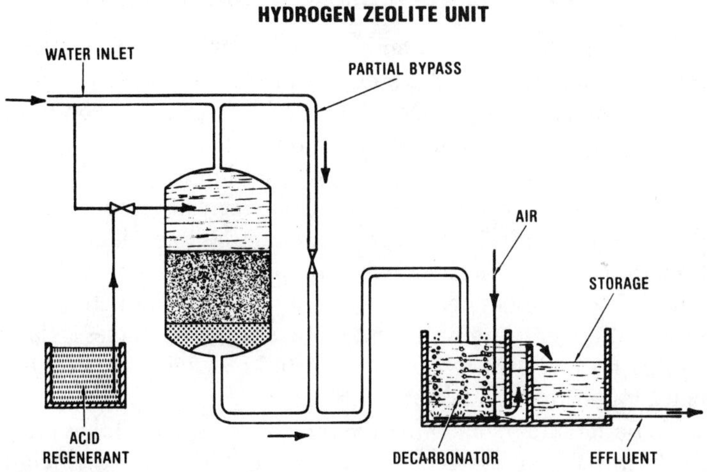

Fig. 16-14. Hydrogen zeolite de-alkalizer and decarbonator.

does not remove any of the anions. Hydrogen zeolite goes a step further by removing alkalinity, as well as the hardness, but it leaves the other anions (sulfates and chlorides) unreduced in the treated water. Demineralizing advances an additional step by removing these anions (sulfates and chlorides) as well, so that the effluent is free from all electrolytes.

What reaction takes place?

Answer: Carbon dioxide gas is liberated from the water in a special degasification heater for further release of dissolved oxygen.

Where is this process advantageously used?

Answer: With raw water high in carbonates; it is credited with being the only process that can successfully handle such cases.

How about the cost of the equipment?

Answer: The cost of the equipment for complete treatment is not greatly different from that required by the cold process, both processes being put on a comparative balance.

Ion-Exchange Processes for Softening

According to Feller, "There are only two basic pretreatments for softening boiler feed water; both have the primary objective of removing scale-forming calcium and magnesium compounds from the water before it enters the boiler (Fig. 16-15).

One method of treatment involves adding chemicals to form precipitates, as in the lime-soda process. The second method depends upon the ability of certain materials to exchange one *ion* for another, hold it in chemical combination temporarily, and give it up to a strong regenerating solution. One characteristic difference between lime-soda and ion-exchange processes is that the former removes waste products as sludge, while the latter eliminates impurities as a liquid solution.

To better understand *ion exchange*, let's see what takes place in a solution. When dissolved in water many substances dissociate to form positively and negatively charged particles known as *ions*. Such substances are called *electrolytes*. The positive ions are

FEED-WATER TESTING AND TREATMENT

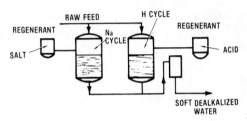

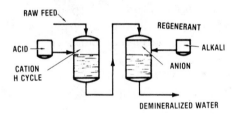

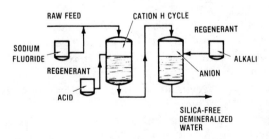

Fig. 16-15. Basic hookups of the ion exchangers used to treat water.

named *cations* because they migrate to the negative electrode (cathode) in an electrolytic cell and negative ones are called *anions* because they are attracted to the anode. These ions are present throughout the solution and lead nearly independent existences. For example, magnesium sulfate ($MgSO_4$) dissociates in solution to form positive magnesium (Mg^{++}) ions, and negative sulfate (SO_4^{--}) ions. In general, all natural waters contain dissolved electrolytes in varying concentrations. Some may impart undesirable properties to the water while others may not be objectionable.

Describe softening by sodium exchangers.

Answer: In this process, the water is brought into contact with the finely divided zeolite particles. As the water passes over the zeolite, it "exchanges" calcium and magnesium ions to the zeolite for sodium ions. This continues until the zeolite runs out of sodium ions, which stops the softening process.

What must be done then, and how?

Answer: The bed must be regenerated by backwashing; that is, reversing the flow of water through the bed to remove any dirt and loosen and regrade the zeolite. Also, a strong salt solution must be introduced downward through the softener.

What does the last operation do?

Answer: It reverses the exchange reaction. That is, sodium from the brine goes back into the zeolite to displace the calcium and magnesium. The water required for backwash and rinse ranges from 100 to 150 gallons (378-567 liters) per square foot ($929 cm^2$) of bed.

In the sodium (base exchange) process, carbonates and bicarbonates of calcium and magnesium are changed to sodium carbonate and bicarbonate, which remain in the feed water. While they do not form scale, they do contribute to the total solids present in the boiler and require greater amounts of blow-down than would be necessary if they were removed.

With zeolite capable of exchanging hydrogen for calcium, magnesium, and sodium ions, the salts are converted into equivalent amounts of their corresponding acids. Carbonates, sulfates and chlorides become carbonic, sulfuric and hydrochloric acids, respectively, and the treated water is a solution of these acids.

Carbonic acid is a combination of water and CO_2 and, because the latter can be expelled by aeration, the total dissolved-solids content and alkalinity of the treated water are reduced by an amount equivalent to the original carbonate content.

How do hydrogen zeolite units compare with sodium units?

Answer: They are similar to sodium units, except that acid-resistant materials are employed. Other differences are the regenerating acid tank and a decarbonator or degasifier to remove carbon dioxide from the effluent of the unit (Fig. 16-16).

How about the mixed-bed method?

Answer: It is possible to use a mixed bed of hydrogen and sodium zeolites in one tank and mix it with a subnormal amount of acid or acid and sodium chloride.

What is the fault of this mixed-bed method?

Answer: It does not produce as uniform an effluent as can be obtained with separate tanks and requires more skill in its operation.

What is another method?

Answer: Another method is to run the hydrogen-zeolite effluent through an anion exchanger, where the sulfuric and hydrochloric acids are absorbed, and then through a degasifier to expel the CO_2.

Silica Removal

According to Feller, sodium aluminate has been suggested for use in removing silica at a controller pH of 8.5. Experiments indicate a greater degree of silica removal than with ferric sulfate with the added advantage of less increase in dissolved solids. The possibility of residual alumina and silica being carried over into the boiler to form aluminum-silicate scale may be avoided, it is claimed, by proper control.

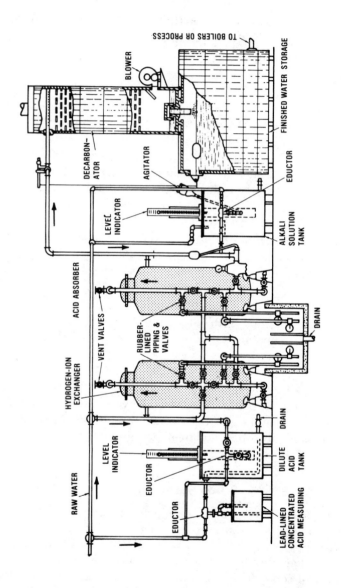

Fig. 16-16. Hydrogen exchanger combined with an acid-absorber unit (anion exchanger) that removes acids from its effluent and a degasifier.

How about prevention of silica scale?

Answer: It is primarily an external treatment problem. Maintenance of high residuals of phosphate and alkalinity in the boiler are an aid in prevention but can cope only with low amounts for high pressures.

How about most waters containing some silica?

Answer: Silica can also be introduced by sand filters, siliceous zeolites or even from concrete storage tanks. Also, certain old-time boiler compounds contained silicates of soda for corrosion prevention. Unless special pretreatment is provided, the silica finds its way into the boiler where it can cause scale, assist caustic embrittlement, and vaporize to pass out with the steam.

If silica is a problem, what should be done?

Answer: Local sources just mentioned should be eliminated before installing a costly removal process. According to Feller, in the boiler silica combines with caustic soda to form sodium silicate which remains soluble if enough caustic and no calcium is present. Calcium combines with silica to form scale. When phosphate is fed, however, calcium phosphate is precipitated, leaving silica to concentrate as sodium silicate.

If circulation is not rapid enough to keep the heating surfaces wet, what happens?

Answer: The sodium salts are thrown out of solution to bake onto the tube walls.

What results under this condition?

Answer: The sodium silicate combines with aluminum salts to form insoluble scale. Magnesium oxide added internally reacts with the silica to form magnesium silicate, a harmless sludge if phosphate is not present. As with all treatments, the sodium scales fall down under some conditions. Several high-pressure plants have discarded them in favor of the potassium family to prevent silica scale and carryover and to eliminate other troubles.

Potassium-silicate compounds have increasing solubilities with increase in temperature and pressure while those of sodium-silicate compounds decrease.

Corrosion

Give one preventive measure against corrosion.

Answer: Maintain enough alkalinity in the boiler water to hold its pH above 10.5.

What causes corrosion and how can it be slowed down?

Answer: Corrosion is caused by such factors as acidity and dissolved oxygen and carbon dioxide. It can be slowed down in the boiler and feed lines by increasing the alkalinity and eliminating oxygen.

CHAPTER 17

Condensers

What is a condenser?

Answer: An appliance designed primarily for removing the back pressure on a steam-generating or -using engine or turbine.

How is this usually accomplished?

Answer: By cooling exhaust steam and converting it into water.

What is the object of reducing the back pressure on a steam engine or turbine?

Answer: To obtain better economy.

Explain why.

Answer: A single pound of steam under an absolute pressure corresponding to one-half-inch vacuum has a volume of 1208 cubic feet (33.8 cubic meters) while a single pound of water has a volume of 0.016 cubic foot (0.00044 cubic meter). Thus, when steam at that pressure is condensed, the reduction in volume is more than 75,000:1. This is what reduces the pressure.

What is a vacuum?

Answer: A space devoid of matter; that is, a space in which the pressure is zero absolute.

ECONOMY OF CONDENSING

What is the economy due to condensing?

Answer: It is popularly held that the economy of condensing is, in round numbers, 25%. This percentage usually relates to simple engines and refers to the economy as measured by the difference in the coal or fuel comsumption produced by a condenser. The evidence of some of Barrus' tests show that "this belief is not well founded except in special cases."

What shows that this belief is not well founded?

Answer: If the feed water is heated by the exhaust steam of the noncondensing engine to a temperature of 100°F., which is that of the ordinary hot well, to a temperature of 210°F., the noncondensing engine can be credited with about 11% less coal consumption, which should be considered in determining condenser economy.

What did Barrus' tests show?

Answer: The average of a number of Barrus' tests gives a saving produced by condensing 22.3%. "If we allow for the steam or power used by an economical condenser, we see that the net economy of condensing is at best not much over 20%, based on steam consumption. If furthermore we allow for the difference produced by heating the feed water to the extent mentioned, the saving of fuel would be reduced to about 10%."

What is the gain in power due to condensing?

Answer: According to C.H. Wheeler, "Changing over a single-

CONDENSERS

expansion, reciprocating engine from noncondensing to condensing operation amounts to 20 to 30%."

What does Wheeler mean by "single-expansion engine"?

Answer: Single-stage-expansion engine. In fact, in engines of this type, there are three to four expansions noncondensing and five to seven condensing.

Since better economy is gained by expanding steam, why isn't the expansion carried further in single-stage-expansion engines?

Answer: Because of the higher range of temperature, which causes additional loss by condensation.

How does the higher temperature range increase condensation?

Answer: Re-evaporation between cut-off and prerelease is increased and its effect is to cool the cylinder walls, which in turn increases initial condensation.

AIR-EVACUATION PUMP

What auxiliary must be connected to a condenser to make them work?

Answer: An air pump.

What does the air pump do?

Answer: It should be understood that the air pump does not create the vacuum but only maintains it. In fact, any first-class, surface condensing equipment in good operating condition can be operated without any air pump for a considerable length of time without substantial loss of vacuum.

Why is the air pump necessary to maintain the vacuum in some systems?

Answer: The inherent reason is that water contains mechanically mixed with it 5% of its volume of air at atmospheric pressure. This air must be removed as fast as it collects or the accumulation of air would gradually build up a pressure which in time would rise

Power Plant Engineers Guide

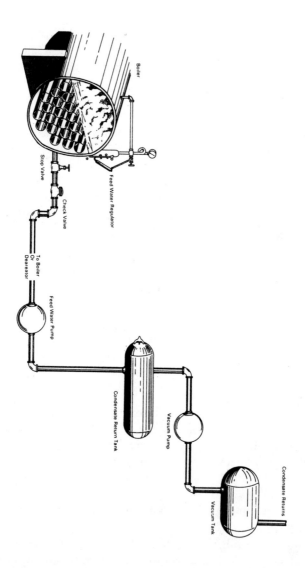

Fig. 17-1. Condensate system showing the position of the air-evacuation (or vacuum pump).

CONDENSERS

to atmospheric and higher, thus destroying the vacuum created by the condenser. See Fig. 17-1.

Give another reason for an air pump.

Answer: In practice, there is always more or less air leakage through stuffing boxes and joints, and this air as well as the air of condensation must be removed.

Why does hot air flow into the condenser?

Answer: Because of pressure difference; that is, the pressure in the condenser is less than atmospheric pressure.

Why shouldn't an air pump be called a vacuum pump?

Answer: Because no pump can "pump" a vacuum.

What then does the pump pump?

Answer: It pumps out most of the air (or other gas) from the condenser, maintaining a (partial) vacuum.

What is this vacuum ordinarily called?

Answer: A vacuum, regardless of its pressure.

Why does an air pump not extract all the air and obtain a perfect vacuum?

Answer: Each stroke of the air-pump piston or plunger removes only a certain fraction of the air, depending upon the percentage of clearance in the pump cylinder and resistance of valves. Theoretically, an infinite number of strokes would be necessary to obtain a perfect vacuum, not considering resistance of the valves.

Give another reason.

Answer: Inherent imperfection of the machine. The exhaustion is limited to the point where the remaining air has not sufficient elasticity to raise the valves.

How is the degree of vacuum measured?

Answer: In terms of inches of mercury. See Chapter 1 for full information of vacuum, absolute, and gauge pressures.

DIRECT-CONTACT (CLASS 1) CONDENSERS

Name two important types of Class 1 condensers.

Answer: Low-level, or jet condensers, and high-level, or barometric condensers.

What is a jet condenser?

Answer: A closed chamber within which exhaust steam comes in direct contact with a spray or jet of cold water and is condensed. See Figs. 17-2 and 17-3.

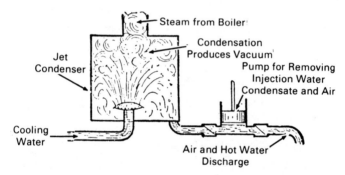

Fig. 17-2. Elementary jet condenser, showing essential parts.

What is the cold water called and why?

Answer: The injection water, because it is "injected" into the condenser.

What happens when the exhaust steam comes into contact with the finely divided injection water?

Answer: It is almost instantly condensed.

What happens when condensation takes place?

Answer: Since each cubic foot of exhaust steam shrinks to about one cubic inch of water when condensed, a partially empty space or partial vacuum is created in the condenser.

Upon what does the actual shrinkage depend?

Answer: The degree of vacuum in the condenser.

CONDENSERS

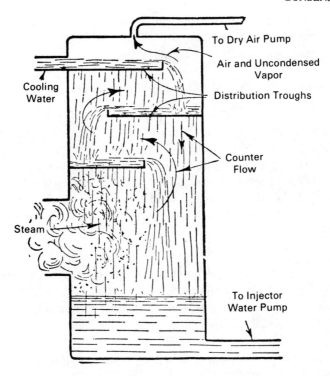

Fig. 17-3. Counterflow jet condenser. In this arrangement, the steam and water flow in opposite directions. That is, the entering steam encounters the warmest water and condenses as it rises, passing through successive curtains of water obtained by suitable arranged overflow trays. Thus, the temperature of the vapors is gradually reduced as they approach the top of the condenser due to the proximity of the incoming injection water. Ultimately, the mixture entering the pipe to the dry-air pump consists of air of relatively high density compared with that of the residual water vapors.

What causes the injection water to enter the condenser?

Answer: When the condenser is not too high, atmospheric pressure forces it in; when higher than the barometric column, a pump is required.

What is a "barometric column"?

Answer: See barometers, Chapter 1.

What must be removed from the condenser in addition to the condensate and injection water?

Answer: A small amount of air.

What means is provided to remove the water and air?

Answer: A wet-air "removal" pump.

What is a "wet-air removal" pump?

Answer: One which pumps both water and air from a condenser. For a jet condenser, the wet-air pump removes (1) condensate, (2) injection water, and (3) air. For a surface condenser, it handles only (1) condensate, and (2) air. It is called a "removal pump" with jet condensers and a "wet-air pump" with surface condensers.

Why is a pump for a jet condenser considerably larger than a pump for a surface condenser?

Answer: Because of the large amount of injection water that must be removed.

On large-size jet condensers what is sometimes used to remove the air?

Answer: A dry-air pump.

What is a dry-air pump?

Answer: A pump designed to handle air and gases only, which permits construction that will give a higher vacuum than that possible with a wet-air pump.

What are the essential features of construction?

Answer: The valves are exceptionally light and the valve spring no stronger than required to ensure proper seating of the valve. Moreover, the clearance is very small.

What do you mean by "the clearance"?

Answer: The volume between the piston and valves when the piston is at the end of the discharge stroke.

Condensers

How is the pump designed for minimum clearance?

Answer: The valves are placed in the cylinder heads, and in some special construction the heads themselves form the valves.

What is frequently used in place of dry-air pumps and why?

Answer: Steam ejectors. They are adapted to service where relatively large quantities of condensable vapors can be condensed in a "precooler" ahead of the condenser.

What important automatic device is necessary for safety with jet condensers?

Answer: A vacuum breaker. See Figs. 17-4 and 17-5.

What is a vacuum breaker?

Answer: An automatic device to protect the main engine or turbine from flooding.

Why?

Answer: It is necessary if the water-removal pump fails.

What do you mean by a water-removal pump?

Answer: A wet-air pump connected to a jet condenser which removes (1) the condensate, (2) the air, and (3) especially the injection water.

Why is a vacuum breaker of such importance?

Answer: At the usual rate of flow of the injection water, a jet condenser would be filled with water in a few seconds should the pump slow down, unless provision is made to break the vacuum and thereby stop the incoming water.

What do the numerous types of vacuum breakers depend upon for automatic action?

Answer: (1) Reduced contact surface, and (2) air admission.

How does the reduced-surface breaker work?

Answer: It consists simply of a constricted neck at the upper part of the condensing chamber which, with undue rise of the injection water, causes the condensing surface to rapidly diminish so that it is inadequate to condense the steam. Thus, the pressure rises within the condenser.

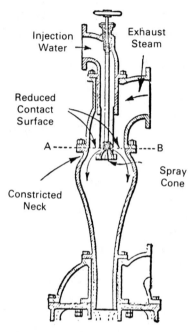

Fig. 17-4. Low-level, parallel-flow jet condenser, showing the method of reduced-contact, surface-vacuum breaker. In construction, the neck of upper part of the condenser chamber is made quite small and the cross-sectional passage area is further constricted at this point by the cooling-water pipe. In operation, rapid condensation is due only to the large surface exposed by the cooling water as it passes through the large section of the condensing chamber. Due to the constricted neck, any accumulation of water rapidly diminishes the condensing surface until the spray cone itself is submerged, leaving only the small annular ring of water at AB to act on the large volume of entering steam. The surface of this ring being far too small to condense the steam, the pressure immediately rises, causing the relief valve between the engine and condenser to open and allow the engine to run noncondensing. In the absence of a relief valve, the exhaust steam will blow out through the cooling-water pipe and pump valves, thus forcing all the water out of the condenser.

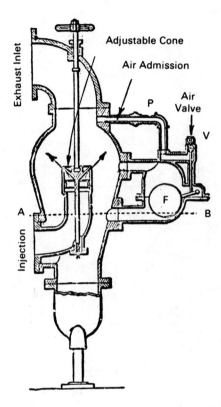

Fig. 17-5. Low-level, combined counter and parallel-flow condenser, showing a typical example of an air-admission vacuum breaker. It consists of a separate and communicating chamber with the float operating an air valve which admits air into the condenser. In operation, when the water rises in the condenser to the level AB, it lifts the float F, which in turn lifts the air valve V from its seat, admitting air into the condenser through pipe P, thus breaking the vacuum.

How does the air-admission—type work?

Answer: It consists usually of a ball float placed either in the condenser proper or in an adjoining and communicating chamber. Upon flooding of the condenser, the float will operate a valve and allow air to enter the condensing chamber, thus destroying the vacuum.

Where should the air be admitted for most efficiently breaking the vacuum?

Answer: In the water injection line.

What is a barometric condenser?

Answer: A high-level jet condenser. See Fig. 17-6.

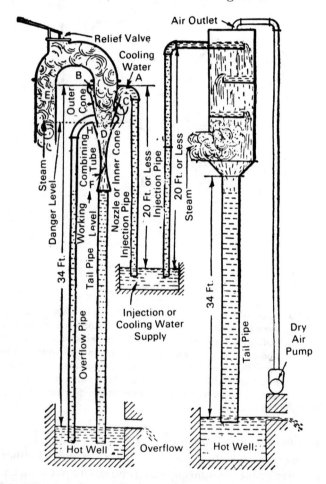

Fig. 17-6. Parallal-flow, barometric condenser or so-called injector condenser (left), and counterflow barometric condenser or dry-air-pump type (right).

What takes the place of the removal pump?

Answer: A *tail pipe* of comparatively large diameter and over 34 feet long attached to the condenser and submerged at its lower end in a hot well.

How does it remove the water without a pump?

Answer: The length of the tail pipe and the weight of the column of water in it overbalances the pressure of the atmosphere (at maximum barometer reading) and accordingly, the injection water flows out of the condenser.

Name two types of barometric condenser.

Answer: (1) The parallel-flow, or so-called "injector" type, and (2) the counterflow, or dry-air-pump type.

What name is sometimes given to the parallel-flow type?

Answer: The ejector barometric condenser.

SURFACE (CLASS 2) CONDENSERS

What is a surface condenser?

Answer: A device for condensing steam in which the steam and cooling water do not come into contact with each other, but are separated by metal surfaces. See Figs. 17-7 to 17-9.

What name is given to the cooling water and why?

Answer: It is called the *circulating* water because it is "circulated" on the water side of the cooling surface.

Into what two classes are surface condensers broadly divided?

Answer: (1) Wet and (2) dry.

What is the distinction between a wet and a dry surface condenser?

Answer: A wet condenser has a common opening for the discharge of the condensate and air. A dry condenser has separate openings for the discharge of condensate and air.

Describe the water- and steam-flow in standard condensers.

Answer: The water flows through the tubes while the exhaust steam flows over the outside surface of the tubes and is there condensed.

How about the flow in a water works condenser?

Answer: It's reversed; that is, the water flows over the outside of the tubes and the steam flows through the tubes and is condensed therein.

What are the advantages of the surface condenser as compared with jet condensers?

Answer: It permits the use of impure or salty cooling water in marine practice without bringing the impure water into contact with the condensate. The condensate is then available for use as boiler feed.

What is the usual assembly of pumps for a marine wet condenser?

Answer: Wet-air pump, circulating pump, and steam unit all direct-drive, tandem-connected; that is, the three pistons threaded on one connecting rod.

What is essential for adequate circulation of the cooling water?

Answer: Fast speed.

On condensers of this type, why should an auxiliary circulating pump be provided?

Answer: To circulate sufficient water through the condenser when the vessel moves slowly into port.

What is the feature of this arrangement?

Answer: Since the air pump is underneath the condenser, the condensate simply drops by gravity into the air pump. In fact, there is a head of condensate (negative lift) on the inlet valves so that the maximum vacuum within the range of the air pump is easily obtained.

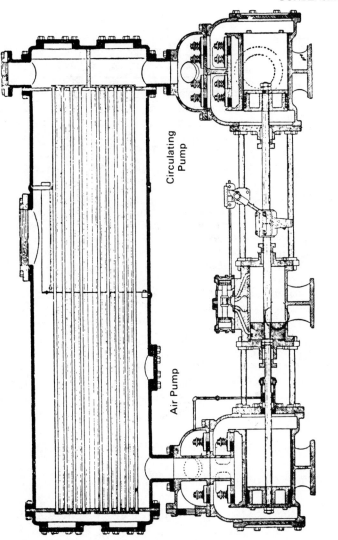

Fig. 17-7. Assembly of condenser, wet-air pump, and circulating pump.

What is an evaporative condenser?

Answer: A condenser in which the cooling surface is kept cool by the evaporation of the cooling water which is sprayed over the

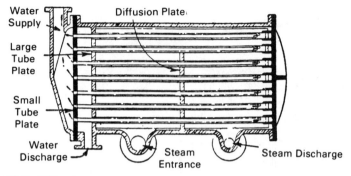

Fig. 17-8. Miller double-tube condenser. In construction, small tubes are placed inside of large ones. The water first passes through the inner tubes and returns through the outer tubes and, after absorbing the heat from the steam, is discharged into the air pump. This type was extensively used at one time, but in recent years the single-tube has represented the prevailing practice.

outer surfaces of the tubes. The rate of evaporation is usually increased by an air blast. See Fig. 17-10.

Condenser Tubes

What materials are used for surface condenser tubes?

Answer: The material most commonly used is Muntz metal for fresh water and Admiralty metal for salt water, pure copper being used only for exceptional conditions.

What should be considered in selecting the proper metal for the tubes?

Answer: Cost, life, and thermal conductivity.

What is the latest practice with respect to tube diameter and thickness?

Answer: The smallest outside diameter of the tube usually is ⅝-inch and the largest, 1 inch. Lengths run approximately from 4 to 22 feet, and the thickness varies from No. 16 to No. 20 Bwg.

How are the tubes connected to the tube sheets?

Answer: By stuffing-boxes or by expanding.

CONDENSERS

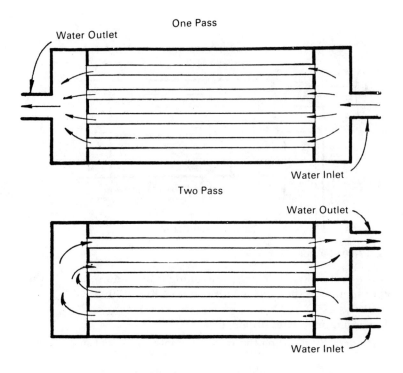

Fig. 17-9. One- and two-pass condensers. Note that the word "pass" relates to the water circuit and not the steam circuit. The steam circuit is not shown in these diagrams.

What provision *must* be made?

Answer: Provision for expansion and contraction.

Water and Steam Circuits

How are condensers classified with respect to arrangement of the water circuit?

Answer: (1) Single-pass, (2) two-pass, and so forth.

How does the two-pass arrangement work?

Answer: The entry box is divided into two sections. The circulating water is admitted to one of these sections, passes to the

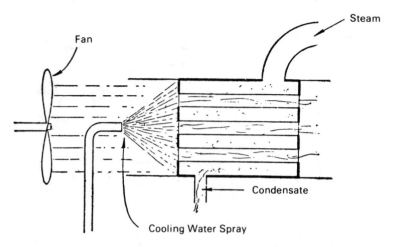

Fig. 17-10. Evaporative condenser.

second water box, enters the remaining tubes, and returns to the other section of the entry-water end. See Fig. 17-9.

How are more passes obtained?

Answer: By dividing the water box into more sections.

What are the various arrangements for steam flow?

Answer: (1) Single-flow, (2) divided (dual) flow, (3) multiflow, (4) crossflow. See Figs. 17-11 through 17-13.

How would you take out a tube of a surface condenser?

Answer: Remove the handhole or manhole plate, whichever is used, loosen the ferrule, and pull out the tube as far as you can. Plug the tube hole until such time as you can renew it. It will be necessary to saw the tube into short pieces to get it out.

CONDENSER OPERATION

Name several ways that a vacuum may be lost.

Answer: (1) The cooling water may be too hot, (2) there may not be sufficient water, (3) some of the valves in the pump may be

CONDENSERS

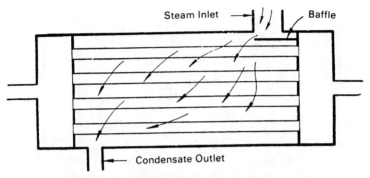

Fig. 17-11. Single-flow condenser. The steam finds its own path from the steam inlet to the condensate outlet with the result that the condensing process and vacuum obtained are not as efficient as might be otherwise. This arrangement is sometimes used on small condensers.

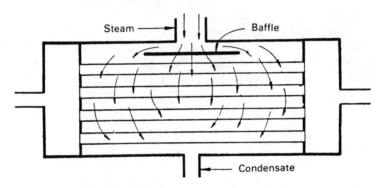

Fig. 17-12. Divided-flow condenser. Note placement of the baffle plate, which causes steam to flow toward the two ends, then downward, converging toward the condensate outlet.

broken, (4) the spray cone may have become plugged, clogged up, or not properly adjusted, (5) the packing of the air pump may leak, (6) may be a leaky joint in the exhaust pipe, or (7) there may be a combination of these causes.

If you were running a surface condenser and the vacuum dropped to 6 or 7 inches within a week, what would you think was the trouble?

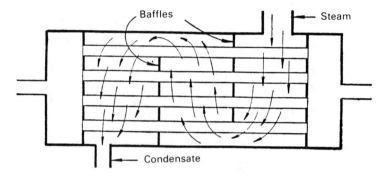

Fig. 17-13. Crossflow condenser in which any number of lanes may be obtained by baffling. The object of this arrangement is to completely control the steam flow, thus preventing short-circuiting. Moreover, the accumulation of condensate is discharged to the bottom of the condenser as it is formed in each alternate lane.

Answer: Water is probably bypassing instead of getting through the tubes. The partition plate on the head is eaten through, preventing the water from passing through the tubes and returning through the other half-bank of tubes. This will occur around salt water.

What is the result of leaky tubes?

Answer: They would lower the vacuum if leaking badly. The air pump will have to handle the leakage water.

How and when is the cylinder of an engine liable to become flooded from a condenser?

Answer: When the throttle valve is closed and the condenser stops.

TYPICAL MODERN CONDENSATE SYSTEMS

Are there condensation systems that return condensate directly to the boiler?

Answer: Yes. Some plants are using a zero-flash, closed-loop system that processes high-temperature condensate directly to the boiler, bypassing the feed-water system. Condensate flows from the process system (see Fig. 17-14) to a vapor seal where it is lifted by differential pressure to a pump surge chamber. A level control regulates the pump discharge to the boiler. Steam and noncondensable gases are removed continuously from the process equipment. Gases flow through the pump surge chamber and are dispelled. The steam is used to preheat makeup water in the feed system. Eliminating the flash loss saves (manufacturers say about 30%) in the fuel needed to generate steam. Some systems flash the condensate at high-pressure discharge condensate at temperatures up to 250°F. (121°C.) at 175 psig.

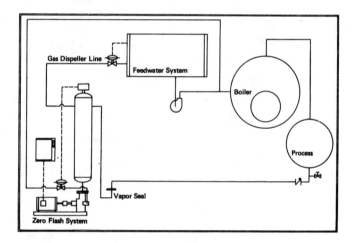

Courtesy Industrial Steam Div., Kewanee Boiler Corp.

Fig. 17-14. Zero-flash, closed-loop condensate system.

Do any condensate return systems operate without a pump?

Answer: Yes, the pumpless unit shown in Fig. 17-15 uses high-pressure steam to transfer the hot condensate. Condensate enters the upper receiver (Fig. 17-15A) and flows to the lower pump tank through the interconnecting piping. As the level in the lower tank rises (Fig. 17-15B) a switch closes, opening the steam valve and closing the equalizing valve. Steam in the lower tank pushes the condensate out the discharge check valve until the level drops to the lower switch position (a check valve in the condensate pipe closes during the pumping cycle and condensate accumulates in the upper receiver). When the switch opens (Fig. 17-15C), the steam valve closes and the equalizing valve opens, which equalizes pressures in the two tanks and allows condensate to enter the lower tank once more. The steam in the lower tank at the end of a pumping cycle is displaced into the upper tank through the equalizing line, as condensate enters the lower tank. The discharge check valve will be closed by discharge line pressure, and the inlet check valve will allow condensate to enter the lower tank (Fig. 17-15D). The system is designed to work in any application where the steam pressure is at least 20 psig higher than the required discharge pressure. For applications where the unit may discharge to a lower pressure than the vapor pressure of the collected condensate (such as in a flash tank), a solenoid valve or back pressure valve must be installed after the discharge check valve to prevent steam from blowing straight through the unit.

Spray-Flow Deaerator

In the system illustrated in Fig. 17-16, makeup water is modulated through stainless steel spring-loaded spray nozzles in the deaerating section where it is blended with condensate returns and recycled deaerated water. The preheated blend of deaerated water is pumped into the scrubbing section at a constant rate of 125% of deaerator capacity. This water is sprayed in the scrubbing section through stainless steel wide-angle, full-cone unrestricted nozzles and provides a continuous vent condensing action. Blended water remains in the scrubbing section until it is heated to 211°F. Then it is either pumped to the boiler if required, or recycled into the deaerating section through the compartment overflow where it partially deaerates incoming makeup and is constantly rescrubbed. The pneumatic temperature control system admits steam to the scrubbing section through a perforated heater tube which provides jets of steam to scrub the water to full saturation temperature. Though this process is externally quiet, a complete heating of the water separates the noncondensible gases which exit through the unrestricted vent.

CONDENSERS

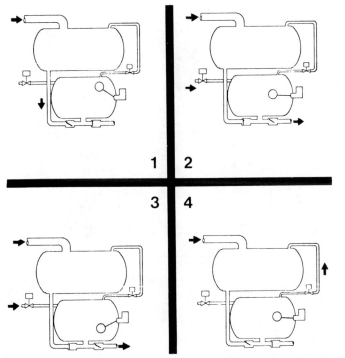

Fig. 17-15. High-temperature condensate return system.

Steam-Flow Deaerator

In the system pictured in Fig. 17-17, makeup water is sprayed through stainless steel spring-loaded nozzles into a stainless steel internal vent condenser located in the deaerating section. This incoming water is heated instantly by direct contact with steam. Returned condensate is also sprayed and deaerated in the same section.

The deaerated water is then pumped into the scrubbing section where it is blasted through stainless steel wide-angle, full-cone unrestricted nozzles. Remaining oxygen is shaken out at the source of the purest steam. The pumped transfer rate is approximately 125% of deaerator capacity, which enables the deaerator to furnish the boiler with deaerator water from startup.

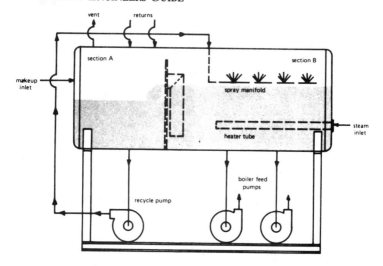

Courtesy Industrial Steam Div., Kewanee Boiler Corp.

Fig. 17-16. Spray-flow condensate-spray-flow deaerator system.

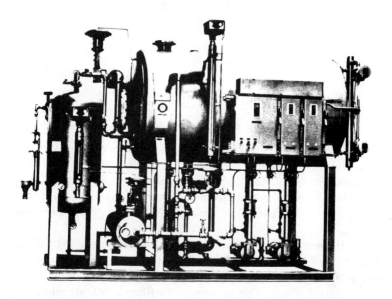

Courtesy Industrial Steam Div., Kewanee Boiler Corp.

Fig. 17-17. Pressurized steam-flow deaerator.

Excess deaerated water, which is not required by the boiler, recycles into the deaerating section through the compartment overflow. This deaerated water is blended with makeup water and is constantly rescrubbed. Noncondensible vapors are expelled from the top of the deaerator through the internal vent condenser.

CHAPTER 18

Cooling Ponds and Cooling Towers

COOLING POND

What is a cooling pond?

Answer: A shallow reservoir having a large surface area for removing heat from cooling water used to condense steam in large-system condensers. See Figs. 18-1 to 18-7.

What is the adaptability of cooling ponds?

Answer: In sparsely settled districts where land is cheap, and when cooling water is scarce, expensive, or rendered unfit for use in condensers by pollution from waste products of manufacturers, cooling ponds are used to advantage as the cooling water may be used over and over again.

What loss takes place with cooling ponds and how is it made up?

Answer: There is a small loss of water by evaporation, which can be made up from an outside source.

Power Plant Engineers Guide

How much cooling water is used where no means is provided for recooling?

Answer: From approximately 25 to 35 times the feed water, depending upon conditions. The following example will give an idea of the very large amount of water required where no means is provided to cool it so that it can be used again.

Example: A 100-horsepower engine runs on 30 pounds of feed water per hour, per horsepower. If the cooling water for the condenser is 27 times the feed water, how many gallons of cooling water are required per 10-hour day?

Total feed water = $30 \times 100 \times 10 = 30,000$ pounds (13,500 kg) daily;

$$\text{Total cooling water} = \frac{30,000 \times 27}{8.33} = 97,200 \text{ gallons}$$

(367,416 liters) approximately daily.

At usual city-water rates, the total expense would largely offset the savings made by condensing, and could make such usage prohibitive because of cost.

Name a few types of cooling ponds.

Answer: (1) Natural, (2) spray (single-deck and double-deck), (3) natural-flow, (4) directed-flow, (5) shallow, (6) deep, (7) open, (8) louvered-fence.

Upon what does the type to be used depend?

Answer: Upon the ground available.

What is a natural pond?

Answer: A natural-flow pond having no baffle walls or spray nozzle. See Fig. 18-1.

COOLING POND OPERATION

How is the cooling accomplished?

Answer: By (1) Radiation, (2) convection, and (3) evaporation.

Cooling Ponds and Cooling Towers

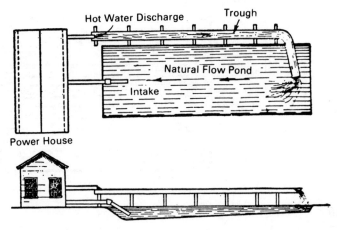

Fig. 18-1. Nondirected-flow, natural-cooling pond suitable for long and narrow lots.

What effect has the weather on the cooling?

Answer: In cold weather, cooling is obtained mostly by radiation and convection; in warm weather, by evaporation.

How much cooling surface is required?

Answer: According to Fernald and Orvok, under the conditions prevailing in northeastern United States, a natural cooling-pond (Fig. 18-2) surface of 250 square feet (23 square meters) is sufficient to cool the condensing water required for a boiler horsepower at 26-inch (660.4 mm) vacuum.

Describe a spray pond.

Answer: A pond of this type is provided with spray apparatus so that the hot water from the condenser is sprayed over the surface of the pond. See Figs. 18-3–18-7.

What is the action of spraying?

Answer: The hot water passing through a multiplicity of jets enters the air in a finely divided state, increasing its available surface area and intensifying the cooling by evaporation. Accordingly, for a given cooling capacity, the size of the pond is much less than that of a natural pond.

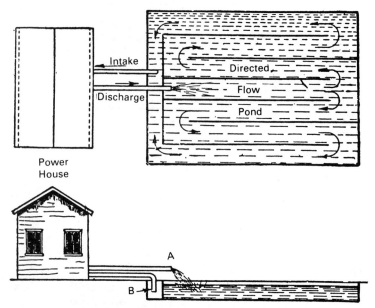

Fig. 18-2. Directed-flow, natural cooling pond. In operation, the hot water enters the middle channel at A and, on reaching the far end, divides into two currents, which are directed by the baffle walls so as to traverse the pond several times before uniting at the intake point B.

In spray ponds, water from the spray can be carried away by the air currents. How can this loss be prevented?

Answer: By extending the sides of the pond or by providing an enclosing fence. See Fig. 18-6.

What is the usual range of cooling?

Answer: Twenty to 40°, depending upon conditions.

How much cooling surface is required for a spray pond?

Answer: About 4 square feet (0.36 square meters) of surface is required for a boiler horsepower to condense steam at a 26-inch (660.4 mm) vacuum.

Cooling Ponds and Cooling Towers

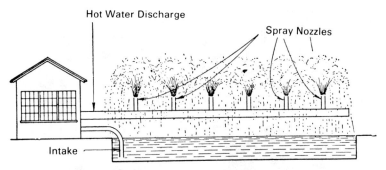

Fig. 18-3. Single-deck spray pond.

What is a double-deck spray pond?

Answer: One having spray nozzles arranged at different elevations.

According to Spray Engineering Company, tests show that the average amount of heat dissipated from the surface of a natural cooling pond with directed flow is 3.5 Btu per square foot per hour per degree.

Is feed water taken from the pond?

Answer: No.

Where is it taken and why?

Answer: It is taken from the discharge-pipe line between the condenser and the pond in order to save heat.

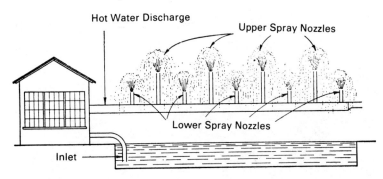

Fig. 18-4. Double-deck spray pond.

 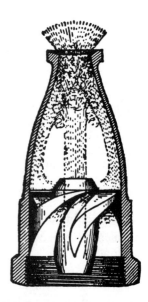

Fig. 18-5. Turbine-type spray nozzle (left), showing a removable turbine center for imparting the rotary motion to the liquid. At the right is a turbine-type spray nozzle with center jet nozzle. In operation, some of the water to be sprayed passes through the outer turbined passages and is gradually given a rapid, rotating motion. The nonrotary central jet strikes this rotating mass of water at a point just below the orifice in the space called the mixing chamber, resulting in a mixing of the rotary and nonrotary jets and issuance of the water from the orifice in the form of a fine, flaring spray.

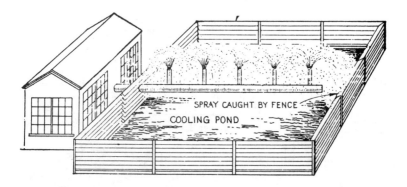

Fig. 18-6. Lower fence for preventing loss of spray water by air currents.

COOLING PONDS AND COOLING TOWERS

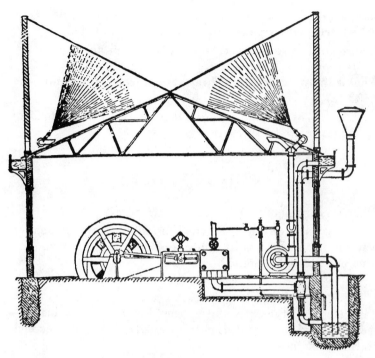

Fig. 18-7. Cooling pond with spray system on roof.

What is the action with a spray cooling system when there is no breeze?

Answer: An effective current of air is created in an upward direction around each nozzle due to the movement of the spray as well as to the heating effect which the spray has on the air which comes in contact with the water.

What is the result of such action?

Answer: The warm, moist air produced is rapidly carried away and replaced by cool, dry air brought in from all sides over the surface of the pond.

When are spray ponds most efficient?

Answer: In extremely hot weather when the humidity is high.

Since a pond 6 inches deep will usually give as good results as one 10 feet deep, why are deep ponds sometimes provided?

Answer: To provide water storage for fire protection.

What is the water loss in a spray system from evaporation and drift?

Answer: The average loss from these causes is from 1 to 2% of the amount of water sprayed.

COOLING TOWERS

What is a cooling tower?

Answer: An apparatus designed to remove from condensing water as much heat as can possibly be abstracted, per unit of space occupied by the apparatus.

Where are they used?

Answer: In large cities where ground is extremely valuable—the expense of buying extra land is avoided by placing the cooling apparatus on the roof.

Describe the construction of a cooling tower.

Answer: Essentially, it consists of a tower or stack from the top of which the heated circulating water is sprayed over a cellular construction of brushwood, earthenware pipes, wire mats, wooden checker-work or other baffles designed to expose the water to the cooling action of the atmosphere while in a film or fine-rain state. Figs. 18-8 and 18-9 show distribution decks and feeder troughs.

How is the process assisted?

Answer: By evaporation.

How are assisting counter-currents provided?

Answer: By side ventilation, natural draft enclosing the tower and using it as a chimney, or by a fan blast (forced-draft). See Figs. 18-10–18-16.

Cooling Ponds and Cooling Towers

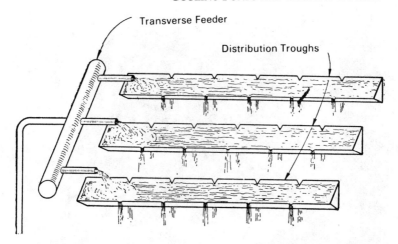

Fig. 18-8. Distribution deck consisting of transverse feeder and distribution troughs.

To what is the cooling effect due?

Answer: (1) Radiation from the sides of the tower, (2) contact of the water with the cooler air, (3) evaporation of the water.

Example: A certain condenser requires 100 pounds (4.5 kg) of water per minute, which is discharged at 110°F. (43.3°C.) and must be cooled to 70°F. (21.1°C.). What will be the rate of evaporation in the cooling tower?

The total heat to be abstracted from the water, per minute, is

$$100 \times (110 - 70) = 4000 \text{ Btu.}$$

If about 20% of the cooling effect comes from radiation and convection or contact of the water with the cooler air, then the heat to be removed by evaporation is

$$4000 \times (100\% - 20\%) = 3200 \text{ Btu.}$$

At 110°F., the latent heat of evaporation is 1030 Btu. So, the rate of evaporation is

$$3200 \div 1030 = 3.1 \text{ pounds (0.139 kg) per minute.}$$

The latent heat absorbed by the cooling water while condensing one pound of steam in the condenser must equal the latent heat

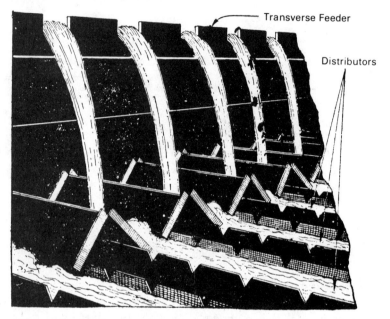

Fig. 18-9. Typical distribution deck.

extracted in the tower when evaporating one pound of water Therefore, the quantity of water evaporated will equal the quantity condensed, less the percentage of heat removed by convection and direct radiation. In other words, the cooling tower has to evaporate a quantity of water equaling 75% to 85% of the weight of steam (corresponds to the feed water) passing through the turbine or engine. This loss must be replaced by a fresh supply.

How much heat does the air absorb?

Answer: Each pound of free air absorbs 2.375 Btu while its temperature is raised 10 degrees. Thus, the temperature difference between the water and the entering air limits the heat-transfer by convection. For every 1000 Btu of heat transferred in this manner, 422 pounds (about 5600 cubic feet or 156.8 cubic meters) of air must be brought in contact with the water and warmed 10 degrees.

The same volume of air will absorb an additional and much larger quantity of heat through evaporation. Each pound of air

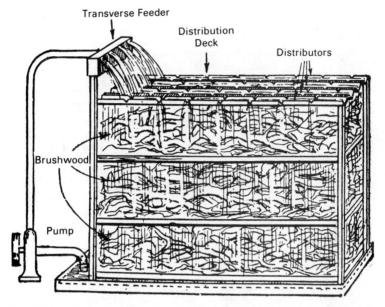

Fig. 18-10. Open, natural-draft brushwood cooling tower. This is about the simplest form of tower, being of very ordinary construction.

entering the cooling tower at 72°F. (22.2°C.) with 70% saturation, and leaving saturated at 102°F. (38.9°C.) will absorb only 7.2 Btu by its rise in temperature, but 28.7 Btu by the water it evaporates.

What is this called?
Answer: The specific heat of air at constant pressure.

How about the location of cooling towers?
Answer: They may be located either at the ground level or on a roof or other elevated structure, depending upon the space available and other local conditions.

What are the advantages of ground-level location?
Answer: (1) Simplicity of foundation and reservoir construction, (2) shorter pipelines, resulting in lower initial and operating costs, and (3) localization of possible spray during high winds.

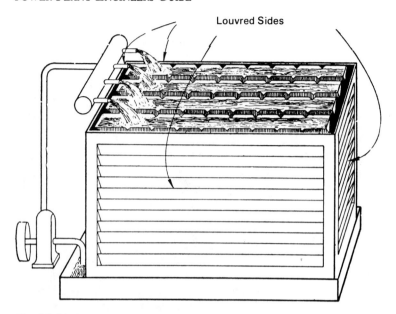

Fig. 18-11. Louvered sides or slatted enclosure to prevent loss of water due to lateral air currents.

Why is an elevated location for a natural-draft tower preferred?

Answer: Because of unimpeded circulation of the air currents and utilization of the otherwise unoccupied space.

What must be considered with respect to elevated towers?

Answer: The cost of pumping against the additional head.

What new design of cooling tower is currently being built?

Answer: The hyperbolic tower. See figs. 18-17 and 18-18.

How did this design evolve?

Answer: Natural-draft towers have been square, rectangular, cylindrical, and even pairs of truncated cones atop one another. The hyperbolic design has been found to be the best scientific design thus far.

Cooling Ponds and Cooling Towers

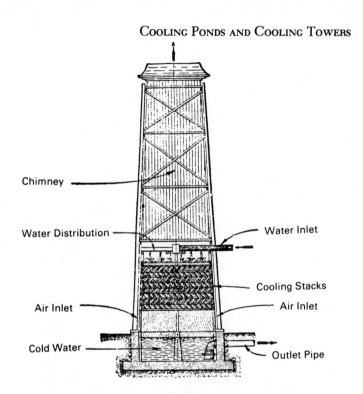

Fig. 18-12. Induced-draft cooling tower with zigzag cooling stacks.

What advantages does the hyperbolic tower have?

Answer: There are three major pluses: (1) It is efficient when operating conditions consist of low wet-bulb temperature and high relative humidity; (2) it has a broad cooling range and long cooling approach, as found in a combination of low wet-bulb and high inlet and exit water temperatures; (3) it is often selected over the mechanical-draft type when heavy, winter heat-load is possible.

What sizes are these towers?

Answer: They may be 500 feet (152 meters) high and 400 feet (121.6 meters) in diameter at the base.

Are hyperbolic designs sturdy?

Answer: According to the Masley Company of Kansas, these

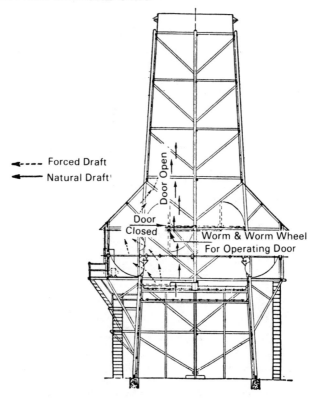

Fig. 18-13. Induced natural-draft tower.

towers can withstand hurricane winds in excess of 100 miles per hour, even though the thickness of the concrete shell is only 7 inches at the waist.

Do these towers have adaptations other than fossil-fuel uses?

Answer: A great number are in use at nuclear steam-electric projects.

Are hyperbolic cooling towers exclusively natural-draft units?

Answer: No. Some are combination induced- and forced-draft units, and some are designed only as forced-draft units because of factors such as humidity, mean temperature, and seasonal loading requirements.

Cooling Ponds and Cooling Towers

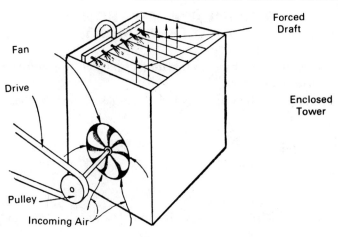

Fig. 18-14. Elementary forced-draft cooling tower, showing fan.

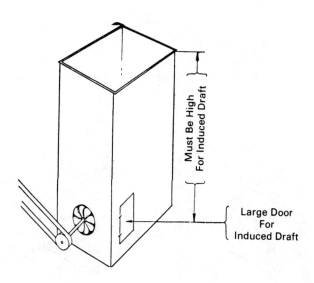

Fig. 18-15. Elementary combination forced- and induced-draft cooling tower.

COOLING-WATER PROBLEMS AND TREATMENT

What are the contaminants that might find their way into a cooling tower or pond?

Answer: Carbon dioxide, industrial gases, dust, bacteria, and oxygen.

What results?

Answer: Fouling, scaling, and corrosion.

How is fouling a problem?

Answer: Fouling causes reduction of both water flow and heat transfer.

What are the causes of fouling?

Answer: The collection of loose debris over pump suction-screens, growth of algae in sunlit areas, and slime formation in dark sections.

How is scaling a problem?

Answer: The deposits of dissolved minerals on interior heat-transfer areas of equipment form an insulation, making the system inefficient.

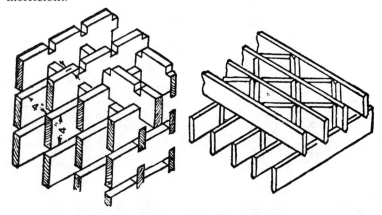

Fig. 18-16. Types of wood checker-work used in cooling towers.

Cooling Ponds and Cooling Towers

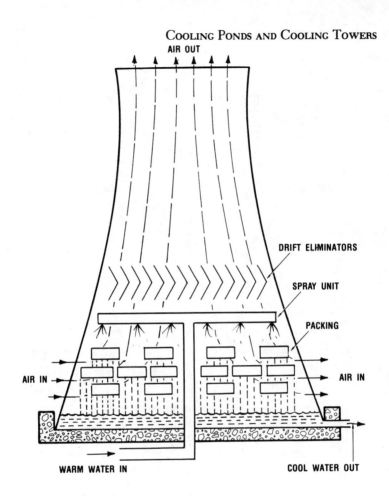

Fig. 18-17. Natural-draft hyperbolic cooling tower.

How is corrosion a problem?

Answer: Airborne chemicals finding their way into the cooling system cause severe damage to the equipment, reducing efficiency and leading to the necessity for costly repair and replacement.

What causes corrosion?

Answer: Five major causes are acidity, oxygen, galvanic action, biological organisms, and impingement.

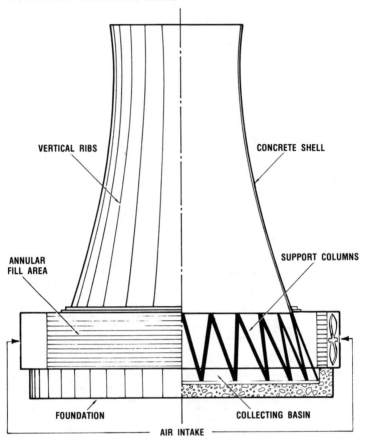

Fig. 18-18. Forced-draft hyperbolic cooling tower.

Explain.

Answer: (1) Acids are formed by chemicals in the water, occurring naturally as dissolved minerals or acquired from the atmosphere; (2) oxygen is most often picked up during the spraying and cooling process; (3) galvanic action is caused by placing dissimilar metals in contact with each other (for example, copper and iron); (4) biological organisms which comprise certain kinds of slime are capable of eating and digesting metal; (5) impingement is the wearing away of pipe walls by the abrasion of entrained gases, minute solid particles, and even the liquid itself as it travels at high speed through the system.

What can be done about cooling-water problems?

Answer: Employ constant analysis to determine the presence of foreign materials.

What can be done to cure the problem when it is detected?

Answer: The addition of proper control agents, such as algaecides, slimicides, chlorines, and water softeners (sodium or phosphorus compounds); the installation of larger-diameter pipes to reduce impingement; the addition of strainers, filters, and settling tanks to remove solids; the reduction of dissimilar metals to reduce galvanic action; and the addition of appropriate chemicals to combat excess oxygen and acids.

CHAPTER 19

Boiler Installation, Startup and Operation

A boiler can be dangerous. Hence, a thorough knowledge of its behavior under varying conditions is necessary for safety. The term operation broadly includes:

1. Installation,
2. Getting up steam,
3. Firing,
4. Repair.

INSTALLATION

Where should a boiler be located?

Answer: As near the engine or load as possible, as in Fig. 19-1A.

Power Plant Engineers Guide

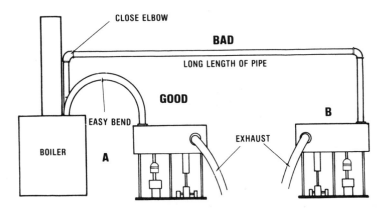

Fig. 19-1. Right and wrong methods of piping steam from a boiler. The arrangement at A shows a short steam pipe with long-sweep fittings, or made in one piece and bent to an easy curve. This arrangement gives the minimum resistance to steam flow and minimum radiation loss. A very objectionable method is shown at B. It will give considerably more pressure drop between the boiler and engine and radiation loss than the arrangement at A.

Why?

Answer: To reduce pipeline condensation to a minimum and, in the case of superheated steam, to minimize the superheat loss.

How about the piping between the boiler and engine?

Answer: It should be as short as possible with a minimum number of elbows and adequate insulation.

Where should the main steam outlet on the boiler be located?

Answer: At the end which will give the shorter pipeline between engine and boiler.

What important provision should be made in erecting piping between boiler and engine?

Answer: It should be so inclined so that the condensate will drain toward the engine (Figs. 19-2–19-4).

Boiler Installation, Startup and Operation

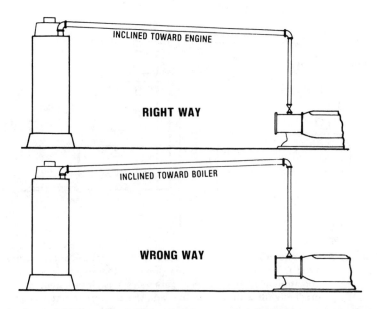

Fig. 19-2. Right and wrong way to pitch the main steam pipe between the boiler and engine.

Why?

Answer: Because if the pipe is filled with condensate on opening the throttle, a solid column of condensate will do damage to the engine.

What mistake is often made in the design of boiler plants?

Answer: No provision for equalized room ventilation in the case of a battery of boilers. See Fig. 19-8.

How should steam pipelines be supported?

Answer: With mounting frames fitted with rollers to allow for expansion and contraction and minimize vibration. See Fig. 19-5.

How should exhaust steam lines be run to the condenser?

Answer: See Figs. 19-6 and 19-7.

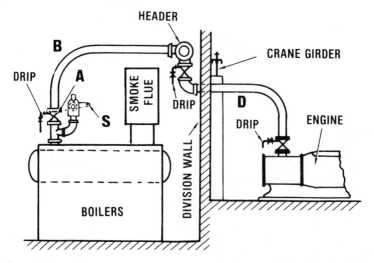

Fig. 19-3. Objectionable arrangement of piping. It was done in order to carry bend D underneath the crane girder in the engine room. It is objectionable because when gate valve A is closed, bend B will gradually fill with condensate, and the column of water will be driven over into the main when A is reopened. If the engine is cut out of service by closing valve E, leaving C open, bend D will fill up with water which will pass into the cylinder when E is reopened. If both E and C are closed, water will collect in the main above C. Whenever a valve forms a water pocket in a steam line, as at A, C, and E, the valves should be drained from above the seat.

RUNNING

The term "running" as used here comprises:

1. Inspection before getting up steam,
2. Getting up steam,
3. Firing methods,
4. Water tending,
5. Meeting fluctuation loads,
6. Banking fires,
7. Blowing off.

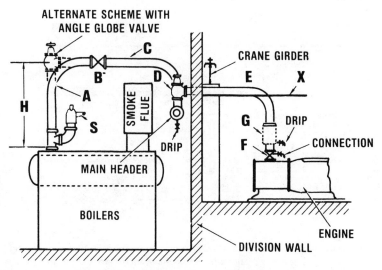

Fig. 19-4. Proper arrangement of piping. In this system, condensation on bend C will drain into the main; the same result is obtained using an angle valve (dotted lines) instead of B. Valves placed at the same distance (as H) above the boiler nozzle should be anchored to prevent vibration. Any leakage through the angle valve or B will fill bend E with condensate; hence, the necessity of separator G, placed close to the cylinder.

What is the first thing an engineer should do on first taking charge of a plant?

Answer: He should trace out all the pipelines, noting the condition of same, carefully examining the safety valve, gauge cocks, and water gauge, and make sure that the gauge column is clear of obstructions by inserting a rod. Attention should also be given to feed line, check valve, feed pump, injector, and blow-off valve. Make sure all auxiliary equipment is working properly: combustion controls, dampers, regulators, blowers, etc. Then close the blow-off valve, water column and gauge glass drains, and gauge or tri-cocks.

What should be done next?

Answer: Fill the boiler and economizer, if any, with water to the proper specified level. Make sure vent valves, water column

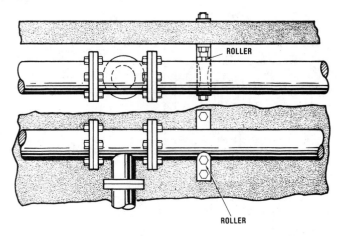

Fig. 19-5. Method of preventing vibration and of supporting pipes. The figures show top and side views of a main header carried in suitable frames fitted with an adjustable roller. While the pipe is illustrated as resting on the adjustable roller, the roller may also be placed at the sides or on top of the pipe to prevent vibration, or in cases where the thrust for a horizontal or vertical branch has to be provided for. This arrangement will take care of the vibration without in any way preventing the free expansion and contraction of the pipe.

shut-off valves, gauge glass shut-off valves, and steam pressure gauge valves are open. If there is an auxiliary feed line, use it during the fill period and hold a water level an inch or so above the bottom of the glass. Water should be fed slowly to a completely vented system.

How much water should be put in a small water-tube boiler and why?

Answer: Fill to the top of the gauge glass because a water-tube boiler holds very little water compared to a shell boiler. Then, in case of trouble in starting the pump or injector, there is a margin of water to work on.

Do the water column and gauge glass need any special attention at this point?

Answer: Yes. Blow down the water column and make sure that

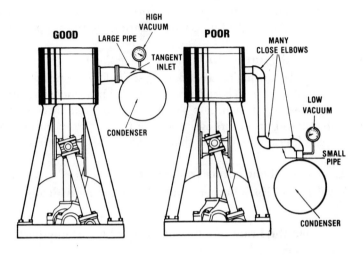

Fig. 19-6. Right and wrong way to pipe from an engine to a condenser.

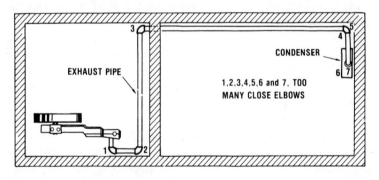

Fig. 19-7. An arrangement of exhaust piping which was expected to give a 24-inch vacuum, but didn't. The only reason for this arrangement was convenience.

the water returns to the proper level immediately. If it doesn't, there could be some obstruction in the connections or column.

How about the economizer and superheater, if they are used?

Answer: All drains and bleed outlets should be open completely, except where special circumstances dictate otherwise.

Follow the manufacturer's procedure precisely in setting bleeding and recirculating controls and valves during the fire-up process.

How should the furnace be ignited?

Answer: According to procedure specified by the firing equipment manufacturer. Where fuels are burned in suspension rather than on a grate, all dampers should be open and forced draft systems turned on before ignition so a free air flow purges the furnace of fuel accumulations that can cause puffs and explosions.

Is it safe to light one burner from another?

Answer: No! Each burner should be lighted individually.

How should the firing rate be regulated where a superheater is used?

Answer: To keep the temperature of the tubes within safe limits and to maintain a safe (according to the manufacturer's specifications) temperature difference between the feed water and the metal which contains it.

Where soot blowers are used, why is it important that they be properly aligned?

Answer: So steam used to blow the tubes doesn't come into contact with any tube surface.

When should the safety valve be checked?

Answer: As soon as possible after pressure builds, by lifting the hand lever or by increasing boiler pressure to the pop-off point. The ASME rules suggest hand testing, and holding the valve open for a short period to blow out any dirt.

What is the proper method of starting a hand-fired fire?

Answer: Spread a thin layer of coal over the grate, then some paper and a moderate amount of wood. It is best ignited by burning some paper in the ash pan, or in the case of a large boiler, some oily waste may be placed in the interstices of the wood, as shown in Fig. 19-9.

Boiler Installation, Startup and Operation

What precaution should be taken upon lighting the fire?

Answer: The draft should not be too strong, as unequal expansion of the metal will cause leakage in the tubes, seams, or rivets.

What should be done as soon as the wood is well ignited?

Answer: Coal should be added, a little at a time.

Why should the coal be added a little at a time?

Answer: This will hold the fire in check and prevent sudden heating of the metal.

After lighting the fire, what should be done?

Answer: Maintain a light fire until the brickwork of the setting is dried out thoroughly. If the entire setting is new, the drying out may require several days; if the lining of the combustion chamber only has been renewed, about 48 hours will be sufficient time for drying out small settings. Larger ones will require a longer time.

What should be done before steam begins to form?

Answer: Open the top gauge cock so that the steam will drive out the air.

What would probably happen if you didn't drain and warm up the header between the main stop valve and the nonreturn valve?

Answer: It would stay cooler than adjoining sections and abnormal expansion would occur when the steam hit it. The result could be leaks and a surge of water in the steam line.

Why?

Answer: The air, if not removed, would work through to the condenser and retard the vacuum. Moreover, its presence would furnish oxygen which would be active in corroding the boiler.

Why open the gauge cock instead of the safety valve?

Answer: Air, being *heavier* than steam, will be next to the water, while the steam will rise. Hence, the air should be expelled at a point as near the water line as possible. See Fig. 19-10.

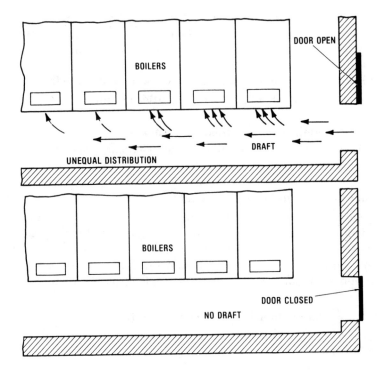

Fig. 19-8. Poor provision for draft in a boiler room. Where the air must come in through a door located at the end of the room, the boiler nearest the door will get most of the air and the others less and less, leaving hardly any air for the most remote boiler. When the door is closed as in cold weather, the draft in all the boilers wil be very poor, the only air available being that which leaks in through cracks.

When steam forms what should be done?

Answer: The safety valve and gauge cocks should be closed, and the steam gauge should be observed at once.

What attention should be given to the water gauge?

Answer: If it appears stationary or "dead," see that the valves are open.

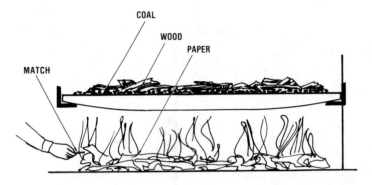

Fig. 19-9. Method of starting the fire. The thin layer of coal protects the grate bars. The mass of wood is best ignited by lighting paper in the ash pan. In this way the wood is ignited in several places at once instead of only at a single point.

What is a good indication that the water gauge is working?

Answer: A "lively" gauge—that is, the water fluctuating up and down, especially in water-tube boilers.

What does a too-lively gauge indicate?

Answer: Priming or foaming.

Why is it important to check the line connecting the steam gauge to the boiler?

Answer: To make sure it is free of obstruction and leaks. Otherwise, the gauge indication won't be accurate.

What should be done when steam is up to the desired pressure and no steam is being used?

Answer: The hinges of the fire door should not be dislocated or broken off by violently throwing the fire door open as is usually done by inexperienced operators.

How is a further rise of pressure and blowing off avoided?

Answer: Close the damper and throw on a little coal, being careful to spread it well.

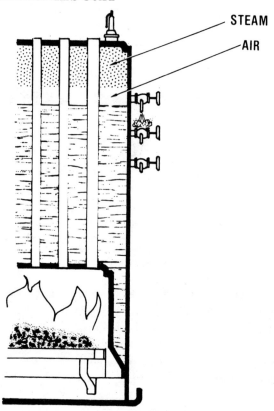

Fig. 19-10. Method of expelling air from a boiler in getting up steam. Air, being heavier than steam, will lie in a layer next to the water line, while steam will rise to the top. Hence, to get out all the air, open the water gauge cock (nearest above the water line) and not the safety valve.

Why not throw open the fire door to check the fire?

Answer: When the door is opened, the sudden inrush of cold air striking the highly heated surfaces causes rapid cooling of the metal with unequal contraction. This, in the case of a shell boiler, subjects it to severe strains. In some types of pipe boilers, it makes no difference.

Why should the ash pit door be left open at all times?

Answer: To prevent the grate bars from becoming overheated, thus prolonging their life.

In general, how is the fire attended after raising steam?

Answer: For power boilers, frequent firing in small amounts; for house-heating boilers, heavy firing with less frequent intervals.

Why is it usually possible to tighten manhole and handhole cover bolts after the boiler reaches operating pressure?

Answer: The gaskets become more pliable and the cover bolts expand more than the boiler steel.

FIRING

There are two kinds of firing:
1. By hand (Fig. 19-11),
2. By specially designed fuel burners and mechanical stokers. These firing methods are covered separately.

What are the various methods of hand firing?

Answer: (1) Spreading methods (even spread, alternate side spread, alternate front and back spread), (2) coking method. See Figs. 19-11–19-21.

Describe even-spread firing.

Answer: The coal is spread evenly, beginning at the back of the grate and working toward the door, as shown in Fig. 19-12.

Describe the alternate side-spread firing.

Answer: Coal is spread on one side of the grate over its whole length, then over the other side, alternately, as shown in Fig. 19-13,

POWER PLANT ENGINEERS GUIDE

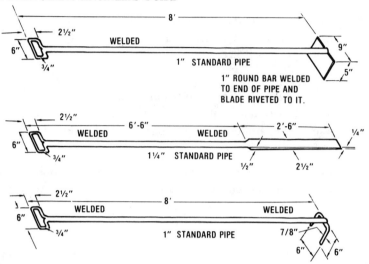

Fig. 19-11. Fire tools as recommended by the Bureau of Mines showing construction and dimensions.

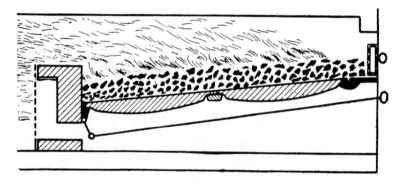

Fig. 19-12. Even-spread mode of firing, the result of which will be a uniform generation of gas throughout the charge.

at equal intervals of time. The firing interval is shortened to one-half the time of the even spread method.

Describe alternate front and back spread firing.

Answer: This method is the same as the alternate side-spread method except that the fresh coal is alternately fired on the front and back halves of the grate. See Fig. 19-14.

498

BOILER INSTALLATION, STARTUP AND OPERATION

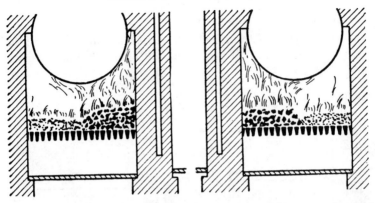

Fig. 19-13. Alternate side-spread firing. Left, fresh coal is put on the right side; right, fresh coal is put on the left side. The coal should be fired in small amounts and well spread. In some cases a boiler that requires a fire on both sides, say every 10 minutes, will require a fire on the right side every 10 minutes, and on the left every 10 minutes, making 5-minute intervals between firings. The gas given off by the fresh coal on one side rises into a hotter furnace or combustion chamber (than it would if both sides were fired simultaneously) due to the hot, clear fire on the opposite side, and the result is that most of this gas is burned if the proper amount of air is admitted through the door. When the last side fired burns clear, the opposite side is fired. Thus, the clear fire on one side serves to ignite the gases baked out of the fresh coal on the other side.

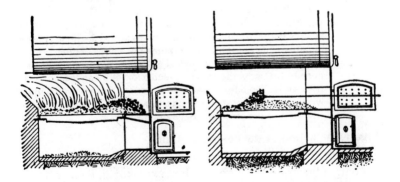

Fig. 19-14. Firing with hard (anthracite) coal. These methods give good results when the fires are properly handled.

What is the coking method of firing?

Answer: In this method the fresh coal is piled on the front of the grate, while the rear half is covered with partially burned coke. See Fig. 19-15.

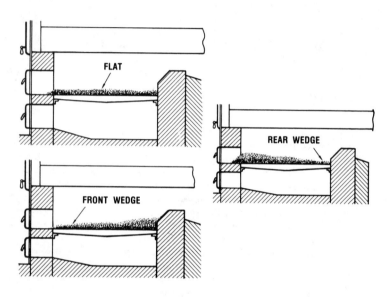

Fig. 19-15. Coking method of firing. Fresh coal is placed just inside the furnace door on the dead plate, covering a few inches of the grate. The draft plate in the fire door is left partly open so as to admit air to the fresh coal. The heat of the fire gradually bakes the volatile matter out of the coal, and if this is mixed with the proper quantity of air entering through the door a considerable proportion of the mixture can be burned. When the fire needs more fuel, the pile just under the door is spread over the incandescent coals with a hoe. The pile of coal is thus reduced to coke which burns without smoke, after spreading a new charge of coal in place on the front of the grate and dead plate.

What is the action of the coking method?

Answer: The gases distilled from the fresh coal pass over the rear half of the grate, through which an excess of air is entering, the air being highly heated as it passes through the bed of coke. The gases and heated air intermingle in the combustion chamber and are completely burned.

BOILER INSTALLATION, STARTUP AND OPERATION

Which is the more efficient, the spreading or the coking method of firing?

Answer: The spreading method.

How should coal be fired in the spreading method?

Answer: In small quantities at short intervals. See Figs. 19-16, 19-17, and 19-18.

Why?

Answer: So that thin places do not burn through and admit a large excess of air.

What are the reasons for light and frequent firing?

Answer: When fresh coal is fired, the volatile matter is immediately distilled. The process is nearly completed in 2 to 5 minutes. Therefore, immediately after firing, large quantities of air should

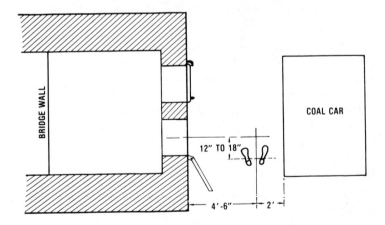

Fig. 19-16. Proper position for firing. According to the Bureau of Mines, the fireman should stand in such position that he can see the thin spots and can throw the coal onto them with the least effort. He should stand 4½ to 5 feet in front of the furnace, and 12 to 18 inches to the left of the center line of the door. He should then be about 2 feet from the coal, which is 6 or 7 feet from the furnace front and should preferably be on a car.

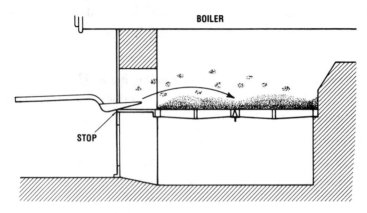

Fig. 19-17. End of the throw. The scoop should travel in a nearly straight line. At the end of the throw, the scoop should suddenly be stopped by laying it on the bottom edge of the door frame. The coal then flies off and is scattered over the proper spot. By thus stopping the scoop, the fireman saves effort and locates the coal better. If he pushes the scoop far into the furnace, he has to jerk it back to get the coal off.

be admitted over the fire and afer 2 to 5 minutes it should be cut down. Such regulation is practically impossible, so large quantities of fuel should not be fired at one time.

Where are thin spots especially likely to occur?
Answer: In the corners at the front of the furnace and between the firing doors.

How does a skilled fireman treat thin spots?
Answer: He gauges the right amount and selects the proper mixture of fine and coarse coal.

When and why is coarse coal used?
Answer: When the spot is burned way down to the grate, the coarse coal will not sift through the grate.

What is the objection to a thin fuel bed?
Answer: Too much attention is required to prevent holes, and

BOILER INSTALLATION, STARTUP AND OPERATION

when holes form, too much air flows through without uniting with the carbon.

With a given draft how can the rate of combustion be increased?
Answer: By reducing the thickness of the fuel bed.

What is the object of wetting coal?
Answer: To cause the coal to burn slower, thereby giving the gases a better chance to mix with the air.

What happens when coal is first thrown on the fire?
Answer: It gives off gas faster than it can be brought into

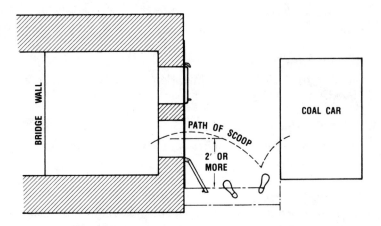

Fig. 19-18. Improper position in firing. If the coal is closer than 6 or 7 feet, the fireman is crowded and will stand away to one side of the door to avoid the intense heat. He cannot see the fire and throws the coal in by guess. His scoop travels in the arc of a circle, scattering coal on the floor and dumping it in a heap directly inside of the furnace door. This results in an uneven fire, low efficiency, and requires raking of the coal onto the back part of the fire. Provide a smooth firing floor or a smooth bottom to the coal car so that the shovel does not hit bumps and rivets. Such items delay the firing operation, keep the door open longer than necessary, and admit excess air.

contact with the necessary amount of air for complete combustion.

Mention an objection to wetting coal.
Answer: The water must be evaporated and this wastes heat.

What difficulty is experienced in burning bituminous slack coal?
Answer: It fuses into a hard, tight crust which admits little air, resulting in a low combustion rate.

What is done to correct this in firing?
Answer: The crust should be broken by lifting with a *slice bar*; and then, level the fire.

What must be done periodically in firing?
Answer: The fire must be "cleaned."

Define the term "cleaning the fire."
Answer: By definition, the operation of removing clinkers and cinders from the burning coal at regular intervals.

Why is this frequently necessary?
Answer: Because the clinker and coarse ash will not pass through the grate.

What tools are used for cleaning the fire?
Answer: The hoe and slice bar and the rake for leveling the fuel bed. See Fig. 19-11.

Upon what does the interval between cleanings depend?
Answer: Upon the proportion of ash in the coal and the type of grate. See Fig. 19-19.

Name two methods of cleaning the fire.
Answer: (1) The side method, and (2) the front-to-rear method. See Figs. 19-20–19-22.

In the side method, one side of the fire is cleaned at a time. The good coal is scraped and pushed from one side to the other. The clinkers may have to be removed from the grates by the slice bar. When they have been loosened and broken up, they are scraped out of the furnace with the hoe. The fireman should gather the clinker on the front part of the grate before pulling it out into the wheelbarrow, as this saves him from exposure to the heat. After the one side is cleaned, the burning coal from the other is moved and scraped to the clean side. It is spread evenly over the clean part of the grate, and a few shovelfuls of fresh coal are added, in order to have enough burning coal to cover the entire grate when the cleaning is done. This adding of coal is important, especially when the cleaning must be done with the load on the boiler. The clinkers are then removed from the second half of the grate. When cleaning is started, there should be so much burning coal in the furnace that enough will be left to start a hot fire quickly, when the cleaning is completed. If a light fire is being carried it may be necessary when starting to clean to put some fresh coal on the side to be cleaned last. During cleaning, the damper should be partly closed. A fireman after becoming familiar with the side method should be able to clean a 200 hp boiler furnace in 10 to 12 minutes.

In the front-to-rear method, the burning coal is pushed with the hoe against the bridge wall. It is usually preferable to clean one-half of the grate at a time. The clinker is loosened and pulled out of the furnace and the burning coal is spread evenly over the bare grates. If the front-to-rear method must be used while the load is on the boiler, the side method should be employed after the day's run is over, so as to prevent the large accumulation of thick and hard clinker at the bridge wall. Some firemen have the habit of pulling the clinkers out of the furnace without scraping and pushing the burning coal against the bridge wall or to one side. This really is not a method of cleaning the fire. They run a slice bar under the clinker to lift it to the surface of the fuel. Then they take a hoe and pull the large pieces out. The small pieces are not easily detected and are left in the fire. These fuse in a few minutes, due to the high temperature near the surface of the fuel bed, and then run into the grates. Thus more masses of clinkers are formed, which are usually worse than those previously removed. This habit should be discouraged.

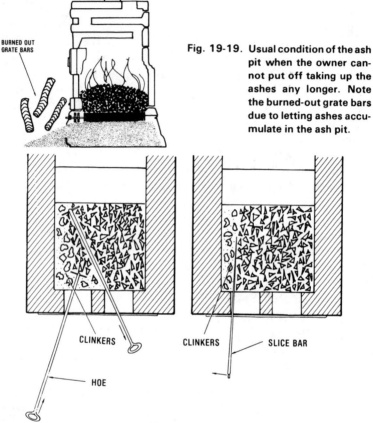

Fig. 19-19. Usual condition of the ash pit when the owner cannot put off taking up the ashes any longer. Note the burned-out grate bars due to letting ashes accumulate in the ash pit.

Fig. 19-20. Side method of cleaning the fire.

What are the requirements for complete combustion of coke?

Answer: Coke needs a greater volume of air per pound of fuel than coal, and therefore requires a stronger draft. The requirement is increased by the fact that coke can only burn economically in a thick bed.

What else should be considered?

Answer: It is necessary to take into account the size of the pieces. For a given combustion rate, about $1/3$ more grate area is required with coke than with coal.

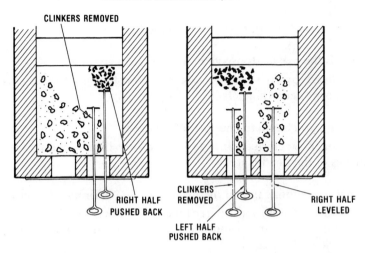

Fig. 19-21. Front-to-rear method of cleaning the fire.

What is the chief difficulty of firing with coal tar?

Answer: It is difficult to get a constant flow of tar into the furnace.

What are the causes of these difficulties?

Answer: Stoppages caused by the regulating cock or other appliance not answering its purpose and by the carbonizing of the tar in the delivery tube, thus choking it up and rendering it uncertain in action.

How may these difficulties be overcome?

Answer: Fix the tar supply tank as near the furnace to be supplied as convenient, and one foot higher than the tar injector inlet. A cock is screwed into the side of the tank, to which is attached a piece of composition pipe $3/8$-inch in diameter, 10 inches long. To this a $1/2$-inch iron service pipe is connected, the other end of which is joined to the injector.

What results are obtained by this arrangement?

Answer: It is found that at the ordinary temperature of the tar well (cold weather excepted) four gallons of tar per hour are delivered in a constant stream into the furnace.

How is the risk of stoppage in the nozzle overcome?

Answer: By the steam jet, which scatters the tar into spray and thus keeps everything clear.

How do you begin to charge the furnace?

Answer: Begin at the bridge end and keep firing to within a few inches of the dead plate.

What precautions should be taken as to timing firing intervals?

Answer: Time your firing so that there is never less than three to five inches of clean, incandescent fuel on the bars, equally spread over the grate area.

Fig. 19-22. The lazy bar. By its use the labor of cleaning the fire is greatly reduced, especially with flat grates and deep fires. When cleaning a fire, the lazy bar serves as the fulcrum for the hoe, instead of the bottom of the fire door. Thus the hoe or rake can be kept in a horizontal position when pushing the live coals back against the bridge wall or drawing clinker forward to the door, and at the same time the hoe does not rest on the fire and drag over it.

Boiler Installation, Startup and Operation

How should the firing be done?

Answer: Keep the bars constantly and equally covered, particularly at the sides and the bridge end, where the fuel burns away most rapidly. Also, see Fig. 19-23.

What should be done if the fire burns unequally or into holes?

Answer: It must be leveled, and the vacant spaces must be filled.

What should be done with large coals?

Answer: They must be broken into pieces not bigger than a man's fist.

What is the procedure with respect to a shallow ash pit?

Answer: It must be more frequently cleaned out. A body of hot cinders, beneath the bars, overheats and burns them.

How about the firing intervals?

Answer: The fire must not be hurried too much, but should be left to increase in intensity gradually. When fired properly the fuel is consumed in the best possible way and no more is burned than is

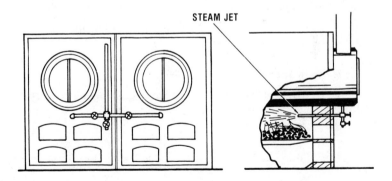

Fig. 19-23. One method of smoke prevention by means of a steam jet. The jet is located just above the fire door as shown. After each firing the jet is opened in a few minutes which prevents black smoke, reducing its density to a haze.

needed for producing a sufficient quantity of steam and keeping the steam pressure even.

How should the draft be regulated?
Answer: By the damper, not by closing the ash pit doors.

How should the fire be checked?
Answer: By the damper, and not by throwing open the fire doors.

On older boilers, how was steam used to help control smoke?
Answer: By releasing a short jet of steam in the furnace. See Fig. 19-23.

BOILER WATER

In taking an examination for engineer's license what is usually the first question the examiner asks?
Answer: "What is the first thing you would do on entering the boiler room?"

Well, what would you do?
Answer: I would find out if there was the proper amount of water in the boiler.

What is the proper amount of water?
Answer: The water level should be high enough to submerge all heating surfaces exposed to intense heat, the exact level depending upon the type of boiler.

Does the water gauge show the true level of the water in the boiler?
Answer: No.

Why not?
Answer: Because the density of the cold water in the glass is

Boiler Installation, Startup and Operation

greater than the hot water in the boiler, causing a thermal equilibrium upset (Fig. 19-24).

What is the indication that the water gauge is not working?

Answer: A stationary water level.

What should be done to correct this?

Answer: Open the drain cock wide until all water disappears from the glass, then close. If the gauge is clear of foreign matter, water will rise at once to proper level. See Fig. 19-25.

What other condition indicates that the gauge is working properly?

Answer: A "lively" water level; that is, the level will fluctuate depending upon the type of boiler. It is especially lively in a water-tube boiler.

How can the water level be expected to vary in a boiler?

Answer: It varies in accordance with the steaming rate of the boiler and changing loads. Water level will rise with heavier demand and fall with decreased demand.

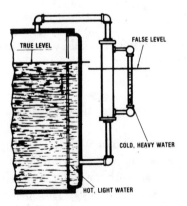

Fig. 19-24. False water level due to difference in density of the cold water in the gauge glass and the hot water in the boiler. The difference between the two levels is considerably exaggerated for clarity; in reality, this difference is very small and may be disregarded.

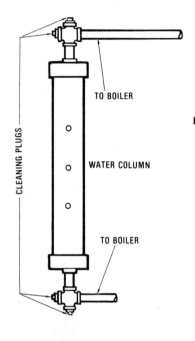

Fig. 19-25. Proper method of connecting a water column to a boiler. By the use of two crosses and plugs, the entire system is made accessible for internal cleaning which is important.

How should steam be raised in a vertical tubular boiler with through-tubes?

Answer: To prevent burning the tubes the boiler should be entirely filled with water, leaving a vent by raising the safety valve or blocking open the whistle valve to allow for expansion of the water as it is heated.

How high should the water be carried in a vertical boiler, and why?

Answer: As high as can be without causing wet steam, because it increases the efficiency of the heating surface and prolongs the life of the tubes. In the author's boiler, he has provided a separator, collector, and dryer to permit carrying an abnormally high water level.

What is the advantage of the abnormally high water level in the boiler?

BOILER INSTALLATION, STARTUP AND OPERATION

Answer: It protects the tubes and increases efficiency of the tubular heating surface.

How should water be fed to a boiler?

Answer: It depends upon the type of boiler. For instance, *constant feed* for a marine boiler and *intermittent feed* for a locomotive. See Fig. 19-26.

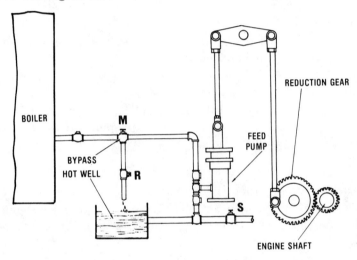

Fig. 19-26. Ideal method of constant feed especially in marine plants. The condensate delivered from the condenser to the hot well is pumped into the boiler by the feed pump. Since the latter of necessity has excess capacity, bypass valve M is provided so that the excess may be returned to the hot well; otherwise, the hot well would soon empty and the pump would force both water and air into the boiler. By close adjustment of valve M, the water is pumped into the boiler at the same rate it is delivered to the hot well by the air pump; thus, the inlet remains covered with water, preventing any air being carried over into the boiler. In time the water level in the boiler gradually falls, due to loss of water through stuffing boxes, whistle, and safety valve, and this may be made up by closing M, and opening makeup valve S. When the water rises to the normal level, S is closed and M again adjusted. By providing a small valve, (R) in the bypass, it is possible to shut off the bypass so that when the correct adjustment of M has been found, it need not be again disturbed. Valve M is used when regrinding is necessary for the needle valve R.

Why intermittent for a locomotive boiler?

Answer: In descending a long grade with throttle closed, the boiler may be filled to the top gauge cock and then shut off, so that during the ascent of the next grade where maximum amount of steam is needed, the evaporation is not reduced by the admission of cold feed water.

In marine practice what means are provided to prevent loss of feed water by the safety valve blowing off?

Answer: A "bleeder," or connection between boiler and condenser, as shown in Fig. 19-27.

How is the bleeder operated?

Answer: The bleeder valve is opened, allowing steam to blow into the condenser; the condensate is then pumped into the hot well by the air pump, thence back into the boiler by the feed pump.

What is necessary for bleeder operation?

Answer: Independent air and feed pumps, or in case of engine-driven pumps, the engine must be run slowly to relieve the condenser of condensate.

What are the indications that a feed pump is working?

Answer: The check or check valve noise each time the valve seats; also, the low temperature of the pump and feed line as indicated by feel and coating of condensate on the pump barrel and line. See Fig. 19-28.

What are the indications that the feed pump is not working?

Answer: No check valve noise and warm temperature of the feed line.

What does the warm temperature of the feed line indicate?

Answer: If the feed-water line is warmer than the temperature of the feed water, it indicates a backflow of water from the boiler due to check valves not properly seating.

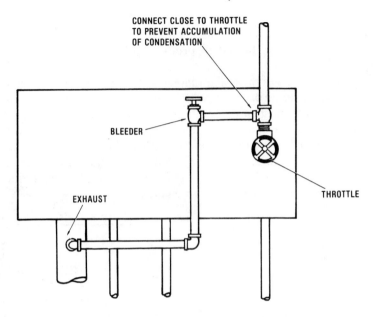

Fig. 19-27. Combination of bleeder and main steam pipe drain. It provides a means of preventing water loss by safety valve blowing, also for draining and warming the steam pipe before starting the engine.

If on entering the boiler room you would find the water out of glass, safety valve blowing off strong, and a good, hot fire under boiler, what should be done?

Answer: First, fuel and air supplies should be cut off or the fire should be smothered as quickly as possible with wet ashes, earth, or coal, closing the ash pit doors and leaving the furnace doors and damper open. If the water has not fallen below the level of either the crown sheet or other extended area of heating surface, the feed pump may be started with perfect safety (Fig. 19-29). If the water level is unknown, the boiler must be cooled down completely, carefully inspected, and repaired if necessary. If no part of the exposed metal is heated to redness, there is no danger except from a rise in the water level sufficient to flood the overheated metal. Hence, care should be taken that the safety valve is not raised so as to produce a priming that might throw the water over the overheated metal, and that no change is made in the working of either engine and boiler that shall produce priming or an increased pressure.

If any portion of the boiler plate is red hot, an additional danger is due to the steam pressure, which should be reduced by continuing the engine in steady operation while extinguishing the fire. If the safety valve is touched at such a time it should be handled very cautiously, allowing the steam to issue steadily and in such quantity that the steam gauge does not show any sudden fluctuactions while falling. The damping of the fire with wet ashes will reduce the steam pressure very promptly and safely.

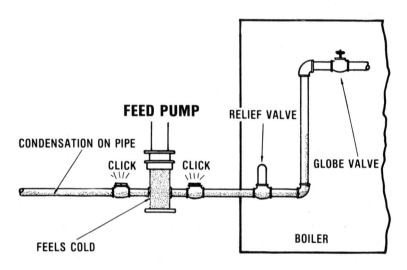

Fig. 19-28. Feed-pump piping and indications of proper operation.

What is priming, and what causes it?

Answer: A boiler "primes" when it lifts the water level and delivers steam containing spray or water, as in Fig. 19-30. It is usually caused by forcing a boiler too hard or by a too-high water level, or a combination of both these causes. When a boiler primes violently, it may be necessary to close all outlets to find the true water level.

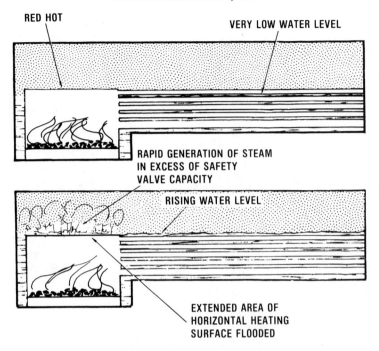

Fig. 19-29. Sectional view of a locomotive boiler with a dangerously low water level, illustrating why the feed pump should not be turned on. With the feed pump in operation the water level gradually rises. When it reaches the elevation of the crown sheet (now red hot), steam is generated quicker than can be discharged by the safety valve. Hence, the pressure rises, bringing more strain on the crown sheet already weakened by the excess heat, and therefore increasing the chances of an explosion.

What is foaming, and what causes it?

Answer: Foaming is severe priming or agitation of the water level due to dirty or impure water, as shown in Fig. 19-30.

In case of fire in the building, what should be done?

Answer: Haul the fire from under the boiler, start the fire tank pump and abandon the boiler room.

If fire has gained such headway that there was no time to haul fire from under boiler, what should be done?

Answer: Open the furnace doors, start the feed and tank pump full speed and abandon the boiler room. If the building has its own electric plant, that engine should be left running.

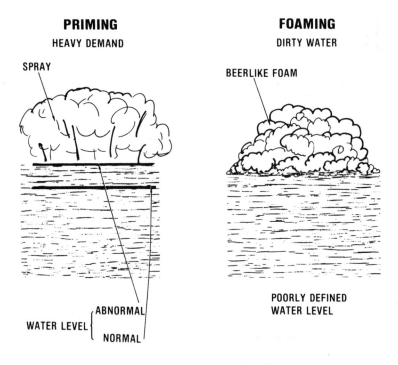

Fig. 19-30. Elementary boilers illustrating the difference between **priming** and **foaming** and the conditions which produce these effects.

ROUTINE OPERATION AND CARE

If the soot deposit on the heating surface is not removed, what happens?

Answer: The increasing accumulation of soot interferes with the draft and considerably reduces the amount of heat transmitted to the water.

Should a soot blower be operated in banked or idle boilers?

BOILER INSTALLATION, STARTUP AND OPERATION

Answer: No. Blowing soot can increase the chance for a furnace explosion, and parts of the boiler can become overheated.

Heavy soot deposits are an indication of what?

Answer: Incomplete combustion. Soot is formed from unburned combustion gases as they strike the relatively cooler metal.

How is the accumulation of soot indicated?

Answer: By an increase in the temperature of the chimney gases. Soot is almost the best insulator known, so keep the heating surface clean.

In furnaces operating on fuel oil, should soot blowers be operated just before securing a boiler after shutdown?

Answer: Yes, to remove unburned carbon from heating surfaces. The higher the sulfur content of the oil used, the more damage will result to boiler metal from the corrosive effect of the carbon, which in combustion with moisture forms sulfuric acid if allowed to remain in the furnace. Soot blowers are much less important on furnaces fired with natural gas; some don't even have them.

What is corrosion?

Answer: Chemical action which causes destruction of the surface of a metal by oxidation; rusting. See Figs. 19-31–19-34.

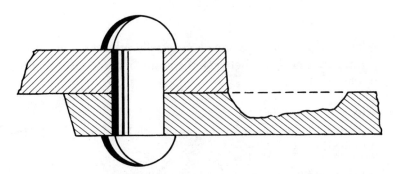

Fig. 19-31. Section of boiler plate at a riveted joint, showing the effects of corrosion of the metal. The normal surface of the plate is indicated by the dotted line across the opening of the corroded gap.

What part of the boiler is especially susceptible to corrosion and why?

Answer: Corrosion occurs especially along the water line because the water-borne air given up during evaporation, being heavier than the steam, collects in a layer between the water and the steam and attacks the metal at this point. See Fig. 19-32.

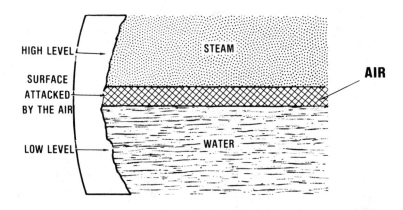

Fig. 19-32. Corrosion along the water line due to air. Even the most nearly pure water, when containing air, will cause corrosion. Air, since it is heavier than steam, forms a layer between the water and steam and rapidly corrodes the plate along the water line. Accordingly, care should be taken to prevent the feed pumps drawing in air.

What part of the furnace shell is especially subject to corrosion?

Answer: The space between the grate or burner area and the furnace shell.

Why?

Answer: Dampness and the lye of solid-fuel ash attacks the furnace shell.

What should be done?

Answer: Clean thoroughly with a wire brush on laying up and give it a coat of red lead mixed with raw linseed oil.

BOILER INSTALLATION, STARTUP AND OPERATION

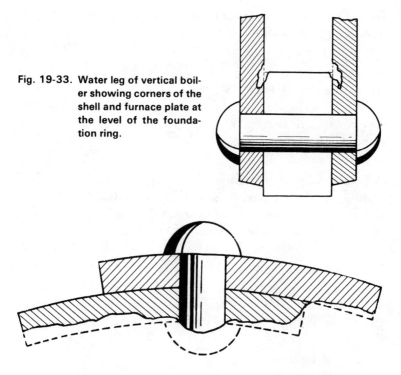

Fig. 19-33. Water leg of vertical boiler showing corners of the shell and furnace plate at the level of the foundation ring.

Fig. 19-34. Corrosion of the boiler shell along a seam. This eating away of plates is due to the chemical action of impure water. Gases absorbed by water, such as sulfurated hydrogen and carbon dioxide, are very active in the corrosion of boiler plates. Grease and organic matter also promote corrosion.

What is pitting?

Answer: A form of corrosion resulting in a series of minute holes or "pits" eaten into the surface of the metal to a depth of sometimes one quarter of an inch. Pitting is the more dangerous form of corrosion.

What condition especially causes pitting?

Answer: When a boiler is merely warm, pitting and corrosion occur to a much greater extent than when it is under pressure.

What is frequently used in boilers to prevent the corrosive action of water on the metal?

Answer: Zinc.

How is the zinc placed?

Answer: Slabs of zinc are suspended in the water by means of wires which are fastened to the upper part of the shell so as to make electrical connection.

Explain the action of the zinc.

Answer: The zinc forms one element of a galvanic battery and the metal of the boiler the other, with the result that the zinc is eaten away and the iron is protected.

Give a test for corrosiveness.

Answer: Fill a tumbler nearly full of water and add a few drops of methyl orange. If the water is acid and corrosive it will become pink, but if alkaline and harmless it will turn yellow.

What is incrustation?

Answer: A coating-over, the coating itself being commonly known as scale.

Describe the formation of scale.

Answer: Water, on becoming steam, is separated from the impurities which it may have contained, and these form sediment and incrustation (see Figs. 19-35–19-36). High concentrations of dissolved solids can cause foaming and priming.

What is the effect of oil carried into the boiler?

Answer: Minute globules of oil coalesce to form an oily scum on the surface of the water. Some globules come in contact with solid scale, forming particles and when they have the same specific gravity as the water, they rise and fall with the convection currents and stick to any surface with which they come in contact.

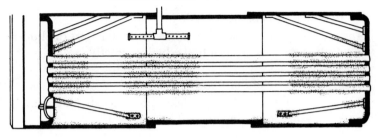

Fig. 19-35. Horizontal tubular boiler showing where scale accumulates most rapidly.

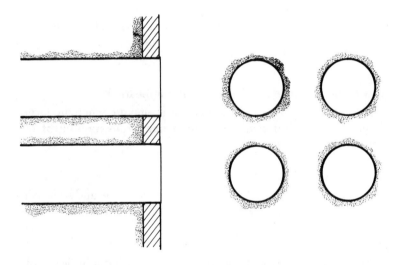

Fig. 19-36. Accumulation of scale between the ends of tubes and on the surface of tube sheets. One patch of thick scale in the proper place can retard the flow of water as thoroughly as a much larger area of very thin scale. It acts as a dam or baffle between the tubes of a fire-tube boiler and in the tubes of a water-tube boiler.

What is the proper method of eliminating scale?

Answer: Some reagent should be used which will precipitate the scale-forming ingredients and soluble salts, and convert them into insoluble salts without increasing the total amount of solids.

What two methods are employed?

Answer: By treating the feed water before it enters the boiler, or by putting the chemicals into the boiler directly.

What about the use of kerosene in boilers?

Answer: It prevents the particles of scale sticking closely together or adhering to the heating surface.

How is an accumulation of deposits prevented?

Answer: By a frequent use of the bottom blow.

What is a scum scoop?

Answer: Apparatus for blowing out water from the surface to remove fine particles of scale-forming foreign matter. See Fig. 19-37.

How is the presence of oil in the boiler indicated?

Answer: It shows in the glass water gauge when in large amounts, but a boiler may contain a dangerous amount of oil without this indication.

How may the oil be reduced?

Answer: By boiling out the boiler with kerosene and soda ash, or by scraping and scrubbing, or a combination of these two methods.

What precautions should be taken when treating a boiler with kerosene?

Answer: Keep all lighted candles, lamps or other fire away from the boiler openings, both when applying the kerosene and upon opening the boiler again.

What is "blowing off"?

Answer: The act of letting out water and steam from a boiler to carry off accumulated mud and scale.

When should the boiler be blown off and why?

Boiler Installation, Startup and Operation

Answer: The blow-off valves should be opened when pressure is lowest or in the morning while the fires are still banked, because a considerable amount of sediment will have settled down during the pressure drop.

Suppose the boiler is used both night and day?

Answer: Blow off at the end of the noon hour shut-down period or at a time when boiler pressure is lowest.

How should the blow-off valves be opened and closed and why?

Answer: Gradually, so as to avoid sudden shock. The valve should be opened fully. If the blow-off valve is preceded by a quick opening valve, it should be opened first, so the blow-off throttling valve is subject to the wear.

What precaution should be taken in closing the blow-off valve?

Answer: Observe the end of the blow-off pipe and see that the valve is tightly closed.

How much and how often should a boiler be blown down?

Answer: At least once a day depending upon the amount of sediment forming. Blow-down controls high chemical concentrations and high water and is also used for emptying the boiler.

Describe the most common blow-down procedure.

Answer: Check the water level. If the blow-down valve is pre-

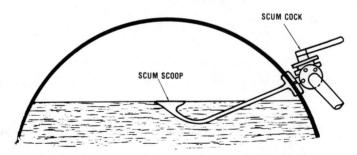

Fig. 19-37. Scum scoop or surface blow for blowing out fine particles of foreign matter floating on the surface of the water.

ceded by a quick opening valve, open it first. Then, slowly open the blow-down valve all the way and close it slowly. Never let the blow-down valve open until the water in the gauge glass drops out of sight, and never walk away from an open blow-down valve.

When pressure is being taken off a boiler during shutdown, why should the vent valve be opened when the pressure reaches 15 to 20 pounds?
Answer: To avoid a vacuum buildup.

What should be done before cleaning a boiler?
Answer: Drain boiler by opening the blow-off valves and the safety valve, allowing water to run out by gravity. Never drain a boiler while hot and not until ready to clean.

Describe a good method of washing a boiler.
Answer: Allow the boiler to gradually cool to disintegrate the scale and give time for most of it to settle on bottom. Remove manhole and handhole plates and play strong stream of water between the tubes and around the shell. See Figs. 19-38 and 19-39.

What precaution should be taken?
Answer: Don't place a lighted candle or lamp inside the boiler until after boiler is partly washed out.

Why?
Answer: If boiler compound or kerosene had been used in the boiler, gases from these substances might cause an explosion.

What should be done after the boiler is washed out?
Answer: It should be entered for inspection and all braces and stays should be carefully tested.

In looking for scale in horizontal boilers where should the lamp be placed, and why?
Answer: Beneath the tubes so that any scale which may be lodged between the tubes can be easily seen (Fig. 19-37).

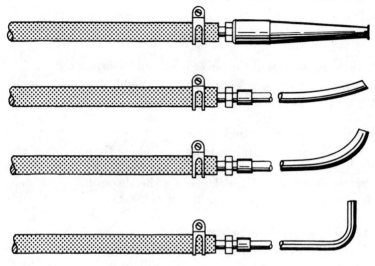

Fig. 19-38. Forms of nozzles for washing out boilers.

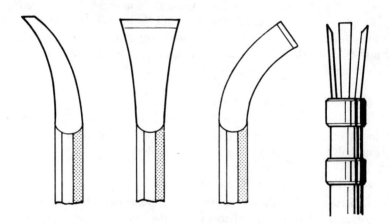

Fig. 19-39. Forms of chisels used in boiler cleaning.

What should be done after closing the boiler?

Answer: Pour several gallons of kerosene into the boiler and allow water to flow into the boiler very slowly. The slower the boiler is filled, the more opportunity the kerosene will have to attack the scale.

How is the interior of the boiler prepared for laying up?
Answer: After being cleaned it is thoroughly dried and must be kept dry to prevent corrosion.

How may the interior be kept dry?
Answer: Some engineers make use only of the natural circulation through upper and lower man- and handholes. Others place flat vessels filled with quicklime, for the absorption of all moisture, into the boiler and then close it up tight.

How do you "boil out" a new boiler?
Answer: A new boiler may be cleaned by boiling out with a mixture of soda ash and caustic soda. Fill the boiler with water to about the middle line of the water glass and add one pound of soda ash and one pound of caustic soda per 1000 pounds of water held within the boiler. Dissolve the chemicals thoroughly before introducing into the water. Close the boiler and start a light fire—sufficient to carry 5 pounds pressure in the boiler. Continue boiling two or three days. Then empty the boiler and wash it thoroughly with fresh water.

What is the procedure after a weekend shutdown?
Answer: A light fire should be carried for about one hour. Only in an extreme emergency should steam be raised in less than one-half hour, even with water-tube boilers.

Describe how to "cut in" a boiler.
Answer: When the steam pressure approaches 90% to 95% of the working pressure and before cutting the boiler into service, blow the water down to the proper level, if necessary. Before cutting in a boiler, open and leave open until the boiler is on the line, all drains in the connections between the boiler and the main header, especially the open drains between the two stop valves.

In general, in cutting in a boiler to a steam header already in service, the steam line between the boiler and the header is usually

Boiler Installation, Startup and Operation

a bypass line

warmed up by backfeed through the drip line or by means of the bypass valve, and the header valve then fully opened to allow the boiler to cut itself in automatically with the nonreturn valve. In case a nonreturn valve is not used, the boiler stop valve should, of course, be opened slowly when the pressure in the boiler and the steam line are approximately equal. The firing rate of the incoming boiler may have to be increased as it comes under load.

What is the procedure with two hand-operated stop valves?

Answer: With two hand-operated stop valves, open slowly (to avoid water hammer) to full opening the stop valve nearer the main header. When the pressure in the boiler is nearly equal to the pressure in the main header, open slowly to full opening the stop valve nearer the boiler.

What do you in the case of one hand-operated stop valve and one combined stop-and-check valve?

Answer: When the hand-operated stop valve is nearer the main header, open it slowly and preferably only a small amount at first. When the pressure of the boiler is still 10 to 50 pounds below the pressure in the header, slowly back off the valve stem of the combined stop-and-check valve sufficiently from the check to provide full opening of the check valve. In order to ensure that the check valve functions properly, always use it automatically for cutting in and cutting out boilers, provided that the main header is filled with steam at full pressure.

What is the procedure with one hand-operated stop valve and one automatic nonreturn check valve?

Answer: Open the hand-operated stop valve slowly when the pressure in the boiler is still 10 to 50 pounds below the pressure in the main header.

How do you raise steam on a boiler not connected to a common steam header?

Answer: In general, it would be more advisable to raise the steam pressure on the whole steam line at the same time, all drips being open.

What should be done after cutting in a boiler equipped with nonreturn valves having two independent outlets?

Answer: If the nonreturn valve is a combined stop-and-check valve, operate the rising stem after the steam pressure in the boiler has reached its working pressure, in order to ascertain whether or not the valve has opened properly.

What should be done in case of foaming?

Answer: Close or partly close the steam outlet valve long enough to determine the true level of the water in the water glass. If the level of the water in the glass is sufficiently high, blow down some of the water in the boiler and feed in fresh water. Use surface blow-off if installed. Repeat the alternate blowing down and feeding several times. If the foaming does not stop, bank the fire and continue the alternate blowing down and feeding. Determine positively the cause of foaming and adopt measures to prevent its recurrence.

What precautions should be taken in operating blow-off cocks and valves?

Answer: They should be opened and closed carefully and slowly; care should be exercised to make sure that when the valves or cocks are closed, they are closed tightly.

How about the frequency of blowing down?

Answer: The volume and number of blow-downs for a boiler should be governed by the condition of the water in the boiler and its effect on the quality of steam, and unless controlled by analyses made at intervals, the minimum blow-down shall be once in 24 hours.

Where the feed water is high in solids, if the load is heavy or fluctuates and wet steam is delivered, the blow-downs must be regulated so that concentration of the suspended matter and dissolved solids is kept below the point at which priming or foaming begins or wet steam is delivered.

What should be done when frequent blowing down does not overcome wet steam?

Boiler Installation, Startup and Operation

Answer: The boiler carrying the highest concentrations should be dropped, cooled, emptied, and refilled with fresh water. If this does not correct the condition and more than one boiler is in operation, follow the same procedure with the other units, taking them down one at a time in order of the highest concentration.

How often should automatic controls and instruments be checked?

Answer: In accordance with the manufacturer's instructions. They must be kept accurate and in good working order.

What are two methods of laying up an idle furnace?

Answer: Dry and wet. With the dry method, after the boiler is cleaned thoroughly, a drier or desecant is used to fill the boiler. So, the boiler must be sealed airtight and kept filled with the drying agent during the entire lay-up period. With the wet method, the boiler is filled with water and the proper chemicals, then fired to thoroughly mix the chemicals in the water. Obviously, this method can't be used if the boiler will be subjected to freezing temperatures during lay-up.

CHAPTER 20

Boiler Maintenance and Repairs

What should *not* be done in repairing boilers?
Answer: The work should not be entrusted to the local handyman or jack-of-all trades.

What should be done before making repairs?
Answer: If the boiler is insured, first notify the insurance company to send an inspector who will recommend how the repairs should be made.

What breakdowns are most likely to occur in a boiler?
Answer: (1) Cracks (fire cracks), (2) bulges (bags or blisters), (3) leaky or split tubes, (4) stay bolts or braces, (5) defective fittings.

SHELL REPAIRS

How may air leaks be found in settings?
Answer: Place the larger end of a funnel-shaped metal cone against the setting and hold a lighted candle at the opposite end.

Fig. 20-1. Air-leak detector cone and method of using.

Any air leak in the brick setting will draw in the candle flame (**Fig. 20-1**).

What is the cause of cracks?

Answer: Original defects of the material or faulty methods in manufacture.

Where do cracks usually occur?

Answer: They may show around rivet holes, at flanged corners, or between tube openings.

What are fire cracks? Are they dangerous?

Answer: Fire cracks run from the end of the boiler plate to a rivet or between rivets. When a crack runs from rivet to rivet, it is dangerous.

What is "grooving"?

Answer: Surface cracking of boiler plates.

Boiler Maintenance and Repairs

What causes grooving?

Answer: Expansion and contraction of parts too rigidly connected.

How should cracks be repaired?

Answer: If possible, the cracks should be drilled off at the ends by small holes, which will prevent further extension, and patches put over them, either on the inside or on the outside.

What is a "bag"? A "blister"?

Answer: A bag appears when metal is softened by heat. Pressure in the boiler pushes out the soft spot. Bags usually occur where scale buildup or oil has weakened the metal. Blisters are the result of defective steel, in which slag or some foreign substance has been rolled into the sheet during manufacture.

How may a bag or bulge be forced back into place?

Answer: If the plate is not burned, and not drawn thin, heat it red-hot by rigging up a gas furnace under the bulge and drive it up into its original position.

What should be done if the plate is burned?

Answer: The burned part must be cut out and a patch put on.

What may be resorted to if the bulged plate is not in convenient reach from rigid parts of the boiler for stay bolt connection?

Answer: Girders.

How should bulged circular furnaces be treated?

Answer: If not altogether collapsed, they may be put in such condition that they will safely carry a reduced steam pressure by attaching closely spaced circular rings of angle or bar iron, fitted in halves. Fig. 20-2 shows such a furnace being removed from a boiler for such repair.

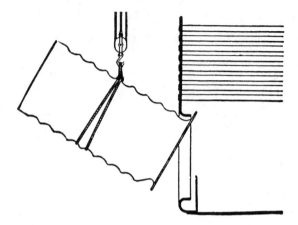

Fig. 20-2. Method of removing a cylindrical furnace from its position in the boiler.

BOILER PATCHES

What is a boiler patch?

Answer: A small piece of metal used to cover and strengthen a weak spot.

Name two kinds of boiler patches.

Answer: Soft and hard.

What is the difference between a soft and a hard patch?

Answer: A soft patch is a covering over a leak or defect which is fastened with bolts, as distinguished from a hard patch which is riveted. The difference between a soft and a hard patch is a favorite question with examiners, hence the applicant for license should understand this thoroughly.

How is a soft patch put on so that no caulking is necessary?

Answer: It may be fitted as closely as possible by heating and forging. The holes may then be drilled and those in the shell beneath the patch tapped or drilled large and through bolts used.

Boiler Maintenance and Repairs

Describe how to make a caulking tool.

Answer: Select a very heavy cold chisel (one made from ⅞-inch to 1-inch octagon steel is the best) and forge or grind the end off until it is about ¼-inch thick. Then round the end of the tool until it is a perfect half-circle ¼-inch in diameter and at least 1-inch wide.

How is caulking done?

Answer: Place the tool against the slope of a rivet about $1/16$-inch from the shell plate, and strike the tool with a hammer, keeping the tool inclined more or less parallel with the shell of the boiler. See Fig. 20-3.

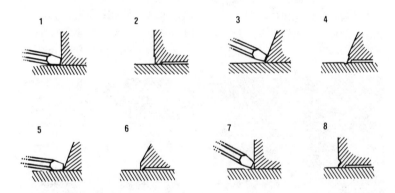

Fig. 20-3. Opinions differ as to the proper shape of the caulking tool and proper shape of the plate edges to obtain the best results in caulking seams. These drawings show the shape (1) of the caulking tool and proper way to hold the tool when the edge of the plate is square; result of caulking (2) a seam in this manner, giving fair results; the same tool improperly held (3) when the edge of the plate is beveled; the bad result (4) of caulking a seam in this manner; the proper way to hold (5) this tool against the beveled edge of the plate; the result obtained (6) after caulking—although the tool is held properly the result is not good; a better method (7), or proper way to hold a roundnose caulking tool against a square-edged plate; result after caulking (8). In caulking with this shape of tool, care must be taken that it is not too small, as it will then act as a wedge and separate the plate. Care must also be used in grinding the flat-nosed tool; if the tool has too much bevel, the lower edge will bite into the lower plate.

GASKET REPAIRS

What are gaskets?

Answer: Rings or discs placed between flanges to make a line leakproof.

What materials are used?

Answer: Composition fibers, cork, and rubber.

What purpose do they serve?

Answer: These materials, being softer than the metal flanges, compress under pressure of the bolts and compensate for slight variations in the flange face, making a leakproof seal. In Fig. 20-4, note that the repairman has carefully cleaned both flange faces to remove any trace of the previous gasket material. Gaskets must be checked to ensure that they fit properly and will not bind or

Courtesy Spaulding Fibre

Fig. 20-4. Gasket replacement on a bolted flange.

BOILER MAINTENANCE AND REPAIRS

wrinkle when the mating flange is fitted in place. It might be possible for a poorly fitted gasket to restrict the flow within the pipe, so a careful check there is also important.

How are they selected?

Answer: Gasket materials are designed and manufactured for specific purposes, such as sealing water or air, and one material should not be substituted for another as a makeshift repair.

Are they available in precut forms?

Answer: Yes, gaskets are produced in a wide variety of shapes. Gasket material is also available in sheet form for the skilled person to cut to the need of the job (see Fig. 20-5).

Courtesy Spaulding Fibre

Fig. 20-5. Gaskets are available in a wide variety of shapes.

TUBE REPAIRS

How do you make a temporary repair to a split tube?

Answer: By plugging with wooden or rubber plugs, or by a metal plug expanded by an explosive detonation inside the tube, using a process developed by Foster Wheeler called Detnaplugging™. See Figs. 20-6 and 20-7.

How is a tube stopper applied?

Answer: The fire must be drawn from the furnace and the boiler cooled so that a person can enter the combustion chamber to insert the plug or the tube sheet can be reached.

How does a wooden stopper stop a leak?

Answer: When pushed in the tube to the point of leakage the escaping steam and water cause the plug to swell, thus stopping the leak.

Where does grooving and cracking of tubes usually occur and why?

Answer: In the lower tubes at the combustion chamber end,

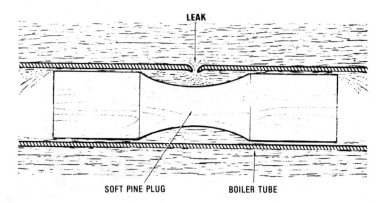

Fig. 20-6. Easy method of repairing a leaky boiler tube. A plug, shaped as shown, is pushed into the flue until it reaches the point of leakage, where the escaping steam and water cause it to swell, thus stopping the leak.

BOILER MAINTENANCE AND REPAIRS

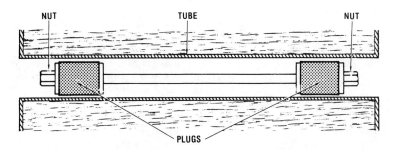

Fig. 20-7. Emergency device for stopping a leaky boiler flue. An iron rod having a thread at both ends is fitted with a nut and washer, then a rubber plug, another washer, and a length of gas pipe against the second washer. The opposite end of the rod is similarly fitted, after which the act of screwing up the nuts causes the rubber plugs to bulge, filling the tube.

caused by pumping comparatively cold water into the boiler in such a way that it comes in contact with the hot tubes.

In water-tube boilers, the process of removing tubes is about the same as in tubular boilers, but the tubes are taken out with less difficulty. Since the ends are not so easily reached a ripper (Fig. 20-8) is almost a necessity for splitting the tubes. The ends can then be folded in by the means of the long slender wedge shown in Fig. 20-9. A good plan is to cut two deep notches in the end of the tube, about an inch apart and then with a bar force the part of the projecting end of the tube between these notches, away from the tube sheet or header, using the edge of the handhole as a fulcrum for the bar. This produces a slight space between the tube and the tube sheet, into which the wedge may be started, and when it is driven home there is plenty of space left, so there is no danger of scoring the tube sheet with a ripper. If the defective tube is in the lower row, as is usually the case, its removal may be facilitated by cutting the tube into two or three sections with a three wheel pipe cutter, and removing the ends from the tube sheets from inside the furnace. From this point the operation is the same as that described for the horizontal tubular boiler, except that they are flared. See Fig. 20-9.

In ordering new tubes, the length should be accurately measured on a strip of wood or a piece of pipe passed through the old ones, allowing from $1/8$ to $3/16$ inch on each end for beading over.

Fig. 20-8. Ripper or plough chisel. It consists of a flat chisel about 1½ inches wide and $5/16$ inch thick. It is made convex or crescent shape on the end, and is ground square or blunt similar to a caulking tool.

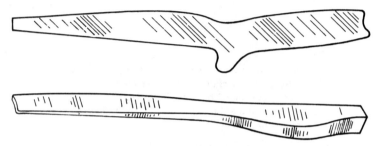

Fig. 20-9. Beading tool and long wedge used to fold tubes.

If more than one boiler is to be retubed, measure each one, as it will probably be found that there is a difference although they are all supposed to be the same. The ends of the new tubes are supposed to have been annealed before delivery, but it is doubtful if this has been done, so it is a good plan to heat them to a dull red and bury them in lime, allowing them to cool very slowly. This will make them much easier to expand and bead, and will reduce the liability of their cracking during these operations.

In retubing, describe how to remove a tube.

Answer: Cut a long slot in the tube from the front extending back about 8 inches to contract the tube. Cut the back end loose and pull out the tube. See Figs. 20-10 to 20-13.

Conventionally, tubes are replaced by using the tapered roller, tapered mandrel, step method expansion technique. Recently, a new, explosive method was introduced by Foster Wheeler for on-site tube replacement. Called the Detnaform® process, it employs explosives sized to achieve full contact between the tube and the hole surface through the depth of the tubesheet. Each charge consists of a length of detonating cord. That portion of the cord which is inserted into the tube is encased in a polyethylene

Boiler Maintenance and Repairs

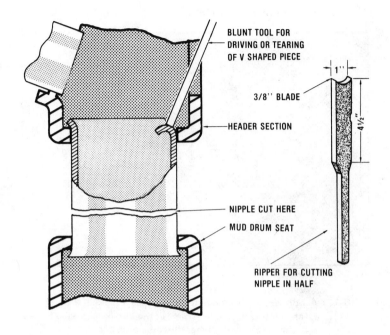

Fig. 20-10. Method of replacing nipples in water-tube boilers, B. and W. type.

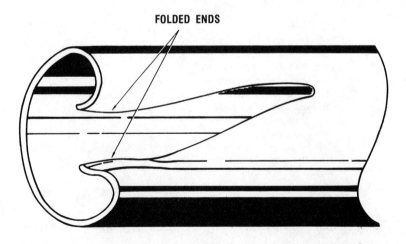

Fig. 20-11. Cutting and folding tube end to permit removal.

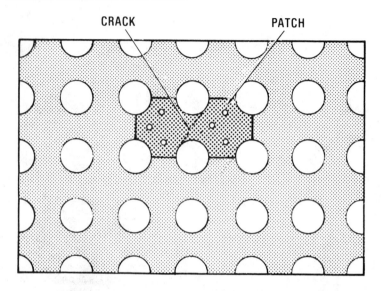

Fig. 20-12. A convenient method of repairing a cracked tube sheet. A patch, such as is shown at base of the cut, is shaped to fit neatly over the crack and between the tube ends. After smearing both surfaces with red lead, it is secured in place by tap bolts.

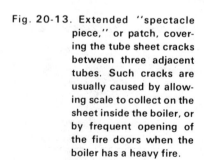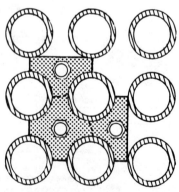

Fig. 20-13. Extended "spectacle piece," or patch, covering the tube sheet cracks between three adjacent tubes. Such cracks are usually caused by allowing scale to collect on the sheet inside the boiler, or by frequent opening of the fire doors when the boiler has a heavy fire.

cylinder. The plastic cylinder transmits the expanding force from the explosive to the tube wall. See Figs. 20-14 and 20-15. A lip at the end of the cylinder and its predetermined length prevents expan-

BOILER MAINTENANCE AND REPAIRS

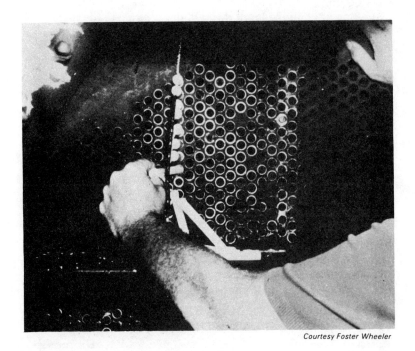

Courtesy Foster Wheeler

Fig. 20-14. Detnaform™ charges are inserted into the water tubes.

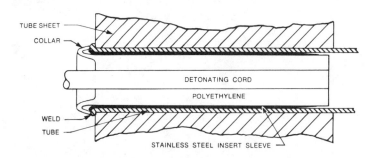

Courtesy Foster Wheeler.

Fig. 20-15. Drawing (not to scale) showing a Detnaform™ explosive insert positioned in a tube and tube-sheet assembly.

Power Plant Engineers Guide

sion of the tube beyond the tubesheet. The detonating cord extending beyond the lip of the cylinder is taped to the remaining lengths of the other explosive charges and to a detonator cap for simultaneous detonation.

The sequence of the Detnaform® operations is as follows: install charges into tube ends, attach the blasting cap, sound the personnel warning, ignite the fuse, detonate, ventilate the chamber, examine for possible misfires, then repeat the cycle if necessary. After the tube is installed, felt pellets are blown through it with compressed air to clean debris and the expanded plastic insert. See Fig. 20-16. The sound and shock waves associated with the process are contained by absorption boxes for open face tubesheet units, and by merely closing the manway on units having elliptical or hemispherical head closures.

TUBE CLEANERS

Tubes in heat-exchange equipment become fouled with different kinds of accumulations. Since any accumulation of foreign

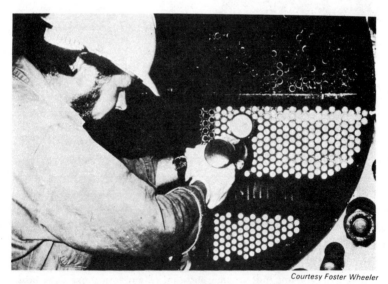

Courtesy Foster Wheeler

Fig. 20-16. Felt pellets are blown through the tubes after forming to remove debris.

Boiler Maintenance and Repairs

matter on the tube surface decreases efficiency, it is important that tubes be kept clean. Very soft deposits of mud and silt can often be washed out or removed by shooting rubber plugs through the tubes. Other mechanical devices are used as well.

How are deposits removed which are thin and not too hard?

Answer: By small tube cleaners which enter the tube.

What type cleaner is used for small tubes which have become scaled with a heavy and hard deposit?

Answer: A powerful, heavy-duty cleaner must be used.

Describe the cleaner.

Answer: The cleaner consists of a powerful motor—air, steam, or electric driven—which turns a shaft carrying a cutting tool. Since only the shaft and cutting head enter the tube, the motor can be made large enough to develop plenty of power.

How is the motor supported?

Answer: By a self-locking differential-rope block which is hooked to an overhead trolley by a hanger, as shown in Figs. 20-17 and 20-18.

Describe the operation of shifting the motor into position.

Answer: The trolley rolls on a length of pipe supported on a framework. In operation, the operator pushes the cleaner into the tubes.

Describe the various types of drives.

Answer: There are four types of driving motors. The direct-drive motor in Fig. 20-17 is powerful and adaptable. It may be operated on either air or steam at 80 to 100 pounds pressure. Speed may be varied by throttling. A two-speed, geared motor is air operated at 80 to 100 pounds air pressure. At low speed, sufficient power is developed to clean the heaviest scale, while high speed can be used to do a fast job on medium and light scale. There are also two types of electric drive, one of which is shown in Fig. 20-18.

Power Plant Engineers Guide

Courtesy Elliott Co.

Fig. 20-17. Elliott tube cleaner direct-drive air or steam motor. The rotor has four paddles, two of which are always under pressure. A ball thrust is used. Air consumption of the motor is about 150 cubic feet of free air per minute at 80 to 100 pounds pressure.

In the case of the direct-drive air or steam motor, water is introduced through a connection in the rear of the motor and in the front for the geared type.

What duty is performed by the water?

Answer: The water serves to cool the cutting tool and flush out removed scale.

What is the construction of the shafting?

Answer: It is made from extra-heavy, seamless tubing. The diameter of the shaft depends upon the size of the tubes to be

BOILER MAINTENANCE AND REPAIRS

Courtesy Elliott Co.

Fig. 20-18. Elliott tube cleaner with direct drive. Geared-drive types are available.

cleaned. All shafting is hollow so that the water can flow through it and out through the drill, brush, or cutter head. The shafting is made either in one piece or in sections, adapting it for use in tubes of different lengths or where there is not sufficient room in front of the unit to handle shafting in one length. Figs. 20-19–20-23 show the various types of drills, cutter heads, and brushes available for the cleaning job at hand.

For what are drills recommended?

Answer: For cleaning heavy encrustations from small tubes.

What is used for similar service in large tubes?

Answer: Solid cutter heads.

What use is made of the various stiff wire brushes?

Answer: They are intended for removing soft, light accumula-

Fig. 20-19. Tube cleaner brushes and cutting tools. *Courtesy Elliott Co.*

tions only. Brushes are fitted with couplings having a hole for the escape of flushing water. Drills and cutting heads are bored to permit the flushing and cooling water to flow through them.

How does it work?

Answer: Shooting rubber plugs through condenser tubes is speeded up by the use of the "jiffy gun." Pushing its nozzle into the tube end, as in Fig. 20-24, automatically applies air or water pressure which shoots the rubber plug through.

BOILER MAINTENANCE AND REPAIRS

Courtesy Elliott Co.

Fig. 20-20. Elliott 1300 series motor equipped with a vibrator head for use in removing scale from the outside of fire tubes.

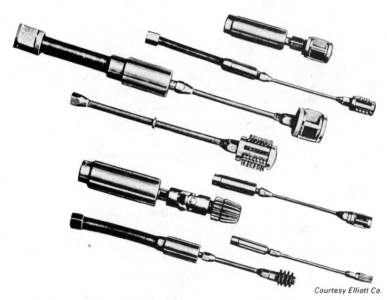

Courtesy Elliott Co.

Fig. 20-21. Various tube cleaner cutters, brushes, scrapers, and universal joints.

What is provided to stop the flow of air or water when the gun is not in use?

Answer: When not in use, a piston is held in closed position by a spring.

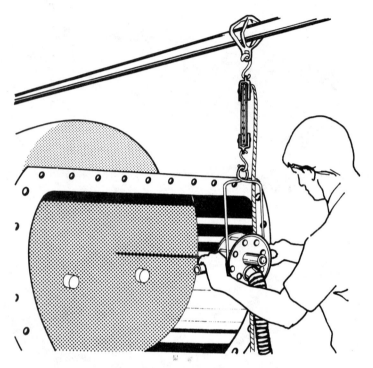

Fig. 20-22. Tube cleaner in operation showing the position of the operator and supporting gear.

What happens when the nozzle is pressed against the tube end?

Answer: The piston slides back, allowing water or air to flow through ports in the piston.

What is the working pressure?

Answer: The gun may be operated at pressures as low as 50 pounds.

LUBRICANTS

By definition, a lubricant is a substance used to reduce friction by preventing direct contact of rubbing surfaces, the substance being pressed out into a thin film on which the moving parts rub.

BOILER MAINTENANCE AND REPAIRS

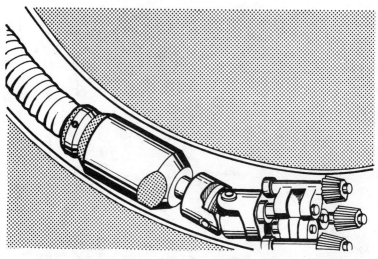

Fig. 20-23. Lagonda cleaner for straight and curved tubes. The 1300 series for curved tubes has a bullet-shaped body designed to slide around the tube turns with minimum interference and motion by use of a universal joint between motor and cutter head and by the use of a short length of hose between the main operating hose and the motor.

What should be noted in the selection of a lubricant?

Answer: A lubricant which would make a large shaft run smooth and cool in its bearings might be quite unsatisfactory if applied in some other place, as, for example, a light, high-speed spindle.

What is friction?

Answer: The resistance existing between two bodies in contact, which tends to prevent their motion on each other.

What causes friction?

Answer: It is partly due to the natural adhesion of one body to another, but chiefly to the roughness of the surfaces in contact.

Why is direct contact objectionable?

Answer: Because metal surfaces, although they appear smooth

Fig. 20-24. "Jiffy gun" type cleaner for shooting rubber plugs through condenser tubes. Standard nozzle fits all tubes from ½ inch to 1 inch.

to the eye and to the touch, are made up of minute irregularities which are visible when magnified.

How do these irregularities act?

Answer: When two metal surfaces are brought into contact, these minute irregularities interlock, retard the motion, and tear off the projecting particles.

What is the duty of a lubricant?

Answer: Its duty is to reduce friction.

How does a lubricant accomplish this?

Answer: By keeping the parts separate. It is pressed out into a thin film on which the moving parts rub, thus preventing direct contact.

What term is applied to the tearing off of small metal particles?

Answer: Wear.

What is the final effect of this?

Answer: If not remedied in time, it will result in freezing, or the adhesion of the surfaces to each other.

Desirable Qualities of a Lubricant

Name several important requirements a lubricant should possess.

Answer: (1) Body, (2) fluidity or viscosity, (3) freedom from gumming, (4) absence of acidity, (5) stability under temperature changes, (6) freedom from foreign matter.

Define body.

Answer: The body of a lubricant indicates a certain consistency in substance which prevents it being entirely squeezed out from the rubbing surfaces.

What is fluidity of a lubricant?

Answer: This term refers to a certain lack of cohesion between its different particles, which reduces the fluid friction.

What happens when a lubricant gums?

Answer: It loses its fluidity.

How about a lubricant that holds free acid?

Answer: The acid attacks the bearing surface, destroys its smoothness, and as a result, increases friction.

How about stability under temperature changes?

Answer: Lubricants should retain their good qualities, even when used under high temperatures, as in a steam cylinder, or when used under low temperatures, as in ice machines, or on exposed bearings. They should not evaporate, be decomposed by heat, or congeal by cold. They should retain their normal body and fluidity as much as possible in all temperatures.

Cold, Flash, and Burning Points

These are three critical temperatures of a lubricant which limit its application and which partly determine the conditions to which it is best suited.

Define cold point.
Answer: The temperature at which any given grade of oil will freeze or become cloudy.

Define flash point.
Answer: The temperature at which the oil gives off flammable vapors.

What is the burning point?
Answer: The temperature at which oil ignites.

Classes of Lubricants

According to form or state, lubricants may be classified as solid or liquid. With respect to composition, they may be termed animal, vegetable, or mineral.

What are the solid lubricants?
Answer: Graphite, soapstone, and various lubricating greases.
Graphite exists in both *crystalline* (or flake) and *amorphous* forms. It is also known as *black lead* and *plumbago*. Black lead usually refers to inferior grades of graphite; plumbago, to the Ceylon product; graphite, to the American product. Graphite may be used alone or in combination with oil.

The action of graphite is to fill the pores of the metal, making the rough surfaces smooth, rather than to form an intervening film to prevent contact. Strictly speaking, graphite is not a lubricant, but in filling the pores of the metal, it greatly reduces friction. One

Boiler Maintenance and Repairs

desirable quality of graphite when used in a steam cylinder is that its presence in a boiler does not produce any injurious effect.

Soapstone, also called *talc* or *steatite*, is used as a lubricant in the form of powder, or can be mixed with oil or fat. Mixed with soap, it is used on surfaces of wood working against either iron or wood.

The various lubricating greases are well adapted for heavy pressures under slow speed, but not for high speed, as their internal or fluid friction is considerable. The lubricating quality of grease may be improved by mixing it with graphite. An advantage of grease is that it does not run, so the machinery can more easily be kept in a clean condition.

What are the applications of liquid lubricants?

Answer: Liquid lubricants are used extensively for both internal and external lubrications.

What animal oils are used?

Answer: Animal oils such as sperm, whale, fish, lard, and neatsfoot oils are used to some extent.

How are they obtained?

Answer: By boiling or melting from the raw animal parts. As acid is sometimes used in the process of manufacture, animal oils are liable to have an acid reaction, and are then undesirable.

Sperm oil is an excellent lubricant. It does not become rancid or dry up, has good body, and is fluid with little internal friction. It is used for rapid-running parts where a high-grade lubricant is desirable without much regard to price. Whale oil is frequently used for external lubrication; it is a good lubricant at a moderate price. Fish oil is also employed to advantage by some engineers for external lubrication, while lard oil is used chiefly for mixing with other oils. Neatsfoot oil, on account of high price, is used in small quantities only for improvement of oils of poorer quality.

How are vegetable oils obtained?

Answer: By pressing the raw materials and cleaning out the cloudy, suspended fibers by treatment with acids. The color of the refined oils is from water white to light yellow.

Under heat, vegetable oils evaporate easily and are therefore employed only for external lubrication. They are gradually decomposed by the oxidizing influence of the atmosphere, and dry up; they are also inclined to gum. Olive, cottonseed, peanut, castor, and rape oil are all used to some extent.

Olive oil is a good lubricant; it neither dries up nor gums, but generally contains acid. On account of its high cost, it is frequently adulterated with cheaper oils. Cottonseed oil dries up less easily than others, and is consequently used sometimes as an admixture to olive oil. Certain grades frequently show an acid reaction, and are undesirable for lubricating purposes. Linseed oil dries up easily, and is therefore undesirable for lubrication; it is often found as an adulterant in other oils on account of its low cost.

How are mineral oils obtained?

Answer: By the distillation of petroleum. These oils are the most important lubricants, and with modern methods of manufacture, their price is relatively low. They retain their qualities well in the air, and if pure, do not gum or dry up.

Describe a test for clearness.

Answer: A sample of the oil is taken from a barrel that has been well-rolled and shaken. The glass containing the sample should be transparent, and the oil, if very cold, should be warmed. If of good quality, the oil will be clear. The amount of suspended matter in a light oil is determined by mixing and shaking it with a relatively larger quantity of gasoline.

How is the purity of an oil indicated?

Answer: By shaking a small quantity in a bottle with a quick, jerking motion, so as to produce air bubbles. If the oil is pure, the bubbles soon burst and disappear; if the test oil has been mixed with other oils, the bubbles will rise to the surface and collect.

How may animal matter be detected in oil?

Answer: About 1 ounce (29.5 cm^3) of the oil is placed in a bottle along with 2 teaspoonfuls of powdered borax. If, on shaking, a soapy deposit should form, the oil contains animal matter.

Boiler Maintenance and Repairs

Describe the acid test.

Answer: A small quantity of oil is mixed with warm water or alcohol, and tested with blue litmus paper. The paper will turn red if any free acid is present.

SAE Viscosity Numbers

These numbers constitute a classification of lubricants in terms of viscosity or fluidity, but without reference to any other characteristics or properties. The refiner or marketer supplying the oil is responsible for the quality of its product. The SAE viscosity numbers have been adopted by all oil companies and no difficulty should be experienced in obtaining the proper lubricant to meet requirements.

Choice of a Lubricant

There are several conditions that determine the choice of a lubricant for any given purpose, and the principal things to be considered are rubbing pressure, rubbing velocity, and temperature.

What is the requirement of a lubricant with respect to pressure?

Answer: For heavy pressure, it should have a good deal of body; for lighter pressures, there should be less body.

What is the requirement with respect to speed?

Answer: For high speed, a lubricant should possess good fluidity, while for slow speed, less fluidity is desirable.

What about temperature?

Answer: The heavier and nearer constant the load, the greater the amount of fuel burned, and consequently, the higher the engine temperature. If high temperatures are to be expected, a heavy-bodied rich lubricant, refined to meet severe heat conditions, would be desirable. If the operating temperatures are moderate, oils of greater fluidity will provide entirely adequate lubrication, and in fact, may be absolutely necessary to meet other conditions. See Fig. 20-25.

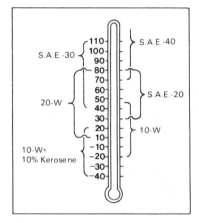

Fig. 20-25. Oil viscosities recommended for various Farenheit temperatures.

Discuss crankcase dilution.

Answer: By definition, this term means a thinning of the crankcase oil on account of certain portions of the gasoline or fuel leaking by the pistons and rings and mixing with the oil. Leakage of fuel or fuel vapors into the oil reservoir mostly occurs during the warming-up period when the fuel is not thoroughly vaporized and burned.

What about lubrication in general?

Answer: The subject of lubrication should receive the special attention and study of every engineer. It is quite important that the engine be properly oiled to avoid excessive friction, wear, and trouble. Owing to its importance, the subject is treated at length next.

LUBRICATION

In the modern power plant, electric motors and some other types of equipment have "permanently" lubricated bearings, which means that oil or grease need never be applied. On the other hand, some equipment must be cared for religiously if unnecessary downtime is to be avoided. It is wisest, of course, to follow the instructions of the manufacturer to the letter, using the type and

Boiler Maintenance and Repairs

grade of lubricant applied as recommended. Quick-disconnect pressure grease fittings are found on some equipment. Hand-operated grease guns or air pressure systems are used to force lubricants to points needing them. Again, the manufacturer's instructions should be followed.

A small amount of a well-selected oil or grease, properly applied, will go further in reducing friction than a much greater amount of an unsuitable lubricant improperly applied. Oiling an engine involves internal lubrication and external lubrication. The former includes oiling the cylinder and valves, and the latter, the external bearings. The lubricators described in this section are what could be described as traditional types found only on certain types of equipment.

Internal Lubrication

There are several kinds of lubricators for introducing oil into the cylinder. They may be classified with respect to their principles of operation as: gravity, and hydrokinetic, and force-feed.

Gravity Lubricators

Those working on this principle are called "plain lubricators," and there are two types: the invisible feed and the sight feed. The action of gravity lubricators depends on the displacement of the oil from a reservoir by condensation and its movement downward by gravity. In operation, steam passes through a central tube to the upper part of the oil reservoir, where it condenses. The water thus formed, being heavier than the oil, sinks to the bottom and displaces a corresponding amount of the oil.

Hydrokinetic Lubricators

The operation of lubricators of this class depends on two well-known principles of physics:

1. If a body (in this case, the oil) is acted upon by two unequal pressures, it will move in the direction of the greater force.
2. The specific gravity or the weight of a certain quantity of oil is less than the same quantity of water; so, the oil will rise to the top.

With these facts in mind, the operation of any hydrostatic lubricator may be easily understood. There are two forms of hydrokinetic lubricator known as the upflow and downflow types. In the former, the oil is seen rising drop by drop in the sight glass, while in the latter, drops of water are seen descending to the bottom of the oil reservoir.

How does an upflow lubricator work?

Answer: Steam from the main steam pipe passes into the connecting pipe above the lubricator and condenser, filling the condenser and part of the pipe above it with water. When the steam valve is opened, the sight-feed glass is also filled with condensation.

In operation (Fig. 20-26), when the condenser and steam valves are open, water from the condenser will pass down the central tube to the lower part of the reservoir; being heavier than the oil, it will stay at the bottom while the oil floats above. On account of the head pressure in the condenser tube, the water will continue to flow until the oil fills the upper part of the reservoir. When the feed valve is opened, the pressure due to the head of water will force the oil, drop by drop, through the nozzle in the sight glass. As soon as a drop of oil leaves the nozzle, it is no longer acted upon by this pressure, but rises because it is lighter than the surrounding water in the sight glass.

Describe the downflow lubricator.

Answer: In this type (Fig. 20-27) there are no internal pipes connecting with the condenser and sight feed. The sight glass consists of two glass discs inserted in the upper part of the reservoir.

How does the downflow lubricator work?

Answer: The operation is quite simple. When the condenser needle valve is opened, the water from the condenser flows through the passage, and, as can be seen through the sight discs, leaves the nozzle drop by drop. Being heavier than the oil, it descends to the bottom and displaces an equal amount of oil into the main steam pipe.

BOILER MAINTENANCE AND REPAIRS

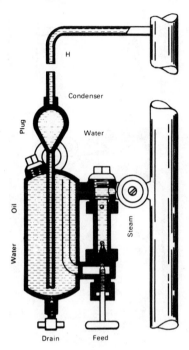

Fig. 20-26. Traditional upflow hydrokinetic lubricator.

What should be done before starting either type of lubricator?
Answer: Time should be allowed for the condenser and sight-feed glass to fill.

Mention a common fault.
Answer: Fouling of the sight glass.

What causes this fouling?
Answer: The nozzle frequently becomes covered with dirt and sediment from the oil, which makes the surface rough and causes the drops to adhere too long to the nozzle. This condition makes the drop become so large that it strikes the side of the glass in rising, which gradually covers the glass with particles of oil from the drop at each contact. This may be overcome by removing the glass and cleaning the nozzle both inside and out, rubbing it smooth with crocus cloth.

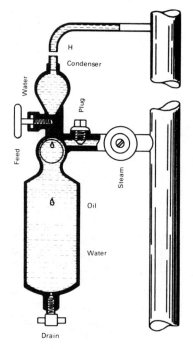

Fig. 20-27. Traditional downflow hydrokinetic lubricator.

Why does the nozzle sometimes become covered with dirt and sediment from the oil?

Answer: Sometimes the orifice in the nozzle is large for the kind of oil used. This causes large drops to form, which tend to foul the glass.

What periodic attention should be given to a lubricator?

Answer: It should be blown out occasionally so as to remove any dirt or sediment that may have accumulated in the small tubes and passages.

What attention should be given to the lubricator when the engine is shut down for a short time?

Answer: Close the feed valve, but the condenser valve in the upflow type should be left open. If both valves are shut, there will be no outlet; if the temperature of the oil should then rise, it will expand and possibly cause the reservoir to bulge or burst.

Force Feed: Oil Pumps

Hydrokinetic lubricators are affected by changes in temperature, causing them to feed too slowly in cold weather and too rapidly in warm weather. In an effort to overcome this defect, oil pump lubricators have been designed and put on the market. The device shown in Fig. 20-28 is typical of those used for years.

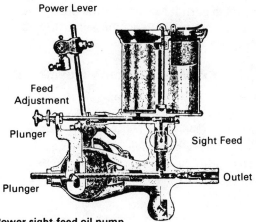

Fig. 20-28. Power sight-feed oil pump.

Name two kinds of force feed (oil pump) lubricators.

Answer: (1) The hand pump, which is used as an auxiliary to the main lubricator, and (2) the power pump operated by the engine, which is employed as the regular feed.

Describe the ordinary hand oil pump.

Answer: It has an oil reservoir with a removable strainer inserted in the central tube, and the filling hole is covered by a cap to keep out dust and impurities. The pump is of the single-acting, plunger type with ball valves. By reversing the positions of the plug and shank, the lubricator may be adapted to horizontal connection.

How does the power force-feed pump shown in Fig. 20-28 work?

Answer: Oil is drawn from the reservoir and forced through the sight glass by an upper plunger; it is then forced on through the check valves and outlet by a lower plunger. The amount of oil supplied with each stroke is regulated by the adjustable upper plunger. The lower plunger is made slightly larger than the upper to avoid any possibility of oil remaining in the sight glass.

Motion is imparted to the plungers by means of a ratchet wheel and cam, which in turn are moved by a lever connected to some reciprocating part of the engine. There is a hand attachment on the ratchet wheel to permit hand operation before starting the engine, or when more oil is needed momentarily while the engine is running.

External Lubrication Systems

The successful lubrication of the bearings of an engine depend in part upon the character of the appliances used to convey the lubricant to the wearing surfaces. There are several systems of external lubrication, the choice of which is governed by the type of engine and conditions of service. They may be classified as gravity, capillary, inertia, centrifugal, pressure, compression, and splash systems.

Gravity Systems

In this method of oiling, the lubricator is placed at a sufficiently high elevation to permit the oil to gravitate or flow to the bearing. Many of the sight feed cups work on this principle.

These cups are made in single or multiple units. A single-cup unit is shown in Fig. 20-29.

Capillary Systems

These include wick-feed lubricators and the chain and collar devices used in self-oiling bearings.

Describe the wick-feed lubricators.

Answer: Wick-feed lubricators are provided with one or more small tubes tapped oil-tight into the bottom of the reservoir and reaching to the top, as shown in Fig. 20-30.

Boiler Maintenance and Repairs

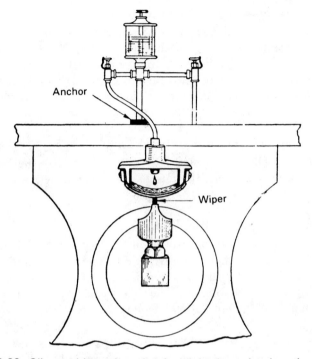

Fig. 20-29. Oil cup with two branches for lubricating wrist pin and guide.

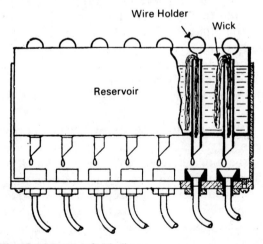

Fig. 20-30. Multiple wick-feed oil cup.

How does the wick-feed lubricator work?

Answer: To lift the oil automatically over the top of the tube, a wick is inserted with one end dipping into the oil in the reservoir. The end of the tube and wick must project below the bottom of the reservoir and a wire is attached to the wick to hold it in place. The wick, due to capillary attraction, becomes saturated with oil, which is then drawn from the reservoir drop by drop.

What may be said of wick-feed lubricators?

Answer: The system is used extensively on marine engines, and although very reliable, the rate of feed cannot be regulated so easily as the gravity feed.

Describe the endless-chain oiler.

Answer: An endless chain or a collar is used on what is known as a "self-oiling" bearing, as shown in Fig. 20-31. The length of the chain is such that it dips into an oil reservoir directly under the bearing, and in rotating with the shaft the ascending side carries with it a small quantity of oil which lubricates the shaft.

Does a ring oiler work on the same principle?

Answer: Yes. A pedestal, or bearing standard, is cored out to form a reservoir for the oil. The rings are in rolling contact with the shaft and dip into the oil at their lower extremity. Oil is then brought up by the rings and distributed along the shaft, where it lubricates the bearing. It returns to the reservoir by gravity and is used over and over. A drain cock is provided in the pedestal so that the oil may be periodically removed and strained to eliminate foreign matter and minimize consequent wear on the bearing.

What about crank-disc oiling systems used on compressors?

Answer: See Fig. 20-32. Oil is carried on the outer rim of each crank disc to the top, where it is diverted into an oil boat by a wiper. The boat, which is bolted to the main bearing cap, has four compartments; all oil enters the main compartment and overflows into three "supply" compartments. One supply compartment feeds the main bearing through a cored opening in the boat; another is piped to a sight-feed oiler at the crosshead guide and

BOILER MAINTENANCE AND REPAIRS

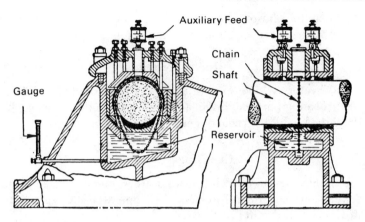

Fig. 20-31. Endless-chain oiler.

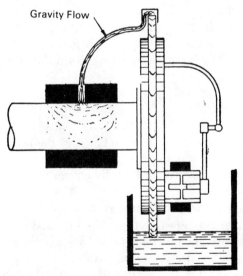

Fig. 20-32. Crank disc "pump" lubricating system for a compressor.

lubricates the guides and crosshead pin bearing; the last feeds oil to a centrifugal oiler which lubricates the crankpin bearing. The system is reliable, and splashing or agitation of the oil is reduced to a minimum if the proper oil level is maintained. Since the boat, the wiper, and the piping are all free from the crankcase oil guard, they may be adjusted with the guard removed. Handhole covers are provided for inspection while the machine is in operation.

Inertia System

Lubricators which operate by inertia are adapted to oiling a reciprocating part moving in an up-and-down direction, such as the wrist pin of a vertical engine.

Centrifugal System

For oiling crankpins, lubricators are sometimes used which depend on centrifugal force for their operation. A typical construction is shown in Fig. 20-33.

Describe this oiler.

Answer: In Fig. 20-33 a grooved ring is attached to the crank to distribute the oil from the sight feed. There is a connecting passage or duct through which the oil may flow from the groove to the crankpin bearing.

How does it work?

Answer: In operation, centrifugal force tends to throw the oil away from the center of rotation and against the bottom of the groove, forcing it through the duct to the bearing and lubricating the pin. This type of oiler is used mostly on center-crank engines.

What kind of oiler is used for a side-crank engine?

Answer: The pendulum-bob type. See Fig. 20-34. The hollow arm which supports the oil cup is bolted solid to the crankpin at one end and is journaled to the oil cup at the other. Since the oil cup is positioned concentric with the axis of crankshaft rotation and is weighted with a pendulum bob, it remains upright and in a steady position as the crankshaft rotates.

How does it work?

Answer: Oil drips from the cup, travels through the journal, and enters the hollow arm, where centrifugal force propels it toward the crankpin end. The bolt holding the arm to the crankpin has

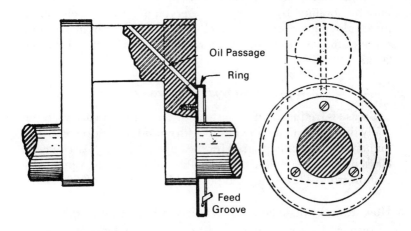

Fig. 20-33. Crankpin centrifugal oiler.

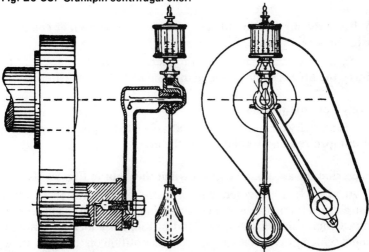

Fig. 20-34. Side and front views of pendulum-bob oiler.

been drilled both axially and radially to allow oil passage from the hollow arm to the crankpin, where additional passages conduct it to the bearing surface. The amount of oil-feed is regulated by a valve located on top of the cup.

Pressure Systems

The necessary pressure for forcing the lubricant to the bearings may be obtained by placing the reservoir at a suitable elevation, by employing air pressure in a closed reservoir, or by means of a force pump. An example of a pressure system is shown in Fig. 20-35.

Compression System: Grease Cups

Grease is frequently used instead of oil on some bearings, especially those which run slowly and with considerable pressure. The numerous kinds of cups used for the application of grease may be classed as hand operated or automatic. All grease cups operate by compression. Three simple types are shown in Fig. 20-36.

How does the plain cup shown at left in Fig. 20-36 work?

Answer: By screwing down the cap over the stationary bottom part, the grease is forced through the outlet to the bearing.

What is the application of the "marine type" shown at center in Fig. 20-36?

Answer: It is a desirable cup in places where it is necessary to force the grease some distance to the part to be lubricated.

How does the marine cup work?

Answer: By turning the handle attached to the piston, the grease is compressed and forced out of the cup.

Describe the automatic cup shown at the right in Fig. 20-36.

Answer: It is provided with a leather piston which is easily raised by a thumb nut whenever refilling is necessary. The thumb nut, which controls the spring and piston, has a locking arrangement which prevents it jarring from its position on the main stem.

How do you fill the cup?

Answer: To fill the cup, turn the thumb nut to the right, drawing the piston to the top; the cup is then unscrewed from the base and filled with grease. After reassembly, pressure is put upon the

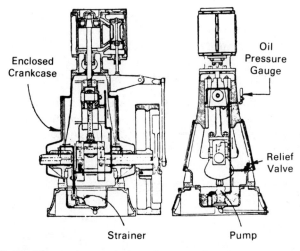

Fig. 20-35. Views of a vertical engine, showing the pressure system of lubrication.

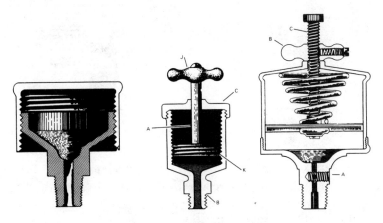

Fig. 20-36. Hand and automatic compression grease cups.

grease by screwing thumb nut to the top of the stem, thus allowing the piston to be pressed downward by the spring.

How is the rate of feed regulated?

Answer: By means of the screw plug valve, which is drilled to register with the feed passage according to the position of the

573

plug. If it is desired to stop lubrication while preserving the rate of feed, turn the thumb nut down to the top of the cup, thus preventing the further advance of the piston.

Splash System

With this method of lubrication, an enclosed frame is necessary. A quantity of oil is placed in the frame and maintained at such a level that the end of the connecting rod comes in contact with it at the lower part of the revolution, splashing it upon the working parts. This system is frequently used with high-speed, horizontal engines, as shown in Fig. 20-37.

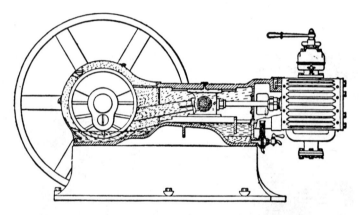

Fig. 20-37. Sectional view of horizontal engine, showing splash system of lubrication.

Upon what does the amount of oil splashed depend?

Answer: It depends upon the rate of speed and the depth of the oil in the frame. As with the pressure system, the oil should be frequently filtered and renewed.

CHAPTER 21

Mechanical Stokers

The basic reason for the wide adoption of mechanical firing has been the need for higher combustion efficiency. Today it is more important than ever before to get maximum steam power for every pound of coal or solid fuel fired. The common aim of all builders and operators of coal-burning equipment is to obtain more heat from a given amount of fuel. Various manufacturers have approached this problem from different angles and with different results. Consequently, some firing units are better than others. In seeking the best over-all operating economy, a vital need is the equipment design which most effectively carries out the basic principle of high combustion efficiency, the intimate mixture of coal and air. The fundamental problem of all firing is to bring together the right amount of fuel and the right amount of air at the right time, so that all combustible elements are burned. The most efficient stoker, then, may be defined as the one in which this intimate mixture of coal and air takes place continuously and most effectively.

What is a mechanical stoker?

Answer: A device constructed to automatically feed fuel to a furnace.

What is the advantage of a mechanical stoker?

Answer: Its use results in more efficient combustion owing to constant, instead of intermittent, firing.

What is the trouble with hand firing?

Answer: The frequent opening of the doors allows a large excess of air to enter which chills the flame, and the dumping of a quantity of fuel at each firing results in a smoke period until normal combustion conditions are restored.

Name the principle types of mechanical stokers.

Answer: (1) Overfeed, (2) underfeed, (3) rotary or sprinkler feed, (4) chain or traveling-grate feed.

OVERFEED STOKERS

What are the two types of overfeed stoker?

Answer: The front-feed and the side-feed types.

Describe a typical overfeed stoker of the front-feed type.

Answer: It has a step grate consisting of a series of stepped grate bars slightly inclined from the horizontal, and a dumping grate at the bottom which receives and discharges the ashes.

Describe its operation.

Answer: The grate bars are given a slow rocking or swaying motion by means of a small engine or motor. This motion gradually carries the fuel as it is burned toward the rear and bottom of the furnace. There is a flat ash-table at the bottom of the inclined grates.

What happens during the process of combustion?

Answer: The fuel is coked on the upper portion of the grates

MECHANICAL STOKERS

and the volatile gases, driven off in this process, are ignited and burned in their passage over the bed of burning carbon lower on the grates.

What are the features of the side-type as compared to the front-feed stoker?

Answer: Fuel is fed from the sides of the furnace for its full length or on the upper part of the grates which are inclined toward the center.

How are the inclined grates moved?

Answer: By rocking bars.

What does the motion do?

Answer: The fuel is gradually carried to the bottom and center of the furnace as combustion proceeds, where a clinker breaker grinds out and removes the refuse.

For what kind of fuel are overfeed stokers adapted?

Answer: For caking coal.

How is caking prevented?

Answer: The fires ordinarily carried are comparatively thin and the movement of the grate bars keeps them broken up and open, thus preventing caking.

UNDERFEED STOKERS

What is an underfeed stoker?

Answer: One in which the fuel is fed upward from underneath. See Fig. 21-1.

How are they classed with respect to the grates?

Answer: Horizontal and inclined.

How does the horizontal type work?

Answer: The action of a screw or worm carries the fuel back

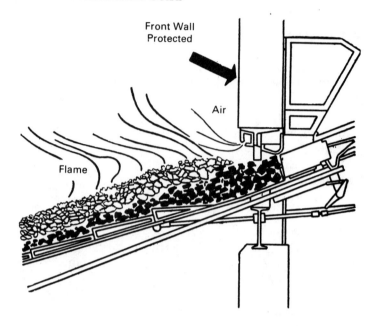

Fig. 21-1. Detail of an underfed, link-grate stoker. The link-grate motion reduces resistance to air flow by keeping the fuel bed porous. Air flowing through the grates is not smothered, but is allowed to support combustion by thoroughly permeating all parts of the fuel bed.

through a retort, from which it passes upward as the fuel above is consumed. The ash is deposited on dead plates on either side of the retort, from which it can be removed.

Describe the inclined type (see Fig. 21-2).

Answer: A piston or ram forces broken coal up through the fire bars into the fuel bed; tuyere boxes covered by perforated, cast-iron blocks are provided, through which air is forced by a fan. A series of rams are driven at variable speed by gearing. As the coal drops behind them, it is pushed upward and backward, working to the surface. Eventually, it moves past the extension grate to the dump plate, on which hot refuse accumulates and is dumped every three to six hours, depending upon the rate of working.

MECHANICAL STOKERS

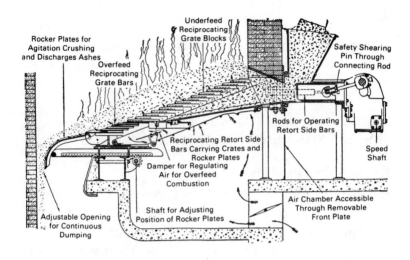

Fig. 21-2. Self-dumping, underfeed, inclined-grate stoker. The stoker is a multiple-retort, underfeed stoker with an incline of 20°. Below the retorts, the grate surface is continuous so as to burn fuel completely before it is pushed off the ash-supporting plates. The distinctive feature of this stoker is that the sides of the retorts reciprocate relative to the bottoms. This provides a means of moving the fuel uniformly along and out of the retort. It also provides a moving grate surface onto which the fuel is passed as it leaves the retort. The same movement serves to push the refuse across the rocker ash-sumping plates where it continuously discharges through the adjustable opening next to the bridge wall. To prevent the possibility of a breakdown due to foreign matter in the fuel, each plunger connecting rod has a safety device consisting of a standard ½ inch rivet so placed that it is double-sheared when the plunger strikes an obstacle.

What are two causes of excessive outage and maintenance?

Answer: (1) Sustained or frequent overloading of stoker and (2) operating with insufficient draft.

What causes fire in the retort?

Answer: It results from too thin a fuel bed, banking with insufficient fuel, or from running with an empty hopper. (See Fig. 21-3.)

What attention should be given to the ends of the grate bars?

Answer: Operate so that the ends of grate bars adjacent to the dump grates are always covered.

What should be done if the ends of the bars become bare?

Answer: Speed up the stoker until they are covered.

What attention should be given to the fuel?

Answer: Be sure that fuel distributes uniformly over grate.

What precaution should be taken when banking?

Answer: Feed in sufficient fuel. In case of long banking periods, renew the supply if necessary.

What should be done when a stoker is being shut down?

Answer: Sufficient ash should be fed to keep fire out of retort.

What are the points with respect to depth of fuel bed?

Answer: This is important. If too thin, fire may burn down into retort and damage retort sides. If too heavy, poor air distribution will result, causing spotty, uneven fire-holes in fuel bed, smoke, and reduced efficiency.

What is the correct depth of the fuel bed on a single-retort, underfeed stoker?

Answer: Correct depth above the top of the retort may be from 4 to 8 inches, depending upon analysis and burning characteristics of coal used.

What is the correct depth of the fuel bed on a multiple-retort,

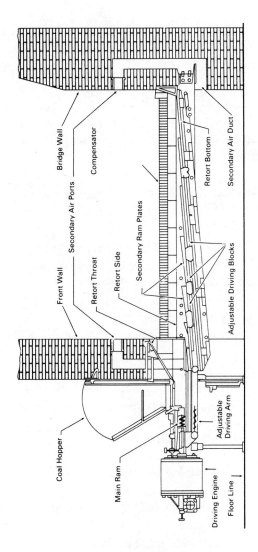

Fig. 21-3. Section through single-retort, link-grate stoker, showing parts.

underfeed stoker?

Answer: Correct depth of the fuel bed above the top of the tuyeres may be anywhere from 16 to 20 inches, depending upon analysis and burning characteristics of coal used.

What precaution should be taken with the retort?
Answer: Keep fire out of it.

What should be avoided in operating underfeed stokers?
Answer: Working of the fire should be avoided as much as possible. If fuel bed requires leveling off, a light rake or T-bar should be used on the surface of the fire. Never slice the fire, as is done in hand firing, by pushing a bar under the fire and raising it through the fuel bed.

What two instruments are essential for good operation?
Answer: A draft gauge and a CO_2 recorder.

What should be the draft for a single-retort stoker?
Answer: Operate with a slight draft just above the fuel bed—preferably 0.1-inch; not less than 0.05-inch.

What does plus pressure cause?
Answer: Excessive temperature at the grates and lower wall areas.

What should be the draft for a multiple-retort stoker?
Answer: Always operate with some draft above the fuel bed, preferably not less than 0.1-inch. The entire furnace, however, should be kept under negative pressure.

How often should a stoker be inspected?
Answer: All accessible parts of a stoker should be inspected daily. Inaccessible parts should be inspected at least twice a year.

What should be noted about lubrication?
Answer: The use of the right lubricants at sufficiently frequent intervals at all points requiring lubrication is essential if unnecessary outages and excessive maintenance costs are to be avoided. There should be a definite schedule for lubrication, regularly adhered to.

MECHANICAL STOKERS

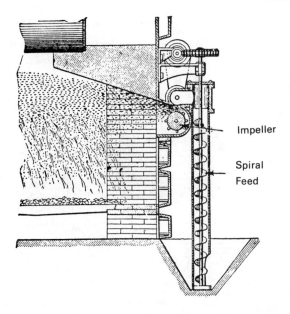

Fig. 21-4. Rotary or sprinkler stoker. In operation, the impeller catches the coal as it is fed and throws or sprinkles it upon the fire.

ROTARY OR SPRINKLER STOKERS

How does a rotary or sprinkler stoker work?

Answer: The revolving member catches the coal as it is fed from the hopper and throws or sprinkles it upon the fire, imitating the operation of a fireman in throwing fuel on the fire with a shovel, but doing it in a more efficient manner. See Fig. 21-4.

TRAVELING-GRATE STOKERS

What is a traveling-grate or chain stoker?

Answer: A type of overfeed stoker consisting of an "endless grate" composed of short sections of bars passing over sprockets at the front and rear of the furnace. See Fig. 21-5.

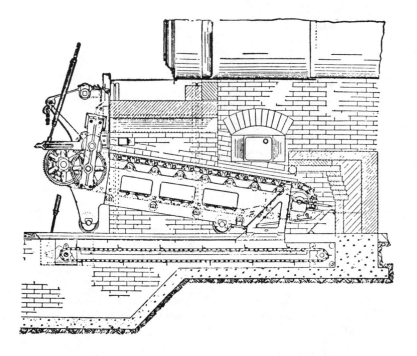

Fig. 21-5. Typical inclined, chain-grate stoker applied to a tubular boiler. Ratchet ash-drags which are recommended when the ash storage pit under the rear of the chain grates is inaccessible. An accumulating pit for storage of ashes is sealed by a trap door. The ash-drag consists of angle members riveted to a series of endless sprocket chains which ride on sprockets on front and rear shafts.

How is the coal fed?

Answer: It is fed by gravity onto the forward end of the grates through suitable hoppers.

How does the grate work?

Answer: The movement of the grate through the furnace is continuous. It is driven through a worm wheel keyed to the front sprocket shaft.

MECHANICAL STOKERS

Describe the combustion.

Answer: Fuel is ignited under ignition arches and is carried with the grate toward the rear of the furnace as its combustion progresses.

For what class of fuel is the chain stoker well adapted?

Answer: For burning low grades of coal running high in ash and volatile matter.

Mention one outstanding feature of chain stokers.

Answer: Cleaning of the fire is continuous and automatic, so no periods occur when smoke will necessarily be produced.

What is the difference between a traveling-grate and a chain stoker?

Answer: The traveling-grate stoker differs in structural design from the chain-grate stokers but functionally, it is the same.

How is the correct fuel-bed thickness, grate speed, and air pressure determined?

Answer: By experimental operation.

How about the supply of coal to the hopper?

Answer: Keep hopper not less than half full. Wherever possible, use mechanically operated swinging-spout, traveling lorry, or nonsegregating spreader chutes.

What happens if hot ashes are allowed to gather close to the rear of the stoker?

Answer: They may cause warping of the grate surface and rear shaft.

What precaution should be taken in banking?

Answer: Never bank a hot fire. Always burn it down first.

What are the points relating to draft?

Answer: In general, the same as for underfeed stokers. A proper supply of air should be maintained at all times.

What attention should be given to the grate?

Answer: Distorted bars or rods may cause excessive damage.

How about worn grate-bar ends?

Answer: Badly worn grate-bar ends and driving chains, burned ledge-plates, or the combination of these conditions, can cause excessive air leakage along the grate line.

What is the result of this condition?

Answer: The resulting "blowtorch" action is very destructive to parts in the area.

How may the life of chains be prolonged?

Answer: By reversing them when badly worn.

Fig. 21-6. Detail of multiple mechanical/pneumatic spreader-stoker.

SPREADER OR SPRINKLER STOKERS

How does a spreader stoker work?

Answer: It spreads coal over a fire in such a way that the larger pieces are distributed over a grate, while the finer particles burn in suspension. This theory utilizes the best features of both stoker and pulverized coal burner operation to burn coal like oil. This stoker

MECHANICAL STOKERS

mechanically carries the coal into the furnace and then sprays the fine particles into suspension while spreading the larger pieces of coal uniformly over the grates. (See Figs. 21-6 to 21-8.)

What are the two main causes of excessive outage and maintenance?

Answer: Sustained or frequent overloading of the stoker and operating with insufficient draft, resulting in positive furnace pressure.

How should the fuel bed be carried?

Answer: Keep it level and thin, with little more than a minute's supply of coal on the grate.

How can this be checked?

Answer: Easily. By shutting off the coal feed with the forced draft on, the fire should be ready to dump in about one minute.

What happens with a heavy fire?

Answer: A heavy fire will smoke and form clinkers. Clinkers frequently cause burned and broken grate sections.

What adjustments are made for an even flow over the entire grate?

Answer: Adjust distributor speed and circular tray.

What happens with a heavy fire at the bridge wall?

Answer: It will not only form clinkers, but will cause excessive erosion of brick work.

What happens with a heavy fire at the front?

Answer: It tends to overheat the arch, press, and distributing mechanism and may injure the stoker front.

What should be done when fuel is changed in type, size, or moisture content?

Answer: Always check the feeder speed adjustment. Adjust the air supply to suit every change in fuel supply.

What size lumps of coal should be used?

Answer: From ¾ to 1¼ inches—not over.

What is the objection to using the coarser sizes?

Answer: They cause excessive wear on the feeder and distributor mechanism and tend to overload and clinker the fuel bed.

How about cleaning?

Answer: Clean fires and ash pits at regular intervals, preferably twice each shift. This keeps fires in good condition and ready to handle load swings.

What happens with accumulated refuse in the ash pit?

Answer: It may cause damage to grates and the grate mechanism.

What should be done when banking a spreader-stoker—fired unit?

Answer: Be sure to leave a layer of ash in the grate to protect it. Cut down on supply, reduce distributor speed, and feed in coal for banking on the front end of the stoker only. Then cut off the forced draft and stoker feed. Maintain a slight draft over the fire.

What draft should be carried?

Answer: In general, the same as for single-retort, underfeed stokers.

What precaution should be taken with respect to the shear key?

Answer: Use only the recommended shear key.

Why?

Answer: The substitution of other material may result in overloading or breakage of stoker mechanism.

STOKER OPERATION

How to Start the Fire

When starting up a new installation, it is well to start with a wood fire or a slow coal fire to dry out the brick or plastic lining of the firebox. When ready for operation, the coal feed is turned on at the control panel, along with the coal blower. (The forced-draft fan is not turned on.) After about an inch-thick layer of coal has been distributed over the grate, both coal feed and coal blower are stopped and a wood fire is built on top of the fuel bed. The forced draft is then started and, after a short interval, the coal feed and coal blower are also started. All of the switches on the control panel are then set to "automatic" and the stoker is allowed to run under the direction of the pressure control or combustion control, as the case may be.

How to Bank the Fire

When shutting down overnight, the fire may be banked by placing four or five shovels of coal inside each fire door. Coal feed, coal blower and forced-draft fan are all shut off, and the stack damper is closed. This will conserve the furnace heat and simplify starting up again next morning.

When ready to start operations again, the coal piled up inside the doors is spread out with a hoe, the stack damper is opened, and the stoker started up. The coal remaining from the banked fire along with the heat from the refractory lining of the furnace should be enough to cause instant ignition of the fresh coal being fed into the furnace by the stoker. It is possible to get up steam pressure quickly by this method.

Regulation and Drafts

Stoker equipment is supplied either for intermittent operation (governed by a pressure switch) or modulating (full floating) operation (governed by a combustion control system and variable speed drive on the stoker coal feed.)

With intermittent operation the pressure switch is set at the desired operating pressure. The stoker then starts when the pressure drops below this point and stops when the pressure reaches a point a few pounds above this. A time switch is also supplied to maintain a fire during periods of light load when the pressure switch may possibly not call for the operation of the stoker for a length of time sufficiently long for the fire to go out. In such a case the time switch will cut in every 3 minutes and will operate the stoker for adjustable periods up to 1½ minutes, permitting the feeding of just enough coal to keep the fire going.

An electric damper regulator is often used with this type of operation. When the pressure switch on the control panel closes due to a drop in steam pressure below the predetermined point, the damper regulator motor starts, opening the stack damper. Ten seconds later, after the damper is fully opened, switches on the control panel start the coal-feed, coal-blower, and forced-draft fan motors.

When steam pressure reaches the desired point, the pressure switch opens, stopping the coal-feed, coal-blower, and forced-draft fan motors and again starting damper regulator motor.

This runs for ten seconds, closing the stack damper, conserving furnace heat, and saving much valuable fuel. Notice that there is no possibility of stoker or forced-draft fan starting *before* the damper is wide open. Thus, there is no possibility of "flare back" through the fire doors.

When the stoker is governed by an automatic combustion-control system, the power cylinder of the master regulator is connected to the variable-speed pulley platform of the stoker, thus changing the speed of the coal-feed drive according to the demand for steam.

It is also connected to the control louvers of the forced-draft fan so that the amount of air fed under the grates is in proportion to the amount of coal being burned. The proper overfire draft is usually maintained by means of an overfire air-regulator which maintains a fixed draft-setting in the firebox by adjusting the boiler damper.

The maintenance of proper undergrate pressure and overfire draft is extremely important if efficient operation is to be obtained. The louvers of the forced-draft fan should be adjusted so that the pressure in the plenum chamber under the grates will be between 0.3 and 0.5 of an inch. The overfire draft should be set at between 0.03 and 0.06 of an inch by means of the boiler damper. These settings are to be made when the stoker is operating at normal load. Once set, the stoker will maintain approximately these conditions because the design of the grates is such that the grates themselves furnish about 80% of the total resistance of the grates and fuel bed; thus, any change in thickness and consequent increase in resistance through the fuel bed means little in the overall resistance of grates and fuel bed.

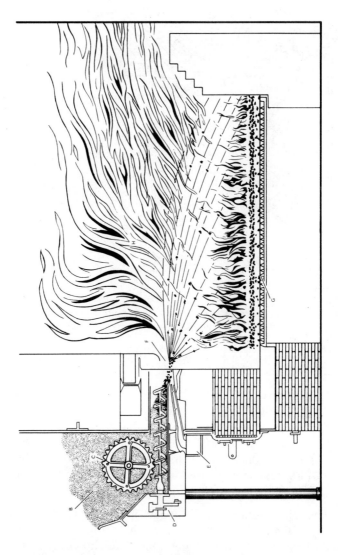

Fig. 21-7. Sectional view of multiple mechanical/pneumatic spreader-stoker showing its large coal hopper capacity (B), coal agitator (C), coal feed ratchets (D), overfire air control (E), ignition zone (F), grate bars (G), fire overfire (H).

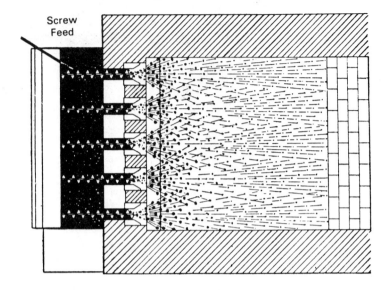

Fig. 21-8. Plan view showing detail of hopper and multiple feeders of mechanical/pneumatic multiple spreader-stoker.

For most efficient operation of stoker equipment, the minimum equipment required is:

1. Draft gauge having a scale capable of registering from zero to at least 0.1 inch of water.
2. Pressure gauge having a scale capable of registering from zero to at least 1.0 inch of water for each section of the plenum chamber. On multisection wind box installations, one gauge may be used with piping and valves which makes it possible to switch from one section to the other.

The coal-feeding capacity of each individual feed screw may be easily changed by merely changing the setting of the control knob at the side of the feeder ratchet. This knob, shown at D in Fig. 21-7, is pulled out and turned to the desired setting. Each number on the dial refers to the number of teeth that the ratchet will pick up at each stroke of the drive bar; thus, the setting may be arranged so that the ratchet pawl picks up from 2 to 8 teeth per stroke.

MECHANICAL STOKERS

How to Clean Fires

When cleaning the fire with a single-section grate, that is, a grate that does not have any division plates or dampers in the plenum chamber, the proper procedure is to first stop the coal feed on the feeders on one side of the stoker by throwing back the ratchet pawls on these feeders.

Allow the machine to operate for about three to five minutes, permitting all of the coal on this side of the grate to be burned out. Then open the fire door and, with a hoe, weed out the ash from this side right down to the bare grate. Then close the door and throw the ratchet pawls back down, starting the coal feed to this side of the grate again.

The heat from the fire on the other side of the grate, plus the heat from the refractory walls, will cause quick ignition of the fresh coal being fed into the firebox.

It is not necessary to spread or work the fire in any way. If the job is equipped with dumping grates, the procedure is the same except that the grates are dumped after the coal has been burned down, instead of cleaning with a hoe.

When cleaning the fire on the multisection grate, that is, one having two or more sections in the plenum chamber, the coal feed is stopped on the feeders opposite one section of the grate. After a few minutes, the forced-draft air to this section is also cut off by means of the plenum chamber damper-control rod which protrudes through the front furnace wall. The first section of the grate is then either cleaned with a hoe or dumper, as the case may be, and the coal feed and forced-draft air again turned on in this section. The same process is then repeated for the other sections.

The cleaning of fires is simple, for it should not be necessary to use any tool but a light hoe. There should be no need for working the fire in any way. It is well to clean the fire when the thickness of the fuel bed (firebed plus ash) reaches a height of approximately four or five inches.

CHAPTER 22

Pulverized Coal Burners

It is agreed by all those who are familiar with combustion that radiant heat is most desirable and that it is about eight times as effective as heat by convection in a combustion chamber. This is the important reason for the use of pulverized fuel, as its luminous flame gives off radiant heat.

What is radiant heat?
Answer: As an example, the sun gives off pure radiant heat, but when clouds intervene, the heat transfer is materially reduced. Therefore, the greatest care should be taken to prevent smoke in a furnace, for it obscures the radiant heat of the flame and retards proper heat-transfer in the furnace.

In a furnace, what part of the heating surface absorbs radiant heat?
Answer: Every surface exposed to light of the flame.

How has the design of boiler furnaces been influenced by the use of radiant heat?

Answer: In power plant work, this was strikingly demonstrated by the introduction of furnace water-walls and by rebaffling arrangements to expose a greater amount of heating surface in the furnace to the radiant heat such as that which comes from the use of pulverized coal.

Give another reason for the use of water walls.

Answer: They are for the protection of refractory walls and have been made necessary by the use of pulverized coal with its radiant heat and high combustion rates. Combustion rates of 40,000 to 200,000 Btu per cubic foot have been attained without difficulty with all refractory surfaces protected by water walls.

What is the nature of a nonradiant or "blue flame," such as a gas-stove flame?

Answer: It does not radiate heat waves, but must be in actual contact with a surface to transfer heat efficiently. Moreover, it delivers very little heat except to the surface it is touching.

POINTS ON COMBUSTION

It is conceded by all that gas is the best fuel for obtaining nearly perfect combustion. Accordingly, the nearer we approach the gaseous state in coal pulverization, the nearer we approach complete combustion. It must be evident, therefore, that coarse coal is not as efficient in either a small or a large furnace.

In the combustion of coal, there exists on the surface of the hot carbon a film of gas through which oxygen must pass before it can react. The rate of combustion and, therefore, the efficiency of the furnace, depends upon the rate at which oxygen diffuses through this stationary gas film which covers the carbon surface. It is also evident that the finer the pulverization, the greater the number of carbon surfaces exposed to the action of the oxygen, thus producing complete combustion in the least time.

What is necessary to secure complete combustion?

Answer: All of the carbon must combine with the oxygen before the coal particles are drawn by gravity out of the path of the flame.

A cubic inch of coal ground to a fineness so that 60% to 75% of it will pass through a 300-mesh screen increases the surface exposure from one piece having six surfaces to more than 27,000,000 particles with innumerable sides or surfaces. Careful estimates indicate that the area of the surface exposed is beyond 5000 square inches. Therefore, by fine pulverization, we closely approach a true gaseous state, and thus secure almost complete combustion.

The advantage of using coal when pulverized to an impalpable powder, and the possibility of using the inferior grades that could not otherwise be used, has brought out many ideas as to the best means of utilizing pulverized fuel.

FIRING PULVERIZED COAL

The firing of coal in pulverized form involves primarily the functions of:

1. Pulverization,
2. Mixing of coal and air,
3. Delivery of coal-air mixture to burners,
4. Combustion.

The equipment for performing these diverse functions is commonly referred to as a *pulverized coal system*.

Three methods of firing are available: (1) horizontal, (2) tangential (also known as "corner"), (3) vertical.

Oil or gas may be used as alternative fuel or in combination with coal.

How is the pulverized coal fed to the furnace?

Answer: Blowers or compressed air are used to force the powdered coal, along with the proper amount of air, into the burner.

How does pulverized coal compare with oil and gas as far as control is concerned?

Answer: It is as easy to regulate the mixture of fuel and air with pulverized coal as it is to regulate it with oil and gas. Compared to the conventional way of burning coal, it is possible to adjust the fuel/air mixture to a much finer degree with pulverized coal.

What is important to any application?

Answer: The furnace design. Furnaces may provide for the discharge of ash in either dry or fluid form. They should be water-cooled to assure continuity of operation as well as minimum maintenance. Water-cooled furnaces also add substantially to the steam-generating capacity or permit a smaller amount of convection evaporative surface for a given output.

PULVERIZERS

The pulverization of materials may be accomplished by various types of pulverizers or mills, such as:

1. Attrition mill (bowl),
2. Impact mill,
3. Combined impact-and-attrition mill.

Attrition (Bowl) Pulverizer

Crushed coal from the rotary feeder falls to the center of a bowl-shaped section which is keyed to a vertical shaft. As the bowl revolves, the coal is thrown between the grinding surface of the bowl and the rollers which are carried by journals attached to the mill housing. During pulverization, the fine particles rise to the surface and are picked up by the current of air which is blown upward around the bowl by the action of the *exhauster*. The coal-and-air mixture enters the *classifier*, where the direction of flow is abruptly changed, automatically sending the heavy pieces to the bowl for further grinding. The remaining mixture, with coal of the desired fineness, passes on through the exhauster and piping to the burners.

All foreign matter, although supposed to be removed by screens and magnetic separator, sometimes passes through the feeder to the mill. Any tramp iron that does not reach the mill is ordinarily discharged by centrifugal force over the rim of the bowl into the air chamber where revolving sweeps discharge it through an opening to a spout with a counterweighted door. Even where magnetic separators are used for attracting metallic substances, the magnet is under the material flowing to the mill and the steel is usually on top. Furthermore, the magnet has no effect on manganese steel, copper, brass, wood, stone, rock, and similar materials. As the magnet cannot pick up these materials, or all the pieces of steel, it is important to use a machine in which magnets are unnecessary and so constructed that it will not be damaged or wrecked if such foreign materials reach the moving parts.

How is the air supplied?

Answer: The air supply to the mill may be taken directly from the boiler room or may be supplied through a duct from an air preheater.

What drive is used?

Answer: The drive unit may be either a constant-speed motor or a steam turbine with a reduction gear. Its shaft is coupled to the horizontal mill shaft which carries the drive worm gear.

Describe the grinding elements.

Answer: They consist of a cast steel bowl which revolves on the vertical shaft, and rollers which revolve on journals attached to the mill housing. At no time, even after the mill is empty, do the rollers touch the replaceable steel grinding ring in the bowl.

Describe the operation of the classifier.

Answer: After sufficient pulverization, the mixture of coal and air passes through openings in the grinding chamber and into the classifier, where it passes through openings equipped with deflectors or vanes. These are adjustable for variation of the fineness. A spiral on the inner cone of the classifier facilitates the return of oversized particles to the bowl through an opening in the bottom of the classifier. A cone suspended above this opening tends to build up pressure at this point and aids in rejecting the coarse particles.

What duty is performed by the exhauster?

Answer: In the exhauster the pulverized coal is blown through the burner supply pipes. Where there is more than one burner per mill, distributors are required to equalize the flow of coal.

Impact Mill

This consists essentially of a rotor with swinging hammers arranged to rotate inside a suitable casing.

For what service is the impact mill suited?

Answer: It is well-suited for use with all coals except those which have excessive amounts of iron pyrites or other highly abrasive material.

Describe the rotor.

Answer: The drive shaft of the mill extends through the pulverizing chamber and has keyed to it a series of steel discs. Manganese steel hammers are pivoted to rods which pass through the discs near their outer edges. Swinging hammers are used to allow for the passage of foreign matter into the tramp-iron pocket without damage to the mill parts.

How is the fineness regulated?

Answer: The smaller mills have a set of adjustable blades in the conical section between the mill chamber and the fan which regulate the fineness of the coal. The larger mills are equipped with classifiers similar to those on the bowl mills. The action of either the adjustable blades or the classifier causes the larger particles of coal to be returned for further pulverization.

Combined Impact-and-Attrition Pulverization

In this mill, impact is accomplished by forged steel balls. The mill is called a *tube mill* and combines the principles of impact and attrition.

PULVERIZED COAL BURNERS

Describe the construction of the mill.

Answer: Figs. 22-1 and 22-2 show details of construction. It is built in various diameters from 24 inches to 96 inches and in lengths desired for producing a given tonnage. The heads of these mills have trunnions cast integral. A steel shell with angle flanges at either end is bolted to the heads and corrugated liners are keyed to the inside. A cut herringbone gear is mounted on one of the angle flanges.

Describe its operation.

Answer: The barrel of the mill is charged with forged steel balls through the feed end while the mill is in operation. The countershaft is direct-connected through a flexible coupling to a motor, or it may be driven by belt, silent chain drive, or direct-connected to a steam engine. Material is fed into the mill in desired quantities by a disc feeder preferably driven by a ¼-horsepower, variable-speed motor. A sleeve from the discharge end of the tube mill leads into the fan housing. The mill is rotated at a speed which causes the balls to be carried around by centrifugal force to a point where they are thrown down onto the material to be ground, just as you would throw a ball onto a piece of material to be crushed. The result is a continuous cascade of balls delivering thousands of blows per second.

How is the pulverized coal discharged?

Answer: The air-swept tube mill has no mechanical discharge and depends upon the air passing through the mill to float out the finished product. An adjustable opening in the feed-end housing admits preheated air in desired quantities. A fan connected to the discharge-end trunnion and driven by a variable-speed motor draws air through the mill, removing only the impalpable powder through the discharge end of the mill. This controls the velocity of the air through the mill and thereby the fineness of the product. According to a series of tests, the flammability of coal dust increases with fineness and the maximum flammability is reached with coal dust within the range of about 10 to 25 microns in diameter (one micron equals one millionth of a meter). The flammability of dust of larger sizes decreases through the range up to

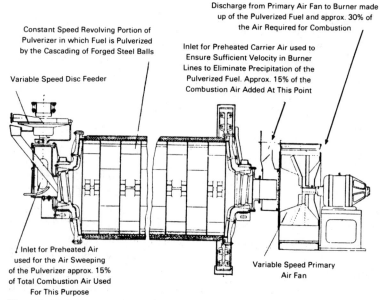

Fig. 22-1. Cross-section of an air-swept tube mill with its auxiliary equipment.

the size of 25 to 74 microns diameter, at which point the flammability has decreased to a very marked extent.

How is additional air obtained?

Answer: It is admitted between the discharge-end of the mill and the fan housing, from which the coal is delivered directly to the burners set in the furnace wall.

BURNERS

In pulverized-coal firing, turbulence through the furnace is essential (1) to bring oxygen and combustible powdered coal into continuous contact, (2) to scrub away the ash from the surfaces of the particles, and (3) to cause the gases to sweep the water-heating surface in the furnace.

Without intensive mixing of coal and air, efficient combustion is

PULVERIZED COAL BURNERS

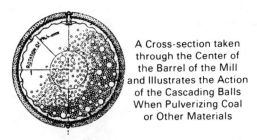

A Cross-section taken through the Center of the Barrel of the Mill and Illustrates the Action of the Cascading Balls When Pulverizing Coal or Other Materials

Fig. 22-2. Cross-section through the center of the barrel of the mill shown in Fig. 22-1.

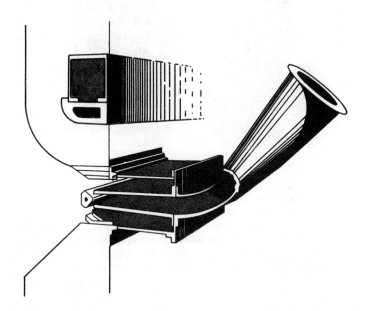

Fig. 22-3. Vortex burner. In operation, the primary air and coal are delivered through a flat orifice, usually horizontal. Coal and primary air impinge on a V-shaped tip, or deflector, placed in front of the burner. By the impingement of the fuel on the deflector, the stream is divided into two parts, one upward and one downward. Secondary air is delivered to the wind box and passes over the curved deflectors shown above and below the burner tip. This

secondary air is delivered at a pressure which drives it into the fuel streams, the result being intense turbulence at the burner tip and a short flame with highly efficient combustion.

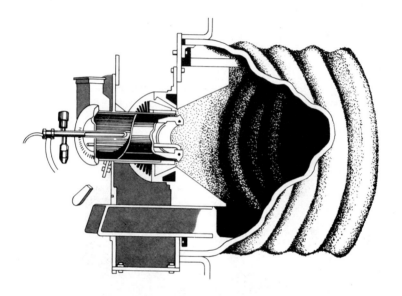

Fig. 22-4. Rotary burner. This burner, which is suitable for coal, gas, or oil, is particularly adapted and recommended for Scotch marine boilers of the return tubular type. In operation, the primary air and coal are delivered at a tangent and pass through an annular orifice with rotary action of a relatively high velocity. The secondary air is also delivered in a rotary manner, on the outside of the coal stream. The combination of the vortex deflector and the rotary action results in probably the most intense form of turbulence it is possible to produce under the given condition. This intense turbulence results in proper mixing of the fuel and a short, highly radiant flame.

impossible. Scrubbing of the coal particles assures contact between the combustible and oxygen, thus promoting rapid combustion and reducing carbon loss. Sweeping of the water-heating surface in the furnace by the gases increases the evaporation rate for the furnace heating surface.

PULVERIZED COAL BURNERS

How is turbulence obtained?

Answer: This may be obtained either by variations in design or arrangement of burners. The most effective method is an arrangement of simple nozzles which provide turbulence by impingement of one flame upon another.

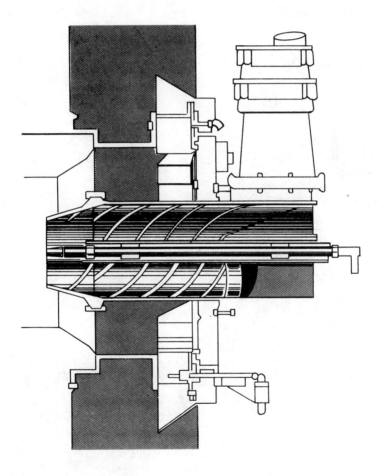

Fig. 22-5. Type R burner for firing pulverized coal and oil. This horizontal firing burner provides for uniform distribution of the coal and air within the nozzle and by means of deflectors and vanes.

Power Plant Engineers Guide

Corner (or tangential) firing is an example of this principle. It is generally conceded that where it can be applied, this type is the most preferable for thorough mixing of coal and air. These burners have horizontally adjustable nozzles arranged in a casing with forced-draft air ports.

Construction details for various burner types are shown in Figs. 22-3 through 22-9. Operation in each case is described in the caption.

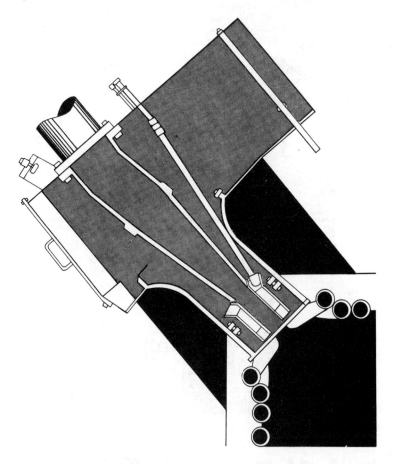

Fig. 22-6. Plan of tangential burner showing the adjustable nozzle.

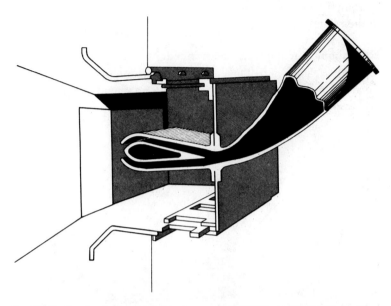

Fig. 22-7. Duplex burner suitable for small installations. With this burner, the secondary air is furnished by natural draft. It will be seen from the illustration that the stream of primary air and pulverized coal is divided into two parts through an upper and lower channel. These two streams impinge on each other as they leave the burner and result in intense turbulence, giving efficient combustion.

AUTOMATIC CONTROL

Response to demands for changes in the rate of coal supply from the pulverizing mill to the burners is obtained by simultaneous changes in proper relation of the coal feed and air supply to the mill.

How is primary air flow controlled?

Answer: By adjusting the exhauster inlet damper.

How is the mill feed controlled?

Answer: By adjusting the speed of the feeder driving mechanism or motor. If these two controls are attached to an automatic

Power Plant Engineers Guide

system, they will respond to the changes in loading pressure acting in the system.

How should the controls be arranged?

Answer: The controllers on the mill feed and primary air should be arranged to permit separate, remote, manual control of each, and also be readily transferable from manual to automatic.

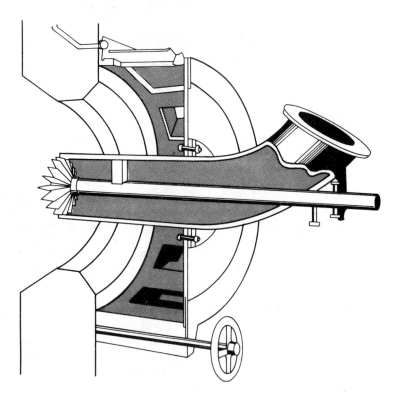

Fig. 22-8. Rotary burner designed for either natural or forced draft. The burner elbow can be bolted to the air housing and burner nozzle at any angle suitable for the particular pipe layout. The elbow carries an attachment for holding the coal diffuser rod and the necessary stuffing box arrangement for making this assembly dust-tight. The air housing for natural-draft burners is a standard, cast-iron,

Pulverized Coal Burners

drum-type air box with an adjustable air register on its periphery to allow for the admission of the necessary amount of secondary air for combustion, which is taken into the air box and furnace by the stack draft. The adjustable air register is equipped with blades which give the secondary air a rotary motion in the same direction as the rotation of the primary air and coal passing through the nozzle and coal diffuser. When using secondary air under pressure from a fan, the steel-plate-type air housing is used, and is equipped with a damper to allow regulation of the proper amount of air to be taken into each burner. All air housings are equipped with two peepholes for observation purposes and for inserting the torch for igniting the coal or oil. The burner nozzle is centrally located in the air housing and discharges the coal from the burner elbow into the furnace. Inside the burner nozzle is the coal diffuser, which consists of a series of curved radial blades giving the stream of primary air and coal a spiral motion. The position of the coal mixer is not fixed and should be adjusted by moving the supporting rod in or out. This adjustment will affect the length and the spreading of the flame. To keep the coal mixer in a central position when outside the burner nozzle, a separate, three-leg support is attached to the coal-diffuser rod or pipe. This keeps the coal diffuser or mixer concentric with the burner nozzle.

What precautions should be taken?

Answer: For the most satisfactory mill performance, the temperature of the primary air and coal mixture should be held at the recommended temperature for the output. Excess temperature may result in pulverizer difficulties. Temperatures which are too low do not produce sufficient drying during pulverization and interfere with normal operation. The use in the mill exit of a recording thermometer with regulator connected to the hot-air inlet damper is recommended. Where automatic control of temperature is desired, a recording temperature controller may be connected to the receiving regulator.

Describe the hot-air control.

Answer: Atmospheric air for tempering the hot air is controlled by a balanced damper in the atmospheric inlet, which is opened by a difference in pressure between the inside and outside of the mill.

Controlled Flow Split Flame Burner

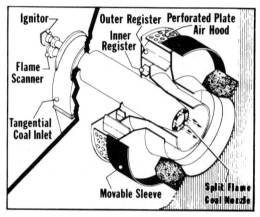

Courtesy Foster Wheeler Energy Corp.

Fig. 22-9 This controlled flow, split-flame coal burner is designed to efficiently reduce nitrogen oxide emissions. Dual series registers control secondary air flow. The outer register is electrically driven, with "closed," "ignite," and "operate" positions. The manually adjustable inner register is optimized during initial shakedown of the boiler. The perforated plate air hood and movable sleeve measures and controls the secondary air flow to each burner. The flow is obtained by measuring the pressure drop across the perforated plate. When the boiler is first placed into operation, burner-to-burner secondary air distribution is optimized by manually adjusting the sleeve, which then remains fixed during subsequent operation. The split flame coal nozzle concentrates the coal into four separate streams to provide four low stoichiometric flames (70% near the burner throat) with normal excess air to the burner (typically 20%). On initial startup, the inner tip of the nozzle is moved manually, to optimize the primary air velocity, and thereafter it remains fixed. Notice the tangential coal inlet and centrally located ignitor and scanner.

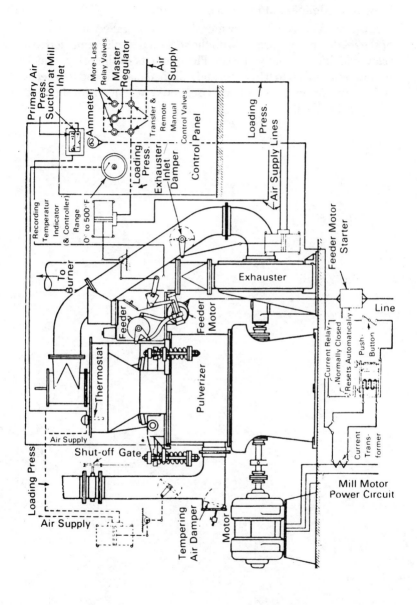

Fig. 22-10. Diagram of an automatic control system for a bowl mill.

How is overloading of the mill prevented?

Answer: A current transformer should be installed in one phase of the feeder-motor circuit, with the secondary coil connected to a relay which is normally closed. The contacts of the relay, which should be adjusted to trip at normal full-load current and to reset at 20% under normal full-load, are connected in series with the feed motor (Fig. 22-10).

CHAPTER 23

Oil, Gas, and Waste-Fuel Burners

FUEL OILS

What should you know about fuel oils?

Answer: Some knowledge of fuel oils is essential for the intelligent operation of oil burners. Fuels are derived from *crude oils* from different fields and vary considerably. See Chapter 2.

How are fuel oils classed?

Answer: Domestic fuel oils are classed as Nos. 1 and 2, with industrial fuel oils classed as Nos. 4, 5, and 6.

Give another classification.

Answer: Sometimes the fuels are referred to as light, medium, and heavy domestic oils and light, medium, and heavy industrial oils.

What determines the grade of oil that can be used?

Answer: The particular grade of fuel that can be used is usually fixed by the design of the burner with respect to the method of spraying and the type of ignition. For instance, gravity-feed burners invariably are designed to burn only the high-grade distillates, whereas some domestic burners use oil as heavy as No. 4.

Upon what does the cost of heating depend to a considerable extent?

Answer: The grade of oil which can be burned.

GRAVITY-FEED OIL BURNERS

Explain the operation of a gravity-feed vaporizing, or gas-type, burner with automatic control.

Answer: In operation, the fuel oil is supplied through a pipe to the vaporizer. See Figs. 23-1 and 23-2. In its passage through the firebox and the vaporizer, the oil is converted into a vapor (or gas) that burns without odor, soot, or residue. From the top of the vaporizer, the gas is conveyed through an elbow pipe to the inlet of the burner, where it escapes through the small orifice and is ignited. The flame is spread in every direction, thus serving the double purpose of generating the gas in the vaporizer and distributing heat to the bottom surface of the vaporizer.

Working below the orifice is a shutoff plunger, which, when raised or lowered, controls the flow of gas. This plunger is connected by means of a rod, counterbalanced rock shaft, bellcrank lever, and connecting rod to another bellcrank lever and to a hollow spring on the outside of the furnace. The weight of these rods is counterbalanced by the rod and ball. The hollow spring is supplied with steam at boiler pressure through the steam port.

The saucer is for fuel oil or alcohol used for generating the proper heat under the vaporizer at starting and until sufficient gas is generated for its own reproduction, usually a matter of from three to four minutes. The burner is furnished with removable plugs for cleaning. The rock shaft is furnished with a stuffing box to prevent leakage.

Oil, Gas, and Waste-Fuel Burners

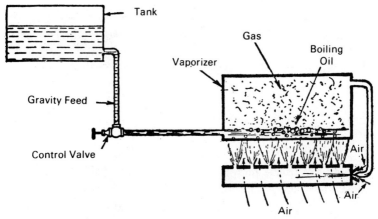

Fig. 23-1. Gravity-feed, induction-mixing, vaporizing burner.

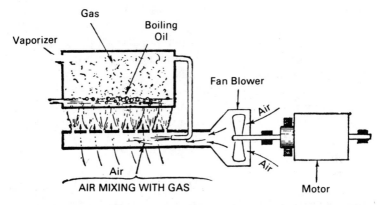

Fig. 23-2. Gravity-feed, mixing-vaporizing burner. Connected to the burner is a blower with outlet pipe surrounding the gas pipe as shown. In operation, air from the blower mixes inside the burner with the gas coming from the vaporizer. Thus the air is mixed with the gas before ignition, resulting in a blue flame and efficient combustion.

For control, the straightening of the spring caused by an increase of pressure in the boiler operates directly on the plunger by means of the adjusting screw, bellcrank lever, and intermediate connections, thus establishing the relation between steam pressure and fire. Should the steam pressure rise, the plunger would close off the flow of gas correspondingly, and vice versa, thereby regulating the heat of the fire. The plunger cannot, however, shut off the flow of gas entirely; a small flow is always left, enough to keep the burner and boiler hot. In this way, the trouble and annoyance of having to relight the fire after every stop is avoided.

PRESSURIZED OIL BURNING

To ignite properly, oil spray must be mixed with air so that it will vaporize and gasify. The higher the temperature and the Baume of the oil at the burner, the easier it is to spray, and the air pressure may be correspondingly lower. The temperature at which it begins to vaporize is the limit to which any oil may be heated.

The lower the pressure at which the oil and air can be used, the less will be the cost for power. Furthermore, a minimum of difficulty will be encountered from foreign matter clogging valves, as the throttle opening will be correspondingly larger and thus permit most of the dirt to pass through.

The air and oil pressure at the burner should be steady. The air blower should be of sufficient capacity, and pipe lines large enough, to deliver the volume of air without more than 10% drop in pressure. All of the oil should be gasified before the flame impinges on any obstruction, refractory or otherwise. If the oil is not thoroughly gasified, the impingement will cause a carbon deposit or the rapid destruction of the refractories; the end of a long, lazy flame, however, is not especially destructive to refractories.

The speed of emission for some gasified oil and air must be less than the speed of propagation of the flame or it will not ignite. For the most efficient combustion, all of the air entering the furnace should pass through the oil spray so that all of the air will be used, and the oil be consumed with the least excess of air.

The length of the flame depends on the fineness of nebulization, or the size of the droplet of oil. It also depends on the temperature of the droplet, temperature of the surrounding air, and the ability of the droplet to absorb the necessary volume of air for complete combustion.

CLASSIFYING OIL BURNERS

What is an oil burner?

Answer: By definition, any device wherein fuel oil is vaporized or atomized and mixed with air in proper proportion for combustion.

How are oil burners classed?

Answer: In numerous ways as to methods of (1) operation, (2) ignition, (3) gasifying, (4) oil feed. See Figs. 23-3 to 23-9. More in detail, they may be classed with respect to the gasifying process as vaporizers or sprayers, with respect to the atomizing agent, as air or steam, and with respect to the method of spraying, as outside-mixing (drooling, atomizer, projector, centrifugal) and inside-mixing (chamber, injector, centrifugal). Other types are rotary, high-pressure, low-pressure gun-type, and pot-type.

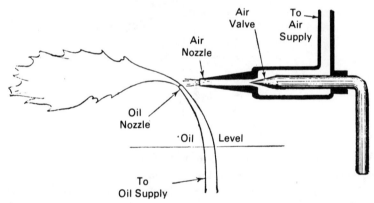

Fig. 23-3. Sprayer burner. The oil is brought through an orifice directly across the path of the jet of air or steam and is "brushed" off by the latter and sprayed.

VAPORIZING BURNERS

What is a vaporizing burner?

Answer: One in which the fuel oil is vaporized by heating in a retort.

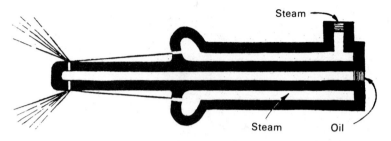

Fig. 23-4. Projector burner. The oil is pumped to the oil orifice and caught by the air or steam jets which are located some distance back.

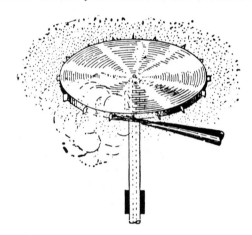

Fig. 23-5. Outside centrifugal burner.

What is meant by the term "vaporize"?

Answer: Fuel is vaporized when a *change of state* takes place, such as a transformation of the fuel from the liquid to the gaseous state. Careful distinction should be made between vaporizing and alleged *atomizing* burners explained later.

Name two types of vaporizing burner.

Answer: Mixing and nonmixing burners.

How does a vaporizing burner work?

Answer: The fuel passes from the source to the retort, or vaporizer, which is a closed vessel heated by the burner underneath.

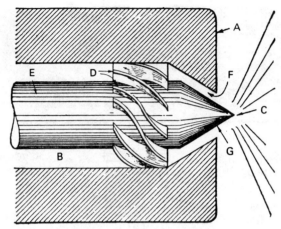

Fig. 23-6. Inside centrifugal burner. In construction, at the end of the pipe (A) that conveys the oil, the oil passage (B) is tapered down to the opening (C) through which the oil is discharged. The series of slanting vanes (D) on the rod (E) deflect the oil and break it up into a number of currents, each of which has a whirling motion as it enters the space (F) around the end (G) of the rod. The centrifugal force due to the whirling motion given by the vanes causes the spray to spread on leaving the burner as shown by the diverging lines.

This causes the oil to boil and supply gas to the burner. In the nonmixing type, mixing takes place when the gas leaves the nozzle of the burner. In the mixing type, the gas from the vaporizer passes into the mixer, is combined with air, and passes out through small holes where ignition takes place. A familiar example of this kind of mixer is the ordinary kitchen gas burner. (See Fig. 23-10.)

ATOMIZING BURNERS

What is an atomizing burner?
Answer: A misnomer, because the fuel oil is not really reduced to component atoms, but made into a fine mist.

What should an "atomizing" burner be called?
Answer: A sprayer.

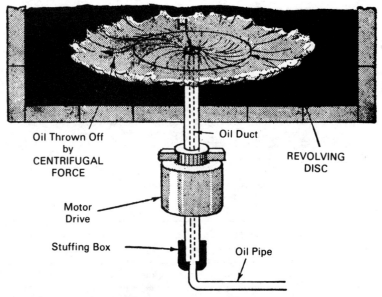

Fig. 23-7. Centrifugal-force atomizing burner. The oil flows through the hollow spindle of a disc which is rotated at high speed by a motor. The oil overflowing onto the disc at its center is hurled off the disc by a centrifugal force, and ignited by a torch or pilot light, producing a ring of flame.

Fig. 23-8. Injector inside-mixing burner.

How does a sprayer work?

Answer: Air or steam blown through a nozzle draws in oil which mixes with the steam or air and passes through the nozzle tip as a very finely divided mixture.

Oil, Gas, and Waste-Fuel Burners

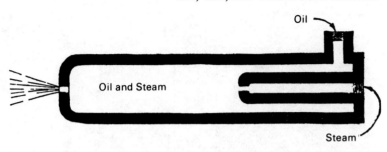

Fig. 23-9. Chamber burner. The oil and steam are mixed before issuing from the burner. With this burner, the oil is heated before leaving the burner.

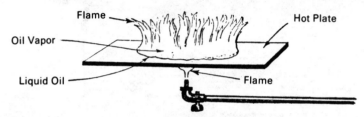

Fig. 23-10. Hot-plate method of vaporizing oil.

Describe the function of a natural-draft register.

Answer: The natural-draft register is designed to give a large range of capacity with a positive and wide variation between maximum and minimum air admission. Air adjustment is accomplished by regulating the small and large ring plates on the air register by means of the handle provided for that purpose. Adjustable stops are also provided so that the best setting can be maintained at all times.

The entire air register is hinged, permitting access to the furnace and all parts of the air register and burner. As with all mechanical pressure atomizing burner air registers, all the air for combustion is admitted at the burner. See Figs. 23-11 and 23-12.

ROTARY BURNERS

Name two types of rotary burners.
Answer: Vertical and horizontal.

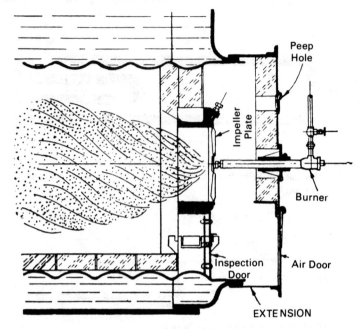

Fig. 23-11. Mechanical burner as applied to a Scotch marine boiler.

How does a rotary burner work?

Answer: The basic principle of operation is centrifugal force. That is, the oil entering at the center of a rotary cup is whirled around very rapidly until the oil is thrown away from the cup. Being mixed with air, it will ignite.

What is meant by vertical and horizontal rotary burners?

Answer: Vertical and horizontal refer to the position of the shaft, not the cup.

What is the application of the vertical type?

Answer: It is used as a domestic burner, and is installed in the boiler with all controls located adjacent to the burner.

How is the horizontal rotary burner installed?

Answer: On the outside of the boiler, with the cup extending into the chamber.

Oil, Gas, and Waste-Fuel Burners

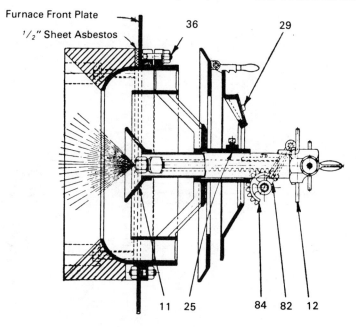

Fig. 23-12. "Twinplex" mechanical pressure sprayer (atomizer) burner and natural draft register.

SPRAY BURNERS

Describe a multiple-spray oil burner for low-pressure air.

Answer: See Figs. 23-13 and 23-14. The burner is provided with a cone-shaped spreader which receives the oil at its pointed end. As the oil issues from the oil passage, the bypass air picks it up and forces it along the surface of the spreader, expanding and thinning it as it goes. The film of oil then meets the second jet of bypass air, which is directed at an angle to strike the spreader at its base and force the oil into the combustion chamber as a cone of very fine spray, the fineness depending on the air pressure, the viscosity of the oil, and the size of the base of the spreader. Less than a sixth of the total air goes through the bypass to form the spray, and it is

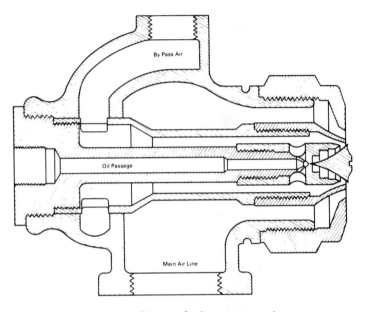

Fig. 23-13. Multiple-spray oil burner for low-pressure air.

always at full air pressure. The greater volume of the air passes through the body cap to support combustion, and its pressure can be varied from zero to full. Having passed through the body cap, this combustion air strikes the spray one or more times in order to change its direction and lower its velocity, thus making a baffle wall or block unnecessary. This burner operates on air pressure as low as 8 ounces with light oil.

How is this burner operated?

Answer: On a cold furnace, it may be necessary at times to start the burner partly throttled until the vents begin to function properly. To widen the spray and shorten the flame, screw the cap further onto the body; unscrewing the cap will squeeze the flame and lengthen it. If a very short, high-temperature flame is desired, it is sometimes necessary to enlarge the taper of the burner hole inside the furnace. If the flame blows away from the tunnelway of the block when the burner is operating at the required maximum, the tunnelway is too small and should be enlarged on the inside of the block or wall. It is often good practice to place an oil shutoff valve back of the oil control valve; the latter can then be adjusted for perfect combustion and left undisturbed while the shutoff can be used to turn the oil on and off. Never place an air control valve back of the burner, as full pressure is needed through the bypass unless the pressure is above 2 pounds. The burner will operate on cold air at any pressure above 12 ounces and on any oil pressure above 5 pounds. Installation of this burner for use with preheated air and automatic temperature control is shown in Fig. 23-14.

LOW-PRESSURE PROPORTIONING BURNERS

Describe a low-pressure proportioning oil burner.

Answer: Such a burner is adapted to industrial heating operations requiring furnaces, boilers, dryers, ovens, kilns, retorts, roasters, stills, and preheaters. The moving of a single lever automatically controls the oil and air supply and simultaneously adjusts both primary and secondary air orifices in the burner, which produces CO_2 readings consistently between 13% and 15% over the full range of the burner rating.

An oil control valve, consisting of a vee-groove in a flat surface, is covered by a rotating cam and bolted to the burner backplate. The oil tube connects the oil valve with the oil nozzle. Around the oil tube and supported by the backplate is the inner air-nozzle operating tube and the inner air nozzle. An operating lever causes the inner air nozzle to move either backward or forward and also causes the oil control valve cam to rotate. Moving the operating lever clockwise causes the inner-nozzle operating pin, which is mounted on the inner air-nozzle operating tube, to move in the curved slot in the backplate tubular section and thus move the inner air nozzle back from the outer air nozzle. This increases the discharge areas, permitting more air to flow out of the burner.

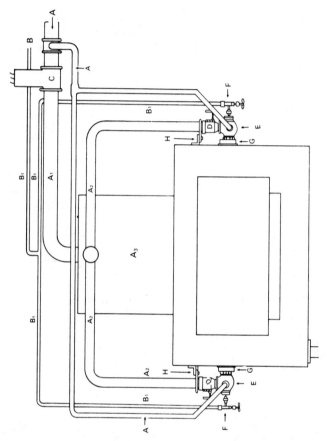

Fig. 23-14. Typical installation of a multiple-spray fuel oil burner in position and piped for using preheated air and automatic temperature control.

Since the oil control valve lever is also connected to the operating lever, the oil valve cam is rotated to regulate the flow of oil into the oil tube and nozzle. Primary atomizing air passes through tangential primary air supply openings in the inner air nozzle, which give it a rotary motion as it enters the space around the oil nozzle. It then leaves the burner through the primary air discharge area, picking up oil from the oil nozzle as it leaves. This high-velocity, rotating, primary air thoroughly breaks up and atomizes the oil. As the air/oil mixture leaves the inner air nozzle in a diverging cone, it is met by air in a converging cone from the secondary air discharge opening, thus further atomizing the mixture and producing a fog which is quickly ignited and easily burned.

What is the difference between low-pressure and high-pressure venturi oil burners?

Answer: Low-pressure venturi burners are used in installations where flexibility of heat is required but where steam or compressed air cannot be used for atomizing purposes. High-pressure venturi burners use steam or air at 30 psig and oil from 5° Baume and up, with the heavier grades requiring preheating. The high-pressure venturi "flat-flame" burners are particularly suited to firing boilers and large heating furnaces where a soft, spreading flame is desired. These operate with steam or air at 40 psig and higher, oils of 19° Baume and lighter without heating, and all heavier oils if preheated.

POT-TYPE BURNERS

What is a pot-type burner?
Answer: A hot-plate burner.

How does it work?
Answer: Fuel oil drops into a plate kept hot by constant heat from either gas or fuel oil permitted to flow slowly into the chamber. The hot plate vaporizes the oil.

GUN-TYPE BURNERS

What are the essential parts of the gun-type burner?
Answer: This sprayer, perhaps called "gun-type" on account of its shape and the required high pressure (100 psi) for its operation, consists of three essential parts: (1) strainer, (2) pump, and (3) pressure-regulating valve.

How does it work?

Answer: The pump forces the oil to the regulating valve at 100 psi, this pressure being maintained by the regulating valve. This pressure is required to force the fuel through the small orifice in the nozzle tip. Surplus oil is bypassed back to either the strainer or the tank, depending upon operating conditions.

What are the three methods of bypass?

Answer: (1) Two-pipe system, (2) internal, and (3) an external loop of copper tubing from the bypass to the strainer, as used when a gravity tank is employed and a separate pump, strainer, and pressure valve is used instead of a fuel unit.

On all three-part units (pump, strainer, and valve), how is the internal bypass constructed?

Answer: The construction is such that the unit may be adapted to operate with either lift or gravity feed.

How is the internal bypass changed to external bypass?

Answer: By means of the ⅛-inch plug, which must be left in position to close the internal bypass.

If internal action is desired, as with gravity feed, what is done and what is it called?

Answer: The small plug is omitted, and the outside bypass plug is inserted so that no external line is needed back to the tank. This is termed a "one-pipe" job.

What should be noted about the internal plug location?

Answer: Location differs with different burners, so it must be placed according to the instructions accompanying the burner.

How is the plug removed on a one-pipe job?

Answer: With a long allen wrench.

What is the construction of the pressure-regulating valve?

Answer: Several types are used, such as piston, diaphragm or bellows.

How is forced draft provided?

Answer: By means of a fan operating in an air tube and driven by an electric motor.

What else does the motor drive?

Answer: The oil pump.

How is the fuel mixture issuing from the nozzle ignited?

Answer: Usually electrically by jump sparks.

What are the essential parts of the ignition system?

Answer: Two electrodes (spark points) spaced for $3/16$-inch air gap and a transformer wound to step up the line voltage to about 10,000 to 15,000 volts.

Describe the flow of oil in a variable-capacity pressure sprayer burner.

Answer: A space is provided between the orifices of both the sprayer plate and the orifice plate; as the oil enters these orifices, the pressure parallel to their axis which is forcing the fluid outward has been converted into velocity in the same direction due to the venturi effect of the whirling chamber and its orifice. At the same time, the rotating oil in the whirling chamber sets up a definite centrifugal pressure perpendicular to the axis of the orifice. As the rotating oil reaches the return annulus, the centrifugal pressure will force some of the oil into this opening if the return-line control valve is open to permit its passage through the return line. Oil which is not returned continues through the orifice in the orifice plate with its energy undiminished and emerges in the form of a hollow, conical spray.

What are the operating instructions for this type of burner?

Answer: The sprayer must be inserted and securely clamped in the quick detachable union connections. Open the oil control valve on the burner return manifold and be sure the return line beyond this control valve is wide open.

The burner is now ready to light off. Insert the torch through the opening in the closure plate. Open air doors and place the torch flame as close to the diffuser as possible. Open the inlet valve. Slowly close down the return line valve until the spray ignites. It may be necessary to quickly close and open the register doors to bring the flame up to the sprayer. Remove the torch after the flame has been properly established. Closing in on the return valve increases the oil flow through the sprayer, producing a larger fire. Adjust the fire to the desired size by means of this valve. The supply pressure to the burners should be held constant. Other fires as required should be lighted off in the same manner and all flames brought to the same size.

It is usually best to put the burners on the main control valve as soon as possible, so that all burners in operation may be controlled from one station. This is accomplished by slowly closing the main return control valve until the size of the flame begins to increase, indicating that the oil control is now on the main valve and not on the individual burner return valves, which should be immediately opened wide so that they do not interfere with the flame control. During these manipulations the air to the burners is, of course, controlled so as to operate smokelessly.

As the pressure on the return line is an index of the quantity of oil being fired for any given burner, the operators will quickly learn to proportion this pressure with the amount of air required to run smokelessly, and adjust the air to the oil flow as required. If automatic combustion controls are being used, they should be adjusted so as to provide the proper quantity of air to maintain a trace of smoke. Most efficient results are obtained when operating in this manner.

While a burner is in use, the oil control valve should never be entirely closed. A small amount of oil circulating through the sprayer is required to prevent carbonizing and overheating at the sprayer. On burners in use, keep valves on supply and return lines wide open—not partly open. Maintain fuel oil pressure to burners at 300 pounds per square inch. This burner is so designed and constructed that the return pressure has no effect on the spraying and regardless of the reading of the return line pressure gauge this does not affect the quality of spraying.

The temperature of the oil at the burner manifold should be that at which the oil has a viscosity of 150 SSU, and should be carefully

OIL, GAS, AND WASTE-FUEL BURNERS

maintained with as little variation as possible. Use enough air pressure for smokeless combustion, but only that much.

Light off burners through lighting holes. Do not attempt to light off from hot brickwork. Close registers for an instant after lighting off; then quickly open again. This assists in properly igniting the oil.

Remember adjustment of sprayer jacket tube, check sprayer-diffuser distance occasionally, and keep sprayer clean and free from grit, carbon or dirt.

To prevent overheating, do not place sprayer in idle register until ready to light off, and as soon as a burner is shut off, remove sprayer. This will prevent caking of oil in the sprayer plate slots and the small passages feeding oil to them.

In shutting off burners, the supply valves should be closed first, followed by the return valve. The register may then be closed. The sprayer should be immediately removed from the register, and care should be taken not to allow oil to drip on the front, piping, etc. The burner should be allowed to cool before cleaning.

When the burner is cool, the sprayer nut should be removed and the sprayer and orifice plates cleaned. Extreme care should be taken so that the surfaces of these parts of the nozzle-body face are kept smooth and free from dents, nicks, or foreign matter. Do not use steel or any other hard material in cleaning these parts.

In storing orifice and sprayer plates, do not place them in bags. Store so that the finished faces are protected against damage. It is important to keep them in good condition.

DUAL- AND MULTI-FUEL SYSTEMS

What is meant by a dual-fuel system?

Answer: There are two sets of units and controls, in some cases with automatic or timed switching from one to the other. An example would be an oil-gun assembly, plus a natural-gas burner, or electricity and one of the three petroleum-based fuels.

What are the advantages of a combination fuel system?

Answer: It allows the user to accommodate changes in the supply or price of either fuel.

Are firing procedures used for gas any different from those used in the oil process?

Answer: Yes—neither the need to preheat the fuel nor the use of auxiliary pumps is necessary in gas firing. Nozzles and burners still need care and cleaning, however.

Are the flame sensors used on gas-fired systems different from those used with heavy oil?

Answer: Ultraviolet is generally used with gas, while infrared is used with heavy oil. Photoelectric-eye systems are also used.

What device protects the boiler in the event of an interruption of gas pressure?

Answer: The pilot gas solenoid, which is also identified as a conductive flame rod.

Can any furnace be fired with solid or liquid fuel alternately?

Answer: Yes. As a result of fluctuating energy prices and availability, some manufacturers are producing boilers that can be fired with a variety of fuels. An example is a tri-fuel system produced by the CNB Tri-Fuel Division of Combustion Service and Equipment Co. This unit, illustrated in Fig. 23-15, offers six firing options: coal only, coal with gas, gas only, coal with oil, No. 2 oil, or No. 6 oil. Listed below are features of the firing system.

The dual-fuel burner includes:

1. Air-atomizing oil burner, capable of burning any commercial grade fuel oil.
2. Gas burner with matching flame, capable of burning a variety of gaseous fuels.
3. Integral blower to supply combustion air.
4. Preburned burner refractory tile for flame stability.
5. Pilot.
6. Observation port for easy viewing of pilot and main flames.
7. Hinged mounting for easy access to nozzles and shutters.
8. Safety switch to prevent firing when the burner is swung open.
9. Individual motors (no belts) for blower and oil pump.

OIL, GAS, AND WASTE-FUEL BURNERS

10. Separate air compressor set that runs only when operated on oil.
11. Full-programming combustion management system to provide enforced low-fire start, supervise the pilot, and monitor the main flame. (If the flame fails, power fails, water is low, or safety limits are exceeded, the main fuel valves close.)
12. Large annunciator lights to show fuel, flame failure, low water, limits, pilot.
13. Controls for modulating input, manual purge, maximum firing rate limiter, on-off switch; large terminal strips.

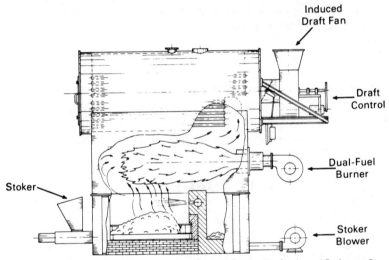

Fig. 23-15. Tri-fuel boiler.   Courtesy Combustion Service and Equipment Co.

The stoker is a single-retort, side-dump (deadplate on smaller sizes), underfeed type, hydraulically operated. It includes:

1. Grate section with retort, tuyeres, live plates, and dump grates selected on the basis of 30 pounds of coal per hour per square foot of grate area.
2. Hydraulic ram cylinder.
3. Hydraulic pump set and controls.
4. Combustion air fan with vortex control on the air inlet.
5. Coal hopper with cover.
6. The overfire air system consisting of a blower, control valve,

manifold, and sufficient air jets to ensure smokeless coal combustion, ready for field installation in the refractory bridgewall.
7. Control motors for stoker feed rate, underfire air, and overfire air adjustment normally are activated by manual positioners at the control cabinet.

The induced draft fan is a centrifugal type designed for combustion gas service. It has a radial blade, and is a paddlewheel type, accessible for inspection and cleaning. The pillow block bearings are self-aligning and located outside the hot gas stream.

The desired negative chamber pressure is automatically maintained by the draft regulator. It actuates an 80% closing damper in the flue gas outlet. The damper is bearing-supported and operated by a screw-jack-type actuator.

Do any multi-fuel burners offer electricity as an alternative?

Answer: Yes. Bryan Steam Corp. produces boilers that may be powered by electricity or one of three petroleum-based fuel options: oil, gas or a gas/oil combination. See Fig. 23-16.

WASTE-FUEL BURNERS

What materials can be used as waste fuel?

Answer: Wood pellets, sawdust, sander dust, hogged lumber, agricultural residue, trash (combustible).

How is it burned?

Answer: In specially designed burners (see Figs. 23-17 and 23-18). Some units mix oil or gas with waste fuel to enhance combustion and switch to either oil or gas if the waste fuel supply becomes exhausted. Waste fuels are burned on a grate or in partial suspension under forced draft. Waste fuel is delivered to the furnace by a pneumatic or mechanical system and injected as needed to maintain the firing rate.

Wood chips and sawdust power and heat a plywood manufacturing plant in Tennessee (Panoply Corp.). Octopus-like collector tubes snake through the plant to every woodworking machine,

Courtesy Bryan Steam Corp.

Fig. 23-16. This multi-fuel heating plant can be fired with electricity, oil, gas, or a gas/oil combination.

sucking up sawdust and wood chips. Pneumatic tubes pump this waste up into a 40-foot-high collector bin. See Fig. 23-18. At the bin bottom, a metered portion of wood waste is dumped onto a conveyor belt that carries it to a hydraulic ram which bites off cubic-yard chunks of the waste and packs it into the incinerator.

To get rid of the smoke and particulates, the Econo-Therm lean-burn incinerator has two chambers. The first uses a gas burner to fire the wood waste; the second has an auxiliary burner that comes on whenever chamber air temperatures drop below 1400 degrees.

The two-stage incinerator works on an unusual principle: Starving the flame to make it burn hotter. Reducing the air (oxygen) supply means higher combustion temperatures and, therefore,

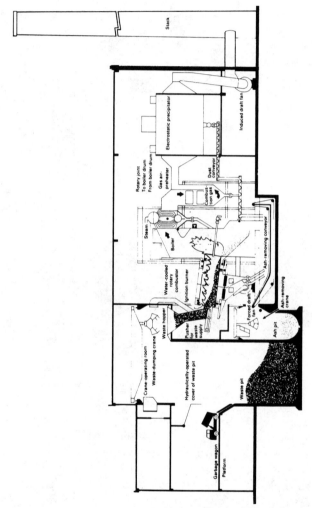

Courtesy O'Conner Environtech Corp.

Fig. 23-17. Solid wastes are used as fuel to fire this furnace, an O'Connor™ Water-Cooled Rotary Combustor. The combustion area comprises a rotating cylinder which stirs and mixes the waste after it is ignited. Heated, forced air promotes burning and drying. The manufacturer says the furnace will consume 25 to 150 tons per day of commercial and municipal waste with a moisture content up to 50%.

Fig. 23-18. Vacuum tubes from Panoply factory (rear) carry sawdust and wood chips to the "lean-burn" Econo-Therm incinerator that turns waste material into steam for heating and processing.

more complete burning. In turn, this assures a smokeless odorless discharge that complies with all federal and local environmental regulations.

The incinerator is sized large enough to handle all Panoply's heating requirements. In winter, heat from the incinerator goes to the boiler to generate steam for heating and processing work; in summer, a damper dumps excess heat up the stack.

CHAPTER 24

Steam Turbines

What is a steam turbine?
Answer: According to Jude, a prime mover in which gradual changes in the momentum of a fluid are utilized to produce rotation of the mobile member. According to Neilson, a turbine is a machine in which a rotary motion is obtained by the gradual change of momentum of a fluid. Graham says a turbine is a prime mover in which a rotary motion is obtained by centrifugal force brought into action by changing the direction of a jet of steam escaping from a nozzle at high velocity.

What are the essential parts of a turbine?
Answer: A disc mounted on a shaft and having attached to its rim many curved vanes; at the side, there is a nozzle (or nozzles) out of which a jet stream is directed to the vanes tangentially or at an angle.

Does the pressure of the steam cause rotation?
Answer: No.

What causes rotation?
Answer: Centrifugal force, created when the direction and momentum of the fast-flowing jet of steam is changed.

Describe this action in more detail.
Answer: In Fig. 24-1, V and V^1 are curved vanes attached to the circumference of a disc or rotor, W. At one side, and placed at an acute angle with the rotor, is a steam nozzle, N. In operation, steam issuing from the nozzle at high velocity strikes the curved vanes and its direction of movement is changed. Centrifugal force is created due to changing the direction of flow of the steam.

Explain more about the centrifugal force.
Answer: It must be obvious that the rotation of the wheel is caused by centrifugal force acting against the forward vanes and not by the pressure of the steam. If the curved passages were closed, thus bringing the steam to rest, it would press equally against vanes V and V^1; since these present equal areas, there would be no excess force tending to cause rotation. See Fig. 24-1.

CLASSIFICATION

There are numerous types of turbines and they are classified in several ways. A common, yet very questionable, division (with respect to the action of the steam) is:

1. Impulse,
2. Reaction,
3. Impulse-and-reaction.

Other classifications are:

1. Depending on whether or not there are one or more revolving vane discs, separated by stationary reversing lines, a turbine is said to be simple or compound.

Steam Turbines

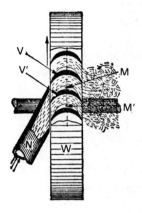

Fig. 24-1. Diagram showing principle of turbine operation.

2. With respect to the number of step reductions in steam momentum, a turbine is either single-stage or multistage.
3. With respect to the direction of steam flow, a turbine can be an axial, radial, tangential, mixed helical, or the re-entry type.
4. With respect to the source of the steam, a turbine is classified as an extraction or accumulator type.
5. With respect to pressure of the steam, the designation is high-pressure, mixed-pressure, or low-pressure.
6. With respect to the terminal pressure, it is a noncondensing or condensing type.

According to Croft, the terms *impulse* and *reaction* have specific meanings in turbine engineering parlance. These specific meanings are different from the meanings of the same words as they are employed in physics, mechanics, and ordinary usage. French says these terms are somewhat misleading because all practical turbines operate by the action and reaction of the working fluid and it would be clearer to designate the two types in some other way. A German writer uses equal pressure for impulse and unequal pressure for reaction; Croft considers these terms more appropriate and so does the author. Perhaps the real reason for all this confusion is the acceptance of loose, misleading definitions for the terms impulse and reaction.

Power Plant Engineers Guide

Define the term "impulse."

Answer: The act of impelling, or suddenly driving forward in the same direction as the applied force. The definition means that the force is applied perpendicular to the surface of the object acted upon, causing it to move in line with the force.

Define the term "reaction."

Answer: Reverse action; a force acting in opposition to, or balancing, another force.

What is an impulse turbine?

Answer: The term is a misnomer.

What causes a so-called impulse turbine to rotate?

Answer: It is not impulse, but the reaction due to centrifugal force—specifically, the reactive or centrifugal force component of the kinetic energy, dynamic inertia, or momentum (whatever you want to call it) of the fast flowing steam.

Give an illustration of impulse.

Answer: If a stream of water is directed at right angles to a board attached to a cart, the force of the stream will move the cart in the same direction as the flow of the water.

In what type turbine could impulse possibly occur?

Answer: Mathematically, but not physically, in a turbine having flat, radial vanes. Even in this primitive arrangement, there is only one instantaneous blade position when there is impulse.

Explain.

Answer: In Fig. 24-2, when the flat blade is at 90° to the axis of the jet or nozzle, impulse takes place. Since the time interval of this action is zero, there is physically no impulse, even in this position.

Give an illustration of reaction.

Answer: This is illustrated by the scotch mill shown in Fig. 24-3. The real force that causes this device to rotate is centrifugal force.

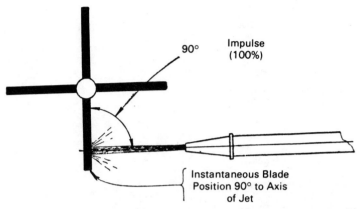

Fig. 24-2. Elementary diagram illustrating the only instantaneous position of the blade of a so-called impulse turbine for 100% impulse.

It is called "reaction" because the curved shape of the pipe gets in the way of the flowing water, forcing it to change its direction. Centrifugal force presses against one side of the curved pipe, which reacts, causing rotation.

Strictly speaking, what causes rotation?

Answer: The tangential component of the centrifugal force. See Fig. 24-4.

Give an illustration.

Answer: In Fig. 24-5, let the nozzle point in some direction between radial and tangential. If P is drawn to length to represent the centrifugal force introduced by the curvature of the pipe, then tangent T, of the right triangle, represents the force available for causing rotation. In amount, this equals the tangential component.

What happens when the blade of the "impulse turbine" shown in Fig. 24-2 points in any other direction?

Answer: The changed angularity of the jet with the surface of the blade introduces centrifugal force and a resulting reaction.

Explain in detail.

Answer: In Fig. 24-6, let the blade be at some angle to the jet

POWER PLANT ENGINEERS GUIDE

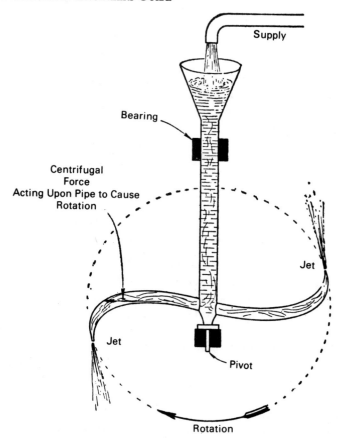

Fig. 24-3. Elementary reaction turbine.

other than 90°. Let AB equal the kinetic energy of the jet and draw AC normal (perpendicular) to the blade. In the triangle, ABC, the kinetic energy of the jet is made up of two components: AC, which is the component available to cause rotation when the wheel is in the position shown, and BC, which represents nonactive or wasted energy.

What does the component AC represent?

Answer: The reaction of the blade due to the centrifugal force, which is what causes rotation at that instant.

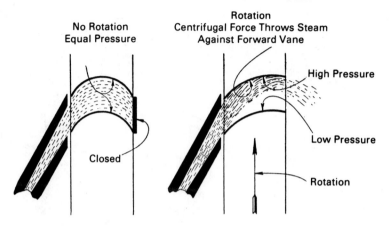

Fig. 24-4. Diagram showing condition for no rotation and for maximum rotation.

VELOCITY ("IMPULSE") TURBINES

What are the distinguishing features of turbines of this type?

Answer: Steam, after expansion in the nozzle to generate velocity, enters and leaves the passages between the vanes at the same pressure. Fig. 24-7 shows essential principles of the turbine.

Describe the nozzle.

Answer: It is of the diverging type, as shown in Fig. 24-8. Note that the velocity of the steam increases as it passes through the diverging section, that is, from A to B. This is due to expansion.

What is the actual construction of nozzles?

Answer: A leading manufacturer specifies nozzles made of monel metal accurately formed over special dies. Each nozzle is individually designed for proper expansion of steam at the pressure and temperature specified. Steam passages to the nozzles are outside the casing. A secondary nozzle with hand-valve which, when closed, gives full-load economy at partial loads. If the main nozzle carries the normal load, the secondary nozzle may be

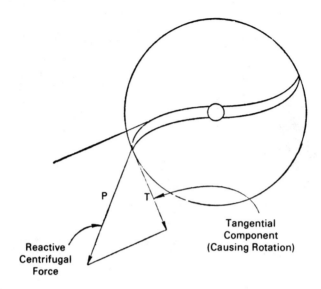

Fig. 24-5. Diagram of the reaction turbine, showing the tangential reactive component, which causes rotation.

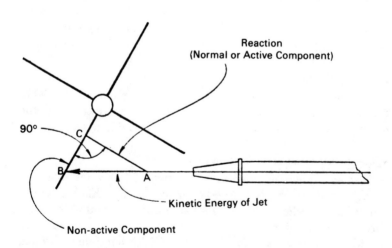

Fig. 24-6. Diagram illustrating the reaction component.

Steam Turbines

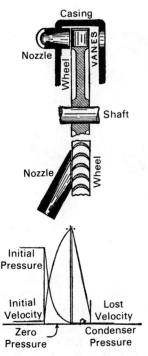

Fig. 24-7. Simple impulse turbine and pressure/velocity diagram. In operation, the steam is completely expanded in the nozzles, being reduced to condenser pressure at the outlet. The velocity falls in passing through the wheel, a certain amount of energy being lost by the inability of the wheel to absorb all the velocity generated in the nozzle.

opened either for an increased load or to compensate for lower steam pressure.

Describe the shape of the rotor vanes or blades.

Answer: The steam-walls of the vanes are of circular form, the distance between the walls at all points from inlet to outlet being the same, so that no expansion will take place in the passage between adjacent vanes.

What is the construction of the vanes?

Answer: They are usually made of stainless or special alloy

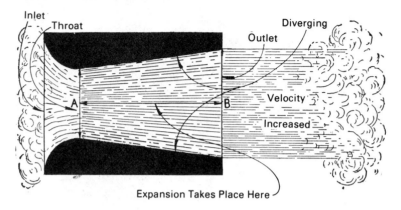

Fig. 24-8. Diverging-type steam nozzle as used in velocity ("impulse") turbines.

steel. In the design shown in Fig. 24-9, they are held in machined slots in the wheel by steel crosspins, in some machines two pins in each blade. Blades are shrouded by inner and outer bands of stainless steel or other metal which confine the steam to a full working passage. The inner bands also protect the wheel from steam erosion. The outer tip of each blade is machined to form a rivet, which extends through and is pressed over the outer band, thus forming a rigid assembly. Vanes are usually removable and renewable.

Fig. 24-9. Single-stage, velocity-turbine wheel and blade assembly.

Steam Turbines

Describe the casing.

Answer: The casing is made of cast-iron or steel, depending upon service conditions, and is suitable for back-pressures up to 50 psi (3.5 kg/cm^2) or more depending on design. It is horizontally split, and the casing is subject to exhaust pressure only. A relief valve is provided. Larger sizes have two exhaust openings, one on each side.

What kind of bearings are provided?

Answer: On older units the bearing shells are babbitt-lined, or solid, lead-bronze and are ring-oiled. They are horizontally split and may be rocked out of position and replaced after removing caps. Cartridge-type water cooling coils are provided. Deflector rings seal the housings and prevent entrance of dirt, moisture, or steam. Oil level is determined by glass oil-level gauges. More recent turbines have self-aligning sleeve, ball or roller bearings.

What are the two governors provided?

Answer: The constant-speed governor and the emergency overspeed governor.

How is the constant-speed governor controlled?

Answer: By flyballs. Sophisticated electrical and electronic controls are used in some power plants.

How does a constant-speed governor work?

Answer: A varying load on the turbine will cause a momentary decrease or increase in speed, and the centrifugal force of the flyballs, acting against the governor spring, moves the governor plunger in or out. This movement, when transmitted to the governor valve by the governor level, opens or closes the valve, increasing or throttling the flow of steam commensurate with the load, which returns the rotor to normal speed.

How is the governor lubricated?

Answer: All vital governor parts are automatically oil- or grease-lubricated. In oil systems, oil from the main bearing is fed into the the governor housing, picked up by the governor hub, sprayed over all governor parts, and then returned to the main oil reservoir.

What is the function of the emergency overspeed governor?

Answer: To stop the flow of steam at a predetermined overspeed, usually, about 10% above normal.

Describe its construction and how it works.

Answer: A spring-loaded plunger is mounted in the governor hub. When overspeed occurs, the centrifugal force on this plunger overcomes the spring tension, allowing the plunger to move outward and strike the trip lever. This action releases the trip latch and closes the trip valve, shutting off the steam supply and stopping the machine.

Describe the speed adjustment.

Answer: The adjusting screw at the lower end of the governor lever is loosened for increased speed and tightened for lower speed. Additional speed-adjustment of about 5% may be made at the governor nut on the end of the shaft after shutting down the unit. Adjustment for lower speeds down to 50% below normal can also be made while in operation by further tightening the adjusting screw, but the governor will be inoperative over this range.

Describe the shaft packing.

Answer: One type of shaft packing uses carbon-rings mounted in bronze rustproof split housings which have two compartments. The inner compartment contains two carbon rings, one ring being a four-piece, self-adjusting type, having $\frac{1}{16}$-inch clearance between the ends of the segments. As wear occurs, the garter spring causes the segments to close in around the shaft. The second ring is a two-piece, split-type, which seals the four-piece ring and aids in preventing leakage along the shaft. The outer compartment contains one two-piece, split-ring which diverts any leakage into the drain. Stainless steel garter springs secure the rings around the shaft and side springs hold them against the sealing faces.

What provision is made on condensing turbines?

Answer: Each box containing four carbon rings and top half-

STEAM TURBINES

housing has a tapped connection for admitting seating steam under 2 to 3 pounds pressure (0.14–0.21 kg/cm^2).

COMPOUND TURBINES

What is a compound turbine?

Answer: One having two or more wheels keyed to one shaft, with reverse-curve, stationary vanes interposed.

What is the object of the stationary vanes between the moving vanes?

Answer: They act to deflect the steam after it leaves the vanes of the first wheel and to cause it to enter the vanes of the second wheel at the proper angle.

Describe in more detail how the compound turbine works.

Answer: Steam, after leaving the nozzles, passes through the vanes of two or more moving wheels separated by stationary redirecting vanes. As with the simple turbine, the steam is completely expanded within the nozzles, issuing therefrom at exhaust pressure. The velocity generated in the nozzles is absorbed in "steps" (not stages) as the steam flows through the moving vanes of the wheels, and remains constant during the passage through the stationary vanes. The essentials and operation of the compound turbine are shown in Fig. 24-10.

What is the object of compounding?

Answer: To reduce the speed of rotation. That is, the speed of rotation would not have to be as high as with a single wheel in order to absorb the same amount of energy from the steam.

Why are the vane passages gradually increased in size in successive wheels?

Answer: The velocity of the steam, as indicated in the diagram in Fig. 24-10, falls during its passage through the wheels, but remains constant in passing through the stationary vanes. Since the velocity is gradually decreased as it passes through the several

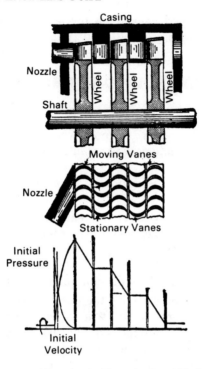

Fig. 24-10. Compound impulse turbine wheel and blade assembly.

wheels, the passages must gradually increase in size if the same quantity of steam is to flow through successive wheels in the same interval of time. The passages through the stationary vanes are of uniform cross-section because the velocity here is constant. However, it should be noted that the second set of stationary vanes is larger than the first since the velocity of the steam is less at the exit of the second wheel than at the first.

MULTISTAGE TURBINES

What is a simple multistage turbine?

Answer: One in which pressure energy of the steam is progressively transformed into kinetic energy in two or more pressure stages. There is only one wheel to each stage.

STEAM TURBINES

What is the distinguishing feature of each stage?

Answer: There is a separate compartment for each stage, as indicated in Fig. 24-11.

How does a simple multistage turbine work?

Answer: The word "stage" relates to working pressure. In a multistage turbine, steam passing through the turbine works at several pressures. There is a separate compartment and set of nozzles for each stage; the pressure of the steam is constant in passing through a compartment, but drops successively as it passes through the several sets of nozzles, regenerating velocity at each reduction of pressure. The essential features and pressure/velocity relations are shown in Fig. 24-12.

Since there is only one wheel in each compartment, the velocity of the entering steam must be proportioned to the speed of the wheel and must be such that it will be reduced to the exhaust velocity in passing through the vane passages. Since the entrance

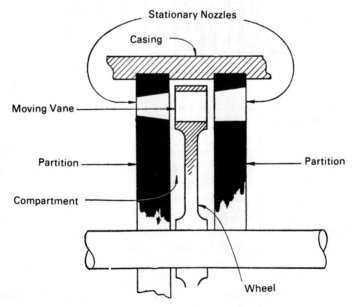

Fig. 24-11. Detail of multistage turbine, showing the partition between stages.

POWER PLANT ENGINEERS GUIDE

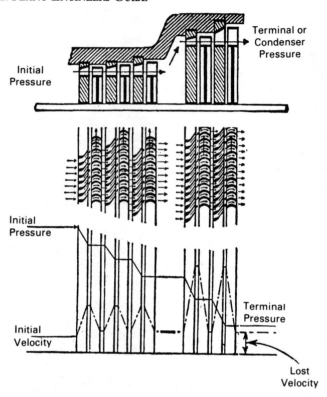

Fig. 24-12. Simple multistage impulse turbine and pressure/velocity diagram.

velocity depends on the amount of pressure reduction at the nozzles, any excess velocity after leaving the wheel represents an undue pressure loss, which otherwise would be available for generating velocity in the second set of nozzles for the second wheel (as the wheels run at slow speed the entrance velocity should be moderate to avoid this loss). With the proper number of stages, the reduction of pressure at each set of nozzles may be so proportioned as to give the correct entrance velocity to avoid loss.

What should be noted with respect to the pressure/velocity diagram?

Answer: That the velocity falls to that of the exhaust during

Steam Turbines

each stage, being regenerated in passing through each set of nozzles by successive reductions of pressure.

Name one inherent defect in the multistage arrangement.

Answer: It is subject to leakage at points where the shaft passes through the walls of the compartments and requires special packing which is usually inaccessible.

What is a compound multistage turbine?

Answer: One in which the pressure energy of the steam is progressively transformed into kinetic energy in two or more stages, with compound working in each stage.

Give another definition for this type of turbine.

Answer: A compound multistage turbine is one which has two or more compartments, each containing two or more wheels and a set of nozzles (Fig. 24-13).

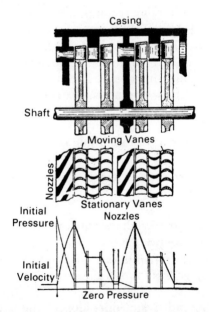

Fig. 24-13. Compound multistage turbine and pressure/velocity diagram. It is virtually two or more compound turbines joined in series.

What is the object of this arrangement?

Answer: To reduce the number of stages that would be necessary in a simple multistage turbine, especially for high-pressure working. That is, by placing two or more wheels in each compartment, a greater pressure reduction may be made between stages, thus reducing the number of stages.

How does the compound multistage turbine work?

Answer: In the pressure/velocity diagram of Fig. 24-13 note that the pressure remains constant in each compartment, being reduced as the steam flows through the nozzles. Also, the velocity falls in two steps in each compartment, remains constant in the stationary vanes, and rises in the nozzles.

UNEQUAL-PRESSURE (REACTION) TURBINES

What are the distinguishing features of turbines of this type?

Answer: In turbines of this type, the nozzle has parallel sides (Fig. 24-14) so that the steam passes through at practically constant pressure; expansion takes place during the steam-flow through the wheel. The passages between adjacent moving vanes are of diverging cross-section to permit this expansion, thus reducing the pressure and increasing the velocity.

How about the pressure on each side of the vanes (that is, each side of the wheel)?

Answer: The pressure is less on the exit side than on the entrance side.

How does this compare with the equal-pressure, or velocity, turbine; that is, the impulse-type?

Answer: The pressure is the same on each side of the wheel on these turbines since there is no vane expansion.

Describe the essentials and operation of the unequal-pressure (reaction) turbine.

STEAM TURBINES

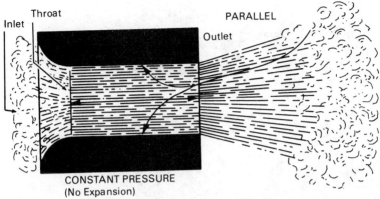

Fig. 24-14. Sectional view of nozzle with parallel sides.

Answer: The principal parts consist of numerous rows of moving vanes separated by alternate rows of fixed guide-vanes. The passages through the latter are of increasing cross-section, so part of the expansion takes place there and is completed in the wheel.

What should be noted about the fixed vanes?

Answer: They perform the function of nozzles.

Give a further description of turbine operation.

Answer: Fig. 24-15 shows essentials and steam distribution. The diagram indicates the increase of velocity as the steam flows through the guide vanes and the decrease as it passes through the adjacent moving vanes. The pressure falls gradually from the inlet to the outlet; it is maintained higher in the turbine passages than the exhaust pressure, this being a characteristic of the reaction principle.

Give an illustration of "reaction."

Answer: In Fig. 24-16, water-flow through opening ABCD relieves pressure on area ABCD but leaves pressure on corresponding area A'B'C'D' unchanged, forcing cart to the left away from the barrier; that is, opposite to the direction of water-flow. Pressure at ABCD is due to friction of the water passing through the short nozzle, while pressure at A'B'C'D' equals the area A'B'C'D' times pressure due to the head of water.

Power Plant Engineers Guide

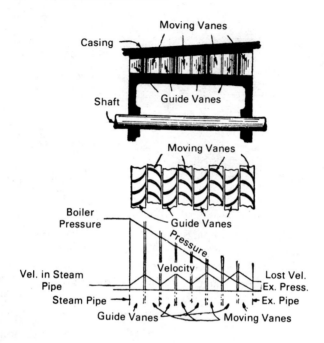

Fig. 24-15. Compound reaction turbine.

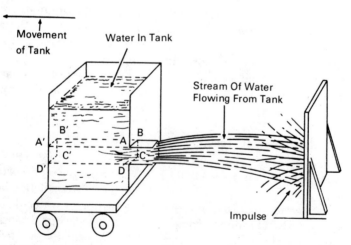

Fig. 24-16. Elementary diagram illustrating reaction.

Describe the effect of steam in passing through the moving vane passages.

Answer: See Fig. 24-17. Take any point as A and draw tangent AB equal in length to the kinetic energy of the steam; complete the right-triangle ABC. Then, the conditions for the point A are BC equal to the nonactive component and AC the reactive component due to centrifugal force.

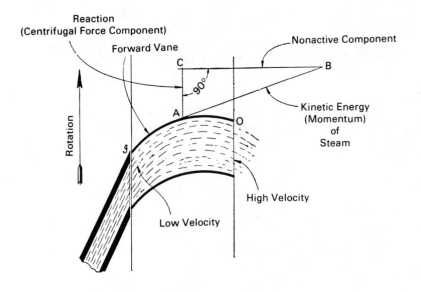

Fig. 24-17. Diagram showing the effect of steam in passing through the moving vane passages.

Does AC in Fig. 24-17 represent the entire turning force? Explain.

Answer: No, the value of AC should be determined for all points from I to O, and the average taken; it is different at each point because of the curvature of the vane. See Fig. 24-18. Divide the vane passage into any number of zones and draw force triangles for each zone, obtaining, for instance, reaction or centrifugal components a, b, c and d. The resultant will be the average of $abcd$. Of course, the more zones taken, the more accurate the calculation.

How is kinetic energy measured?
Answer: In foot-pounds.

Just what is kinetic energy?
Answer: The kinetic energy of a *moving* body is the work which the body is capable of performing against a resistance before it is brought to rest; that is, it equals the work which has brought it from its state of rest to its actual velocity.

How is the value of kinetic energy obtained?
Answer: The measure of kinetic energy is equal to the weight of the body multiplied by the height from which it must fall to acquire its actual velocity.

The kinetic energy of a body is expressed by the formula $E = \frac{1}{2} mV^2$, but since $m = W/g$, we can substitute and obtain

$$E = \frac{1}{2} \times \frac{W}{g} \times V^2 \text{ or } \frac{WV^2}{2g}$$

where,

m is the mass,
E is the kinetic energy in foot-pounds,
W is the weight of the body in pounds,
g is the acceleration due to gravity, or 32.2, and
V is the velocity in feet per second.

This formula will express the kinetic energy of the fluid passing through a turbine if W is used to represent the weight of fluid per second and V its velocity at entrance.

The smaller W may be, the larger V must be in order to develop the same power. So, if the fluid *leaves* the turbine with the velocity V_a, then $WV_a^2/2g$ represents the energy *not* absorbed by the turbine. If V_a is small, the wasted energy will likewise be small.

Does the construction in Fig. 24-18 give the total reaction?
Answer: No, only the reaction of one vane. The result must be

Steam Turbines

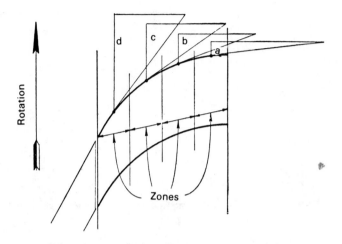

Fig. 24-18. Diagram for obtaining the average value of turning force.

multiplied by the number of vanes to which steam is supplied simultaneously; that is, the number of passages covered *at one time* by the nozzle or nozzles.

What must be considered with respect to reaction?

Answer: The speed of the steam and the tangential speed of the wheel.

Explain.

Answer: In the velocity diagram shown in Fig. 24-19, let AB equal the tangential speed at point A on the vane, and AC the component in direction of rotation. If the speed of the vane represented by A'C' equals AC, the reaction will be zero because the vane will move in the direction of rotation as fast as the steam will move in this direction; that is, its component of motion in this direction. If the speed of the vane equals A"C", the reactive velocity component will be AC−A"C", which equals aC, or the factor V of the formula $WV^2/2g$.

661

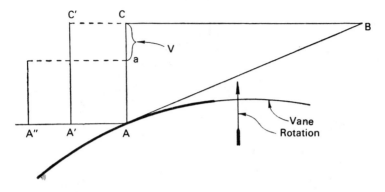

Fig. 24-19. Diagram for obtaining the reactive velocity component or resultant.

MISCELLANEOUS TURBINES

What is a helical-flow turbine?

Answer: A single-pressure, multiple-velocity-stage machine designed for helical flow.

How does it work?

Answer: The steam is expanded to exhaust pressure in the nozzles, the steam jet being redirected back into the turbine rotor from the stationary reversing buckets until its kinetic energy is practically absorbed.

What is a bleeder turbine?

Answer: One in which steam is extracted at one or more intermediate stages for industrial use, often at comparatively high pressures.

What is a low-pressure turbine?

Answer: One operating at approximately atmospheric pressure and expanding to condenser pressure.

What is a mixed-pressure turbine?

Answer: One designed to work on two or more pressures.

STEAM TURBINES

ERECTION OF TURBINES

Instructions given are for a simple, single-stage machine, but they can be applied generally to all turbine installations.

How is the turbine usually mounted?

Answer: Usually on a continuous cast-iron or steel baseplate which extends under the driven machine. The two shafts are connected by a flexible coupling.

What precaution should be taken?

Answer: Any baseplate, no matter how heavily constructed, may be sprung in shipment. Also, it may be distorted by an uneven support on the foundation, uneven tightening of foundation bolts, or by the pull from improper pipe connections.

After the unit is set on the foundation, it is necessary to check alignment and make certain that the shafts of the turbine and of the driven machine are precisely in line. No machine will operate satisfactorily if the shafts are out of line, and serious troubles such as bent shafts, worn bearings, and worn coupling parts may develop. The flexible coupling will *not* compensate for misalignment.

How should the foundation be constructed?

Answer: It may be of either concrete or structural steel. Consult the certified dimension drawing furnished with the machine for size of baseplate and size and location of foundation bolts. When setting the unit on the foundation, use wedges under the baseplate so that there is at least ½-inch (1.27 cm) clearance between baseplate and top of foundation for grouting.

Describe the method of aligning.

Answer: Adjust the wedges so that the baseplate is level, and then check alignment of coupling hubs. Pin-and-bushing–type couplings have wide flanges and are easily checked. Couplings of the gear- or spring-type have enclosing covers, and these covers must be disconnected and slid back along the shaft before alignment can be checked.

When cold, the turbine shaft should be set low to compensate for upward expansion when the casing becomes heated. Allow 0.001-inch (0.0025 cm) for each inch (2.54 cm) in height from turbine feet to centerline of shaft. However, when the driven machine is also heated, such as a hot-oil or hot-water pump, the allowance need not be so great. Therefore, it is necessary to check alignment when cold and also after both members of the unit have been heated to operating temperature.

What checking should be done after setting the unit on the foundation?

Answer: Check alignment by placing a straightedge on the coupling flanges or both faces. Check at top, bottom, and both sides. Also, insert calipers or a feeler gauge between the coupling halves to make certain that both faces are equidistant in any position of the shaft.

How are corrections made?

Answer: Corrections may be made by "throwing" the base or shimming under the turbine or driven machine. Both the drive and driven shafts must run freely with and without the coupling bolts or covers in place. Pull down the foundation bolts, dowel turbine and driven machine to the base, and again check for alignment; make corrections if necessary. The base may then be grouted in. When the turbine is mounted separately on its own foundation, with or without soleplate, the method of installation and alignment is the same as for units with continuous baseplates.

How should the inlet line be installed?

Answer: The inlet line should be anchored at the header and must have loops or offsets to provide flexibility so that expansion of this inlet line will not exert undue strain on the turbine. Install a drain in the inlet line at its lowest point between the header and the turbine. Install a throttle valve in the steam line near the turbine. Close this valve and turn steam into the line; when hot, check the face of pipe flange with turbine flange. Make sure that the faces are in line and parallel and that no force is necessary to bring them together or to match the bolt holes.

STEAM TURBINES

How should the exhaust line be installed?

Answer: The exhaust line should have a flexible expansion joint placed close to the turbine or should otherwise be arranged to compensate for expansion. Provide a drain at the lowest point in this line. Exercise the same care in lining up the pipe flange with turbine flange as described for the inlet line.

Do not depend upon the sentinel relief valve to protect the turbine casing against excessive back pressure. It is advisable to install a large relief valve at some point in the exhaust line. This can be large enough to protect all equipment which may be exhausting into a common header.

How about the miscellaneous pipe connections?

Answer: Each packing box has a drain connection in the lower half. These can be piped separately or in series, but in either case, they must exhaust to atmosphere. For condensing turbines, there will be a tapped opening in the upper half for admitting sealing steam under two or three pounds pressure. Each bearing has a cartridge-type, water-cooling coil. Inlet opening is at the top and outlet at the bottom. These may be piped separately or in series. Place a valve in the inlet line. Water pressure should preferably be 25 to 30 pounds (1.75–2.10 kg/cm^2) and must never exceed 100 pounds (7.0 kg/cm^2).

The inlet flange is tapped on the side for bleed-off line. Install a valve at this point and arrange blow-off to atmosphere. After the overspeed governor valve has closed, either due to over-speeding or hand-tripping, close the throttle valve in the steam line and then open this bleed-off valve to relieve pressure before attempting to reset the overspeed trip valve. A drain connection is provided at the lowest point of the casing. Install a valve at this point, with or without a steam trap.

How may a turbine be installed to provide automatic standby power?

Answer: Pumps, fans, blowers, and other plant auxiliaries normally driven by constant-speed motors can be equipped with a double-extended shaft for connecting a turbine opposite the motor. If this is impossible, the motor shaft can be extended and the turbine coupled behind the motor so that it can drive the auxiliary through the motor shaft. On dual-driven units, the standard speed-governor can be easily adjusted so that the turbine will automatically pick up the load in case of electric power failure. With the unit in operation (driven by the electric motor), the adjusting screw at the lower end of the turbine governor lever is tightened, closing the governor valve and shutting off steam-flow. The screw is then locked in that position, and the governor valve will remain closed as long as normal electric-motor speed is maintained. Upon motor-power failure, the unit immediately slows down, flyballs close in, governor valve opens, and steam is admitted to the nozzles. The load is automatically picked up by the turbine at a speed 3% to 5% below normal, with full-speed operation possible by loosening the adjusting screw to open the governor valve. No other automatic controls are required.

TURBINE OPERATION

What should be done before attempting to start the turbine?

Answer: Clean off any dirt which might have accumulated during erection, flush out bearing cavities, and fill with new oil. Then, work the governor and trip valves by hand to make certain that they are free.

What should be done when starting for the first time?

Answer: Check alignment after the turbine has become heated. Coupling halves must be in line within 0.004-inch (0.010 cm). If found to be out of line, disconnect the inlet and exhaust flanges and realign. Be sure that the flanges of these lines are parallel with the turbine flanges before reconnecting them.

How would you start the turbine?

Answer: I would proceed as follows:

1. Check the oil level in the bearings
2. Open the drain from the turbine casing
3. Open drains in the inlet and exhaust lines
4. Open the exhaust valve

Steam Turbines

   5. Crack the throttle valve
   6. Close the drain valves after condensation ceases to flow and the turbine has become heated
   7. Open the throttle valve sufficiently to start the turbine
   8. Examine the oil rings—make sure they are turning
   9. Open the throttle valve gradually until the machine comes up to speed and the governor has taken control
  10. Open the throttle valve wide.

When the operator becomes familiar with the turbine, it may be brought up to full speed within 30 to 60 seconds.

What important precaution should be taken after starting the turbine?

Answer: Test the overspeed trip governor by actually over-speeding the machine. If not advisable to over-speed the driven machine, disconnect the flexible coupling.

How do you over-speed the turbine?

Answer: Push on the top governor lever.

What does this do?

Answer: It opens governor valve, increases the supply of steam to the nozzle, and speeds up the turbine. The overspeed governor should trip at a speed of 10 to 15% above the designed speed shown on the nameplate.

How do you shut down the turbine?

Answer: Proceed as follows:

  1. Trip the overspeed governor valve by disengaging the trip latch. This will test the valve mechanism.
  2. Close the throttle valve in the steam line.
  3. Reset the overspeed governor.
  4. Open all turbine drains.

What care should be given to the turbine?

Answer: There are numerous items that should be attended to, such as:

  1. Watch the oil level in the bearings;
  2. Examine the oil at least once each week to make sure that it is

in good condition. Evidences of deterioration are an increase in viscosity, development of acidity, and the deposit of sludge or sediment;
3. Clean out and flush the oil reservoirs at least every two or three months. Refill with new oil;
4. Keep the valve stems clean and properly packed;
5. Trip the overspeed governor occasionally to see that it is in proper condition;
6. Watch for endwise movement in the shaft of the driven machine. Make sure that this movement is not transmitted to the turbine shaft;
7. Inspect the bearings frequently. On sleeve bearing turbines the governor end bearing is set for 0.008-inch (0.020 cm) total clearance between thrust faces. See that this clearance is maintained. Keep the bearings in good condition and the turbine will operate indefinitely.

What kind of oil should be used for lubrication?

Answer: In order that the oil rings may carry the oil to the top of the journals and distribute it freely over the bearing surfaces, the oil itself must have the proper body under the existing operating temperatures. It must not be so heavy as to retard the action of the rings. An oil that is too light, on the other hand, will not be able to maintain the necessary film between the surfaces. As temperature has a positive influence on the body of an oil, it is necessary that an oil be selected which will attain the proper body at the operating temperatures to which it will be subjected in actual service.

How long will the oil last?

Answer: The time that an oil will last in a turbine varies according to operating conditions. The continuous circulation of oil in a bearing results in a gradual deterioration of the oil itself, due to the action of heat, moisture, and air. Evidences of this deterioration are development of acidity, increase in viscosity and the deposit of sludge or sediment.

The length of service that may be expected from the oil is dependent upon many variables, such as stability, operating temperature, and rate of circulation. Consequently, it is impossible to make any definite statement in regard to the frequency of oil changes. Since the quantity of oil required for ring-oiled bearings is small, good advice is to change the oil frequently.

TURBINE MAINTENANCE

How are the two halves of the casing sealed?

Answer: With Permatex or other liquid gasket such as oil and graphite. Top half casing is provided with tapped holes for jack screws which are used to break the joint. There are no high-pressure joints inside this casing.

How is leakage of steam along the shaft prevented?

Answer: By carbon-ring packing. In those so equipped, each packing box consists of a bronze split housing, three carbon rings, bronze diaphragm, and cover. Garter springs which hold rings together are stainless steel.

How do you renew the carbon rings?

Answer: Remove the top half of the turbine casing and loosen the cover. Slide it back along the shaft; then, lift off the top half of the packing box. This exposes the rings, which can easily be removed. Make certain that the shaft and packing housings are free from rust or scale.

How do you insert new rings?

Answer: Line them up so that the slots for the retaining pins are at the top and on exact center with the shaft.

What should be done next?

Answer: Replace the top half housing, using oil and graphite at the split. Next, replace the top half casing and then pull up the cover. When the packing box is properly assembled with rings seating and pins in slots, very little pressure is required to pull up the cover. When in proper position, there is $1/16$-inch (1.58 mm) clearance between cover and housing. If clearance is greater, the pins are riding on the carbon rings and it will be necessary to dismantle and reassemble. The joint of the housing with casing is made with Permatex or oil and graphite. Carbon-ring clearance on shaft is 0.002-inch (0.005 cm).

Describe the bearings.

Answer: In this type turbine they are babbitt-lined, sleeve-type, ring-oiled, and horizontally split, mounted in cast-iron, split brackets. They may be rocked out of position and replaced after removing the bearing caps. Governor end bearing is babbitt-faced at both ends and serves as a thrust bearing as well as radial bearing. This bearing has radial oil grooves in both ends for lubricating the thrust faces. Running clearance between each thrust face is 0.004-inch (0.010 cm), or 0.008-inch (0.020 cm) total for both faces. Coupling end bearing has oil collector grooves in both ends with drain holes in lower half.

What do replacement bearings require?

Answer: Scraping and fitting to the shaft, so that the running clearance will be 0.003-0.004-inch (0.007-0.010 cm). Also the governor end bearings should be scraped on both ends to fit into the space occupied by the old bearing. If necessary, the locating collar, which is set-screwed to the shaft, can be shifted either way to match the new bearing.

What precaution should be taken in fitting the bearings?

Answer: Make certain that the rotor is equidistant between the nozzle and the reversing nozzle. Clearance at the nozzle and also the reversing nozzle is $\frac{1}{32}$-inch (0.794 mm).

What provision is made on bearing caps?

Answer: Each bearing cap has a tapped opening at the top for filling the oil reservoir. Through these openings, the operator can view the oil rings to make certain that they are turning with the shaft.

How is the oil cooled?

Answer: By cartridge-type, water-cooling coils.

Are they necessary?

Answer: Not always.

Steam Turbines

What should be done when it becomes necessary to renew blades?

Answer: The rotor should be returned to the factory for assembly, balance, and inspection.

What should be done when the rotor cannot be returned to the factory?

Answer: Order a complete new wheel.

How would you remove a nozzle?

Answer: Unbolt the steam-chest assembly. The nozzle-stand flange is then exposed, and this assembly slides out of the casing.

How would you replace?

Answer: A dowel pin permanently fixes the position of both nozzle and stand with respect to blades. When replacing the stand, make sure that the nozzle outlet lines up with the blades and then redowel.

How do you remove the overspeed valve assembly?

Answer: To remove the overspeed valve assembly, first disconnect the trip link and then remove capscrews from the valve cage. The joint with the steam chest is broken by two jack screws which are inserted into tapped openings provided for this purpose. The basket-type steam strainer is removed with this assembly.

How do you dismantle this assembly?

Answer: Remove the strainer by taking out two screws. Unscrew the trip-steam knob and nut, both right-hand threads. Lift off the spring, loosen the packing nut, and push the valve stem through the packing. Unscrew the recessed pipe plug from the lower end of the trip valve and push out the valve stem. Reverse this procedure when reassembling. Clearance between trip lever and governor hub is $1/16$-inch (1.58 mm). Adjust at the trip link.

How do you the remove governor valve?

Answer: Take off the governor lever bracket and loosen the capscrews in the flange of the steam-chest head. The steam-chest head and governor valve may then be pulled out. This exposes the valve cage, which may also be pulled out. The pin is set for $1/64$-inch (0.397 mm) clearance at steam-chest head.

How do you remove the steam-chest assembly?

Answer: The steam-chest assembly can be removed as a unit simply by loosening the nuts which hold it to the casing. On two-nozzle turbines, the auxiliary valve stand and hand valve are part of, and removed with, this assembly.

How do you renew the trip valve packing?

Answer: Unscrew the knob and nut (right-hand threads), lift off the spring, unscrew the packing nut, and then pull old packing. Replace with packing ordered from the factory. Work in by pushing stem down by hand. Make certain that the stem works freely before starting the turbine.

How do you renew the governor valve packing?

Answer: Take off the bracket, unscrew the nut, unbolt the cradle, and then pull the old packing. Replace the packing ordered from the factory. Never fill the box completely full. After assembling, work the stem by hand to make certain that it is free. Never use standard square packing. Use only conical packing.

How do you dismantle the speed governor?

Answer: Disconnect the bracket, withdraw the governor plunger, remove the top half governor housing, and unscrew the governor nut (right-hand thread). Then the governor spring with spring seat, plunger seat, and flyballs can be lifted out. Before unscrewing governor nut, be certain to note its exact position on the shaft so that it can be replaced in the same location.

To inspect or renew the governor plunger or ball bearing, simply disconnect bracket and draw out plunger assembly. No other parts need be disturbed.

How do you make the speed adjustments?

Answer: Speed may be changed while turbine is in operation.

Turning the governor adjusting screw to the right reduces speed, while turning to left increases speed.

A total adjustment of about 20% or approximately 10% above and below normal, may be made in this manner. Additional speed adjustment of about 5% may be made, after shutting down the turbine, by tightening or loosening the governor nut on the end of the shaft. Governor nut is accessible through inspection window after removing the cover.

Adjustment for speeds to about 50% below normal can also be made, while in operation, by further tightening the adjusting screw, but the governor will be inoperative over this range. Operation of the turbine will be satisfactory, provided the load of the driven machine is fairly constant and the steam pressure does not fluctuate to any great extent.

How does the overspeed governor work when over-speeding occurs?

Answer: Trip plunger moves outward, striking a lever, which releases a latch and permits spring to close trip valve. Do not attempt to reset the valve until steam pressure has been relieved. To relieve pressure, close the throttle valve in steam line and open the hand valve which was installed at turbine inlet flange.

How do you raise the trip speed?

Answer: Remove the inspection cover, loosen allen-head screw, and then tighten adjusting nut. A slight turn of this nut will increase speed considerably. Be sure to tighten the setscrew after any adjustment.

Check this overspeed valve at least once each week by hand-tripping. Simply disengage the trip latch, which allows the valve to close. Also check trip plunger periodically to make certain that it is free to move outward. This is done by inserting a blunt-nosed rod through the hole in the adjusting nut and pressing down against spring tension.

TURBINE TROUBLESHOOTING

Here are a few troubles that may be encountered from time to time, together with their probable causes.

Vibration

1. Bent shaft,
2. Worn bearings,
3. Coupling out of balance,
4. Driven machine out of balance,
5. Misalignment between turbine and driven machine—check when turbine case is hot.

Stuffing Boxes Leaking Steam Along Shaft

1. Clogged drain line,
2. Carbon rings improperly installed,
3. Worn carbon rings—check and replace if necessary; clean and polish shaft surface at same time.

Governor Hunting

1. Worn or bent valve stem,
2. Worn or fouled governor plunger,
3. Lost motion in governor lever,
4. Governor valve sticking,
5. Governor valve or cage worn,
6. Worn governor ball-bearing.

Turbine Will Not Come Up to Speed

1. Speed governor out of adjustment,
2. Load of driven machine too great,
3. Nozzle-throat fouled by foreign matter,
4. Low steam-pressure at turbine inlet,
5. Low pressure at nozzle—caused by partially closed valve in steam line, clogged steam line, or clogged strainer. Check nozzle pressure at tapped opening in top of steam chest, close to turbine case.

Turbine Speed Too High

1. Governor out of adjustment,
2. Worn governor valve or cage.

Bearing Failures

1. Misalignment,
2. Inferior or improper oil,
3. Improper fitting or oil grooving,
4. Thrust transmitted to turbine shaft from driven machine.

Bearing temperatures above 200° are not alarming on sleeve- or ball-bearing machines. However, these higher temperatures require closer attention to oil selection and more frequent oil changes.

CHAPTER 25

Air Compressors

What is free air?

Answer: Air at atmospheric condition at the point where a compressor is installed.

What is compressed air?

Answer: Air forced into a smaller space than it originally occupied.

What happens when air is compressed?

Answer: Both its pressure and temperature rise.

What is the object of compressing air?

Answer: The power available from compressed air is used in many applications as a substitute for steam, electricity, or other forces, as in operating shop tools, rock drills, and so forth. In some cases, compression is used to provide air for combustion, and to produce and maintain reduced pressure levels.

What happens when the space occupied by a given volume of air is enlarged?

Answer: The volume of air expands, which reduces both its pressure and temperature.

USING BOYLE'S AND CHARLES' LAWS

State Boyle's law.
Answer: At a constant temperature, the volume of a gas varies inversely with absolute pressure.

What is compression of a gas at constant temperature called?
Answer: Isothermal compression.

State Charles' law.
Answer: At constant pressure, the volume of a gas is proportional to its absolute temperature; at constant volume, the pressure is proportional to its absolute temperature.

What is the ratio of compression?
Answer: The ratio of the final volume to the initial volume. For example, if the final volume is 10% of the initial volume, what is the ratio of compression?

$$\text{Ratio} = \text{Initial volume} \div \text{Final volume}$$
$$= 100 \div 10$$
$$= 10$$

How is the final pressure of compression obtained knowing the initial pressure and the number of compressions?
Answer: The rule is that the final pressure of compression is equal to the initial pressure times the number of compressions. For example, if air is compressed from an initial pressure of $33\frac{1}{3}$ psi absolute to $\frac{1}{3}$ volume, what is the final pressure?

$$\text{Number of compressions} = 1 \div \frac{1}{3}$$
$$= 1 \times \frac{3}{1}$$
$$= 3$$

Air Compressors

$$\text{Final pressure} = 33\frac{1}{3} \times 3$$
$$= 100 \text{ psi absolute } (7 \text{ kg/cm}^2)$$

How do you find the initial pressure of compression?

Answer: The rule is that the initial pressure of compression equals the final pressure divided by the ratio of compression. For example, if the final pressure is 100 psi absolute, and the number of compressions, or compression ratio, is 3, what is the initial pressure?

$$\text{Initial pressure} = 100 \div 3$$
$$= 33\frac{1}{3} \text{ psi absolute } (2.33 \text{ kg/cm}^2)$$

What precaution should be taken in making compression calculations?

Once and for all, remember that the pressures should be *absolute* pressures and not gauge pressures.

Please summarize the formulas.

Answer: (1) Compression ratio equals initial volume divided by final volume; (2) Final pressure equals initial pressure times compression ratio; (3) Initial pressure equals final pressure divided by compression ratio.

What happens when air is compressed?

Answer: Heat is generated, as illustrated by the familiar operation of pumping up a bicycle or automobile tire.

What is the big problem in the design of compressors?

Answer: To get rid of the heat of compression which results in a loss by boosting the pressure abnormally.

What is compression without receiving or giving up heat called?

Answer: Adiabatic compression.

EFFECT OF BOYLE'S AND CHARLES' LAWS COMBINED

In the ordinary process of air compression, what two elements are at work toward the production of a higher temperature?

Answer: (1) The reduction of volume by the advancing piston, and (2) the increasing temperature due to the increasing pressure corresponding to the reduced volume.

Describe the application of the two laws.

Answer: Suppose that a cylinder is filled with air at atmospheric pressure (14.7 psi absolute, or 1.03 kg/cm²) and the piston is moved to compress the air to ⅓ volume. According to Boyle's law, the pressure will triple, becoming 44.1 psi absolute (3.08 kg/cm²). In reality, here is where Charles' law comes in, because actual measurement would show a pressure greater than 44.1 psi due to the increase in temperature of the air from the compression process. This is called *adiabatic* compression (Fig. 25-1), and extra work is required from the piston in order to overcome excess pressure due to the temperature rise.

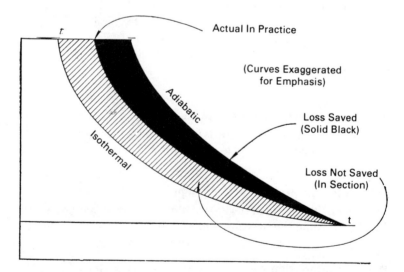

Fig. 25-1. Isothermal, adiabatic, and actual compression curves.

Why is this extra work lost?

Answer: Because after the hot compressed air leaves the cylinder, it cools and the pressure drops to what it would have been if compressed at constant temperature; that is, compressed according to Boyle's law.

Is all the extra work lost in the operation of a compressor? Why?

Answer: No, because in compressors where working efficiency is considered, some means of cooling the cylinder is provided, such as projecting fins or jackets for the circulation of cooling water.

How much work is lost?

Answer: About half, more or less, depending upon the design of the compressor. See Fig. 25-1.

TYPES OF COMPRESSORS

What is a compressor?

Answer: A machine (driven by any "prime mover") which compresses air into a receiver to be used later.

What is a prime mover?

Answer: An apparatus or mechanism whereby motion and force are received directly from some natural source of energy and transmitted into some form of motion by means of which the power may be conveniently applied.

Name some prime movers.

Answer: Steam engines, gas engines, and electric motors.

Mention some different types of compressors.

Answer: (1) Single-acting, (2) double-acting, (3) single-stage, (4) double-stage, (5) air-cooled, (6) water-cooled, (7) single-cylinder, (8) multicylinder.

Compressors generally fall into two categories: *positive displacement* types where quantities of free air or gas are confined or trapped within a closed space and compressed to a higher pressure, and *dynamic* types in which a high-speed rotating element increases the movement of air as it passes through it and converts the speed of movement into head pressure.

Positive displacement compressors include reciprocating, rotary, sliding-vane, liquid-piston, two-impeller straight-lobe, and helical or spiral-lobe types. Dynamic compressors are the centrifugal, axial, and mixed-flow compressors.

What is a single-acting compressor?

Answer: One in which compression takes place every other stroke.

What is a double-acting compressor?

Answer: One in which compression occurs every stroke. See Fig. 25-2.

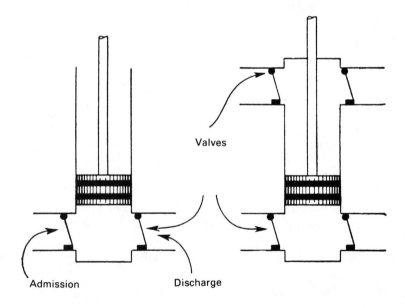

Fig. 25-2. Single-acting (left) and double-acting compressors.

What is a single-stage medium-pressure compressor?

Answer: One in which the compression cycle takes place in a single cylinder (Fig. 25-3).

What is a two-stage high-pressure compressor?

Answer: One in which the compression begins in one cylinder and is completed in a second cylinder.

What is the object of two-stage compression?

Answer: It divides the temperature range between the two cylinders and permits cooling between the cylinders (Fig. 25-4).

What is a reciprocating compressor?

Answer: One having a piston arranged to move forward and backward. See Fig. 25-5.

What is the proper name for a centrifugal compressor?

Answer: A low-pressure blower.

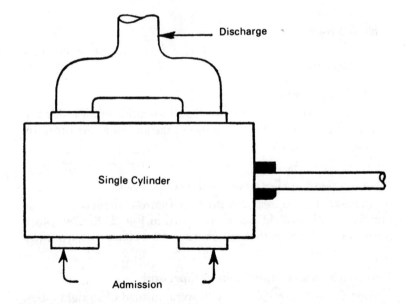

Fig. 25-3. Single-stage compressor.

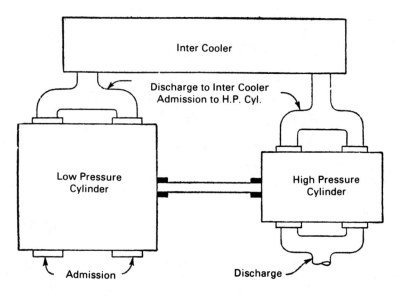

Fig. 25-4. Two-stage compressor.

What is a rotary compressor?

Answer: One having a vane rotor, or its equivalent, mounted in a starionary casing (Fig. 25-6).

What is a liquid-piston compressor?

Answer: A rotary compressor that traps a ring of air inside the casing. The shape of the casing forces the air out as the pressure increases. See Fig. 25-7.

What is a two-impeller straight-lobe compressor?

Answer: A compressor with two figure-8 shaped rotors. or impellers with straight lobes, as shown in Fig. 25-8. The rotors push air from the inlet to the discharge where compression occurs.

What is a helical- or spiral-lobe compressor?

Answer: A two-rotor unit with spiral instead of straight lobes. See Fig. 25-9.

AIR COMPRESSORS

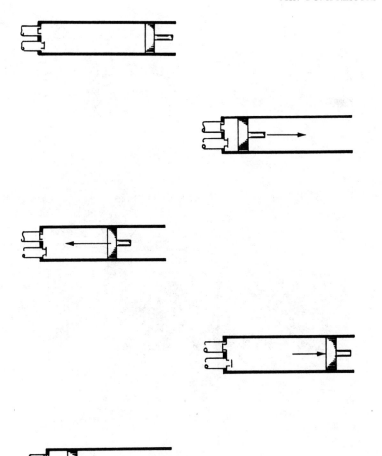

Fig. 25-5. Steps in the reciprocating compressor cycle.

What is an axial-flow compressor?

Answer: A high-speed type where each stage has two rows of blades, one rotating and the other stationary. As the rotor turns, it forces air through the stationary blades, thus compressing it.

What is a direct-connected compressor?

Answer: One in which the prime mover is attached direct to the

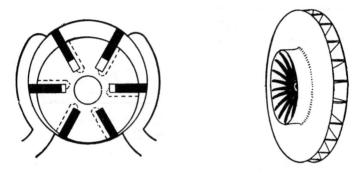

Fig. 25-6. Centrifugal (left) and rotary compressors.

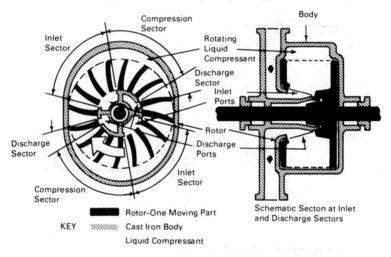

Fig. 25-7. Liquid-piston compressor.

compressor without any interposed transmission such as a chain or belt.

What is the difference between vertical, horizontal, and vee cylinder placements?

Answer: This relates to the position of the axis of the cylinder or cylinders with respect to the horizontal.

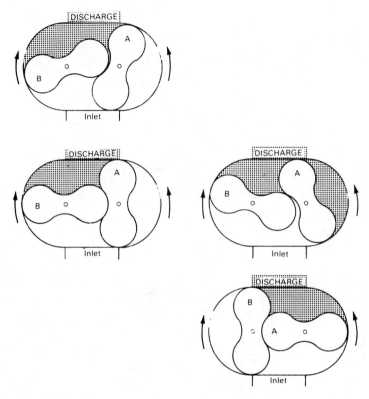

Fig. 25-8. Two-impeller straight-lobe compressor showing four "stages" of operation.

What is the point about single-cylinder and multicylinder compressors?

Answer: This relates to the power end. Where a steam engine is a prime mover, one cylinder (or two, if compound) does the job. Of course, when a gas engine or diesel engine is the prime mover, as many as eight cylinders are required to approach the performance of a steam engine.

What is the point as to compound steam-engine drive?

Answer: It is not a case of getting a smoother turning effect, but to get better economy.

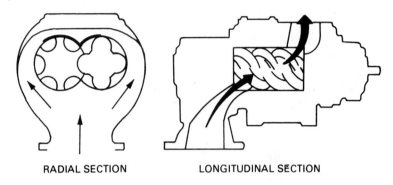

Fig. 25-9. Helical-lobe rotary compressor.

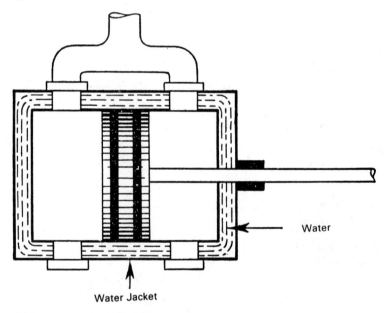

Fig. 25-10. Water-cooled compressor.

COOLING

What is a water-cooled compressor?

Answer: A compressor whose cylinder is water jacketed (Fig. 25-10).

AIR COMPRESSORS

What duty is performed by the cold water in the jacket?

Answer: It functions as a transmission medium to carry off some of the heat of compression.

What other medium is employed to carry off the heat of compression?

Answer: Air.

How is the cylinder constructed for air-cooling?

Answer: It has numerous thin fins cast integral to form a large cooling surface. A draft of cool air then forms the medium to carry off some of the heat of compression (Fig. 25-11).

What is a power-driven compressor?

Answer: They are all power-driven, but this means a compressor whose prime mover is a separate unit connected by a transmission such as a belt or chain.

What are "en bloc" or unibloc cylinders?

Answer: Two or more cylinders cast integral; that is, all in one casting.

Describe the hopper cooling system.

Answer: This is a nonexternal-circulating system. The cylinder

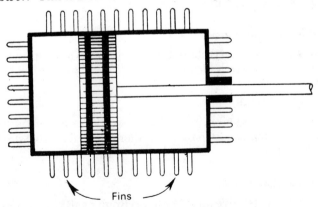

Fig. 25-11. Air-cooled compressor.

has an open water jacket of considerable volume. It is filled with water and the volume of water depended upon to cool the cylinder by thermocirculation.

COMPRESSOR VALVES

What type of valves are used on compressor cylinders?

Answer: The various types of air valves in general use may be classed as (1) finger, (2) feather, (3) plate, (4) ring (plain or dual-cushion), (5) channel.

What role do valves play?

Answer: Allow a one-way flow of air into and out of the cylinder. Each must open and close once during each turn of the crankshaft.

What are finger valves?

Answer: Valves made of narrow strips of stainless steel fastened to a seat at one end and free to flex along their length. See Fig. 25-12.

What is the application of finger valves?

Answer: They are suitable for light service, as in small compressor units for garage and service stations.

What is a feather valve?

Answer: A strip of ribbon steel which covers a slightly narrower slot when the valve is closed.

How does it work?

Answer: In opening, a feather valve flexes against a curved guard, allowing the air to pass on either side of the valve strip with very little friction.

What is a plate valve?

Answer: This valve (also known as Rogler valve) consists of flat discs of special steel, clamped upon valve seats and held in posi-

Air Compressors

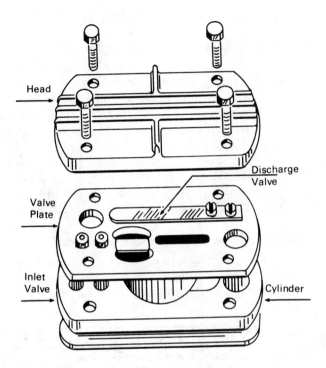

Fig. 25-12. Finger-valve assembly used on small compressors.

tion by spring arms formed in one piece with the valve itself. The stop-plate has spring arms which press upon the valve, limiting the opening movement and closing the valve at the end of the stroke. The stop-plate is backed up by a buffer plate which is identical with the stop-plate except that the spring arms are not cut through.

What is a ring disc-valve?

Answer: The valve proper consists of annular rings stamped out of thin alloy steel. One valve of this class is cushioned by special springs. See Fig. 25-13.

What is a channel valve?

Answer: This is a heavy-service valve. See Fig. 25-14.

PACKAGED COMPRESSOR SYSTEMS

Discuss the operation of a two-stage compressor.

Answer: In compressing air to 100 psi (7 kg/cm²) in a two-stage compressor, the air is compressed from atmospheric pressure to about 26½ psi (1.85 kg/cm²) in the low-pressure cylinder and is delivered to the intercooler at 240°F. (115.6°C.). If all of the heat of compression is taken out at the intercooler, the air is admitted to the high-pressure cylinder at atmospheric temperature. It is then compressed from 26½ psi to 100 psi and delivered to the receiver at 240°F. In a single-stage compressor, the air is compressed from atmospheric pressure to 100 psi in one cylinder and reaches the receiver at 480°F. (about 249°C.).

Mention another type of low-pressure compressor.

Answer: In the impeller-type of rotary compressor (Fig. 25-18), two meshing impellers, called lobes, are gear-driven in opposite directions within a housing. Minimum operating clearance pre-

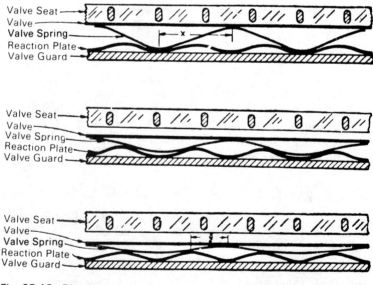

Fig. 25-13. Diagrams showing operation of the ring disc valve with dual cushion springs.

Fig. 25-14. Parts of a cushioned channel valve.

vents leakage of the confined air during compression. Such units can handle large volumes of air per minute at a relatively low pressure and steady delivery.

What is a screw cycloidal, or twist-type, compressor?

Answer: See Fig. 25-19. The axial movement of air through this type of compressor is caused by meshing of the four spiral lobes on the male rotor with the six cavities on its female counterpart (the male rotor is driven by speed-up gears). As the rotors turn, the initial volume of trapped air is forced along each cavity and through a discharge port designed to provide the desired discharge pressure. Because of its relatively simple design, pulsation-free discharge, quiet operation, and low discharge-air temperature, the use of this design of compressor is rapidly increasing.

What are the elements of a "packaged" system?

Answer: A complete air-power system includes an inlet filter and silencer, an aftercooler, a moisture separator, condensate trap, air receiver, motor, motor starter, and controls, in addition to the compressor itself (depending on requirements, an air dryer and an automatic control system may also be necessary). For the sake of efficiency, many manufacturers supply "package systems" consisting of all or most of these units, preassembled and tested, and requiring only the final pipe and electrical connections to put them into operation.

Figs. 25-15 to 25-22 are examples of horizontal and vertical single-stage compressor package units. Fig. 25-15 is a gas-engine- or electric-motor-driven, single-acting air-cooled compressor for automotive and industrial uses. This class of machine is designed to produce up to 200 acfm at 250 psig. Fig. 25-16 shows a heavy-duty, single-acting two-stage, compressor unit, with air-cooled cylinders and intercooler. It is designed to produce up to 550 acfm at 250 psig. A crosshead, two-stage, double-acting water-cooled compressor is shown in Fig. 25-17. It has balanced opposing cylinders and is said to be capable of producing up to 650 acfm. Fig. 25-18 is an L-type cylinder model also designed for continuous use at an output up to 2500 acfm. The compressor shown in Fig. 25-19 is a direct-drive rotary screw packaged unit. Fig. 25-20 is

AIR COMPRESSORS

Courtesy Ingersoll-Rand Co.

Fig. 25-15. Type 30 single-acting, air-cooled compressor unit.

a single-cylinder, water-cooled vertical belt-driven reciprocating type designed to produce up to 487 cfm at 100 psig. A two-cylinder V-type compressor is pictured in Fig. 25-21. It is said to be capable of 1092 cfm at 125 psig continuous duty.

INTERCOOLERS AND AFTERCOOLERS

All free air as well as manufactured gas contains varying amounts of moisture in the form of water vapor. Since it is temperature and volume, not pressure, which determine the moisture capacity of air or gas, the capacity increases as the temperature rises. For example, 1000 cubic feet of saturated air (air at 100% humidity) will contain 1.57 pounds of water at an air temperature of 80°F. At an air temperature of 100°F., it will contain 2.85 pounds. Since one gallon of water weighs 8.33 pounds, it is apparent that air can contain an appreciable amount of water and that

695

Courtesy Ingersoll-Rand Co.

Fig. 25-16. Type 40 single-acting, two-stage, air-cooled compressor unit ready for service connections.

great quantities of water are taken into compressors.

At the higher temperatures developed during compression, the air or gas leaving the compressor can carry this moisture in the form of vapor despite the reduction in volume. However, after

Courtesy Ingersoll-Rand Co.

Fig. 25-17. Class PHE heavy-duty, crosshead two-stage, double-acting water-cooled compressor unit.

AIR COMPRESSORS

Courtesy Ingersoll-Rand Co.

Fig. 25-18. Two-stage, double-acting water-cooled compressor, with L cylinder design.

this compressed air or gas becomes cooled in the pipeline to atmospheric temperature, it will hold only a small percentage of moisture in the form of water vapor and the majority of it condenses into liquid, causing problems in the pipes, lines, receiver, and air-operated equipment.

Intercooling and aftercooling precludes excess-moisture problems by extracting moisture before it reaches the pipeline and receiver. If the air mentioned previously is cooled to 80° with 65° water (both F.) circulating in the intercooler and/or aftercooler, it will hold only 0.20 pounds of water per 1000 cubic feet, or about 13% of the water vapor originally taken into the compressor. In a compressor of 1000 ft^3/min capacity, nearly 10 gallons of water per hour will be removed, so the advantages of aftercooling are evident.

Air cooled to within 15°F. of incoming cooling-water temperature is considered good practice. Attempting to secure a lower air temperature with the same water temperature requires a much larger and more expensive cooler, and increases water consumption. These disadvantages more than offset the extra gain in condensation, for if the air had been cooled to 70°F. instead of 80°F.,

Courtesy Joy Machinery Co.

Fig. 25-19. Twistair rotary screw air compressor.

about 92% of the original moisture would be condensed, which is a gain of only five percentage points in the total condensate.

Intercoolers

What is an intercooler?

Answer: A species of surface condenser placed between the two cylinders of a two-stage compressor so that the heat of com-

AIR COMPRESSORS

Courtesy Joy Machinery Co.
Fig. 25-20. WGAP-9 continuous duty compressor unit.

pression generated in the first-stage cylinder may be removed in part or whole from the air as it passes through the intercooler to the second-stage cylinder (Fig. 25-22).

Why is this necessary for efficiency?

Answer: Because the pressure of a gas increases with a rise in temperature, as previously explained, and this excess pressure results in a waste of power in compressing the gas.

What is the approximate saving?

Answer: Roughly, each 1°F. decrease in temperature between stages results in a 1% saving in power input.

Mention another favorable result obtained by intercooling between stages.

Courtesy Joy Machinery Co.

Fig. 25-21. WN-112 series, two cylinder V-type reciprocating compressor unit.

Answer: Since it reduces the maximum temperature in the second-stage cylinder, better cylinder lubrication is obtained, which minimizes valve troubles.

What objectionable action is brought about by intercooling?

Answer: Intercooling (or *any* cooling) causes condensation of some of the moisture passing through the intercooler. See Fig. 25-23.

How is the condensate disposed of?

Answer: A separator with float-valve ejector is provided to get rid of the condensate.

What would happen if the condensate passed into the second-stage, high-pressure cylinder?

Answer: It would wash away lubrication and induce rapid wear.

Air Compressors

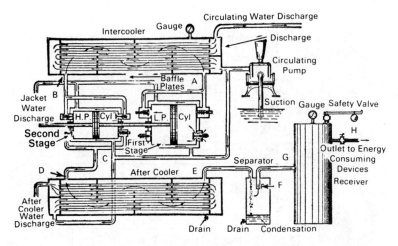

Fig. 25-22. Free air is moderately compressed in the first-stage cylinder and discharged into the intercooler (A), where the heat of compression is removed by the circulating cooling water. Air leaves the intercooler (B), enters the high-pressure cylinder, is further compressed, and is discharged (C) into the aftercooler (D), where the heat of the second-stage compression is removed in the same way. The highly compressed air leaves the aftercooler (E), passes to the separator (where condensate is removed), leaves the separator (F) and enters the receiver (G), where it is stored at working pressure.

What should be noted about temperatures?

Answer: The final temperature in each cylinder will be the same if the work has been equally divided and the intercooler properly designed. The final temperature will be much lower than it would be if the total compression were done in one cylinder (single-stage).

What difficulties are encountered in an attempt to cool the air in the cylinder during compression?

Answer: There is not sufficient time during the stroke to thoroughly cool the air by any available means. The old method of injecting cooling water into the cylinder (wet compression) is out.

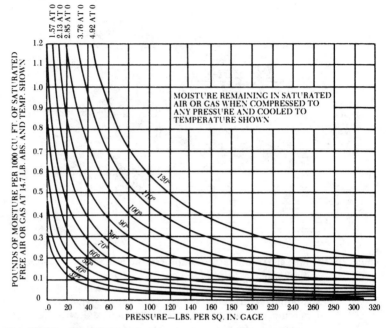

Fig. 25-23. Air moisture chart.

What should be inferred about compressors whose indicator diagrams show compression lines "running away" from (that is, lower than) the adiabatic?

Answer: Such lines should be regarded with suspicion as they probably result from leakage rather than from efficient cooling. Such leaks will explain many isothermal compression lines.

Describe the construction of a typical intercooler.

Answer: It consists of a nest of special metal tubes expanded into corrosion resistant tube plates at each end. Baffles effectively direct the air or gas across the tubes many times, thus affording perfect contact with the cool tubes. The entire nest of tubes may be withdrawn as a unit for cleaning. Provision is made for tube expansion.

What stabilizing effect has the intercooler?

Answer: The large receiver volume formed by the intercooler and the connecting piping results in a nearly uniform discharge pressure from the low-pressure cylinder.

How are intercoolers mounted?
Answer: Compressors of 1200 ft^3 capacity (33.6 m^3) and larger have their intercoolers mounted directly over the air cylinders. The intercoolers of smaller machines are usually arranged underneath the cylinders.

Aftercoolers

What is an aftercooler?
Answer: Like an intercooler, a species of surface condenser, in which compressed air is cooled after compression. See Fig. 25-22.

Describe the construction of a typical aftercooler.
Answer: When water is used as the cooling medium, air may flow (1) through the tubes, or (2) outside the tubes. In the first type, air passes through numerous tubes, being divided into thin streams. Proper air velocities are developed for high heat-transfer to the surrounding water.

What is placed at the end of the aftercooler?
Answer: A separator which catches the condensate. Fig. 25-23 is an air-moisture chart.

How are pipelines affected by moisture?
Answer: Moisture has various harmful effects in pipelines. It (1) produces water hammer, (2) reduces the carrying capacity, (3) causes loss of power by accumulation at low points, and (4) may freeze pipelines in cold weather.

What bad effect is encountered in pipelines in the absence of an aftercooler?
Answer: Where hot air is passed directly from the compressor into pipelines, the heat causes the lines to expand; this is followed by contraction when the compressor is shut down. The repetition

of this cycle induces leaky joints and consequent air losses even when expansion joints are provided.

How do separators separate the condensate?

Answer: The action depends upon centrifugal force, which is brought into effect by suddenly changing the direction of flow of the gases, usually through 180°.

What is the object of a receiver?

Answer: To provide a storage reservoir for the compressed air. This gives a supply to work on, thus avoiding sudden pressure fluctuations on load demands.

When is a large receiver capacity recommended?

Answer: When automatic start-and-stop control is used on the compressor.

SPEED AND PRESSURE CONTROLS

Why is some method of control necessary for a compressor?

Answer: So that the compressor can maintain a supply of air at a predetermined pressure in the receiver to meet any demand.

Name two general control methods.

Answer: Regulation may be secured by providing suitable devices at the air ends or at the power ends.

What kind of regulating devices are used at the power end of steam-driven compressors?

Answer: Throttling or variable cut-off governors. See Figs. 25-25 and 25-26.

What is the peculiarity of these governors for air compressors as compared with other service?

Answer: The usual function of a governor on a steam engine is to maintain a constant speed; however, as applied to a compressor,

Air Compressors

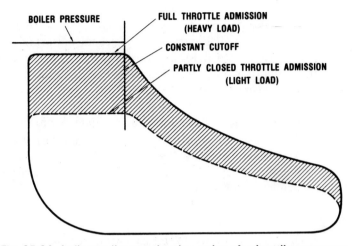

Fig. 25-24. Indicator diagram showing action of a throttling governor.

it acts to maintain a predetermined constant pressure by varying the speed of the machine.

What is the actuating force for these governors?

Answer: The pressure of the air in the receiver by acting upon a

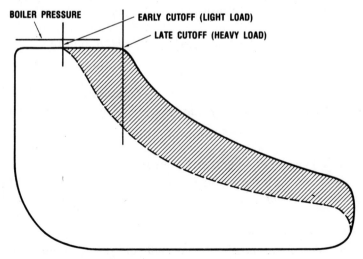

Fig. 25-25. Indicator diagram illustrating regulation by variable cut-off.

705

diaphragm or piston and with proper connection operates the governor to vary the amount of steam admitted to the cylinder as in the case of the throttling governor or to vary the cut-off of the cut-off engine. This acts to increase or decrease the speed accordingly as the variation in cut-off increases or decreases the mean effective pressure (mep), respectively.

How does the cut-off governor operate to facilitate starting?

Answer: It is necessary to lengthen the cut-off and thus obtain the necessary torque for starting. It automatically adjusts the cut-off for starting, then returns it to the proper operating position as soon as the machine is up to speed.

What provision is made to prevent over-speeding?

Answer: In case of damage to the governor, over-speeding is prevented by an independent safety stop valve.

When is the automatic "stop and start" method of control used?

Answer: When the demand for air is intermittent.

Explain the stop and start method.

Answer: The compressor either operates at full capacity or is shut down, the motor being automatically started and stopped by a pressure switch. When no air is being used the compressor is at a standstill and consumes no power.

What important device operates as the compressor slows down?

Answer: A regulator *unloads* the cylinder and shuts off the cooling water.

How long is this condition maintained?

Answer: Until the motor is started and running again at full speed. This results in a saving in electric current and cooling water.

What is dual control?

Answer: A combination of *constant* speed and automatic *start and stop* control.

AIR COMPRESSORS

What does it provide?

Answer: A constant speed control where there is a steady and large demand for air or gas which permits switching to automatic start and stop control during periods of intermittent or small demand.

UNLOADERS

What is a free-air unloader?

Answer: An automatic device used on some compressors that varies the amount of air or gas being pumped.

How does it work?

Answer: If the pressure in the receiver built up, due to the compressor supplying air when none is being used, a relief valve must blow, thereby resulting in a loss of air, and consequently of power. See Figs. 25-26 and 25-27.

Name the various methods of unloading.

Answer: (1) Inlet line unloading valve, (2) inlet and discharge

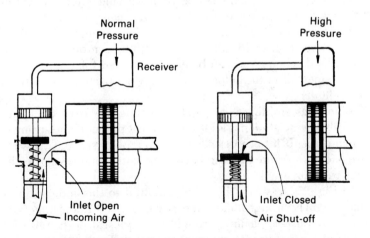

Fig. 25-26. Elementary free-air inlet unloader showing operation of the automatic control valve.

POWER PLANT ENGINEERS GUIDE

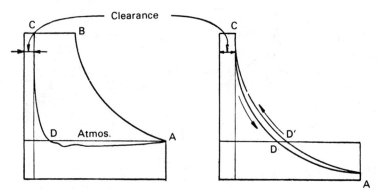

Fig. 25-27. Diagrams corresponding to the operation in Fig. 25-26 explained in the text.

unloader, (3) inlet valve kept closed, (4) inlet valve kept open, (5) discharge valve kept open, (6) adjustable compression stroke, (7) variable clearance volume.

What is the application of unloaders on constant-speed compressors?

Answer: Constant speed compressors require some form of unloading device which will permit the driving unit to run at full speed without delivery or more air or gas than is required.

The free-air unloader assembly includes the inlet valve subassembly. See Fig. 25-28. The inlet valve subassembly is bolted to the unloader body and can be easily removed for inspection or replacement after the unloader assembly has been removed from the cylinder.

When reassembling a valve to the unloader where used, be sure the valve center-bolt nut is locked and that pusher B and spring C are in place. If it becomes necessary to replace spring C, always use one of the same size and having the same compression as the one originally used.

In Fig. 25-28, when the air is admitted to the under side of diaphragm E through hollow inlet cover setscrew K, the fingers on the valve pusher, passing through the valve seat ports, raise the inlet valve off of its seat until the pressure under the diaphragm is released. When the inlet valves are held off of their seats, the air or

Air Compressors

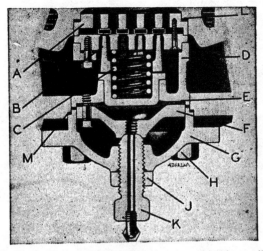

Fig. 25-28. Channel valve free-air unloader assembled in a cylinder.

gas cylinder is vented to the inlet passage and the air or gas passes alternately in and out of the cylinder without being compressed.

Valve cover G has a machined recess into which fits inner cover F of the unloader assembly. Thus, when the unloader assembly is being put into the bottom of a cylinder, unscrew valve cover setscrew K and set the whole assembly in outer cover G with the unloader assembly in position. After the valve seat and cover come to rest in the cylinder, insert and tighten the valve cover tap bolts. Now tighten valve cover setscrew K, forcing valve seat A solidly onto its gasket, and then lock the setscrew into position by means of locknut J.

For what service are unloaders used in addition to their regular function?

Answer: To facilitate starting a compressor, permitting temporary operation without doing the full amount of work at the start.

With what kind of drive is this especially important and why?

Answer: With certain types of electric motors that do not develop full torque until full speed is reached.

What is an *inlet line* unloader?

Answer: An unloader that automatically opens and closes the inlet line under pressure variations in the receivers.

How does an inlet line unloader work?

Answer: In operation, when the pressure is low in the receiver, a spring overcomes the air load on a piston and forces the valve open. With the inlet line open, free air is drawn into the cylinder. Again, when the pressure reaches a predetermined point, it in acting upon the piston pushes the valve shut, which shuts off the air supply and accordingly no more air is pumped into the receiver until the pressure in the receiver drops below the predetermined working pressure.

How does actual construction vary from the arrangement just described?

Answer: In the actual unloader there is a small piston which is actuated by the pressure in the receiver against the resistance of a spring. When the pressure in the receiver becomes too high, the piston is moved sufficiently to open a small port which admits air to a plunger valve of larger area. This plunger valve is forced against its seat, thus preventing further admission of free air to the cylinder until the pressure in the receiver drops below the maximum pressure for which the mechanism is set.

Name a standard unloading method.

Answer: Usually free-air unloaders are placed on each inlet valve.

What is the application of free-air unloaders controlled by an auxiliary valve?

Answer: This type of regulation is standard where the compressor is driven at a constant speed from any source of power, and where the demand for air is fairly constant.

Air Compressors

What happens when the receiver pressure rises to a value for which the auxiliary valve has been set?

Answer: The auxiliary valve operates to admit live air to the free-air unloaders.

Describe the action of a safety stop valve.

Answer: In the starting position, the shut-down weight must be raised and latched by raising the trigger latch handle. The red spot indicates a lack of oil pressure, as is the case when the compressor is idle. As the compressor is started and the oil pressure rises, the trigger latch handle drops into the running position and the red warning dot disappears. The trigger will now function if the oil pressure falls below a preset level.

When the compressor stops, or if the oil pressure drops for any other reason, the trigger latch handle releases the shut-down weight and closes the shut-off valve. It must be reset to the starting position before the compressor can be restarted.

What is the effect of this?

Answer: The free-air unloaders push all the inlet valves off their seats, allowing air and gas to pass in and out through the inlet valves while the compressor piston continues to move back and forth.

How is the compressor unloaded for starting when there is pressure in the receiver?

Answer: Screw on the wing nut at the bottom of the auxiliary valve.

What is the effect of this adjustment?

Answer: Air will be admitted to the free-air unloader and thereby vent the compressor cylinder to the intake passage.

COMPRESSOR OPERATION

Startup

What is the preliminary procedure on taking charge of a newly installed plant?

Answer: Clean up everything, especially the compressor, its foundation, and the floor around it. Blow out all oil piping, passages and pockets with a jet of air. Thoroughly inspect the compressor and give whatever attention that may be necessary.

Then what should be done?

Answer: Turn on a full supply of cooling water to the cylinder jacket. Open the necessary valves in the unloader, discharge, and regulator lines. Be sure the intake filters and screens are in place and charged with oil as specified by the manufacturer.

How about the lubrication?

Answer: Follow the manufacturer's instructions. See that there is plenty of oil in all lubricators and that all are functioning properly.

What precautions should be taken in starting?

Answer: Make sure that the tops of the wheels rotate toward the cylinder.

How do you start a compressor?

Answer: Rotate or bar the compressor over a few times by hand to see that it works freely and that everything is clean.

What should be done next with a new compressor?

Answer: Run the compressor without any pressure or only partial load for an hour or so.

How are wheels or pulleys installed or removed from the shaft?

Answer: In order to secure a tight fit in some cases, the bore of the wheel hub is a few thousandths of an inch smaller than the shaft diameter. Therefore, the plug which is furnished must be used

Air Compressors

when installing or removing wheels. To use it, remove the hub clamping bolt and place the plug in the slot so that the plug boss fits in the bolt hole. Hold a nut in the slot and turn the bolt until the end touches the plug and the nut rests against the opposite wall of the slot. By using a wrench on the bolt head, the inner end can be made to exert force against the plug, spreading the split hub so that the wheel can be placed on, or removed from, the shaft. Wheels are properly placed when the end of the shaft projects $\frac{1}{8}$-inch beyond the hub face on 7-inch-stroke machines or $\frac{3}{16}$-inch on 9-inch-stroke and larger machines.

How do you start a steam-driven compressor?

Answer: Open the drain cocks on the steam cylinder and steam chest, and drain the steam pipe above the throttle valve until it is warmed up. After draining, open the throttle very little and let steam blow into the steam cylinder until it is thoroughly heated, turning the compressor over very slowly until all condensate is worked out. Then close the cylinder cocks and bring compressor gradually up to speed.

What should be done after the compressor attains full speed?

Answer: Unscrew the wing nut and the compressor will begin to operate normally.

What is used in place of the wing nut on gas compressors?

Answer: A three-way valve in the line between the auxiliary valve and the unloaders, with one connection of the three-way valve piped to receiver.

Explain the method of control by variable clearance.

Answer: In this method, a number of clearance pockets are thrown into communication with the ends of each air cylinder when the pressure in the receiver has reached the required pressure.

What is its principle of operation?

Answer: When the compressor piston returns, the air in the clearance pockets expands and partly fills the cylinder, thus preventing the opening of the automatic inlet valves until later in the admission stroke. When no air is used from the receiver, air from the clearance pockets may completely fill the cylinder and prevent the opening of the inlet valves, admitting no free air.

What is the object of five-step, variable-compression regulation?

Answer: In this method, there are five approximately equal steps from "full" to "zero" which regulate the degree of compression.

What provision is made in the cylinder to obtain variable clearance?

Answer: Each cylinder is equipped with four clearance pockets.

Describe the operation of the five-step regulator with push-button starting station.

Answer: In operation, a slight rise in the receiver pressure causes the first step of control to operate and reduce the compressor capacity. A further reduction in the demand for air results in a slightly higher receiver pressure and causes the next step to operate and so on to no-load, if necessary, or until the compressor output equals the demand for air or gas. The total increase in receiver pressure from full-load to no-load is very small. When desired, the discharge pressure may be varied by a screw adjustment on the top of the regulator.

What provision is made to accommodate the standard two-stage, duplex compressor to the starting characteristics of the synchronous motor?

Answer: Relief valves are provided on the high-pressure cylinder, and an unloader is provided on the intercooler.

How is remote-control obtained?

Answer: To obtain remote-control, a magnetic unloader is employed which unloads the compressor during the time it is being started or stopped.

Air Compressors

Describe the bypass system.

Answer: In some special compressors, the cylinders are provided with bypass passages, which connect the two ends of the cylinder through the clearance pockets and allow air to pass back and forth without being compressed. In these, clearance valves are used to open or close the passages. The valves may be hand-operated, automatic, or a combination of both.

What is this bypass valve used for on single-stage compressors?

Answer: To unload for starting.

What should be noted about the closing of bypass valves?

Answer: They will not close automatically when there is no pressure in the receiver. Therefore, it is necessary to close the bypass valves by hand until enough pressure is built up in the discharge line to hold them closed.

How do you regulate the cooling water?

Answer: Adjust the supply through the cylinder jackets so that the discharge-water feels warm to the hand (about 100°-110°F. or 38°-43°C.).

What precaution should be taken after running the compressor without any pressure for an hour or so?

Answer: If the compressor shows no signs of heating in the bearings or stuffing boxes, the pressure may be increased gradually up to the working point.

What is the next precaution that should be taken?

Answer: Watch the regulators (and governors on a steam-driven machine) and never leave the machine until satisfied that they control the compressor properly.

What should be done after the compressor has run for an hour or so and is thoroughly warmed up?

Answer: Go over the exterior of the machine and tighten any loose nuts. Gaskets squeeze up a little when new. Do not use too much wrench pressure on the valve-cover setscrews as this may spring the valve seats and cause leakage and excessive heat.

What should be done next?
Answer: Adjust the flow of cooling water to the proper temperature so that the discharge feels warm to the hand.

Are there limits to the number of starts and stops with an automatic system?
Answer: Yes, especially, electric-motor-driven machines; motor manufacturers usually place a limit on the number of starts and stops per hour.

Running

What general attention should be given when the compressor is running?
Answer: The compressor should be watched closely at the start, particularly the crankcase. Even though the compressor has been carefully cleaned, there is sure to be some dirt remaining which will be washed out by the circulation of the oil. All of this dirty oil must be removed.

What should be done with the oil when removed?
Answer: The oil removed may be filtered and used again.

How often should the oil be removed?
Answer: Once the compressor is in operation and the interior thoroughly cleaned, it should not be necessary to change the oil more often than once a month or at intervals recommended by the manufacturer.

What attention should be given to safety valves?
Answer: Test the safety valves on the receiver and other locations occasionally.

AIR COMPRESSORS

How much oil is required for compressor cylinders?

Answer: Usually 1 to 5 drops of oil per minute, according to size.

What precaution should be taken?

Answer: Too much oil will form carbon on the valves.

What attention should be given to belts?

Answer: For belt-driven machines (1) inspect the belts regularly, (2) keep belt surfaces clean, (3) apply dressing to a leather belt which is too dry, (4) apply proper amount of dressing—enough dressing in winter is too much in summer, (5) take grease out of belting which is too oily, (6) do not expect too much from a leather belt when subjected to temperature in excess of 100°F. (38°C.).

Which are the major wearing parts of the compressor?

Answer: (1) Piston rings, (2) cylinder or liner, (3) valve, (4) piston rod.

How should the main bearing be adjusted with liners?

Answer: The bearing cap should be adjusted by means of these liners so that there is about 0.004-inch (0.01 cm) clearance between the cap and the top of each side box.

What should be done before crossheads have been worn in?

Answer: Care should be taken to allow enough clearance to keep the gibs from heating.

Describe the clearance setting.

Answer: Start with about 0.012-inch (0.03 cm) clearance, and when the bearing surfaces have been worn in, take up the clearance to approximately 0.008-inch (0.02 cm).

What happens when wear occurs at the piston and cylinder wall?

Answer: The piston naturally sets in a relatively lower position.

What should be done to compensate for this wear and reestablish the original alignment?

Answer: The crosshead should be lowered an amount equal to the wear.

What attention should be given to piston-rod wiper rings?

Answer: Clean the wiper rings periodically.

What should be done in the case of broken springs?

Answer: They should be replaced at once.

How are wiper rings assembled?

Answer: In assembling, stagger the joints.

What should be noted about piston rings?

Answer: They tend to rotate when in operation.

In this connection, what is the important procedure in installation?

Answer: When installing, make sure that each three-piece ring has one right-hand and one left-hand outer ring so rings will rotate in opposite directions. This avoids the possibility of the splits lining up and allowing leakage through the joint.

What is the most important requirement to get the maximum efficiency out of any compressor cylinder?

Answer: The admission and discharge valves *must* be tight.

What inspection and attention should be given at regular intervals?

Answer: Inspect the valves, valve passages, and cylinder bore and remove any accumulation of foreign matter.

What precaution should be taken with respect to cleaning a cylinder?

Answer: Never use kerosene or gasoline in an air cylinder to clean it out. This is a very dangerous practice and should be prohibited.

What should be noted about free-air unloaders?

Answer: A complete understanding of the installation and operation of the free-air unloader should be obtained by the operator in order to maintain an efficient regulating system.

What about cold weather?

Answer: Whenever the compressor is exposed to a freezing temperature while not in operation, drain the water from the jackets and intercooler in order to prevent injury from freezing. Drain valves will be found at the side of the cylinder for draining the jackets. The intercooler can be drained through the valve provided for that purpose at the water inlet connection. If the compressor is to stand idle for any length of time, remove piston rod packing. If this is not done, the packing will cause rods to rust very rapidly, causing trouble later.

CHAPTER 26

Refrigeration Basics

In some cases, power plant engineers are responsible for the operation and maintenance of refrigeration equipment for air conditioning and cold storage facilities. Therefore, for those interested we include in this chapter the basics of refrigeration. Obviously, for a complete course in refrigeration, a text or study guide is recommended. A rather detailed study aid is available from the NAPE, a guide used as a reference in the preparation of this chapter.

REFRIGERATION SYSTEM COMPONENTS

What is refrigeration?
  Answer: The removal of heat from air or any material.

What is a refrigerant?

Answer: A liquid that boils or vaporizes at a low temperature, thus absorbing heat from the air or material to be cooled.

What are the basic components of a refrigeration system and what part does each play in the refrigeration process?

Answer: A *refrigerant* is contained in a closed system comprising a *compressor, condenser, receiver, control valve,* and *evaporator* (Fig. 26-1). The liquid refrigerant, at low pressure, absorbs heat from the area to be cooled. In so doing, it eventually boils. The vaporized refrigerant with the absorbed heat is drawn by the compressor from the evaporator and compressed. As the refrigerant is compressed, the heat is concentrated until the temperature of the refrigerant is higher than that of the surrounding air or water, whichever is used for cooling, in the condenser. The hot, high-pressure refrigerant vapor passes into the condenser where the cooling medium absorbs heat from it. As the refrigerant cools, it changes to a liquid and flows into a receiver, a storage area for high-pressure liquid. (In some systems the condenser is also the receiver; there is no separate receiver unit.) At this point it is ready for "recycling" through the system. The control or expansion valve

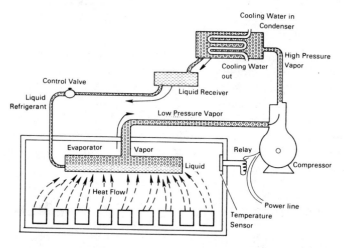

Fig. 26-1. Basic components and functions of the mechanical refrigeration cycle.

Refrigeration Basics

meters the flow of liquid refrigerant into the evaporator. As the refrigerant passes through the valve, the pressure drops and part of the liquid instantly flashes to gas, thus absorbing the latent heat in the liquid and lowering the temperature of the low-pressure liquid refrigerant flowing into the evaporator, where it continues to absorb heat and begin the cycle again. When the desired temperature is reached, a thermostat actuates a relay which stops the compressor, which stops the refrigeration cycle until the temperature in the cooled area begins to rise above the predetermined point. When this point is reached the compressor starts and the cycle is repeated.

What is commonly known as the "high side" in a refrigeration system?

Answer: All parts of the system where the refrigerant is under high pressure—the top half of the compressor, the condenser, the receiver if there is one, and the line up to the control or expansion valve. Pressure is measured in pounds per square inch above atmospheric by a gauge.

What is the "low side"?

Answer: The other half, beginning at the output of the expansion valve, the evaporator and the bottom half of the compressor. Low-side pressure, which is less than atmospheric, is measured in inches of vacuum. At sea level 30 inches of vacuum is a perfect vacuum when the barometer reads 30 inches. A 15-inch vacuum would be a barometer reading 15 inches below atmosphere or one half of a perfect vacuum or 7.5 psi.

How is the refrigerant transferred from one part to another?

Answer: Through pipe or tubing. Usually, the lines from the expansion valve to the evaporator and from the evaporator to the compressor are insulated so a minimum of unwanted heat is picked up from the surrounding air.

What is the pressure-temperature relationship of the refrigerant?

Answer: The evaporator pressure needed so that the liquid refrigerant vaporizes at the desired temperature. The temperature control in effect varies the pressure-temperature relationship. See the tables for various refrigerants in Chapter 30.

How is the capacity of refrigeration units rated?

Answer: Tons of refrigeration or Btu per hour. A ton of refrigeration is equal to the amount of cooling that would be produced by a ton of ice melting in 24 hours. If one pound of ice has a latent heat of 144 Btu, a ton will have a latent heat of 288,000 Btu (2000 × 144). This would equal 12,000 Btu per hour (2000% ÷ 144). This would equal 12,000 Btu per hour (288,000 ÷ 24).

REFRIGERANTS

What characteristics must a liquid refrigerant have?

Answer: It must be suitable for heat absorption by evaporation. The freezing point should be well below the normal evaporator operating temperature and its critical temperature should be well above the condensing temperature. Likewise, pressures required for evaporation and condensing should be reasonable (as near atmospheric as possible) for the type of system under consideration. A refrigerant should be nontoxic and nonflammable, and leaks should be easy to detect. The liquid should be stable and have no effect on metal or oil. For economy a refrigerant should produce maximum refrigeration per cubic foot of vapor and be readily available at reasonable prices.

Which refrigerant is used in many systems where an evaporator temperature well below 0°F. is needed?

Answer: Ammonia is one of the most common of all the refrigerants. It is a chemical compound of nitrogen and hydrogen (NH_3) which has a boiling temperature at atmospheric pressure of -28.0°F. Ammonia is somewhat flammable and with the proper mixture of air is explosive. Accidents from this are rare, however. Ammonia gas is not classed as poisonous, but its effect on the respiratory system is so violent that only very small quantities of it can be breathed.

What are some of the popular refrigerants?

Answer: Ammonia, Freon 12. Listed below are additional liquids used in commercial systems with the boiling point of each at 0 psi:

REFRIGERATION BASICS

Ammonia	(NH_3)	$-28.0°F.$
Carbon dioxide	(CO_2)	$-109.3°F.$
Carbon tetrachloride	(CCl_4)	$+170.2°F.$
Ethyl chloride	(C_2H_5Cl)	$+54.5°F.$
Freon 11 (F-11) trichloromonofluoromethane		$+74.7°F.$
Freon 12 (F-12) dichlorodifluoromethane		$-21.7°F.$
Freon 21 (F-21) dichloromonofluoromethane		$+48.0°F.$
Freon 22 (F-22) monochlorodifluoromethane		$-41.0°F.$
Freon 113 (F-113) trichlorotrifluoroethane		$+117.9°F.$
Freon 114 (F-114) dichlorotetrafluoroethane		$+38.4°F.$
Isobutane	(C_4H_{10})	$+13.6°F.$
Methyl chloride	(CH_3Cl)	$-10.7°F.$
Methyl Formate	$(C_2H_4O_2)$	$+89.3°F.$
Methylene chloride	(CH_2Cl_2)	$+103.6°F.$
Sulfur dioxide	(SO_2)	$+13.86°F.$

	Freon Symbols	trade name
Freon 11	(CCl_3F)	Carrene #2
Freon 12	(CCl_2F_2)	
Freon 21	$(CHCl_2F)$	Thermon
Freon 22	$(CHClF_2)$	
Freon 113	$(C_2Cl_3F_3)$	Carrene #3
Freon 114	$(C_2Cl_2F_4)$	

C = carbon, S = sulfur, O = oxygen, H = hydrogen, Cl = chlorine, F = fluorine, and N = nitrogen

What is the most common synthetic refrigerant?

Answer: Freon 12 is a colorless, almost odorless gas with a boiling point of -21.7°F. at atmospheric pressure. F-12 has no effect on clean metal. However, as a powerful solvent, it will loosen any grease or scale that may be on the inside surface of castings, pipes, or tubes and plug the orifices in expansion valves and cause wear in compressors. Its solvent action will also dissolve natural rubber, so none of this can be used in Freon system gaskets or packing. Freon 12 will dissolve in oil in any portion and thin the oil. It is nontoxic and nonflammable, and almost odorless; a halide torch must be used to detect leaks.

COMPRESSORS

Describe a simple compression system.
Answer: A single-stage system where the refrigerant is compressed from evaporator to condenser in a single stroke of the compressor piston. (The system in Fig. 26-1 is a single-stage system.)

What is the normal evaporator pressure in a single-stage compression system?
Answer: Zero pounds gauge.

At zero pounds evaporator pressure, what must the condenser pressure be?
Answer: High enough to condense the refrigerant.

What are the minimum temperatures attainable with some of the popular refrigerants in simple compression systems?
Answer: Ammonia, -28.8°F.; Freon 12, -21.6°F., Freon 22, -41.6°F.

How is the cooling capacity of a refrigeration system regulated?
Answer: (1) Stopping and starting the compressor automatically by room thermostat or manually; (2) controlling the flow of refrigerant; and (3) control of water (if used) flow through the condenser.

Describe a compound compression system.
Answer: A multistage system employing at least two stages of compression, which may involve two or more individual compressors or by a single-body compressor where the refrigerant is compressed in a series of cylinders with the gas flowing from one to the other. In other words, the pressure of the refrigerant from evaporator to condenser is increased in several steps. The drawing in Fig. 26-2 illustrates the operation of a two-stage system. Also shown are pressures and temperatures of the ammonia refrigerant at various points in the cycle. With this system it is possible to hold temperatures at -60° to -85°F.

REFRIGERATION BASICS

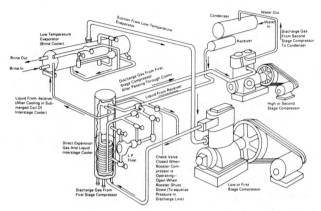

Fig. 26-2. Drawing of a compound or multistage compression system.

What refrigerant is usually used in low-temperature industrial systems?

Answer: Ammonia, for the reasons given above.

What happens during interstage cooling?

Answer: The refrigerant gas is precooled before entering the second stage to reduce its superheat and keep the discharge temperature within operating limits. The liquid refrigerant is precooled ahead of the evaporator expansion valve (increases the heat-absorbing capacity of the refrigerant). The liquid refrigerant is held at the condenser evaporator pressure until it enters the low-temperature evaporator.

A drop in the workload on a compound system results in what consequence?

Answer: Unloads the first stage and loads the second (and higher) more heavily. The opposite effect occurs with an increase in workload.

When putting a compound compressor on line, which stage goes first?

Answer: Top stage(s) drawing gas directly from the evaporator (bypassing the first stage). The first stage goes on line only after the suction pressure is down enough to avoid overload.

Why is adiabatic compression impossible?

Answer: The refrigerant would undergo adiabatic change if it were to neither gain nor lose heat as it is compressed. Because of cylinder radiation, jacket cooling, and other heat losses, adiabatic compression is impossible.

Why is isothermal compression impossible?

Answer: Isothermal compression could occur only if the refrigerant temperature were held constant as the gas goes through a series of pressure and volume variations.

What is a reciprocating compressor?

Answer: A gas pump that pulls refrigerant from an evaporator as it is boiled off and transfers it under the required pressure to the condenser. The piston or pistons of a reciprocating compressor alternately increase and decrease the volume of a cylinder. The cylinder admits refrigerant through inlet ports and discharges it through outlet ports. See Fig. 26-3.

How many cylinders does a reciprocating compressor have?

Answer: One to 16.

What cylinder arrangements are used?

Answer: Vertical, horizontal, radial, or V/W.

What are some of the most common?

Answer: Vertical single-acting, horizontal double-acting, and V/W.

What are the advantages associated with vertical compressors?

Answer: Adaptable to varying load conditions and are slower running. Vertical compressors are available in a power range from one-quarter to 700 hp, and are divided into two classes: fully enclosed and semienclosed.

What is a double-acting horizontal compressor?

Answer: Pistons moving horizontally draw in gas at each end of the cylinder, so it compresses two charges per revolution.

What is the advantage of a horizontal compressor?

Answer: It takes less head room because it is spread out horizontally. It will last longer because a horizontal compressor runs

REFRIGERATION BASICS

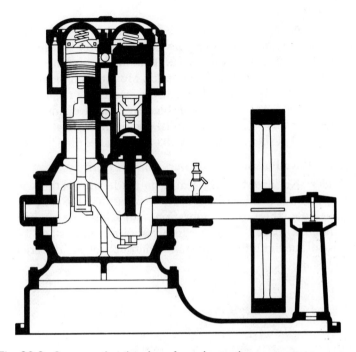

Fig. 26-3. Cross-section drawing of a reciprocating compressor.

slower than any other type. Horizontal compressors are less efficient than vertical types, though.

What are V/W compressors?

Answer: High-speed, single-acting, multicylinder machines in which the cylinders are arranged in a V or W position in respect to the crankshaft axis. See Fig. 26-4.

What are the advantages of a V/W compressor?

Answer: The most efficient due to its automatic partial capacity control during periods of lighter loads, plus unloading valves that unload the compressor during starting. V/Ws can be operated in parallel or intermittently, and are smaller and lighter per ton of capacity. This type of compressor is widely used in air conditioning applications.

729

What is a rotary compressor?

Answer: A rotary compressor has a horizontal rotor which spins inside of a horizontal cylinder. The axis of the rotor is eccentric to that of the cylinder to the degree that the rotor runs in close clearance to the bottom of the cylinder. Mounted on the rotor are blades that move out radially from the axis as the rotor reaches operating speed. As the ends of the blades reach the cylinder walls, each forms a seal which traps (pulls) gas from the inlet (suction) chamber and pushes it into the discharge chamber (See Fig. 26-5).

What are the advantages of a rotary compressor?

Answer: No-load starting because the blades are not thrown out until operating speed is reached. It is smaller than the reciprocating machines, there is less vibration, and it may be connected directly to a motor. Rotary compressors are multistage types because of the limitation of the input/output pressure differential of each stage.

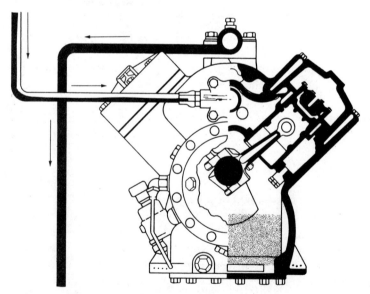

Fig. 26-4. Cross-section of a V-type compressor.

REFRIGERATION BASICS

What is a centrifugal compressor?

Answer: A compressor with a rotating impeller that agitates the refrigerant to a high velocity, thus increasing the kinetic energy of the gas. The kinetic energy in the gas is converted to static pressure in the expanding section of the impeller housing, which is the head pressure needed for the refrigeration cycle. See Fig. 26-6.

How are centrifugal compressors classified?

Answer: Open and closed. In an open type, the driver (motor, turbine, etc.) and compressor are separated, with the driver exposed to air. The closed-type driver and compressor are hermetically sealed in a common housing.

How widely used are centrifugal compressors?

Answer: To a large degree, they've replaced reciprocating types due to many advantages.

What are the advantages of centrifugal compressors?

Answer: They are compact and light in weight for the refrigeration capacity produced. The only major rubbing surfaces are the main bearings. Due to its high speed the compressor can be directly connected to the driver. Since the compressor is designed

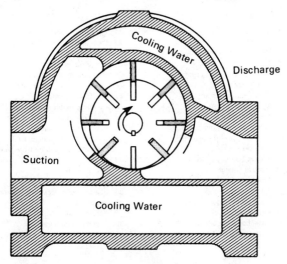

Fig. 26-5. Cross-section of a rotary compressor.

731

Power Plant Engineers Guide

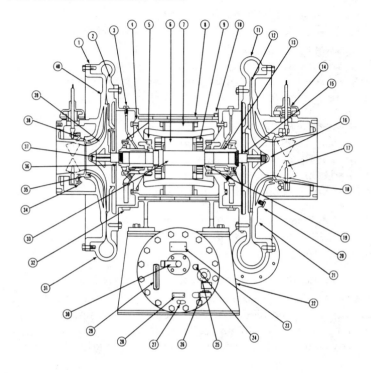

1. Gasket
2. Gasket
3. Oil inlet to bearing
4. Motor bearing bracket
5. Motor bearing end cover
6. Motor rotor
7. Motor stator winding
8. Motor water jacket
9. Motor fan blade
10. Motor jacket water connection
11. First stage volute casing
12. Diffuser passage
13. Sleeve bearing
14. Control linkage
15. Labyrinth seal
16. Shaft nut
17. Variable inlet guide vane
18. First stage impeller
19. Fixed bearing collars
20. Rotation sight glass
21. First stage suction cover
22. Base and oil tank assembly
23. Oil pump motor terminal box
24. Water outlet connection
25. Oil level sight glass
26. Solenoid valve, water inlet
27. Oil charging valve
28. Oil pump motor nameplate
29. Oil temperature thermometer
30. Pressure relief valve
31. Second stage volute casing
32. Oil outlet connection from bearing
33. Shaft
34. Ball joint linkage
35. Impeller labyrinth
36. Bearing collar locknut
37. Impeller key
38. Actuator ring
39. Second stage impeller
40. Second stage suction cover

Fig. 26-6 Cross-section of a centrifugal compressor.

with rotating parts it is easy to balance both statically and dynamically, thus reducing foundation requirements and allowing easier installation on upper floors. Centrifugal compressors are noted for their uniform gas flow at the compressor discharge without any pulsating effect and their ease of control for constant volume or constant pressure. Centrifugal compressors cannot produce excessive pressure, hence no relief valve is required on the compressor. The impeller cannot produce a pressure greater than pumping limit. Internal lubrication is not required, so a centrifugal operates with little oil contamination of refrigerant gas. Centrifugal compressors are available at the speed needed to fit the load requirements. The speed can be reduced to balance reduced load demands.

OTHER REFRIGERATION METHODS

What is the absorption system of refrigeration?

Answer: In an absorption system, cold water absorbs ammonia gas in one part of the cycle and gives it up in the other. It is still used where waste or exhaust steam is available.

Describe an absorption system and the role of each major section.

Answer: It is composed of a condenser, receiver, absorber, liquid pump, heat exchanger, and generator. Ammonia vapor from the evaporator goes to the absorber, where it is absorbed by the water. The heat generated during this absorption process is controlled by a water coil installed in the absorber. The ammonia and water are then pumped to the generator and through a steam coil under thermostatic control. The weak ammonia and water liquor (which falls to the bottom of the generator) are returned to the absorber through a pressure-reducing valve which maintains a pressure difference between the two chambers. The ammonia vapor distilled off in the generator is passed to the condenser, which acts in the same manner as a condenser in a compression system. The liquid ammonia goes to the liquid receiver then through the expansion valve into the evaporator. See Fig. 26-7.

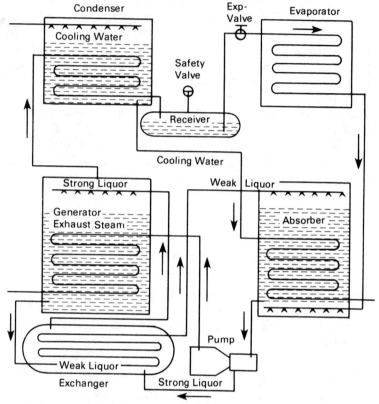

Fig. 26-7. Absorption refrigeration system.

What is steam-jet refrigeration?

Answer: A system operating on the principle that heat is needed to change water from the liquid to the vapor state. It is economical where a large amount of steam is readily available. It has few moving parts (liquid-handling centrifugal pumps) and does not involve odor- or fume-producing chemicals or gases.

How does a steam-jet system operate?

Answer: Water is generally cooled by very rapid evaporation, which in turn requires that a high vacuum be kept in the evaporator. For example, a vacuum of 29.75 inches of mercury is necessary to obtain an evaporator temperature of 40°F. To exhaust the large volumes of water vapor needed by a steam-jet refrigeration system, high-pressure steam is applied to the evaporator through a nozzle arrangement. As the steam expands, its potential energy is converted into kinetic energy. The steam leaving the nozzle at approximately 4000 feet per second maintains a vacuum in the evaporator by carrying entrained water vapor to the condenser. In the evaporator the rapid evaporation cools the water. As water returns from the cooling system it is sprayed through nozzles, creating a fine mist which evaporates very rapidly. Remaining cooled water falls to the bottom of the tank, and a centrifugal pump circulates it through the cooling system. Makeup water from a service main is added to the system to maintain the desired water level in the evaporator. The prime mover of the steam jet refrigerator system is the steam booster ejector. High-pressure steam expands as it leaves the nozzle and moves through the ejector, entraining water vapor as it moves. Low-pressure vapors in the evaporator enter the ejector body. After passing through the ejector and picking up the water vapor from the evaporator, the steam is condensed in a surface condenser or a jet condenser where there is intimate mixing of the steam and water. See Fig. 26-8.

What is a cascade system?

Answer: A multistage refrigeration system used to produce low temperatures where, due to independent system circuits in each stage, it is possible to use different refrigerants in each stage. It operates very similar to reciprocal or centrifugal systems. Cascade systems are used mostly to replace a centrifugal system or to add capacity to an ammonia system. See Fig. 26-9.

HEAT PUMPS

What is a heat pump?

Answer: A reverse-cycle refrigeration system that uses condenser heat to advantage. Such systems are used to cool air in the summer and heat it in the winter. See Fig. 26-10.

Power Plant Engineers Guide

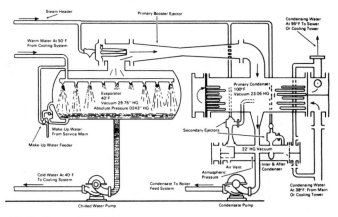

Fig. 26-8. Steam-jet refrigeration system.

Name two heat pump systems in use.

Answer: Air-to-air and water-to-water.

How does an air-to-air system work?

Answer: During the heat cycle, outside air is used as a source of heat, and heat from the condenser is used to heat space. During the cooling cycle, heat is removed from inside air by the evaporator and rejected to the outside air by the condenser. The refrigerant

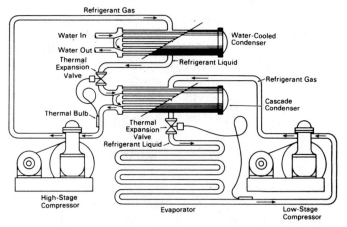

Fig. 26-9. Cascade refrigeration system.

REFRIGERATION BASICS

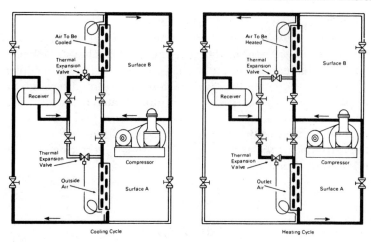

Fig. 26-10. Heat pump or reverse-cycle system.

path is reversed for summer and winter operation by stop valves. During the cooling cycle, the refrigerant follows the conventional path. Some heat pump system designs reverse the air flow past the condenser and cooling coil for summer and winter operation rather than reversing the refrigerant flow. In some systems heat is drawn or expelled into the earth to use the ground as the heat source rather than the surrounding air.

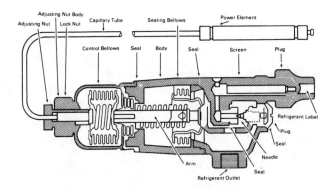

Fig. 26-11. Temperature-controlled expansion valve.

737

How does a water-to-water system work?

Answer: In water-to-water heat pumps, water is used as both the heat source and the heat transfer medium. Such systems have a water-cooled condenser and a shell-and-tube type water cooler. The heat source may be a well with water pumped through the condenser and discharged to the drain. Water from the water cooler is pumped through the air conditioner coil for summer cooling. During winter operation the flow of the well water and the refrigerated water are reversed without interfering with the refrigerant circuit.

CONTROLS

How does a pressure-controlled expansion valve work?

Answer: A flexible bellows is connected to the needle of a needle valve with chilling-unit pressure on the inside and atmospheric or confined gas pressure on the outside. As the pressure in the chilling coil decreases, the difference in pressures will open the needle valve and allow some refrigerant into the coil which keeps the box temperature within design limits. The expansion valve usually opens only when the compressor is running because this is the only time pressure is reduced enough.

How does a temperature-controlled expansion valve work?

Answer: A temperature-controlled expansion valve really is a pressure-controlled valve, but instead of a pressure tap attached to the chilling coil the pressure sample is taken from a power element. This power element is a metal bulb connected to the expansion valve by a small flexible (capillary) tube. The bulb is partially filled with sulfur dioxide or methyl chloride fluid and is attached to the last portion of the chilling coil tubing (the exhaust end). As the chilling unit warms up, the fluid in the power element evaporates, increasing the pressure, which is transmitted by the tubing to the expansion valve. This pressure on the bellows opens the needle valve and allows more refrigerant into the chilling unit. See Fig. 26-11.

Refrigeration Basics

How does a low-pressure side control work?

Answer: It consists of a float which controls the amount of refrigerant in the chilling unit. The float is connected to a needle valve or a ball check valve and calibrated so that the valve will close when the float is at the proper level, when there is enough refrigerant.

How does a high-pressure side float control work?

Answer: A float is located in the liquid receiver tank or in an auxiliary chamber in the high-pressure side of the system. When enough liquefied refrigerant has collected in the float chamber, the float will rise and open the needle valve enough to let the liquid into the low-pressure side.

How does a pressure control work?

Answer: It turns the compressor on as pressure begins to increase above normal in the evaporator and turns it off when normal evaporator pressure is reestablished. The switch that operates the motor control is a mercury-bulb or a contact-point type that is bellows-activated. This type is used now only in the commercial field.

How do temperature controls in home-type units work?

Answer: Two types are used. The oldest of the two (Fig. 26-12) has a thermostat with a power element that is clamped to the evaporator outlet tubing. As it warms, the fluid in the power unit (alcohol and water) melts and the pressure against the bellows (just above the power unit) drops. As the bellows contracts, a switch closes, which starts the compressor. When the evaporator cools, the liquid in the power element freezes, expanding the bellows and opening the switch. Temperature adjustment increases or decreases spring pressure on the bellows. Newer type controls (Fig. 26-13) operate on the same principle but have a safety time fuse and a defrosting switch. The only difference in design is a remote power element connected to the electrical control by a capillary tube.

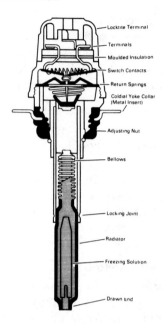

Fig. 26-12. Home-type refrigerator temperature control.

What is a water-regulating valve used for?

Answer: To regulate the flow of water to a water-cooled condenser.

Why is it important to regulate water to a water-cooled condenser?

Answer: To conserve water. A considerable amount can be wasted if water is not automatically regulated.

Where is a regulating valve normally installed?

Answer: In the water supply line to the condenser.

How does a water-regulating valve work?

Answer: See Figs. 26-14 and 26-15. A capillary tube connects a pressure sensor, located in the top of the condenser or connected to the condenser discharge manifold, to the control valve bellows. As head pressure in the condenser rises, it increases pressure in the

REFRIGERATION BASICS

capillary tube which pushes against the valve bellows in opposition to the spring pressure holding the valve closed. As the capillary pressure overcomes the spring pressure, the valve opens and admits water. As the addition of fresh water lowers condenser head pressure, the valve closes in response, cutting off the water. Thus, head pressure in the condenser is held constant within the range determined by the spring tension against the valve bellows. It is adjusted to open and close at the specified condenser pressures.

What is an automatic refrigerant head pressure controller?

Answer: A diaphram-type control valve located in the condenser liquid line. As the head pressure falls under decreasing load, it slows down the flow of refrigerant from the condenser. It is used in air conditioning systems to reduce the load as the outdoor temperature drops or where the air conditioning load decreases for any reason.

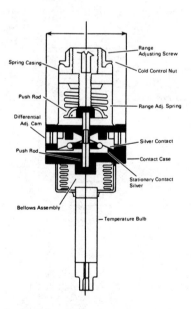

Fig. 26-13. Remote power element type refrigerator temperature control.

741

POWER PLANT ENGINEERS GUIDE

How does it work?

Answer: As head pressure falls, the valve restricts the flow of refrigerant from the condenser by pneumatic and mechanical means. The valve reduces refrigerant flow by closing the inlet port and simultaneously opens the hot gas port. As the liquid accumulation builds in the condenser, the condensing surface area decreases. As a result, the head pressure drops slowly. The valve is adjusted while throttling the inlet gas to keep the minimum head pressure

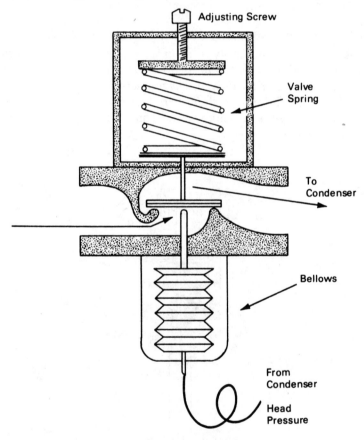

Fig. 26-14. Water-control valve.

REFRIGERATION BASICS

needed to hold the condenser temperature to the level recommended. The refrigerant charge must always be adequate to fill the condenser completely.

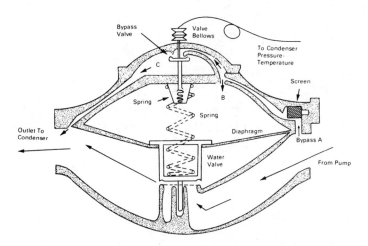

Fig. 26-15. High-capacity water-control valve.

How are solenoid valves used in air conditioning systems?

Answer: To control the flow of refrigerant to the cooling coils. They are usually located ahead of the expansion valve.

How does a solenoid valve work?

Answer: An electromagnet (Fig. 26-16) opens the valve when there's a current in the magnet coil. It usually opens when the compressor starts and closes when it stops, thus keeping liquid refrigerant out of the evaporator when the compressor is shut down. Where the compressor is to be started and stopped automatically, it's best to have the thermostat open and close the solenoid valve. When the liquid supply to the cooling coil is cut, the compressor pulls a vacuum inside the evaporator. When the pressure drops, a low-pressure switch stops the condenser.

What does an evaporator pressure regulator do?

Answer: It keeps the evaporator pressure from dropping below a specified point.

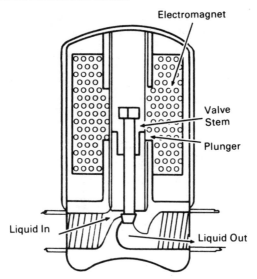

Fig. 26-16. Solenoid valve.

How does it work?

Answer: The evaporator pressure regulator is installed in the suction line from the evaporator coil (or each coil in a large system) or the common line from a multicoil system. It is a spring-loaded pressure-operated valve (Fig. 26-17) connected into the suction line. When the pressure on the evaporator side is above a specified point (37 psi in some systems) the sensor transmits enough pressure through a capillary tube to overcome the spring tension against the bellows, which holds the valve open. When the pressure on the suction side drops, the spring tension forces the valve closed. Some evaporator pressure regulators are operated by a pilot valve internally connected to the evaporator side of the suction line. The pilot valve operates the main valve.

DEFROSTING SYSTEMS

What are the most popular defrosting systems?

Answer: Hot gas, brine spray, warm brine, water, warm air, and electric.

How does each work?

Answer: Hot gas—gas from a normally operating evaporator is applied to the evaporator being defrosted (Fig. 26-18). Here is how it is operated:

1. Close valve A.
2. Open valve B; connects the evaporator suction line to the compressor discharge.
3. Close valve K.
4. Open valve J; connects the evaporator liquid line to the defrosting shell.
5. Close valves H and Q.
6. Open valve G; connects the defroster shell to compressor suction.

Hot gas, pulled from evaporators 2 and 3 by the compressor, and suction from the defrosting shell are forced into evaporator 1 through the suction line. As gas condenses in evaporator 1, it is defrosted. Condensed refrigerant goes to the defrosting shell. After adequate defrost:

7. Close valves B, J and G.
8. Open valves A, K, H, and Q.

The defrosting shell is placed under compressor discharge pressure and refrigerant drains into the receiver (receiver must be mounted lower than the defrosting shell). Evaporators 2 and 3 are defrosted in the same way, using the appropriate valves. Automatic defrosting can be achieved by substituting solenoid valves for the valves in the drawing and operating them by a timing device or other type of frost-sensing control.

Brine spray.—Used only in conditioners operating at room temperature. Above the evaporator coils are the spray headers and beneath them are pans with drain lines to collect the sodium or calcium chloride brine solution. During defrost, the circulating fan is stopped and brine is pumped through the headers to melt the ice from the coils. Diluted brine falls into the collecting pan and drains to a storage tank. The storage tank has an overflow drain to handle water accumulated during defrost and a brine replenishment system. In some systems a heating coil drives off the excess water.

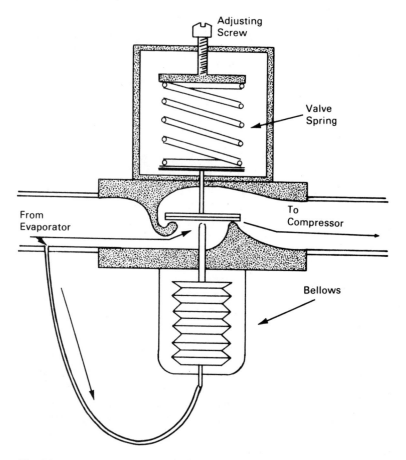

Fig. 26-17. Evaporator pressure regulator.

Disadvantages are high initial cost and corrosion from the salt. Some systems use more expensive brines such as ethylene or propylene glycol.

Warm brine.—Applicable only in air-cooling coils using brine as a coolant. Heat (warm brine) is applied to the inside of the coils, the same as in hot gas defrosting. The defrosting brine must be at least 70°F., and the initial system cost is high.

Water defrosting.—Used only with coils operating at room

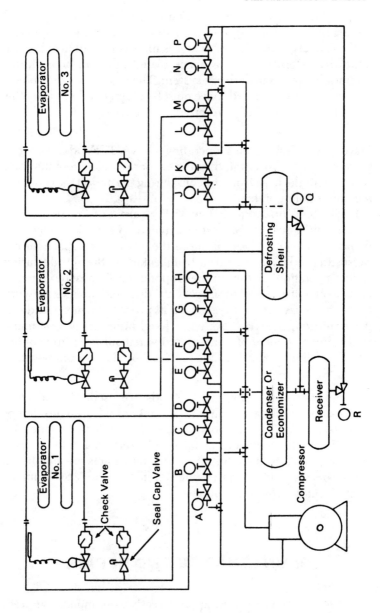

Fig. 26-18. Hot gas defrosting system.

temperature of 0°F. and above. During defrost, the refrigeration cycle and air-circulating fans are stopped and water is sprayed over the coils. If the defrost water is unusually cold, it is heated somewhat. Defrost water is accumulated in a pan beneath the evaporator coils as in the brine system. The water lines and collector inside the refrigerated space must be drained quickly after the defrost cycle to avoid freeze-up.

Warm air.—For systems operating above 35°F., defrosting is automatic. A fan forces air over the evaporator coils continuously, which melts the frost from the coils during the period the condensing unit is not operating.

For systems operating below 35°F., warm outside air is circulated over the coils. These systems involve a network of ducts and dampers to keep outside air out of the refrigerated space during defrosting and to keep cold air from leaking to the outside during normal operation. Therefore, the evaporator coils must be mounted close to an outside wall to simplify the duct and damper network as much as possible. This system has its problems when outside air is humid (longer defrost cycle due to additional frost accumulation at the beginning of defrost) and when outside is too cold to use. Drain lines are susceptible to freeze-up during cold weather.

Electric defrosting.—Several methods are used. One involves strip heaters mounted in the coils or banks of infrared heat lamps mounted around the evaporator coils. In another, a connection from the discharge duct to a cold air inlet recirculates the air over strip resistance heaters. Supply and return ducts are dampered during defrost. Electric defrosting is easily adaptable to automatic control but expensive to operate.

MAINTENANCE AND REPAIR

The following procedures apply directly to ammonia compressors but are generally applicable to any small system.

Refrigeration Basics

Symptom	Probable Cause	Suggested Action
Too high condensing pressure.	Too little or too warm condensing water.	Supply more or cooler water to condenser.
	Condenser coils fouled with scum or scale.	Scrape condenser coils clean.
Very high condensing pressure, trembling of pressure gauge, and condenser pipe. Temperature corresponding to the pressure considerably higher than the discharge temperature of condensing water.	The presence of air or noncondensable gases in the system or too great a charge of refrigerant.	Blow off air or noncondensable gases. Remove ammonia from the system.
Too low condensing pressure.	Too small a charge of gas.	Add ammonia to the system.
Rapid fall of suction pressure.	Expansion valve closed or not open far enough.	Gradually open expansion valve until correct condition is obtained.
Too high suction pressure; discharge connection cold; heavy frost on compressor.	Expansion valve open too wide.	Gradually close expansion valve until correct condition is obtained.
Loud hammering of compressor valve.	Broken spring on compressor valve.	Replace spring.
Irregular action of compressor valve.	Dirty or leaking compressor valve.	Overhaul valves.
Capacity of compressor reduced.	Leaky piston.	Overhaul piston and cylinder.
Stuffing box and discharge pipe hot.	Expansion valve closed or open too little.	Open expansion valve wider very slowly.

CHAPTER 27

Basics of Electricity

Due to the increasing use of electrically operated boilers and the number of electrical devices associated with power-plant operation a knowledge of the basic principles of electricity often can be helpful for the modern power-plant engineer. Therefore, it seems appropriate to include at least enough information on the basics of electricity to help you reach a general understanding of the subject. This chapter is an introduction only. For more detailed engineering theory, see any of the many excellent texts on electrical engineering, including one prepared by Northwest Orient Airlines Technical Training Section and published by the Educational Committee of the Minnesota Chapter of National Association of Power Engineers, a reference used in the preparation of this chapter.

MAGNETISM

Are electricity and magnetism related?

Answer: Yes, closely. Electricity is generated from magnetism. Most electrical units depend on magnetism to operate.

What are magnets made of?

Answer: Steel alloys.

If a straight bar magnet is suspended in the middle and is free to swing, what position will it take?

Answer: The north pole of the magnet will point to the earth's north magnetic pole, which is the principle of the ordinary magnetic compass. See Fig. 27-1.

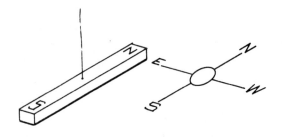

Fig. 27-1. A suspended, freely swinging magnet will align its poles with the earth's magnetic poles.

How do magnetic lines of force (flux) flow?

Answer: From the north to the south poles. If the opposite ends of two bar magnets are placed under a piece of paper sprinkled with iron filings, the filings will take the positions shown in Fig. 27-2.

How do unlike poles (north and south) react when near each other?

Answer: They are attracted to each other (Fig. 27-2). The closer they are, the stronger the attraction. Like poles repel each other. See the filings pattern in Fig. 27-3.

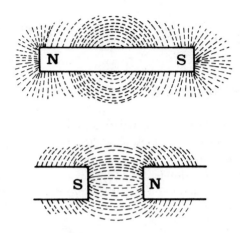

Fig. 27-2. Iron filings show flux lines between magnet poles. Unlike poles attract.

A metal object that is attracted by a magnet and held within the magnet's field for a short period tends to retain some magnetism after the influence of the magnetic field is removed. What is this characteristic called?

Answer: Retentivity.

Soft iron is very receptive to magnetic flux. What is this characteristic called?

Answer: Permeability.

Some materials resist magnetic flux. What is this resistance called?

Answer: Reluctance.

Fig. 27-3. Like magnetic poles repel each other.

What is an electromagnet?

Answer: A temporary magnet created by an electric current, usually a bar or rod mounted inside a coil of wire.

CURRENT, VOLTAGE, AND RESISTANCE

Name the two types of electricity and define each.

Answer: Static electricity is electricity that is not moving nor capable of doing work. The charge that builds on the human body as the result of friction is static (until you touch another person or metal object). *Current* electricity is moving electricity that is capable of doing work.

Similar to magnets, electrical charges exist in two states, positive and negative. How do electrical charges react when near each other?

Answer: The same as magnetic poles; like charges repel; unlike charges attract. When charges attract there is current, similar to magnetic flux.

What causes an electric current?

Answer: Electron movement.

What is an electron?

Answer: The smallest negative electric charge.

How do electrons make a current?

Answer: When there is a connection or path between negative and positive charges, electrons move toward the positive charge.

What is an electrical circuit?

Answer: A connection between a negative and positive charge that provides a path for electron flow and, therefore, current (see Fig. 27-4). Electrons move at 186,000 miles per second (299,460 km per second).

BASICS OF ELECTRICITY

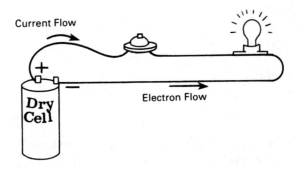

Fig. 27-4. When the button is pushed, the light lights.

In an electrical circuit the pressure that causes current is called:
Answer: Voltage difference, measured in volts, between the positive and negative sides of the circuit.

What is current?
Answer: The rate of electron flow through a circuit, measured in amperes.

How is the amount or quantity of electricity measured?
Answer: Coulomb.

What is resistance?
Answer: The opposition to current flow in a circuit. Resistance is measured in ohms.

What is a conductor? An insulator?
Answer: A *conductor* is any material that offers little resistance to an electrical current. Silver, copper, gold, and aluminum are good conductors, which is why these metals are used to make wire and contacts in switches. A conductor material has an abundance of free electrons in its atomic structure. An *insulator* is any material that offers a high resistance to an electrical current. Rubber, glass, and plastics are good insulators. An insulating material has few free electrons in its atomic structure.

What factors determine resistance?

Answer: Type of material, length, cross-sectional area, and temperature.

What is an electrical load?

Answer: A resistance in a circuit that does work—a motor, lamp, heater, etc.

What is the relationship between voltage, current, and resistance?

Answer: A pressure of one volt will force a current of one ampere through a resistance of one ohm. This relationship is called Ohm's law. If any two of these values are known, the other can be calculated by E (voltage) equals I (current) times R (resistance). Other forms of this formula are: $I = E/R$ and $R = E/I$.

What is electrical power?

Answer: The rate of doing work. Electrical power is measured in watts. One watt is equal to 0.00134 hp or 746 watts equals one horsepower.

How is power in watts calculated?

Answer: Watts = E (voltage) times I (current). One watt is the power expended when a current of one ampere is sustained by a pressure of one volt.

What is the difference between AC and DC?

Answer: Direct current (DC) flows only in one direction, positive to negative (the opposite of electron flow), usually at a steady rate. A battery is a familiar source of DC. Alternating current (AC) flows first in one direction, then reverses and flows in the opposite direction. The alternations usually occur at a steady rate, 60 times a second for normal power line AC.

What are series and parallel circuits?

Answer: In a series circuit, the electrical loads (motors, lights, heaters, etc.) are connected in a series. If we have a power source (battery, for example) with terminals numbered 1 and 10 and four lights with terminals numbered 2 and 3, 4 and 5, etc., in a series

Basics of Electricity

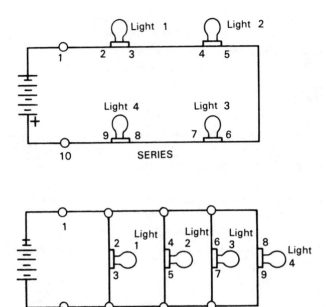

Fig. 27-5. Electrical circuits.

circuit, battery terminal 1 would connect to light terminal 2. Light terminal 3 would connect to light terminal 4, and light terminal 5 would connect to 6 and so on through the other lights until finally light terminal 9 connects to battery terminal 10, as shown in Fig. 27-5. Electrons flow from terminal 1, through each light in turn and re-enter the battery at terminal 10. If any light burns out, none will light, because the circuit will be open. In a parallel circuit each light is connected directly across the battery.

What does a switch in an electrical circuit do?

Answer: It interrupts the current through the circuit load. Switches are used to turn equipment on and off. Some are operated manually, others operate automatically. A thermostat operates an automatic switch, as does a circuit breaker when a load is drawing more current than the wiring is designed to carry.

Are electrical circuitbreaker and fuse ratings critical?

Answer: Yes. They are used to protect power sources from overload and possible fire or other hazards. A breaker or fuse should never be replaced with one of a higher rating.

Is a wire carrying an electric current surrounded by a magnetic field?

Answer: Yes. Whenever a wire carries electricity, a magnetic field is set up around the wire. A compass held near a wire carrying current will detect the magnetic flux.

What happens to the magnetic field if the current-carrying wire is wound into a coil?

Answer: The coil is surrounded by a much stronger magnetic field equal in strength to combined fields of the individual loops. The more loops or turns, the stronger the field. Increasing the current through the wire increases the strength of the field.

What happens if an iron core is inserted into the coil of wire or if the wire is wound around an iron core?

Answer: The iron concentrates the magnetic flux and increases the magnetic effect. This is an electromagnet, acting as a magnet only when current exists. Motors, relays, solenoids, generators and many other electrical devices depend on electromagnetism to operate.

What is an inductor?

Answer: When AC flows in a coil of wire, the magnetic field around the coil resists the change in alternations of current. This resistance to change is called inductance (measured in henries); therefore, a coil is an inductor. With DC applied to a coil, the magnetic field is steady, hence no inductive opposition to current exists. A coil with DC flowing is simply a magnet you can turn on and off.

What is a capacitor (or condenser)?

Answer: Two plates of metal separated by a dielectric (insulator) of some kind. A capacitor stores energy in a static electric field

BASICS OF ELECTRICITY

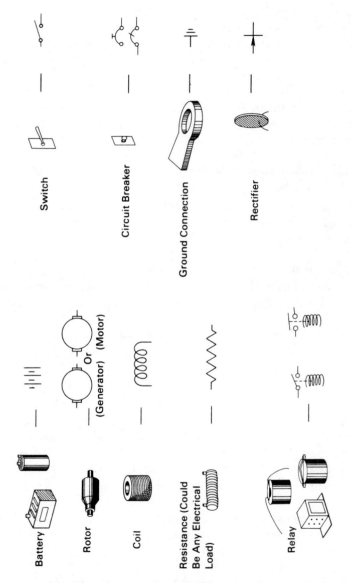

Fig. 27-6. Drawings and schematic symbols of common electrical components and devices.

because of the difference of electrical charge applied to its plates. The unit of capacity (or capacitance) is the farad. Capacitors are used to suppress sparks across mechanical breaker points as they open. Drawings and schematic symbols for the most common electrical components and devices appear in Fig. 27-6.

How is electricity measured?

Answer: With an instrument called a galvanometer. Some instruments are designed to measure voltage (voltmeter), others measure current (ammeter) and resistance (ohmmeter).

Batteries fall into one of two categories. Name each and describe its characteristic.

Answer: Primary cell is a battery that cannot be recharged after it runs down (flashlight or lantern battery). A secondary cell can be recharged many times.

What is the most common type of secondary cell?

Answer: The lead-acid battery used in automotive systems. It has lead plates immersed in a sulfuric-acid-and-water solution.

Do most lead-acid batteries have more than one cell?

Answer: Yes. A 12-volt battery has six appoximately 2-volt cells connected in series.

How is the storage capacity of a lead-acid battery measured?

Answer: By the length of time it will produce a specific amount of electricity. Batteries capacities are usually rated in ampere-hours.

What determines the capacity of a lead-acid cell?

Answer: The surface area of the lead plates.

Most automotive systems use a single-wire electrical circuit. What serves as the other side of the electrical circuit?

Answer: The body or steel frame of the unit.

Identify each of electrical symbols shown in the drawing.

Answer: See Fig. 27-6.

Basics of Electricity

What is the current in the series DC circuit in Fig. 27-7?

Answer: 2 amperes.

What is the total current drawn from the battery in the parallel DC circuit in Fig. 27-7?

Answer: 18 amperes. Notice how resistances (representing loads) combine. In a series circuit, they add (2 + 2 + 2 = 6), while in a parallel circuit, the total is less than the smallest. The reciprocal of the reciprocals method is used to calculate the combined resistance in the parallel circuit. ½ + ½ + ½ = 1.5, then 1 ÷ 1.5 = 0.6666 ohms combined resistance. Since each load is of equal resistance, the total current would divide evenly, or 6 amperes per load.

What voltage would appear across each load in the series DC circuit in Fig. 27-7?

Answer: 4 volts. E = IR or E = 2 × 2. If the resistances were not equal, the voltages would not be equal. In the parallel circuit, the voltages would be the same across each load regardless of

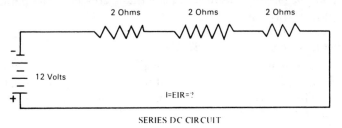

SERIES DC CIRCUIT

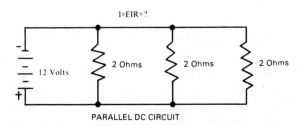

PARALLEL DC CIRCUIT

Fig. 27-7. DC circuits demonstrating **Ohm's law**.

Power Plant Engineers Guide

resistance. From the above we can say that the current is the same through all loads in a series circuit and the voltage across each load depends on resistance. The opposite is true of a parallel circuit; the voltage is the same across each load, but the current through each depends on the resistance.

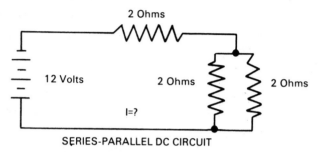

Fig. 27-8. Complex DC circuit.

What is the rate of power consumption in the circuits in Fig. 27-7?

Answer: 24 watts in the series circuit and 216 watts in the parallel circuit. Watts = voltage × current (W = EI). As you can see the rate of power consumption is 9 times greater in the parallel circuit.

What is the current in Fig. 27-8?

Answer: 4 amperes. In this series-parallel circuit, to determine current, the combined resistance of the three loads must be calculated. The net resistance of the parallel loads is 1 ohm (½ + ½ = 1 and 1 ÷ 1 = 1) plus 2 ohms = 3 ohms.

What is the voltage across the series load in Fig. 27-8?

Answer: 8 volts (E = IR).

What is the voltage across each of the parallel loads?

Answer: 4 volts. Remember, voltages across parallel loads are the same. But the current through each parallel load would be only 2 amperes. The total of the voltage drops in a series circuit must equal the battery or source voltage and the total of the individual load currents in a parallel circuit must equal the total current output of the battery or source.

BASICS OF ELECTRICITY

GENERATORS

If a conductor which is part of a circuit is moved through a magnetic field, what happens?

Answer: An electric current is developed in the wire as it "cuts" the magnetic lines of force. See Fig. 27-9. We say the current is *induced* in the wire by the magnetic field.

What would happen to the induced current if the wire stops in the middle of the magnetic field?

Answer: It would disappear. Induction depends on movement of either the wire or the magnetic field.

What determines the amount of induced current?

Answer: (1) The *strength* of the magnetic field, the number of lines of force; (2) the *speed* of the wire through the field; the faster a wire cuts more lines of force, the greater the induced current; (3) the *angle* of the cutting action; a wire cutting straight across generates more current than one moving at an angle; (4) the *number* of wires doing the cutting; the more wires the more current.

Draw a sketch of a simple generator and explain how it works.

Answer: See Fig. 27-10. The induced current flows as indicated by the arrows in Fig. 27-10A, 27-10B and 27-10C. As the wire rotates, the direction of current flow reverses during each half of the wire rotation. Thus the induced current at *a* and *d* in Fig. 27-10 alternates with each rotation of the wire (representing the armature in a generator) and drops to zero twice during each rotation when the wire is not cutting the magnetic field (Fig. 27-10B). Thus, the generator is producing AC, alternating current.

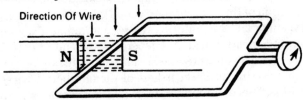

Fig. 27-9. Current is induced in a circuit moving through a magnetic field.

763

POWER PLANT ENGINEERS GUIDE

What's the difference between an AC and a DC generator?

Answer: All generators produce AC. The AC load is connected to the generator as shown in Fig. 27-11A. A *slip ring* is connected to each end of the wire and rotates with the wire. Contact between the slip rings and the circuit is achieved by brushes.

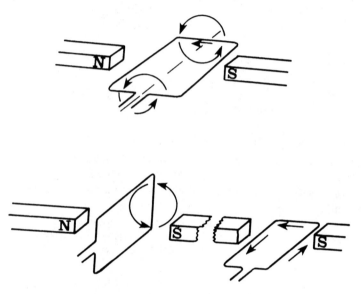

Fig. 27-10. Basic electrical generator.

How do DC generators operate?

Answer: Each end of the armature wire is connected to semicircular segments at the end of the armature. The assembly of semicircular segments is called a commutator. As shown in Fig. 27-11B, current always flows through the external circuit in the same direction but pulsating from zero to maximum twice during each rotation of the armature. An actual generator has many loops of wire and a DC generator has many commutator segments, so the output is steady. In an AC generator, the end of each loop of armature wire is connected to the same slip ring (an AC generator has only two slip rings, regardless of the number of wires in the armature). The commutator in a DC generator is a form of *rectifier*.

BASICS OF ELECTRICITY

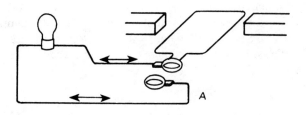

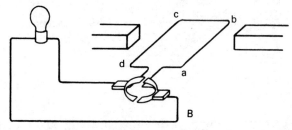

Fig. 27-11. With the addition of slip rings (A), an alternator or generator produces an AC output. A commutator (B) produces a DC output.

What is a rectifier?

Answer: Any device that changes AC to DC.

How is an armature built?

Answer: See Fig. 27-12. Armature loops are made in the form of a coil and the coils are wound on a core which supports them.

How is the core made?

Answer: From laminations of thin metal discs separated by insulation to prevent current buildup in the metal, which would result in heating.

The generators discussed so far show coils rotating in the field produced by permanent magnets, a technique which serves purposes of illustration. But even if the permanent magnets produced extremely powerful fields, such a generator would produce little usable electricity. Therefore, in practical generators, electromagnets are used to produce the magnetic fields, as shown in Fig.

27-13A and schematically in 27-13B. Notice that the field coils are connected in parallel with the armature and the load (the lamp). Electricity to energize the field coils could come from a battery or some other DC source, but it isn't necessary with a ready source in the generator itself.

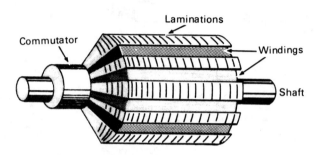

Fig. 27-12. Typical armature showing various parts.

What type of generator does the drawing in Fig. 27-13 represent?

Answer: A *shunt-wound* type, because the field coils are shunted across in parallel with the armature. Current through the field windings is independent of current in the load and the voltage across the field is proportional to generator output.

What type of generator is shown in the drawings in Fig. 27-14?

Answer: A *series-wound* type; notice the field coils are connected in series with the armature and the load.

What is the relationship between the current in the load and in the field coils?

Answer: It's the same.

What happens when the load current increases in a series-wound generator?

Answer: The field-coil current also increases, thus increasing the strength of the magnetic field and the output of the generator. So a series-wound generator automatically increases output as the load on the generator increases and decreases output when the load decreases, a favorable characteristic.

BASICS OF ELECTRICITY

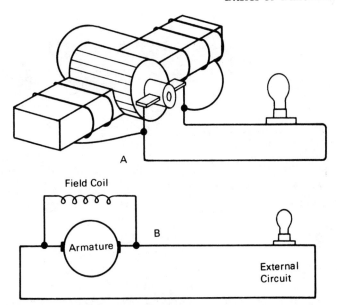

Fig. 27-13. Shunt-wound generator.

What is compound-wound generator?

Answer: A generator with two sets of field coils, a series winding and a shunt winding, as shown in Fig. 27-15.

What is the advantage of two windings?

Answer: The shunt winding tends to keep the voltage output of a compound-wound generator constant, while the series winding varies the output with a varying load current. A compound-wound generator, therefore, is a better self-regulator. A voltage control can be put in series with the shunt field to control the generator voltage output. A variable resistance is shown in the drawing in Fig. 27-16, but the actual control can be any type suitable, including automatic electronic controls.

ALTERNATING CURRENT

What is an alternator?

Answer: A generator that produces an AC output.

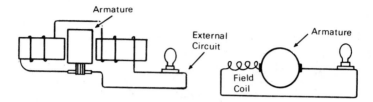

Fig. 27-14. Series-wound generator.

What is the difference between an alternator and DC generator?

Answer: The output of an alternator cannot be used to directly supply the field coils. DC is needed for this purpose.

How are the fields of an alternator supplied?

Answer: A battery or DC source as shown in Fig. 27-17 or a small DC generator on the same armature shaft with the alternator. (Remember, all armatures produce AC. The difference lies in the way current is taken from the armature; a commutator is used in a DC generator and slip rings in an AC generator or alternator.) The variable inductance in Fig. 27-17 varies the DC applied to the field coils as the alternator output varies.

How is the output of an alternator controlled?

Answer: The same way as in a DC generator output. The circuit in Fig. 27-17 shows an adjustable automatic control. The automatic feature compensates for varying loads. As the alternator output voltage tends to drop under increasing load, the automatic control allows the current through the shunt field to increase.

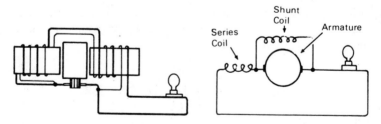

Fig. 27-15. Compound-wound generator.

BASICS OF ELECTRICITY

Draw a graph or a "picture" of current in an AC circuit.

Answer: See Fig. 27-18. Current starts at zero at the left. As the alternator armature turns, the current increases upward in the drawing (usually called a positive direction). As the armature reaches the ¼ point in its rotation, it is producing maximum current. Then, during the next ¼ turn, the current decreases at the same rate to zero at the halfway point in the armature rotation. As the armature begins the third quarter of its rotation, current increases in the opposite (negative) direction. At the end of the third quarter of the revolution, the negative (bottom) peak is reached and current begins decreasing to zero again during the final quarter armature turn, completing one AC cycle, which equals one complete rotation of the alternator armature. (A graph or "picture" of DC would be a steady, straight line at a point above the zero-current line in Fig. 27-18.) Power line AC goes through 60 complete cycles (*a* to *b* in Fig. 27-18) per second. The *frequency* of the power line is 60 cycles per second (referred to now as 60 *hertz*, Hz.)

What would a graph of AC voltage look like?

Answer: Very similar to the drawing in Fig. 27-18. It starts at zero and goes through the same cycle.

How is the work done by AC compared to DC?

Answer: By the heating effect. Experience shows that 1.41 times as much AC voltage is needed to have the same heating effect as DC; i.e., 141 volts AC has the same heating effect as 100 volts DC. Or we can say that *peak* AC voltage is only 0.707 times as effective as DC. So AC voltmeters register the *effective* AC voltage (0.707 of the actual peak). The peak voltage of a 115-volt powerline is over 160 volts.

How does a transformer work?

Answer: As AC flows through the coil, the magnetic field around it builds and collapses 120 times a second, 60 times in each direction. If another coil is located within the rising and falling magnetic field, a voltage will be induced in the second coil or secondary. (See Fig. 27-19). The voltage induced in the secondary

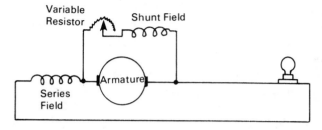

Fig. 27-16. Compound-wound generator with output control.

by the primary or first coil is proportional to the strength of the magnetic field and the relationship between the number of turns of wire on the primary and secondary, as shown in Fig. 27-19. The coils in a power transformer are wound around laminated metal cores to keep heating losses down.

As voltage is stepped up and down, what happens to the current-delivery capability of a transformer?

Answer: Current available from any transformer decreases as voltage is stepped up and increases as voltage is stepped down. For example, if a 1:2 step-up transformer is capable of delivering 115 volts AC at 10 amperes to the primary, the secondary voltage will be 230 volts at 5 amperes. The output capability in watts doesn't change, though; 10 amperes at 115 volts will do the same work as 5 amperes at 230 volts.

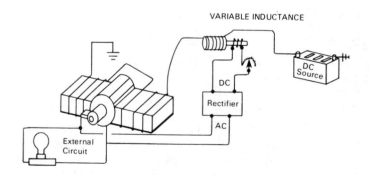

Fig. 27-17. Compound-wound generator with automatic output control.

BASICS OF ELECTRICITY

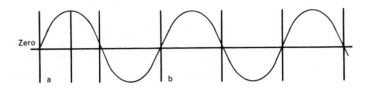

Fig. 27-18. AC current waveform.

How does inductance of a coil affect the flow of AC?

Answer: The magnetic field produced around a coil by an alternating current resists each change or tries to hold back the change in the magnetic field with each AC cycle. So an inductor resists the flow of AC current.

How does the AC resistance of an inductor affect the relationship (called phase) between AC voltage and current?

Answer: Because of the resistance to change in the magnetic field, which is produced by the current, the current lags behind the voltage by one-quarter cycle or 90° (one fourth of a full cycle or circle).

What is power factor?

Answer: Power factor is a ratio that tells us how efficient an electrical device is; i.e., how much actual power is being delivered in comparison to the apparent power. A power factor rating is usually less than one (one would be 100% efficient).

How does the capacitance of a capacitor or condenser affect the flow of AC?

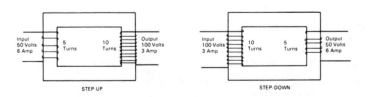

Fig. 27-19. Simple transformer drawings show relationship between step-up and step-down windings.

771

Answer: The exact opposite of a coil. The electrostatic field produced by an AC voltage between the plates of a capacitor resists each change or tries to hold back the change in the static field with each AC cycle. As the voltage alternates it must *charge* the capacitor each time it changes direction. Charging takes time, so the voltage lags behind the current 90° in an AC circuit with a capacitor in it.

If an inductor or coil and a capacitor or condenser are both used in an AC circuit, what is the effect?

Answer: The effect of one offsets the other, bringing AC voltage and current together again, or back in the same phase relationship, theoretically.

What is the resistance to AC introduced by a coil or capacitor called?

Answer: Reactance, inductive and capacitive, respectively.

What happens if DC is applied to a coil?

Answer: A magnetic field builds around it and holds until the DC is turned off. The only resistance is in the energy used to heat the wire.

What happens if DC is applied to a capacitor?

Answer: A capacitor effectively blocks the flow of DC after the initial charge when the current is first turned on.

MEASURING VOLTAGE, CURRENT, AND RESISTANCE

What is a voltmeter?

Answer: A galvanometer or electronic meter designed to measure the voltage present across a circuit. Voltmeters are calibrated in different ranges to measure voltages at varying levels.

What is an ammeter?

Answer: A galvanometer or electronic meter designed to mea-

BASICS OF ELECTRICITY

sure amperes or current in a circuit. Ammeters are calibrated in various ranges to measure differing levels of current. A circuit must be opened and the ammeter leads connected in series with the circuit to measure current (unlike voltage which is measured across a circuit).

What is an ohmmeter?

Answer: A galvanometer or electronic meter designed to measure ohms of resistance. An ohmmeter has its own power supply, therefore, resistance can be measured *only* when no other power source is connected to circuit. An ohmmeter measures resistance by applying a known voltage to a circuit or component and measuring the current in the circuit. An ohmmeter indicator is calibrated ohms.

MOTORS

How similar are a DC motor and a generator?

Answer: They're the same, basically, except we take current from a generator and apply current to a motor. If a current is applied to the wire in Fig. 27-20, a magnetic field will develop around it. The magnetic field around the wire will react with the field of the magnets and spin the wire (like poles repulse). As the wire turns, and like poles begin to line up, the direction of the current is reversed by the commutator and the repulsion of the like

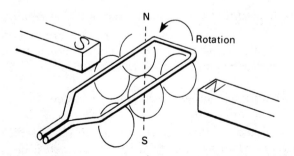

Fig. 27-20. Basic electric motor.

magnetic fields spins the wire. As long as current is applied to the wire, it keeps running.

What is a series-wound motor?

Answer: The same as a series-wound generator. See Fig. 27-21.

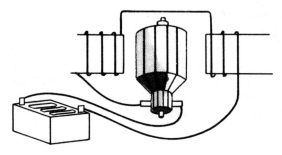

Fig. 27-21. Series-wound motor.

What is a shunt-wound motor?

Answer: The same as a shunt-wound generator. See Fig. 27-22. The only fundamental difference between a series- or shunt-wound motor and generator is that current is applied to a motor to produce mechanical energy and mechanical energy is applied to a generator to produce current.

Is a compound motor the same as a compound generator, too?

Answer: Yes. See Fig. 27-23. The series winding produces a high degree of torque (turning force) at startup and at low speeds.

Are there similarities between an AC motor and alternator?

Answer: Yes. See Fig. 27-24. The loop armature will produce the AC current wave drawn beside it. One complete turn of the wire produces one complete cycle (*a* to *b* in the wave drawing).

What will happen if another armature loop is added at right angles to it, as shown in Fig. 27-25?

Answer: It, too, would produce a complete AC cycle with each turn. So, the two loops would produce two complete cycles (*a* and *b* in Fig. 27-25) with each complete turn of the armature. One cycle would be 90° ahead of the other, of course.

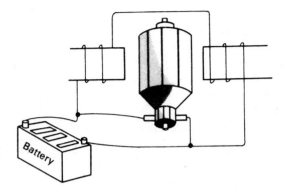

Fig. 27-22. Shunt-wound motor.

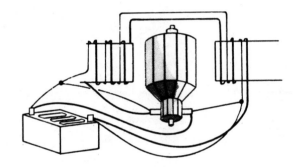

Fig. 27-23. Compound-wound motor.

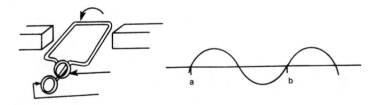

Fig. 27-24. Simple alternator and current flow drawing.

If a third loop were added, as shown in Fig. 27-26?

Answer: Three AC cycles (*a*, *b*, and *c* in the drawing) would be produced with each complete turn of the armature. Each cycle would be 60° apart in time.

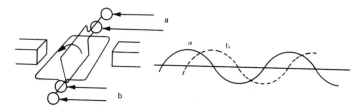

Fig. 27-25. Two-loop armature produces two output waves at different times.

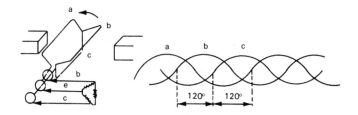

Fig. 27-26. Three-loop armature produces three output waves.

What are the alternators in Figs. 27-25 and 27-26 called?

Answer: The alternator in Fig. 27-25 is a two-phase type and the machine in Fig. 27-26 is a three-phase. Of course, the single-loop alternator is a single-phase type. These same principles are used in AC motors.

Draw a diagram of a simple two-phase motor and describe its operation.

Answer: See Fig. 27-27. A permanent magnet rotor is mounted in the fields of two sets of magnetic poles. One AC phase energizes one set and the other phase energizes the other. As the current changes, the pole pieces are magnetized in turn. Of course, the rotor has to turn in an effort to keep unlike magnetic poles lined up.

Draw a sketch of a single-phase motor.

Answer: See Fig. 27-28. Although this motor would run with only two pole pieces, a second set is shown to help in starting.

BASICS OF ELECTRICITY

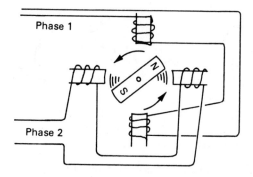

Fig. 27-27. Two-phase motor. Notice the similarity with starting winding.

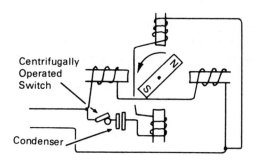

Fig. 27-28. Single-phase motor with starting winding.

When operating speed is reached, the centrifugal force produced by the armature opens a switch in the circuit supplying the second set of poles. The capacitor throws the phase of the AC voltage applied to the second set of pole pieces off by 90°. It also absorbs the surge of voltage which develops when the switch opens, thus keeping the potentially damaging voltage out of the field coils.

How many sets of pole pieces does a 3-phase motor have?

Answer: Three. See Fig. 27-29. Of course, each set is energized in turn by each of the three phases of AC.

Are permanent-magnet rotors used in all motors?

Answer: No, only small ones. Most often, coils of fine wire are

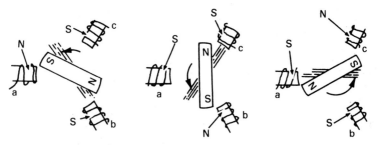

Fig. 27-29. Three-phase motor.

wound on the armature but not connected to any outside circuit. Current is induced in the armature coils, which gives the motor more power. This type is called an *induction* motor.

What is a repulsion-start induction motor?

Answer: It has a commutator with brushes, similar to a generator. Current is applied to the armature windings only during startup.

CHAPTER 28

Diesel Engine Principles

What is a diesel engine?
Answer: A high-compression, internal-combustion engine which depends upon the *heat of compression* for ignition (Fig. 28-1).

Name two general classes of diesel engine.
Answer: Four-cycle and two-cycle.

FOUR-CYCLE DIESEL ENGINES

What are the four strokes of a four-cycle diesel engine?
Answer: (1) Admission or intake, (2) compression, (3) power, (4) exhaust.

What occurs during the admission stroke?
Answer: Air is admitted into the cylinder (Fig. 28-2).

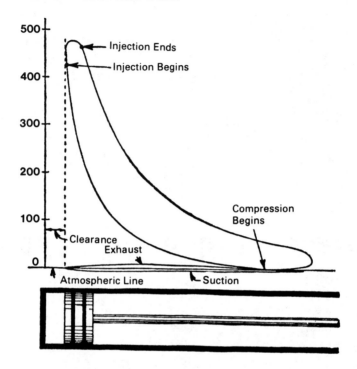

Fig. 28-1. Typical indicator card of diesel four-cycle engine.

What occurs during the compression stroke?

Answer: The air previously admitted is compressed to about 500 psi (35 kg/cm²) which causes its temperature to rise to about 1000°F. (about 538°C.). As this pressure is reached gradually, it does not cause a shock to the engine, such as an explosion to the same pressure would give. See Fig. 28-3.

Describe the events of the power stroke.

Answer: At the beginning of this stroke, as in Fig. 28-4, the fuel is injected into the cylinder. Meeting the highly heated air, it immediately ignites by the heat of compression and burns throughout the period of injection, this period being a small percentage of the stroke. The resulting pressure of combustion forces the piston down on its power stroke.

Diesel Engine Principles

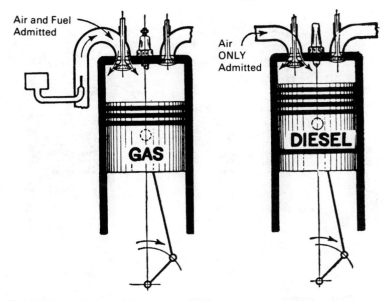

Fig. 28-2. Admission stroke.

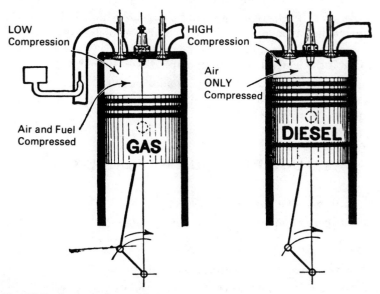

Fig. 28-3. Compression stroke.

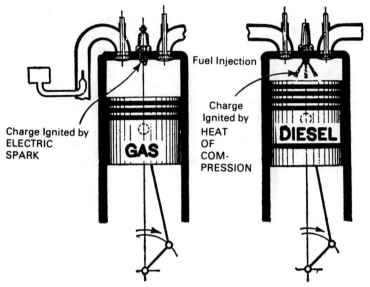

Fig. 28-4. Power or expansion stroke.

What occurs during the exhaust stroke?

Answer: The products of combustion are expelled through the open exhaust valve, as in Fig. 28-5, thus completing the cycle.

What is the general construction of a diesel engine?

Answer: The diesel engine has many parts in common with the gasoline engine, but they are made heavier to adapt them to the more severe cycle.

COMPRESSION PRESSURES

What is the vital feature of diesel engines?

Answer: High compression, much higher than in gasoline engines.

Why is such high compression necessary?

Answer: Without high compression, there would not be enough

DIESEL ENGINE PRINCIPLES

heat to ignite the charge; accordingly, the engine could not operate.

What determines the air temperature at the time of fuel injection?
Answer: The degree of compression.

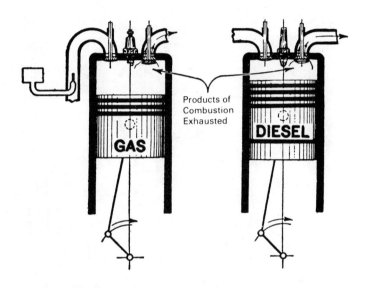

Fig. 28-5. Exhaust stroke.

What are the compression ratios usually employed in the United States?
Answer: They range from 11 to as high as 19 to 1.

FUEL-INJECTION METHODS

How many methods of fuel injection are employed and what are they?
Answer: Two: air injection and airless injection. Nondescript names for the second method, such as solid, direct, mechanical, and so forth should not be used.

What is air injection?

Answer: In this method, as shown in Fig. 28-6, air compressed to several hundred pounds higher than the compression pressure is used to force the fuel charge into the cylinder.

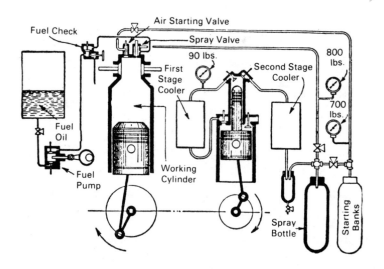

Fig. 28-6. Elementary air-injection diesel, showing two-stage compressor and other auxiliary apparatus.

Describe the method in more detail.

Answer: The highly compressed injection air is, at the proper moment, blown into or over a measured charge of fuel. It forces the fuel into the cylinder and aids combustion by the added quantity of air furnished and the turbulence produced during admission of the charge.

What is turbulence?

Answer: The state of being in violent, disordered commotion.

What is the object of turbulence?

Answer: Designers intend it to more efficiently mix the incoming charge with the air of combustion. The turbulence due to air

Diesel Engine Principles

injection is constituted to smooth and regulate fuel combustion so that such engines usually can use a lower grade of fuel.

What is airless injection?

Answer: In this system, used in most diesels today, fuel is drawn from the main tank by a fuel-supply, or transfer, pump, filtered, and delivered to an auxiliary tank or direct to the injection pump. The injection pump forces the liquid at very high pressures through the nozzle or spray valve and into the cylinder at a predetermined time and quantity as controlled by timing and metering mechanisms.

Name the several systems of airless injection.

Answer: (1) Master pump, (2) individual pump, (3) distributor, (4) injector.

Master pump system.—This is the use of one pump to supply fuel to all the cylinders. The pump maintains a high fuel pressure in a common manifold or "rail" (hence the name, "common-rail system"), as in Fig. 28-7. Connection is made from the rail to each injection nozzle. The injection nozzles are closed by spring-loaded valves which are opened by separate mechanism or valve gear at the proper time.

How is the power output controlled in this system?

Answer: Either by variation in the length of time the injection valve is held open or by variation in the fuel pressure at the nozzle.

How are the injection valves operated?

Answer: Usually by cams, either directly or through pushrods and rocker arms.

Mention one characteristic of the system.

Answer: In common-rail systems, the injection valves are always under full pump-pressure.

What is the chief advantage of the common-rail system?

Answer: Simplicity. All that is required is a single-plunger

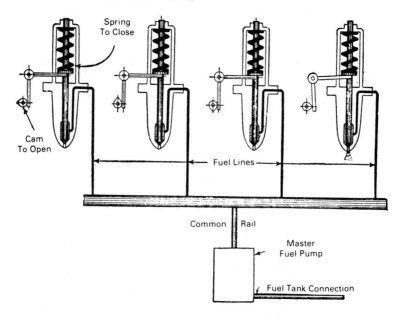

Fig. 28-7. Elementary master pump, or common-rail, fuel system.

pump and spray valve for each cylinder, plus mechanical timing gear for operating the valve.

What pressure is maintained by the master pump?

Answer: From 2000 to 8000 psi (140-560 kg/cm^2), depending upon the design of the system.

Individual pump system.—In this system, a separate pump is used for each cylinder, as shown in Fig. 28-8.

How are the pumps mounted?

Answer: Either in a single housing or individually on each cylinder. They are usually placed in a single housing.

Describe the pumps.

Answer: They are of the plunger type and work at pressures from 1000 to 10,000 psi (70-700 kg/cm^2).

DIESEL ENGINE PRINCIPLES

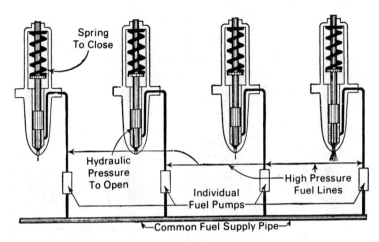

Fig. 28-8. Elementary individual-pump fuel system.

How are metering and timing accomplished?

Answer: Metering of fuel and timing of injection are done within the pump unit, the pump of each cylinder discharging through its connecting line and injection valve at the proper time.

Describe the injection valves.

Answer: They are usually spring-loaded, opening and closing at a definite, predetermined pressure in order to ensure spray characteristics and accurate cut-off of fuel.

Why is this system called the "hydraulic timing" method?

Answer: Because the injection valve merely acts as a hydraulic check valve, its opening depending upon the hydraulic pressure built up by the pump at the proper time.

Name three methods of metering the fuel.

Answer: (1) By variable stroke of the plunger, (2) by throttling, and (3) by variable bypass. The variable bypass system is used on practically all modern, small- and medium-speed engines.

Distributor system.—Here, the fuel is supplied by one pump and switched to each cylinder by a multioutlet, rotating valve, or distributor. It may be compared in principle to synchronous ignition, so far as the switching idea is concerned (**Fig. 28-9**).

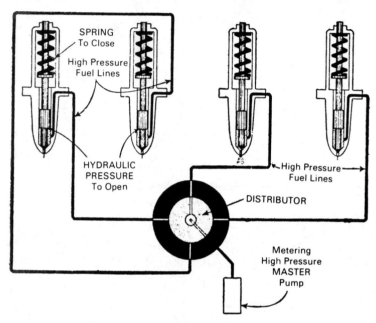

Fig. 28-9. Elementary distributor fuel system.

What should be noted about the distributor system?

Answer: The pump must make as many delivery strokes as there are power strokes. Considering the heavy pressure, the duty is severe and causes considerable wear, especially on high-speed engines. This makes the system more suited to large, slow-speed engines.

Injector system.—This is virtually a distributor-type two-stage pumping system; that is, a modified distributor system.

Why was it introduced?

Answer: To relieve the metering pump of the severe duty of pumping against high pressure and to avoid the need for high-pressure tubing lines.

What does the master, or main, pump do?

Answer: The main pump (sometimes called the distributor pump) meters the fuel and delivers it to the unit injectors at a low pressure of about 10 psi (0.7 kg/cm^2). After the fuel has passed through a strainer, the distributor functions to connect the unit injectors in proper sequence.

Just what are these unit injectors?

Answer: They are small, high-pressure pumps and nozzles combined.

Where are they located?

Answer: In the center of each cylinder head.

Describe the drive.

Answer: The plungers may be driven directly by camshafts, rocker arms, or pushrods, or they may be spring-loaded and ride on the cams (Figs. 28-10 - 28-11). In the operation of the reversed cone nozzle injector, note the lever arm fulcrumed at the center. When the nose of the cam pushes up one end of the lever, the other end pushes down the valve stem, to which is attached the pump plunger and reversed valve of the nozzle, allowing the charge to enter the cylinder.

Describe the Graham injector.

Answer: As the valve closes, the master pump forces a metered charge of fuel into the receiving chamber at low pressure. During the opening of the valve, this charge is transferred to the discharge chamber, where it is preheated and discharged through the multinozzle passages during the closing of the valve. The Graham injector requires no valve adjustment and no high-pressure fuel line.

COMBUSTION

Name three periods relating to combustion.

Answer: (1) Delay, (2) uncontrolled combustion, (3) direct-burning.

What is the delay period?

Answer: When the fuel is injected, it does not start to burn immediately because its temperature must be raised to the ignition level.

What other name is sometimes given to this period?

Answer: The time lag of ignition.

What is the uncontrolled combustion period?

Answer: Ignition begins at the end of the delay period and

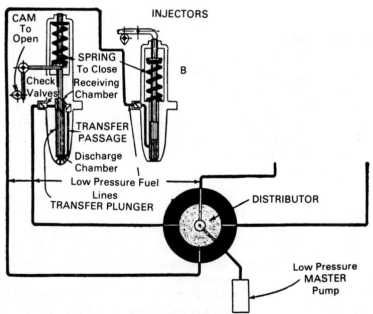

Fig. 28-10. Distributor system with unit injectors illustrating the Graham injector (A) and reversed-cone nozzle injector (B).

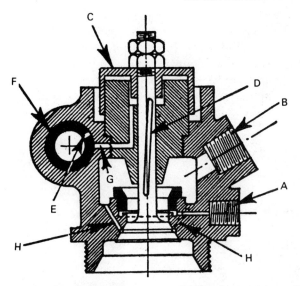

Fig. 28-11. Raabe pneumatic injector. The parts are (A) fuel connection, (B) air connection, (C) valve operating piston, (D) shallow air groove communicating with the valve operating piston, (E) injection timing port in injection valve, (H) air turbulence ports.

spreads to the rest of the charge. This progressive combustion up to the instant of complete combustion is called "uncontrolled combustion."

What is the direct-burning period?

Answer: After the instant of maximum pressure, the fuel still being injected finds the oxygen needed and the temperature is raised to such an extent that it begins burning immediately, or direct-burning.

Name the auxiliary combustion-chamber designs upon which the various combustion methods are based.

Answer: (1) Plain, or "open," (2) turbulence, (3) precombustion, (4) separate, (5) antecombustion, and (6) air cell.

What is a plain combustion chamber?

Answer: A nondivided chamber which is simply an extension of the cylinder itself beyond the upper travel-limit of the piston (Fig. 28-12). This type gives the minimum cooling surface exposed to compressed air and flame. In small cylinders, ignition delay may retard the ignition until a good portion of the fuel has been admitted, resulting in "diesel knock," which is a sudden pressure rise during the power stroke. Injection in a plain combustion chamber is known as airless injection.

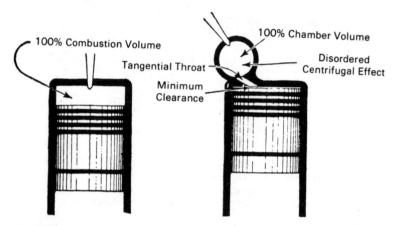

Fig. 28-12. Plain, or "open," combustion chamber (left) and turbulence combustion chamber (right).

Why is the plain combustion chamber undesirable on modern engines?

Answer: Owing to the constant demand for more power and less bulk, the rotative speed has greatly increased. This necessitates something different from the plain combustion chamber for efficient combustion.

What is a turbulence air chamber?

Answer: A chamber designed, as in Fig. 28-12, to provide the turbulence so essential for efficient combustion in small, high-speed engines.

What is the construction and operation?

Answer: The cylinder clearance is reduced to a minimum and the turbulence chamber made large enough to receive practically all of the compressed air charge. In operation, the piston coming up on its compression stroke forces the air through the port and into the chamber, where it is compressed to a pressure of about 550 psi (38.5 kg/cm^2). The throat entering the combustion chamber is tangent to its outer wall, a design which forces the air past the injection nozzle at great speed and sweeps the nozzle tip with air at all times during the compression stroke. The high-speed, swirling action of the air ensures prompt and intimate mixing of the air and fuel; since the airspeed increases with the speed of the engine, it gives automatic compensation that permits high-speed engine operation, a clear exhaust, and high power output.

What is a precombustion chamber?

Answer: This is a partial combustion chamber, as shown in Fig. 28-13, in which combustion of a part of the fuel takes place. By definition, it is a chamber so proportioned with respect to the clearance volume of the cylinder that only about 30% of the combustion takes place within the chamber itself.

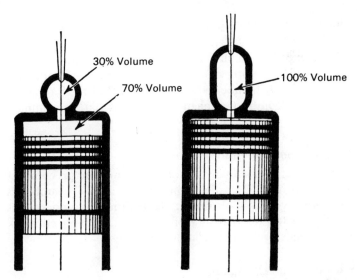

Fig. 28-13. Precombustion chamber (left) and separate combustion chamber (right).

What is the function of the chamber?

Answer: To produce turbulence and higher wall-temperatures near the nozzle.

What is a separate combustion chamber?

Answer: One in which all the combustion takes place (Fig. 28-13).

Describe the construction.

Answer: Cylinder clearance is reduced to a minimum. A small passage connects the separate combustion chamber with the cylinder.

Describe its operation.

Answer: In operation, the entire charge is ignited in the separate combustion chamber before initial expansion forces the burning gases through the connecting passage and against the moving piston.

What is the distinction between a precombustion chamber and a separate combustion chamber?

Answer: It lies in the ratio between the cylinder clearance volume and the volume of each type of combustion chamber. The precombustion chamber contains about 30% of the former, while the separate combustion chamber contains 100% (assuming zero clearance).

What is an antecombustion chamber?

Answer: It is essentially a modification of the precombustion chamber (Fig. 28-14). The objective of such a chamber is to progressively supply air during the combustion period. It is so designed as to extend away from its outlet, with injection taking place directly opposite the latter.

What is an air cell chamber?

Answer: See Fig. 28-14. Because of the position of the injection nozzle, only air is compressed within the chamber and all combus-

DIESEL ENGINE PRINCIPLES

tion takes place outside of it. The practice of making air cell chambers 100% volume can be of doubtful benefit because the extremely small quantity of initial combustion air outside the cell could become exhausted before piston advance produced sufficient pressure drop to supply the additional air required.

Describe its operation.

Answer: In operation, fuel is injected only into the main cylinder during expansion of the burning gases in the main chamber, when the pressure therein drops below that of the air in the cell.

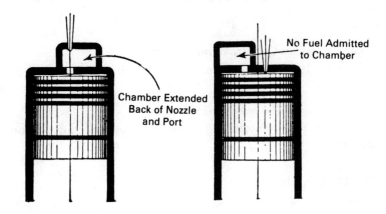

Fig. 28-14. Antecombustion chamber (left) and air-cell chamber (right).

FUEL-INJECTION SYSTEM

Of what does the fuel-injection system consist?

Answer: (1) Fuel transfer pump, (2) sediment trap, (3) strainer, (4) injection pump or pumps, (5) injection valves.

What is the function of the fuel transfer pump?

Answer: The low-pressure transfer pump is used to deliver the fuel from the tank to the injection pump.

What is placed between the transfer pump and the injection pump and why?

Answer: A sediment trap and strainer, because the fuel must be separated from all impurities.

Name two classes of injection pumps and their uses.

Answer: (1) Constant-pressure (used with common-rail system), and (2) metering (used with individual pump system).

What other classification can you give for fuel pumps?

Answer: They are classed as constant-stroke and variable-stroke.

What does a constant-stroke pump do?

Answer: It delivers a definite quantity of fuel on each stroke.

Can the quantity be regulated?

Answer: Yes, by bypassing part of the fuel back to the supply tank.

What is its application?

Answer: Automotive engines.

What is a variable-stroke pump?

Answer: One in which the stroke is lengthened or shortened so as to meter the correct charge to meet varying operating conditions.

Name two general types of airless injection valves.

Answer: Mechanical and hydraulic.

What is a mechanical injection valve?

Answer: One which is lifted from its seat by cam action and closed by the action of a spring, as shown in Fig. 28-15.

DIESEL ENGINE PRINCIPLES

What is a hydraulic valve?

Answer: One held on its seat by a spring, as in Fig. 28-16, and provided with an enlargement on the spindle so that impulse pressure exerted by the metering fuel pump will overbalance the tension of the spring and open the valve.

Name some types of fuel nozzles.

Answer: (1) Single jet, (2) multijet, (3) pintle, (4) conical, as shown in Figs. 28-17–28-19.

What is a glow plug?

Answer: A device which screws into the combustion chamber like a spark plug and has a small heater coil. See Fig. 28-20.

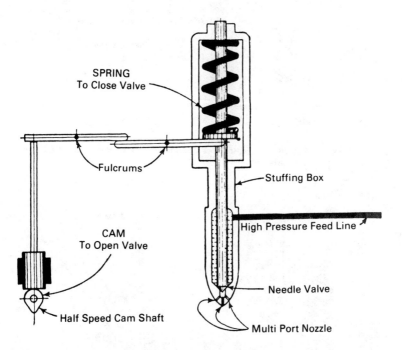

Fig. 28-15. Mechanical injection valve opened by cam drive and closed by spring tension.

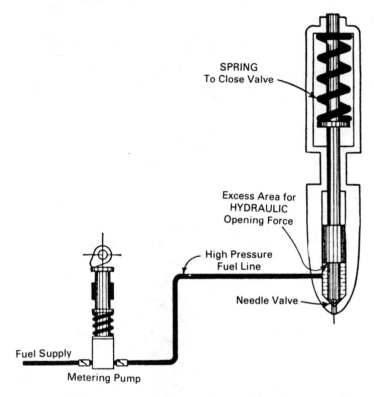

Fig. 28-16. Hydraulic injection valve opened by fuel pressure and closed by spring tension.

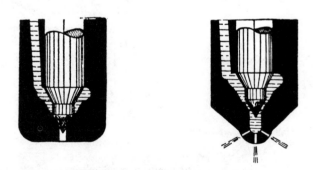

Fig. 28-17. Single (left) and multijet (right) nozzles.

DIESEL ENGINE PRINCIPLES

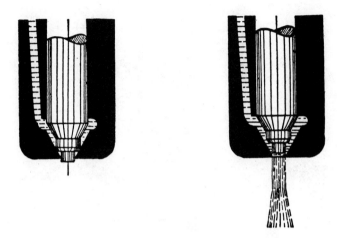

Fig. 28-18. Pintle nozzle closed (left) and open (right).

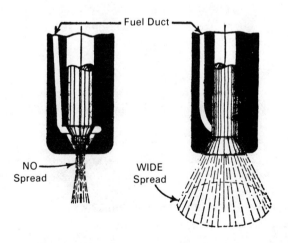

Fig. 28-19. A pintle valve (left) produces a cylindrical discharge, while a conical nozzle has a conical discharge (right).

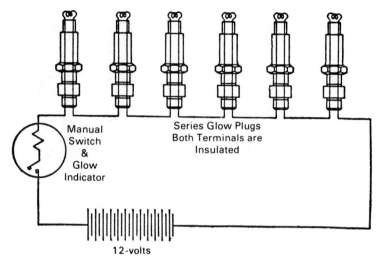

Fig. 28-20. Glow-plug circuit diagram for six-cylinder engine.

SERVICE

What should be noted about poor compression?

Answer: There is a certain critical point of compression below which a diesel will not ignite the charge.

What causes loss of compression?

Answer: Leakage. High compression pressure introduces an increased tendency for air to leak through the smallest crevices.

What precaution should be taken in fitting piston rings?

Answer: The mechanic should be guided by the clearance recommendations of the manufacturer to ensure a leak-free fit.

Name another important cause of compression loss.

Answer: Leakage through the valves.

Give Raabe's method of grinding injection valves.

Answer: This noted authority on diesel engines advises that

DIESEL ENGINE PRINCIPLES

whenever it is necessary to use any abrasive material, be sure it is of extra-fine grain and of uniform consistency. Ground glass mixed with light oil is best, but care must be taken to remove any coarse grains from the powder. The abrasive must be applied to the surface to be ground very sparingly, and very little pressure should be applied while grinding.

When the surface has acquired a strictly uniform, dull color, as revealed by frequent examination, the abrasive must be wiped off very carefully. A drop of light oil is then applied to the surface, and with hardly any pressure, the grinding is repeated. This latter operation tends to remove any broken-down grains of the abrasive that may have been crushed into the pores of the metal. Moreover, the microscopic chips of metal which have been cut from the surface by the abrasive will be removed.

After a few light rubs with pure oil, examine the surface. Repeat the grinding and examining until no further evidence of metallic or abrasive particles is found upon the surface or in the film of oil upon it.

What extra care should be taken?

Answer: Thoroughly remove any abrasive remaining in contact with any bearing surfaces of valve guides or pump barrels and plungers.

What should be done after overhauling and reassembling any of the fuel-handling apparatus?

Answer: Be sure it is cleared of air, as air pockets in the fuel system always cause trouble.

How about the practice of idling the engine?

Answer: Do not allow the engine to idle any more than is absolutely necessary, and do not try to reach the minimum speed at which the engine will idle.

TWO-CYCLE DIESEL ENGINES

What is a two-cycle engine?

Answer: One in which the four events of the cycle are performed during two strokes of the piston, or one revolution of the crankshaft, as in Fig. 28-21. Fig. 28-21A shows a piston in this GM diesel at the lower end of its scavenging and compression stroke; air from the blower is entering the combustion chamber through the cylinder air ports. In Fig. 28-21B, the rising piston has closed off the cylinder ports, stopping the flow of incoming air; the exhaust valves are closed, and the air in the cylinder is being compressed. The piston has reached the uppermost limit of its travel in Fig. 28-21C and is about to begin the power stroke; the fuel charge is being injected into the combustion chamber. Fig. 28-21D shows the piston just after completion of the useful part of the power stroke; the exhaust valves are open and exhaust is taking place. The downward-moving piston is about to uncover the cylinder air ports, when the cylinder will again be swept with clean, scavenging air and the entire cycle repeated. See Fig. 28-22 for a timing diagram of one crankshaft revolution.

What are the advantages of the two-cycle engine as compared with the four-cycle engine?

Answer: Saving in weight and space, and better power-flow.

How is the engine constructed for two-cycle operation?

Answer: A series of ports cut into the circumference of the cylinder wall above the piston on its lowest position admit the air from the blower into the cylinder as soon as the face of the piston uncovers the ports.

What happens as the piston continues on the upward stroke?

Answer: The exhaust valves close and the charge of fresh air is subjected to the final compression. Shortly before the piston reaches its highest position, the required amount of fuel is sprayed into the combustion space. The intense heat generated during the high compression of the air ignites the fine fuel spray and combustion continues as long as the fuel spray lasts.

What occurs now?

Answer: The resulting pressure forces the piston downward

Diesel Engine Principles

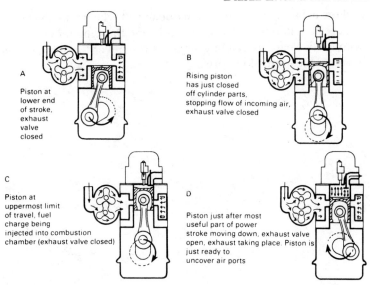

Fig. 28-21. Events in the two-stroke diesel cycle.

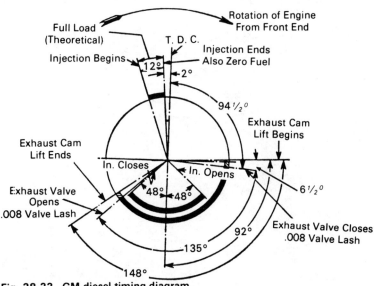

Fig. 28-22. GM diesel timing diagram.

803

until the exhaust valves are again opened, at which time the burnt gases escape into the exhaust manifold and the cylinder volume is swept with clean, scavenging air as the downward-moving piston uncovers the admission ports.

CHAPTER 29

Plant Safety

SAFETY VALVES

What is the most important fitting on a boiler and why?

Answer: The safety valve, because it keeps the steam from rising above the safe working pressure; that is, the pressure at which the safety valve is set.

What is a safety valve?

Answer: A circular valve connecting to the steam space of a boiler and loaded to such an extent that when the pressure of steam exceeds a certain point, the valve is lifted from its seat and allows the steam to escape.

How is the valve loaded?

Answer: Either by a weight or by a spring.

How many safety valves does a boiler need?

Answer: One, if the boiler heating surface is 500 square feet

Power Plant Engineers Guide

(46.45 m^2) or less; two if the heating surface is larger.

What care should a safety valve receive and why?
Answer: It should be kept clean and raised by hand at least once a week to ensure that it is in proper working condition.

Why should it be raised so often?
Answer: So that it cannot stick in its seat through the accumulation of dirt and scale.

Name the different types of safety valves.
Answer: (1) The dead-weight, (2) lever-with-weight, (3) lever-with-spring, (4) spring.

What types are in general use?
Answer: The weighted-lever and the safety spring types.

Where has the dead-weight valve been used?
Answer: It has sometimes been used for very low-pressure apparatus, as in steam heating.

What type valve is required in the "code states"?
Answer: The direct-spring-loaded valve.

Why do the code states prohibit lever or dead-weight valves?
Answer: Because they are too easily tampered with.

What are the characteristics of a lever valve?
Answer: The lever valve has no definite "pop" point; the valve lifts slowly in opening and settles gradually in closing.

What is the difference with spring valves?
Answer: They have a positive opening to practically the full amount possible.

What are the essential parts of the lever safety valve?
Answer: (1) A valve chamber containing the valve seat, inlet,

Plant Safety

and outlet opening, (2) a cover containing the upper spindle and lever guides, and an arm having a pivot hole at its end forming the fulcrum, (3) a valve and spindle, the latter being attached to the valve and the projecting part terminating in a knife edge, (4) a lever pivoted at one end to the projecting arm of fulcrum, in contact with the knife edge of the spindle at an intermediate point, and weighted at the other end with a ball. (See Fig. 29-1.)

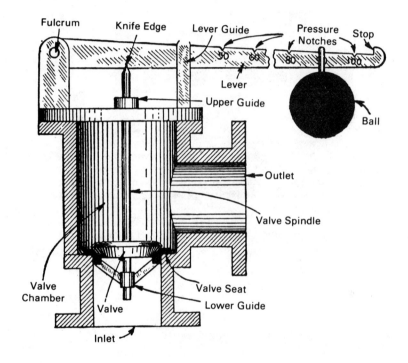

Fig. 29-1. Sectional view of a lever safety valve showing essential parts.

Why is a lever valve objectionable on steam vessels navigating rough water?

Answer: The inertia of the weight produces a variable pressure on the valve, tending to close and open the valve respectively with the rise and fall of the boat on the waves. Moreover, when the boat rocks, the horizontal position of the lever is disturbed and the blowing pressure of the valve is lowered. (See Fig. 29-2.)

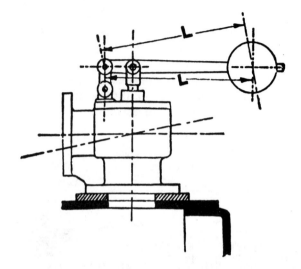

Fig. 29-2. Diagram of a lever safety valve showing the decrease of the weight's effect as the result of incline in a heavy sea. In the diagram, L is the length of the lever arm, the full length being effective when horizontal but when inclined the effective length is reduced.

What precaution should be taken with a lever valve?

Answer: The lever should be raised *frequently*, permitting the valve to blow to guard against the valve sticking to the seat.

What is a pop safety valve?

Answer: One so constructed that it opens very suddenly, like a cork popping out of a champagne bottle, and remains open until the pressure is reduced a predetermined amount.

Describe how the spring resistance acts.

Answer: The force due to the compression of the spring opposes the steam pressure and keeps the valve on its seat until the pressure of the steam on the valve becomes strong enough to overcome the resistance of the spring.

How is it constructed?

Answer: The construction is such that as soon as the valve

PLANT SAFETY

begins to open, an excess area of the disc is presented to the escaping steam and it suddenly opens wide (See Fig. 29-3.)

What is the object of a pop safety valve?

Answer: To prevent too frequent blowing of the valve by remaining open until the pressure is lowered a few pounds below the popping point.

How is the blowing-off pressure regulated on a spring loaded valve?

Answer: By increasing or decreasing the tension of the spring. This style of valve always has an adjustable setscrew for this purpose.

What names are given to the pressure at which the valve opens and at which it closes?

Answer: The blow-off and the blow-down pressures, respectively.

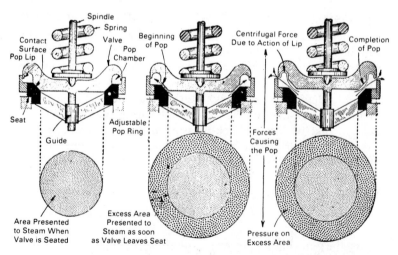

Fig. 29-3. Construction and operation of a pop valve. As the valve begins to lift, steam rushes into the pop chamber, acts on an excess area as indicated by the shaded ring, and causes the valve to suddenly lift or "pop" to its full opening. The main object of the pop ring is to regulate the blow-down or closing pressure.

What kind of a spring should be used with superheated steam?
Answer: One with an outside spring, as in Fig. 29-4.

Why?
Answer: Because the high temperature of the superheated steam would in time destroy an inside spring.

How should a pop-type valve be installed?
Answer: Safety valves must be connected to the boiler independent of any other steam connection. They must be attached as close as possible to the boiler without any unnecessary intervening pipe or fitting. When an intervening pipe or fitting is necessary, it must not be longer than the face-to-face dimension of the corresponding tee fitting of the same size as the valve inlet and of the same pressure rating. The safety valve must be installed so as to stand in an upright position with the spindle vertical. A discharge pipe should be provided for safety valve outlets to protect personnel from injury while the valve is discharging. (See Fig. 29-5.)

On a pop safety valve, how is the intensity of the "pop" regulated?
Answer: By adjusting the pop ring. See Fig. 29-6. At the left, the ring is in high adjustment so that with the valve fully opened as shown, the opening past the ring is restricted, resulting in wire drawing of the steam with a higher back-pressure in the pop chamber than at the right, where the ring is in low adjustment, giving free passage for the steam to the atmosphere. It must be evident that with high back-pressure in the pop chamber, the valve will remain open longer and reduce the boiler pressure to a lower point.

What is the main object of the pop ring?
Answer: To regulate the blow-down, or closing pressure.

When the pop chamber opens to the atmosphere, what two forces tend to keep the valve open?
Answer: The pressure of the steam in the excess area presented by the pop chamber and the centrifugal force caused by the action of the curved pop lip in changing the direction of the steam.

TABLE OF DIMENSIONS

	1½"	2"	2½"	3"	3½"	4"	4½"	6"
A	17 15/16	17 15/16	19 5/8	22 3/4	22 3/4	27 3/8	27 3/8	42 5/8
B	5 5/8	5 5/8	5 1/8	5 3/16	5 3/16	6 7/8	6 7/8	9 3/4
C	5	5	4 11/16	6	6	7 1/16	7 1/16	....
D	6 1/8	6 1/2	7 1/2	8 1/4	9	10 3/4	11 1/2	12 1/2
E	1 1/8	1 1/4	1 3/8	1 1/2	1 5/8	1 3/4	1 7/8	1 7/8
F	2 13/16	3 3/8	4 1/8	4 3/4	4 3/4	5 7/8	5 7/8	7 5/8
G	1 1/2	2	2 1/2	3	3	4	4	....
H	2 7/8	3 5/8	4 1/8	5	5 1/2	6 3/16	6 3/4	8 1/2
I	1/4	1/4	1/4	1/4	1/4	1/4	1/4	1/4
J	4 1/2	5	5 7/8	6 5/8	7 1/4	8 1/2	9 1/4	10 5/8
K	7/8	3/4	7/8	7/8	1	1	1 1/8	1
L	2 1/2	2 1/2	3	4	4	5	5	8
M	5	5	5 1/16	6 1/4	6 1/4	7 1/4	7 1/4	10
N	11/16	11/16	3/4	15/16	15/16	15/16	15/16	1 1/8
O	7	7	7 1/2	9	9	10	10	13 1/2

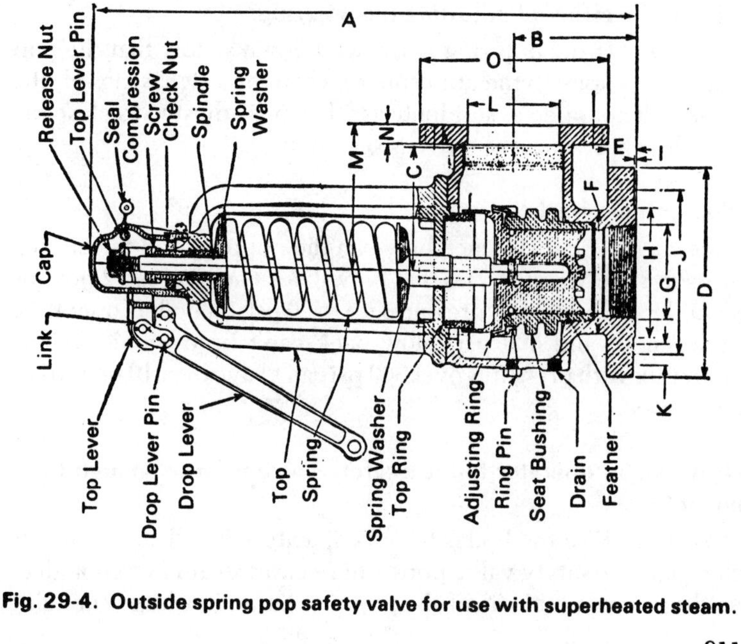

Fig. 29-4. Outside spring pop safety valve for use with superheated steam.

POWER PLANT ENGINEERS GUIDE

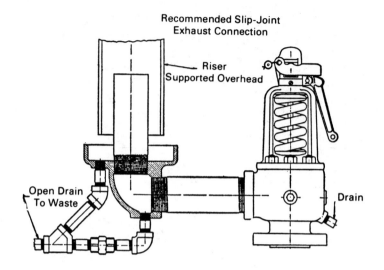

Fig. 29-5. Pop-type safety valve installation.

What is the effect of adjusting the pop ring?

Answer: If the pop ring is screwed down so low that the pop chamber is open to the atmosphere when the valve is closed, the valve will not open as suddenly as when it is adjusted to close the pop chamber when the valve is seated.

What is blow-back?

Answer: The difference between the popping and the closing pressures. ASME specifies that blow-back can't exceed 2 pounds on boilers operating at pressures up to 70 psi; on boilers operating between 70 and 300 psi, blow-back can't be over 3%. Boilers operating with pressures over 300 psi can't have over 10 pounds of blow-back.

How can an operator tell if a safety valve is large enough for a boiler?

Answer: Fire the boiler to full capacity with all steam outlets closed. If the safety valve pops and releases steam fast enough to hold pressure within 6% of the popping point, it is large enough.

PLANT SAFETY

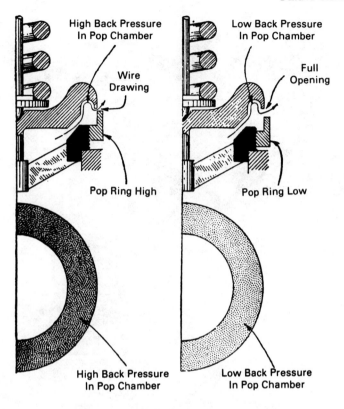

Fig. 29-6. Pop valve with high and low adjustment of the pop ring.

Who, if anyone, should adjust the pressure at which a safety valve lifts?

Answer: The boiler inspector.

LOW-WATER CUT-OFF

What is the purpose of a low-water cut-off?

Answer: A device that will cut off the burner or at least sound an alarm if water in the boiler drops to a dangerously low level. Many different types are used. See Fig. 29-7.

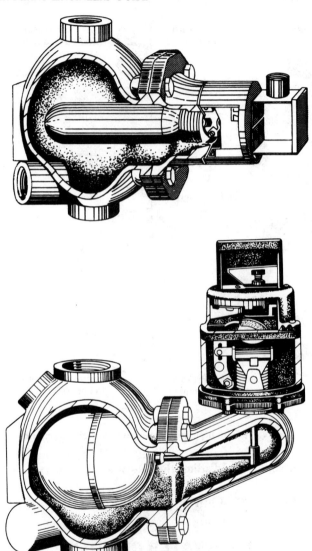

Fig. 29-7. Two types of low-water cut-off controls which stop the burner when the boiler water level falls too low. This prevents damage to the furnace heating surfaces and eliminates the possibility of a boiler explosion.

PLANT SAFETY

How does it work?

Answer: All types operate similarly, regardless where they're mounted. A sensing part of the device (usually a float) drops, opening the burner or firing electrical circuit or closes an alarm circuit. Fig. 29-8 illustrates one type of control.

How should it be tested?

Answer: Blow down the boiler water level while the furnace is firing until the level is low enough to operate the cut-off. It should be done daily. Once a month it's a good idea to cut off the feed water and make sure the low-water cut-off works when the water level in the gauge glass drops.

FLAME DETECTORS

What is a flame detector?

Answer: A device that will shut down an oil or gas burner if the fuel doesn't ignite, thus avoiding a fuel accumulation in the firebox and a probable explosion. See Fig. 29-9.

How does a flame detector work?

Answer: Usually, a lead sulfide cell (Fig. 29-10) is used to "see" the pilot light and then the main burner flame, thus "proving" the ignition and operation of the burner. If the pilot or burner fails to light, the flame detector interrupts the burner timing cycle.

How should a flame detector be tested?

Answer: Follow the manufacturer's instructions.

STEAM GAUGE TEST

Why test a steam gauge?

Answer: To make sure it is reading accurately.

Power Plant Engineers Guide

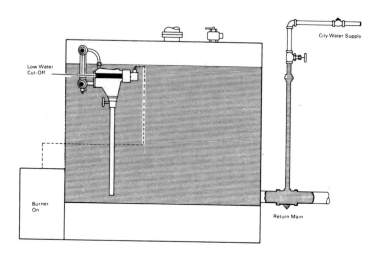

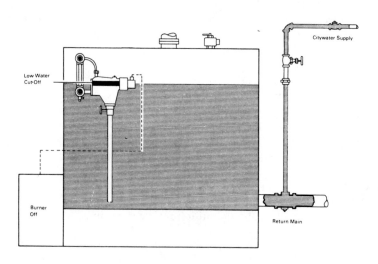

Fig. 29-8. When water level is normal (A), the burner operates normally. But when water drops to an unsafe level, the burner is prevented from firing by the water level control.

PLANT SAFETY

How can a steam gauge be tested?

Answer: By temporarily substituting a gauge of known accuracy. Of course, the boiler must be in normal operation. It can also be checked with a gauge testing set.

If a 150-psig safety valve popped and the steam gauge indicated boiler pressure at only 125 psig, where would you look for the fault?

Answer: It's more likely that the pressure gauge is registering wrong, so a check of the gauge is the most logical step. If it is all right, the safety valve must be defective. If such is the case, it's best to call the safety inspector or the manufacturer for advice.

SAFETY GUIDELINES

A set of rules can't prevent accidents, of course. Only extreme care can. The following guidelines, based partly on recommendations of the National Association of Power Engineers and other sources, are intended to promote the kind of care that precludes all types of mishaps.

A better accident history can only result from the combined efforts of everyone who works in a boiler room.

- Always wear a helmet-type hard hat where required—especially when there is danger of head injury from electric wiring or falling objects. Flying particles or looking at the rays of a welding machine or torch can permanently damage eyes.
- Frayed or loose-fitting clothing might catch in machinery or cause tripping. Never wear shoes that could cause you to slip or shoes with worn soles that are easily punctured.
- Protect your hands from burning, slivers, or abrasions.
- Never use an extension light with broken insulation or a frayed cord or without a cage around the bulb.
- Do not experiment with electrical hookups beyond your knowledge.
- A flashlight that does not give suitable light and is not dependable could leave you in the dark in some remote area of a large vessel or boiler.

POWER PLANT ENGINEERS GUIDE

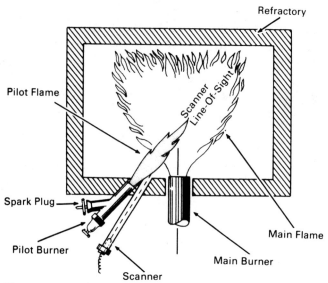

Fig. 29-9. Drawing of a flame detector system.

- Never walk in a dark area if you can't see where you are putting your foot down.
- Never take a handhole or manhole cover plate from a boiler unless you are sure the boiler is empty.
- Do not drain a boiler without opening the proper vents. This will prevent collapse and allow the boiler to drain completely.
- Never venture inside the furnace of an automatically fired boiler unless you make sure that the firing mechanism is in the locked out position and is tagged to this effect. When you go inside the furnace of a water-tube boiler, watch for falling slag.
- Never enter a boiler installed in a battery of boilers without making sure that all stop valves in the connecting pressure lines are securely closed and tagged. If there are bleed-off connections between the stop valves, open them.
- Never enter a boiler without a reliable person stationed nearby in case of personal mishap or weakness.
- Before using a ladder always check for loose or worn rungs and loose side rails. Make sure it's not too short. In trying to reach above the ladder, the angle of the ladder and you causes

Plant Safety

you to lean backward. When trying to enter an inspection opening, make sure the ladder reaches to or above it. Climbing from the top rung can cause the ladder to tip out at the top or kick out at the bottom. Watch out for electric wiring when you're placing a ladder.

- Never try to use a ladder not designed for the job. The bottom ends of the rails should be fitted with spikes or cleats to prevent slipping. If a ladder is not so equipped, or is placed on a metal floor, fasten the bottom to some fixed object. Do not rely on someone to hold the ladder; he or she may be called away or be distracted. Ladders are not designed for use as a bridge.
- Never use a rope ladder or bosun's chair without first inspecting the rope. Also be sure that you are physically able to make an inspection from such flexible equipment.
- You risk your life by attempting to remove fuses from high-voltage electrical circuits with your bare hands. Use tongs or fuse pullers.
- Never go inside a pressure vessel that once contained a toxic agent or that may have *dead air* in it until it is thoroughly purged. If necessary, wash and clean the surfaces.
- Never attempt to inspect a boiler or pressure vessel while it is under a hydrostatic pressure test without due care. Always make sure a vessel is completely vented of air before a hydrostatic test is applied.
- Do not operate soot blowers unless the burners are operating at a high firing rate. Air flow should be adjusted to ensure a high carbon dioxide but low oxygen content in the flue gases.
- If ammonia fumes, sewer gas, natural gas or any other toxic gas is present near a building or enclosed area, never go in without proper clothing and respiratory equipment—and then only with extreme caution.
- Standing in front of the discharge opening of a relief valve is very dangerous—particularly while testing it. Never blow down an appliance on a boiler without being aware of the point of discharge—your foot may be under it.
- Never blow down a boiler under pressure by opening the slow-opening valve first. Always open the quick-opening

valve or cock first and then the slow-opening valve. After blow-down, close the slow-opening valve first. This prevents a sudden back-pressure wave on the connecting piping. In fact, a valve should never be closed suddenly in a pressure line except in an extreme emergency. The sudden change in flow may cause a rupture at some weakened location. Also, never open a valve suddenly (particularly in a steam line) where water hammer might result.

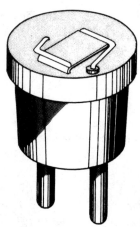

Fig. 29-10. Closeup view of a lead sulphite flame-sensing cell.

- Never apply pressure to a vessel with only part of the head or cover plates, bolts, lugs or clamps in place. Don't try to use any bolt, pin, lug, or clamp after it becomes worn, sprung, stripped, or otherwise weakened.
- Never release the holding mechanism of any quick-opening door or end closure until you're sure that the chamber is relieved of pressure.
- Automatic controls are not dependable as safety devices unless they are kept clean and are checked, tested, and otherwise regularly maintained.
- Never operate boilers, pressure vessels or machines at pressures or speeds above that specifically rated for the unit in question.
- Never take someone else's word for something you should have checked yourself. Avoid any *horseplay* in a boiler room.

Plant Safety

- Never neglect your fire protection. Check fire buckets and extinguishers regularly, and never leave a water line valve to a sprinkler system shut off after maintenance. Red tag it and stay with it until corrections or alterations have been completed; then return all valves to their proper positions for service.

CHAPTER 30

Tables and Data

BASIC MATH CALCULATIONS

What is a digit?

Answer: A single numeral; 5 is a one-digit number; 399 is a 3-digit number.

What is a whole number?

Answer: Any number that represents whole units, rather than parts of units; 8 is a whole number, ½ and 0.5 are parts of a number or a fraction or decimal.

What is a mixed number?

Answer: An expression representing a mixture of whole and partial units: 4½ or 4.5.

How are mixed numbers containing decimals read?

Answer: 0.5 is five tenths; 0.02 is two hundredths; 0.003 is three thousandths; 0.0004 is four ten thousandths; 150.02 is one hundred fifty and two hundredths.

Addition

What operation does a plus sign (+) indicate?

Answer: Addition, combining two or more whole or mixed decimal numbers (called addends) into a sum; $2 + 6 + 8 = 16$.

How is a series of more than two-digit whole or mixed decimal numbers added?

Answer: Arrange the numbers one beneath the other, with the decimal points or the last digit on the right lined up on the right:

```
      384      (addend)
       55.6    (addend)
        1.2    (addend)
    +4891.06   (addend)
     5331.86   (sum)
```

Beginning at the rightmost column, add each column, either up or down; in the above series, a 6 is the only number in the rightmost column, so the total here is 6. The second column, $0.6 + 0.2 + 0.06 = 8$. Column 3 contains $4 + 5 + 1 + 1$, totaling 11; the left 1 of the 11 is carried to the next (4th column from the right): $1 + 8 + 5 + 9 = 23$. This time the 2 is carried to the fifth column, where it is added to the 3 and 8 for 13. After the 3 is placed, the remainder (1) goes to the last column and is added to the 4.

Subtraction

What operation does the minus sign (-) indicate?

Answer: Subtraction, finding the difference between two whole or mixed numbers:

```
     30    (minuend)
    -20    (subtrahend)
     10    (remainder)
```

Tables and Data

The number you are subtracting a quantity *from* is the minuend; the number you are subtracting the minuend from is the subtrahend, and the answer is the remainder. If the minuend is smaller than the subtrahend, the remainder is less than 0.

How is one whole or mixed number subtracted from another?

Answer: The two numbers are positioned as in addition, decimal points or rightmost digits lined up on the right:

$$\begin{array}{r} 384 \\ -55.6 \\ \hline 328.4 \end{array}$$

Though it isn't there, the decimal in the minuend would go after the 4, so the decimal in the subtrahend is positioned under this point. Subtraction, like addition, begins on the right. Since there is nothing or zero above the 6, a 1 must be borrowed from the 4 to make the 0 a 10. Then 6 is subtracted from 10, which is 4. The 4 in the minuend is now a 3 (we borrow a one from it), and the 5 beneath it is larger. So a 1 must be borrowed from the eight to make the 3 a 13. Then, 5 from 13 is 8. Now, the 7 (which was 8) in the minuend is larger than the other 5 in the subtrahend so borrowing isn't necessary; 5 from 7 is 2. The absence of another digit in the subtrahend under the 3 in the minuend equals zero, and zero from 3 is 3. Accuracy is easy to check; add the remainder and subtrahend; the sum must equal the minuend.

Minus Signs

What does a minus sign before a number tell us?

Answer: That the number is less than or below zero. Zero is the borderline between positive (plus) and negative (minus) numbers or quantities.

How are negative numbers added and subtracted?

Answer: If someone owes you $100 (+100) and pays $50 (-50), $50 is still due you. If you were to lend that same person another $50, you would add that to the amount already owed, (-100)+ (-50) = (-150). So, when adding two negative numbers the sum is negative, or the sum given the sign of the largest quantity. If you owed someone $50, then that person borrowed $100 from you

Power Plant Engineers Guide

(+50) − (−100), $50 then would be due. When subtracting a negative number from a positive number, or a larger positive number from a smaller positive number, the remainder is given the sign of the largest quantity.

Multiplication

What operation does the times sign (× or ·) indicate?

Answer: Multiplication, adding a number (multiplicand) as many times as there are units in the number which you are using to multiply (multiplier). The total is the product:

```
   123     (multiplicand)
   ×45     (multiplier)
   615     (subproduct)
   492     (subproduct)
  5535     (product)
```

To multiply 3 × 5, 3 is added to itself 5 times. To solve the above, the operation begins with the rightmost digits in multiplicand and multiplier, 5 × 3 = 15; the 5 goes into position under the multiplier and the 1 is carried. Now 5 × 2 = 10 + 1 (carry) = 11, the right 1 is placed to the left of the first subproduct and the left 1 is carried. Next, 5 × 1 = 5 + 1 (carry) = 6. The 6 goes next to the 1, completing the first subproduct. If the multiplier was a single number the operation would be finished. But, since our multiplier is two digits, the second subproduct must be calculated, just as the first subproduct. Remember, the first digit of the second subproduct is placed under the multiplier. The subproducts are added for the final product. When there's a zero in a multiplier it can be treated as a digit only, not a number.

```
    1024              1024
     105               150
    5120             51200
   1024              1024
  107520            153600
```

How is multiplication of decimal numbers handled?

Answer: The same as whole numbers, initially ignoring decimals, then adding the total number of digits, counting zeroes, to right of the decimals in multiplicand and multiplier, then placing the decimal that number of digits from the right in the product:

```
   1.28         0.128         128
    1.2           12         0.012
   ───          ───          ───
   256          256          256
  128          128          128
  ────         ────         ────
  1.536        1.536        1.536
```

Powers of Numbers

How are powers of numbers used?

Answer: To abbreviate math expressions: $5 \times 5 \times 5$ can be written as 5^3, or 5 cubed or 5 to the third power. The superscript (3 above) tells how many times to multiply the root (5 above) by itself.

What does square or cube root mean?

Answer: A number which has been multiplied by itself; 2 is the square root of 4, 3 is the cube root of 27. Square and cube roots are written: $\sqrt{4}$, $\sqrt[3]{27}$; the absence of a superscript usually means square root. The $\sqrt{}$ is called a radical sign and means root of.

Division

What do these signs mean: $\div$, /, ?

Answer: Division, an operation that determines how many times one number is contained in another; $4 \div 2 = 2$, there two 2s in 4, or 2 will divide into 4 two times.

What is short division?

Answer: Division by a single divisor.

```
                    3117       (quotient)
                   ──────
    (divisor)   4 │ 12468      (dividend)
```

Above, 4 will not divide into 1, so we take the next digit too, making 12. Now 4 will divide into 12 three times. Since we had to bring the 2 in to form our first workable dividend, the 3 goes above the 2. (Technically, a zero goes above the 1, but it's not used since it would be meaningless.) Now, 4 will go into 4 one time, and one time into the 6. However, 2 is left over (called a remainder), making the next part of the dividend 28, which divides 7 times by 4. As you can see division is the reverse of multiplication.

What is long division?

Answer: Division by a divisor with two or more digits:

```
              283.38
         44 ) 12469
              88
              ---
              366
              352
              ---
              149
              132
              ---
              170
              132
              ---
              380
              352
              ---
              28  (remainder)
```

To solve the above problem, there is less than one 44 in 1 and 2, so we have to use the first three digits in the dividend, into which 44 will divide 2 times; so, the first digit of the quotient (2) is placed above the 4. The next step is multiply the divisor (44) by the first digit of the quotient (2) and place the product (88) as shown under 124 in the dividend. Subtracting, we have a remainder of 36 in this first step. Now, bring down the next digit in the divisor (6) and place it to the right of the remainder (36), which is the next dividend, 366. Since 44 will go into 366 8 times (but not 9), the second digit of the quotient is 8, which is placed over the 6 as shown. The divisor, 44, times the second quotient digit, 8, is 352, which is placed under 366. Subtracting leaves a remainder of 14. Now, bring down the 9 from the dividend, to make the next dividend 149, and repeat the division operation. If the accuracy of the quotient is satisfactory for your purposes, you can accept the remainder of 17. Or, you can carry it into decimals by placing the

Tables and Data

decimal after the 3 in the quotient and 9 in the dividend. The decimal goes here because all of the digits in the dividend have been used, but you can add as many zeroes as you want and carry to as many decimal places as accuracy demands. Division with a decimal point in the divisor is not possible. If the divisor above had been 4.4, it would have been necessary to move it one place to the right and, at the same time, move the decimal in the dividend one place to the right, adding a zero to fill in the empty space:

$$4{,}4_x \overline{\smash{)}124690_{\curvearrowleft}}\quad\text{add zero}$$

Instead of 12,469, the dividend would have become 124, 690. Division would proceed normally beyond that point.

How is the square root of a number calculated?

Answer: By division, just the opposite of squaring a number.

Find the square root of 1874.

Answer: Since this is a whole number, we place a decimal to the right of the 4, then mark off the number in 2 digit groups:

		4 3	(quotient)
(power)	$\sqrt[2]{}$	18′74	(dividend)
	4)	16	(product)
(trial divisor)	80)	274	(remainder)
(true divisor)	83	249	(product)
		25	(remainder)

Reading from the left, the first group we'll deal with is 18. The nearest whole-number square root is 4, which is placed in the quotient over the 18 and to the left in the position shown, on the first product line under the radical sign. The first product is the square of the first quotient digit. As in division, the product is subtracted, producing a remainder of 2. Now, bring down the next group, 74, and place it to the right of the 2 in the remainder line. The trial divisor is the product of the power times the first quotient digit, 8, to which a 0 is added. Next, we determine how many times 80 will divide into 274 and place a 3 in the quotient over the 74 in the dividend. Adding the 3 in the quotient to the trial divisor, we arrive at the true divisor for this set, 83, and multiply 83 × 3 to determine the second product which is written under the 274. Subtraction leaves a remainder of 25. So, the square root is 43, approximately, plus a remainder of 25. To carry the root in decimals, the second quotient digit, 3, is added to the true divisor, 83, and a zero added to arrive at the next divisor, 86. Two 0s are added (officially brought down from the dividend as the next group) to the remainder, 25, making 2500. Now, as before, 860 is divided into 2500. The first digit in the decimal is 2. The same procedure (add the last quotient digit to the last divisor and add a zero and add two 0s the new remainder) is used to go farther. The same procedure is used for mixed decimal numbers, observing the position of the decimal. Only one root digit is possible for each group of 2 digits in the quotient.

Fractions

What are the parts of a fraction called?

Answer: The lower digit (the 2 in ½) is the denominator, which tells how many parts the whole is divided into, and the top digit is the numerator, which indicates the number of parts we refer to. Since a fraction indicates division, the denominator is the divisor and the numerator is the dividend.

What is a proper fraction?

Answer: A fraction where the numerator is smaller than the denominator. If the numerator is larger than the denominator, it is an improper fraction, because the value it represents is greater than 1.

Does ¾ represent the same value as 9/12?

Answer: Yes. The numerator and denominator of a fraction can be multiplied or divided by the same number without changing the value of the expression.

Reduce 9/12 and 12/16 to the lowest common denominator.

Answer: Since each digit is evenly divisible by 4, the LCD is 3, or ¾.

TABLES AND DATA

What is $3\frac{1}{2}$?

Answer: A mixed number. Improper fractions can be reduced to mixed numbers by dividing the denominator into the numerator; $\frac{7}{2} = 3\frac{1}{2}$

What is a decimal fraction?

Answer: A fraction with 10, 100, 1000, etc., as the denominator (0.7, 0.07, 0.007, etc.)

How can a fraction be converted to a decimal?

Answer: Divide the denominator into the numerator:

$$\frac{7}{32} = 32 \overline{)7.00000} = .21875$$

How are fractions added?

Answer: Convert all to fractions with the lowest common (LCD) denominator, add the numerators, then reduce to the LCD:

$$\frac{1}{2} + \frac{1}{16} + \frac{3}{8} + \frac{7}{32} = \frac{16}{32} + \frac{2}{32} + \frac{12}{32} + \frac{7}{32} = 1\frac{5}{32}$$

How are fractions subtracted?

Answer: Convert to the LCD, subtract the numerators, and reduce to the LCD:

$$\frac{15}{16} - \frac{1}{2} = \frac{15}{16} - \frac{8}{16} = \frac{7}{16}.$$

How is the LCD of a series of fractions determined?

Answer: To find the LCD of $\frac{1}{4}$, $\frac{1}{3}$, $\frac{1}{9}$, and $\frac{1}{16}$, place the denominators in a row, separated by commas:

```
2 | 4, 3, 9, 16
2 | 2, 3, 9, 8
3 | 1, 3, 9, 4
    1, 1, 3, 4
```

The digits on the left of the vertical line are divisors. To arrive at each, find a number that will divide evenly into at least two of the denominators. In the first line above, 2 will divide evenly into 4 and 16. Where a divisor will not divide evenly, the number is brought down as is. So line 2 above reads: 2, 3, 9, 8, which contains two numbers evenly divisible by 2 again. Therefore, our next divisor is 2. Line 3 reads: 1, 3, 9, 4. Now, two are evenly divisible by 3 and the bottom line reads, 1, 1, 3, 4. To determine the LCD multiply the divisor numbers on the left by the bottom line numbers: $2 \times 2 \times 3 \times 1 \times 1 \times 3 \times 4 = 144$, which is the LCD.

How are fractions multiplied?

Answer: Multiply all numerators by each other and denominators by each other, then reduce to the LCD or lowest form if an improper fraction:

$$7/8 \times 3/16 \times 4/32 = 84/4096 = 42/2048 = 21/1024$$

$$3\,7/8 \times 1/2 = 31/8 \times 1/2 = 31/16 = 1\,15/16$$

$$2 \times 3/8 = 2/1 \times 3/8 = 6/8 = 3/4$$

Cancellation could have been used in the problem immediately above and saved one step:

$$\overset{1}{\cancel{2}/1} \times 3/\cancel{8}\ \ 3/4$$
$$\qquad\quad\ 4$$

How are fractions divided?

Answer: Invert the divisor and multiply:

$$3/8 \div 3/32 = \overset{1}{\cancel{3}/\cancel{8}} \times \overset{4}{\cancel{32}/\cancel{3}} = 4/1 = 4$$
$$\qquad\qquad\quad\ \ 1\qquad\ \ 1$$

SQUARE, RECTANGLE, AND PARALLELOGRAM CALCULATIONS

1. To find the area of a square or rectangle, multiply the width times the length, using the same units of measure.
2. To find the perimeter of a square or rectangle, add two times the width and two times the length or add the dimensions of all

four sides, using the same units of measure.
3. To find the area of a parallelogram (a figure with two pairs of parallel sides), multiply the measurements between each pair.
4. To find the volume of a square, rectangle, or parallelogram, multiply the area of one side by the height.

TRIANGLE, TRAPEZOID, AND CYLINDRICAL CALCULATIONS

1. To find the area of a triangle, multiply the measurement of the longest side by the measurement between the longest side and the opposite point and divide by 2, using the same units of measure.
2. To find the area of a trapezoid (a figure with one pair of parallel sides), add the measurements of the parallel sides, divide by 2, then multiply the quotient by the measurement between the parallel sides.
3. To find the area of a parallelogram (a figure with two pairs of parallel sides) multiply the measurements between each pair.
4. To find the volume of a trapezoid, multiply the area of the trapezoid by the length of the trapezoid.
5. To find the surface area of a tube, multiply the circumference by the length.
6. To find the volume of a cylinder, multiply the circular area by the length of the cylinder.
7. To find the weight of water or any liquid in a cylindrical tank, find the volume and multiply by the weight of the volume unit. The same principle would apply to a rectangular bar or any volume measure. To find the quantity of water or any liquid in a cylindrical tank, find the volume of the tank and multiply by the number of liquid units (gallons or liters) per unit of measure (cubic foot or cubic centimeter or cubic meter).

BOILER HORSEPOWER

The equivalent evaporation of 34.5 pounds (15.62 kg) of water per hour from a feed-water temperature of 212°F. (100°C.) into dry steam at the same temperature is defined as a boiler horsepower. However, since the actual operating conditions of a boiler are seldom from and at 212°F., a factor is needed to convert the rating from the actual conditions to what they *would* be from and at 212°F. This is called the *factor of evaporation* and equals the ratio of heat required to generate one pound of steam under actual conditions to that required to make one pound of steam from and at 212°F.

ENGINE HORSEPOWER

The determination of an engine horsepower is based on the fact that one mechanical horsepower is equivalent to 33,000 foot-pounds of work in one minute or 550 foot-pounds of work in one second.

To express the indicated horsepower of an engine, the following formula is used:

$$IHP = \frac{PLAN}{33,000} \times K$$

where,

P is the mean effective pressure in pounds per square inch acting on the piston, as shown by the indicator diagram,
L is the length of stroke in feet;
A is the area of the piston in square inches,
N is the number of working strokes per minute,
K is a coefficient equal to ½ the number of cylinders in gasoline engines or twice the number of revolutions in double-acting steam engines.

CIRCLE CALCULATIONS

1. To find the area of a circle, multiply the square of the diameter by 0.7854, the square of the circumference by 0.07958, or half

TABLES AND DATA

the circumference by half the diameter. See Table 30-2.
2. To find the circumference of a circle, multiply the diameter by 3.1416 or the radius by 6.283185. See Table 30-2.
3. To find the diameter of a circle, multiply the circumference by 0.31831 or the square root of the area by 1.12838.
4. To find the diameter of a circle equal in area to a given square, multiply a side of the square by 1.12838.
5. To find the radius of a circle, multiply the circumference by 0.159155 or the square root of the area by 0.56419.
6. To find the side of a square equal in area to a given circle, multiply the diameter by 0.8862.

INSCRIBED CIRCLE CALCULATIONS

1. To find the diameter of a circle inscribed in an equilateral triangle, multiply a side of the triangle by 0.57735.
2. To find the diameter of a circle inscribed in a hexagon, multiply a side of the hexagon by 1.7321.
3. To find the side of an equilateral triangle inscribed in a circle, multiply the diameter of the circle by 0.866.
4. To find the side of a hexagon inscribed in a circle, multiply the diameter of the circle by 0.500.
5. To find the side of a square inscribed in a circle, multiply the diameter by 0.7071.

SPHERICAL CALCULATIONS

1. To find the area of the surface of a sphere, multiply the square of the diameter by 3.1416.
2. To find the volume of a sphere, multiply the cube of the diameter by 0.5236.

WATER CALCULATIONS

A US standard gallon of water weighs 8.336 pounds and occupies a volume of 231 cubic inches. A cubic foot of water contains 7½ gallons, occupies 1728 cubic inches, and weights 62.425 pounds at a termperataure of about 39°F. The weights change slightly above and below this temperature.

1. To find the pressure in pounds per square inch (psi) at the base of a column of water, multiply the height of the column in feet by 0.433.
2. If the diameter of the pipe is doubled, its capacity will increase by four times.

BOILER CONVERSION FACTORS

1. To convert boiler horsepower (BHP) to pounds of steam (water) per hour, multiply by 34.5.
2. To convert boiler horsepower to gallons of water per minute (GPM), multiply by 0.069.
3. To convert boiler horsepower to Btu per hour, multiply by 33,479.
4. To convert boiler horsepower to square feet of equivalent direct radiation (EDR), multiply by 139.
5. To convert cubic centimeters per liter of oxygen to parts per billion of oxygen, multiply by 1400.
6. To convert cubic feet of water to gallons of water, multiply by 7.48.
7. To convert cubic feet per minute to gallons per hour, multiply by 448.8.
8. To convert cubic feet per minute into pounds of water per minute, multiply by 62.43.
9. To convert feet of water (head) to pounds per square inch, multiply by 0.4335.
10. To convert gallons of water to pounds of water, multiply by 8.345.
11. To convert pounds of steam (water) per hour to gallons of water per minute, multiply by 0.002.
12. To convert pounds per square inch into feet of water, multiply bty 2.307.
13. To convert square feet of equivalent direct radiation to gallons of water per minute, multiply by 0.000496.

TABLES AND DATA

METRIC MEASUREMENT

The metric unit of length is the meter; the unit of weight, the gram; the unit of volume, the cubic meter; the unit of capacity, the liter; the unit of area, the square meter.

Note: The National Bureau of Standards recommends spelling "meter" as "metre" to avoid confusion with the instrument. Words such as "centimeter" and "liter," however, take the older spelling.

To convert any of these base units to milli-units, multiply by 1000; to centi-units, by 100; to deci-units, by 10. To convert to deka-units, multiply by 0.10; to hecto-units, by 0.01; to kilo-units, by 0.001.

One *kilo-unit* contains 10 hecto-units, 100 deka-units, 1000 base units, 10,000 deci-units, 100,000 centi-units, or 1,000,000 milli-units.

One *hecto-unit* contains 0.1 kilo-units, 10 deka-units, 100 base units, 1000 deci-units, 10,000 centi-units, or 100,000 milli-units.

One *deka-unit* contains 0.01 kilo-units, 0.1 hecto-units, 10 base units, 100 deci-units, 1000 centi-units, or 10,000 milli-units.

One *base unit* contains 0.001 kilo-units, 0.01 hecto-units, 0.1 deka-units, 10 deci-units, 100 centi-units, or 1000 milli-units.

One *deci-unit* contains 0.0001 kilo-units, 0.001 hecto-units, 0.01 deka-units, 0.1 base units, 10 centi-units, or 100 milli-units.

One *centi-unit* contains 0.00001 kilo-units, 0.0001 hecto-units, 0.001 deka-units, 0.01 base units, 0.1 deci-units, or 10 milli-units.

One *milli-unit* contains 0.000001 kilo-units, 0.00001 hecto-units, 0.0001 deka-units, 0.001 base units, 0.01 deci-units, or 0.1 centi-units.

METRIC-TO-ENGLISH CONVERSIONS

One *centigram* equals approximately 0.15 grains. To convert from centigrams to grains, multiply by 0.154323.

One *centiliter* equals approximately 0.61 cubic inches or 0.34 fluid ounces (US). To convert centiliters to cubic inches, multiply by 0.610254; to US fluid ounces, by 0.338149.

One *centimeter* equals approximately 0.39 inches. To convert centimeters to inches, multiply by 0.393700 or divide by 2.54.

One *cubic centimeter* equals approximately 1.0 milliliters, 0.061 cubic inches or 0.034 fluid ounces (US). To convert cubic centimeters to cubic inches, multiply by 0.061023 or divide by 16.387; to US fluid ounces, multiply by 0.033814 or divide by 29.57.

One *cubic decimeter* equals approximately 61.02 cubic inches, which is the same as one liter. To convert cubic decimeters to cubic inches, multiply by 61.023744.

One *cubic meter* equals approximately 35.31 cubic feet or 1.31 cubic yards. To convert cubic meters to cubic feet, multiply by 35.314667; to cubic yards, by 1.307950; to US liquid gallons, by 264.172.

One *decigram* equals approximately 1.54 grains. To convert decigrams to grains, multiply by 1.54.

One *deciliter* equals approximately 6.1 cubic inches or 0.21 pint (US liquid). To convert deciliters to cubic inches or US liquid pints, multiply by these factors.

One *decimeter* equals approximately 3.94 inches. To convert decimeters to inches, multiply by 3.937007.

One *dekagram* equals approximately 0.35 ounces (avdp). To convert from dekagrams to ounces (avdp), multiply by 0.353.

One *dekaliter* equals approximately 0.35 cubic feet or 2.64 gallons (US liquid). To convert dekaliters to cubic feet or US gallons, multiply by these factors.

One *dekameter* equals approximately 32.81 feet or 10.94 yards. To convert dekameters to feet, multiply by 32.808399; to yards, by 10.93613.

One *gram* equals approximately 0.035 ounces (avdp). To convert grams to oun;ces (avdp) multiply by 0.035273; to US fluid ounces, divide by 29.57; to dynes, multiply by 980.665; to grains, multiply by 15.432358.

One *hectogram* equals approximately 0.22 pounds (avdp). To convert hectograms to pounds (avdp), multiply by 0.220462.

TABLES AND DATA

One *hectoliter* equals approximately 3.53 cubic feet or 26.41 gallons (US liquid). To convert hectoliters to cubic feet, multiply by 3.531566; to US gallons, by 26.41794.

One *hectometer* equals approximately 328.08 feet or 109.36 yards. To convert hectometers to feet, multiply by 328.08399; to yards, by 109.3613.

One *kilogram* equals approximately 35.27 ounces (avdp) or 2.20 pounds. To convert kilograms to ounces (avdp), multiply by 35.273962; to pounds, by 2.204622; to long tons, by 0.000984; to metric tons, by 0.001; to dynes, by 980665; to grains, by 15432.358.

One *kiloliter* equals approximately 1.31 cubic yards, 35.31 cubic feet, or 264.18 gallons (US liquid). To convert kiloliters to cubic yards, multiply by 1.307987; to cubic feet, by 35.31566; to US liquid gallons, by 264.1794.

One *kilometer* equals approximately 1093.61 yards or 0.62 miles. To convert kilometers to yards, multiply by 1093.6133; to miles, by 0.621371.

One *liter* equals approximately 61.02 cubic inches, 33.81 ounces (US fluid), or 1.06 quarts. To convert liters to cubic inches, multiply by 61.02545; to US fluid ounces, by 33.81497; to US liquid quarts, by 1.056718; to US liquid gallons, by 0.264179.

One *meter* equals approximately 39.37 inches, 3.28 feet, or 1.09 yards. To convert meters to inches, multiply by 39.370079; to feet, by 3.280839; to yards, by 1.093613.

One *metric ton* equals approximately 2204.6 pounds (avdp), 0.98 long tons, or 1.10 short tons. To convert metric tons to pounds (avdp), multiply by 2204.6226; to long tons, by 0.984206; to short tons, by 1.102311.

One *milligram* equals approximately 0.015 grains. To convert milligrams to grains, multiply by 0.015432.

One *milliliter* equals approximately 1.0 cubic centimeters, 0.06 cubic inches, or 0.27 ounces (US fluid). To convert from milliliters to cubic inches, multiply by 0.061025; to US fluid ounces, by 0.033814.

One *millimeter* equals approximately 0.04 inches or 0.003 feet. To convert from millimeters to inches, multiply by 0.039370; to feet, by 0.003280.

One *square centimeter* equals approximately 0.155 square inches or 0.001 square feet. To convert from square centimeters to square inches, multiply by 0.1550003; to square feet, by 0.001076.

One *square kilometer* equals approximately 0.39 square miles. To convert from square kilometers to square miles, multiply by 0.386102.

One *square meter* equals approximately 10.76 square feet, or 1.19 square yards. To convert from square meters to square feet, multiply by 10.763910; to square yards, by 1.195990.

One *square millimeter* equals approximately 0.0015 square inches. To convert from square millimeters to square inches, multiply by 0.01550.

ENGLISH-TO-METRIC CONVERSIONS

One *cubic foot* equals approximately 0.03 cubic meters. To convert from cubic feet to cubic meters, multiply by 0.028316.

One *cubic inch* equals approximately 16.39 cubic centimeters. To convert from cubic inches to cubic centimeters, multiply by 16.387064.

One *cubic yard* equals approximately 0.765 cubic meters. To convert from cubic yards to cubic meters, multiply by 0.764554.

One *foot* equals approximately 30.48 centimeters. To convert from feet to centimeters, multiply by that factor.

One *gallon* (US liquid) equals approximately 3.78 liters. To convert from US gallons to liters, multiply by 3.785411.

One *grain* equals approximately 0.065 grams. To convert from grains to grams, multiply by 0.064798.

One *inch* equals approximately 2.54 centimeters. To convert from inches to centimeters, multiply by that factor.

One *mile* equals approximately 1.61 kilometers. To convert from miles to kilometers, multiply by 1.609344.

One *ounce* (avdp) equals approximately 28.35 grams. To convert from ounces (avdp) to grams, multiply by 28.349523.

One *ounce* (US fluid) equals approximately 29.57 milliliters or 29.57 cubic centimeters. To convert US fluid ounces to milliliters, multiply by 29.572702; to cubic centimeters, by 29.573730.

One *pint* (US liquid) equals approximately 0.47 liters. To convert from US liquid pints to liters, multiply by 0.473163.

One *pound* (avdp) equals approximately 0.45 kilograms. To convert from pounds (avdp) to kilograms, multiply by 0.453592.

One *quart* (US liquid) equals approximately 0.95 liters. To convert from US liquid quarts to liters, multiply by 0.946326.

One *square foot* equals approximately 0.093 square meters. To convert from square feet to square meters, multiply by 0.092903.

One *square inch* equals approximately 6.45 square centimeters. To convert from square inches to square centimeters, multiply by 6.4516.

One *square mile* equals approximately 2.59 square kilometers. To convert from square miles to square kilometers, multiply by 2.589988.

One *square yard* equals approximately 0.84 square meters. To convert from square yards to square meters, multiply by 0.836127.

One *ton* (long) equals approximately 1.02 metric tons. To convert from long tons to metric tons, multiply by 1.016046.

One *ton* (short) equals approximately 0.91 metric tons. To convert from short tons to metric tons, multiply by 0.907184.

One *yard* equals approximately 0.91 meters. To convert from yards to meters, multiply by 0.9144.

ENGLISH-TO-ENGLISH CONVERSIONS

To convert *cubic feet* to cubic inches, multiply by 1728; to cubic yards, by 0.03704; to US liquid gallons, by 7.4805.

To convert *cubic inches* to cubic feet, multiply by 0.0005787; to cubic yards, by 0.0002143; to US liquid gallons, by 0.004329.

To convert *cubic yards* to cubic inches, multiply by 46656; to cubic feet, by 27.

To convert *feet* to yards, multiply by 0.333; to miles by 0.0001894.

To convert *grains* to ounces (avdp), multiply by 0.002286.

To convert *inches* to feet, multiply by 0.0833; to yards by 0.02778; to miles, by 0.00001578.

To convert *miles* to inches, multiply by 63360; to feet, by 5280; to yards, by 1760.

To convert *ounces* (avdp) to pounds (avdp), multiply by 0.0625; to long tons, by 0.00002790; to short tons, by 0.00003125.

To convert *pounds* (avdp) to ounces (avdp), multiply by 16; to hundredweight, by 0.01; to long tons, by 0.000446; to short tons, by 0.0005.

To convert *square feet* to square inches, multiply by 144; to square yards, by 0.11111.

To convert *square inches* to square feet, multiply by 0.00694; to square yards, by 0.0007716.

To convert *square yards* to square inches, multiply by 1296; to square feet, by 9.

To convert *tons* (long) to ounces (avdp), multiply by 35840; to pounds, by 2240.

To convert *tons* (short) to ounces (avdp), multiply by 32000; to pounds, by 2000.

To convert *yards* to inches, multiply by 36; to feet, by 3; to miles, by 0.0005681.

MISCELLANEOUS CONVERSIONS

To convert *atmospheres* to pounds per square inch, multiply by 14.696; to do the opposite, multiply psi by 0.068046.

To convert *cheval vapeurs* to horsepower, multiply by 0.9863.

To convert *foot-pounds* to newton-meters, multiply by 1.355818; to do the opposite, multiply n/m by 0.737562.

To convert *frigories per hour* into tons of refrigeration, divide by 3023.9.

To convert *grams per cubic centimeter* into pounds per cubic inch, multiply by 0.036127; to do the opposite, multiply lbs/in^3 by 27.679905.

To convert *grams per liter* into pounds per cubic foot, multiply by 0.06243.

To convert *grams per square centimeter* into pounds per square inch, multiply by 0.014223; to do the opposite, multiply psi by 70.306958.

To convert *joules* into foot-pounds, multiply by 0.737684; to do the opposite, multiply foot-pounds by 1.35582.

Tables and Data

To convert *kilogram calories* into Btu, multiply by 3.968320.

To convert *kilogram meters* into foot-pounds, multiply by 7.233.

To convert *kilograms per cheval* into pounds per horsepower, multiply by 2.235.

To convert *kilograms per cubic meter* into pounds per cubic foot, multiply by 0.062427; to do the opposite, multiply pounds per cubic foot by 16.018463.

To convert *kilograms per meter* into pounds per foot, multiply by 0.672.

To convert *kilograms per square centimeter* into pounds per square inch, multiply by 14.223; into feet of water at 60°F., by 32.843. To do the opposite, multiply psi by 0.070306.

To convert *kilowatts* into horsepower, multiply by 1.341.

To convert *watts* into horsepower, divide by 746; into foot-pounds per second, multiply by 0.7373.

Table 30-1. Equivalent Temperature Readings for Fahrenheit and Celsius Scales

Degrees F	Degrees C	Degrees F	Degrees C	Degrees F	Degrees C	Degrees F	Degrees C
-459.4	-273.	-21.	-29.4	17.6	-8.	56.	13.3
-436.	-270.	-20.2	-29.	18.	-7.8	57.	13.9
-418.	-260.	-20.	-28.9	19.	-7.2	57.2	14.
-400.	-240.	-19.	-28.3	19.4	-7.	58.	14.4
-382.	-230.	-18.4	-28.	20.	-6.7	59.	15.
-364.	-220.	-18.	-27.8	21.	-6.1	60.	15.6
-346.	-210.	-17.	-27.2	21.2	-6.	60.8	16.
-328.	-200.	-16.6	-27.	22.	-5.6	61.	16.1
-310.	-190.	-16.	-26.7	23.	-5.	62.	16.7
-292.	-180.	-15.	-26.1	24.	-4.4	62.6	17.
-274.	-170.	-14.8	-26.	24.8	-4.	63.	17.2
-256.	-160.	-14.	-25.6	25.	-3.9	64.	17.8
-238.	-150.	-13.	-25.	26.	-3.3	64.4	18.
-220.	-140.	-12.	-24.4	26.6	-3.	65.	18.3
-202.	-130.	-11.2	-24.	27.	-2.8	66.	18.9
-184.	-120.	-11.	-23.9	28.	-2.2	66.2	19.
-166.	-110.	-10.	-23.3	28.4	-2.	67.	19.4
-148.	-100.	-9.4	-23.	29.	-1.7	68.	20.
-139.	-95.	-9.	-22.8	30.	-1.1	69.	20.6
-130.	-90.	-8.	-22.2	30.2	-1.	69.8	21.
-121.	-85.	-7.6	-22.	31.	-0.6	70.	21.1
-112.	-80.	-7.	-21.7	32.	0.	71.	21.7
-103.	-75.	-6.	-21.1	33.	+0.6	71.6	22.
-94.	-70.	-5.8	-21.	33.8	1.	72.	22.2
-85.	-65.	-5.	-20.6	34.	1.1	73.	22.8
-76.	-60.	-4.	-20.	35.	1.7	73.4	23.
-67.	-55.	-3.	-19.4	35.6	2.	74.	23.3
-58.	-50.	-2.2	-19.	36.	2.2	75.	23.9
-49.	-45.	-2.	-18.9	37.	2.8	75.2	24.
-40.	-40.	-1.	-18.3	37.4	3.	76.	24.4
-39.	-39.4	-0.4	-18.	38.	3.3	77.	25
-38.2	-39.	0.	-17.8	39.	3.9	78.	25.6
-38.	-38.9	+1.	-17.2	39.2	4.	78.8	26.
-37.	-38.3	1.4	-17.	40.	4.4	79.	26.1
-36.4	-38.	2.	-16.7	41.	5.	80.	26.7
-36.	-37.8	3.	-16.1	42.	5.6	80.6	27.
-35.	-37.2	3.2	-16.	42.8	6.	81.	27.2
-34.6	-37.	4.	-15.6	43.	6.1	82.	27.8
-34.	-36.7	5.	-15.	44.	6.7	82.4	28.
-33.	-36.1	6.	-14.4	44.6	7.	83.	28.3
-32.8	-36.	6.8	-14.	45.	7.2	84.	28.9
-32.	-35.6	7.	-13.9	46.	7.8	84.2	29
-31.	-35.	8.	-13.3	46.4	8.	85.	29.4
-30.	-34.4	8.6	-13.	47.	8.3	86.	30.
-29.2	-34.	9.	-12.8	48.	8.9	87.	30.6
-29.	-33.9	10.	-12.2	48.2	9.	87.8	31.
-28.	-33.3	10.4	-12.	49.	9.4	88.	31.1
-27.4	-33.	11.	-11.7	50.	10.	89.	31.7
-27.	-32.8	12.	-11.1	51.	10.6	89.6	32.
-26.	-32.2	12.2	-11.	51.8	11.	90.	32.2
-25.6	-32.	13.	-10.6	52.	11.1	91.	32.8
-25.	-31.7	14.	-10.	53.	11.7	91.4	33.
-24.	-31.1	15.	-9.4	53.6	12.	92.	33.3
-23.8	-31.	15.8	-9.	54.	12.2	93.	33.9
-23.	-30.6	16.	-8.9	55.	12.8	93.2	34.
-22.	-30.	17.	-8.3	55.4	13.	94.	34.4

Table 30-1—*Continued*

Degrees F	Degrees C	Degrees F	Degrees C	Degrees F	Degrees C	Degrees F	Degrees C
95.	35.	134.	56.7	172.4	78.	211.	99.4
96.	35.6	134.6	57.	173.	78.3	212.	100.
96.8	36.	135.	57.2	174.	78.9	213.	100.6
97.	36.1	136.	57.8	174.2	79.	213.8	101.
98.	36.7	136.4	58.	175.	79.4	214.	101.1
98.6	37.	137.	58.3	176.	80.	215.	101.7
99.	37.2	138.	58.9	177.	80.6	215.6	102
100.	37.8	138.2	59.	177.8	81.	216.	102.2
100.4	38.	139.	59.4	178.	81.1	217.	102.8
101.	38.3	140.	60.	179.	81.7	217.4	103.
102.	38.9	141.	60.6	179.6	82.	218.	103.3
102.2	39.	141.8	61.	180.	82.2	219.	103.9
103.	39.4	142.	61.1	181.	82.8	219.2	104.
104.	40.	143.	61.7	181.4	83.	220.	104.4
105.	40.6	143.6	62.	182.	83.3	221.	105.
105.8	41.	144.	62.2	183.	83.9	222.	105.6
106.	41.1	145.	62.8	183.2	84.	222.8	106.
107.	41.7	145.4	63.	184.	84.4	223.	106.1
107.6	42.	146.	63.3	185.	85.	224.	106.7
108.	42.2	147.	63.9	186.	85.6	224.6	107.
109.	42.8	147.2	64.	186.8	86.	225.	107.2
109.4	43.	148.	64.4	187.	86.1	226.	107.8
110.	43.3	149.	65.	188.	86.7	226.4	108.
111.	43.9	150.	65.6	188.6	87.	227.	108.3
111.2	44.	150.8	66.	189.	87.2	228.	108.9
112.	44.4	151.	66.1	190.	87.8	228.2	109.
113.	45.	152.	66.7	190.4	88.	229.	109.4
114.	45.6	152.6	67.	191.	88.3	230.	110.
114.8	46.	153.	67.2	192.	88.9	231.	110.6
115.	46.1	154.	67.8	192.2	89.	231.8	111.
116.	46.7	154.4	68.	193.	89.4	232.	111.1
116.6	47.	155.	68.3	194.	90.	233.	111.7
117.	47.2	156.	68.9	195.	90.6	233.6	112.
118.	47.8	156.2	69.	195.8	91.	234.	112.3
118.4	48.	157.	69.4	196.	91.1	235.	112.8
119.	48.3	158.	70.	197.	91.7	235.4	113.
120.	48.9	159.	70.6	197.6	92.	236.	113.3
120.2	49.	159.8	71.	198.	92.2	237.	113.9
121.	49.4	160.	71.1	199.	92.8	237.2	114.
122.	50.	161.	71.7	199.4	93.	238.	114.4
123.	50.6	161.6	72.	200.	93.3	239.	115.
123.8	51.	162.	72.2	201.	93.9	240.	115.6
124.	51.1	163.	72.8	201.2	94.	240.8	116.
125.	51.7	163.4	73.	202.	94.4	241.	116.1
125.6	52.	164.	73.3	203.	95.	242.	116.7
126.	52.2	165.	73.9	204.	95.6	242.6	117.
127.	52.8	165.2	74.	204.8	96.	243.	117.2
127.4	53.	166.	74.4	205.	96.1	244.	117.8
128.	53.3	167.	75.	206.	96.7	244.4	118.
129.	53.9	168.	75.6	206.6	97.	245.	118.3
129.2	54.	168.8	76.	207.	97.2	246.	118.9
130.	54.4	169.	76.1	208.	97.8	246.2	119.
131.	55.	170.	76.7	208.4	98.	247.	119.4
132.	55.6	170.6	77.	209.	98.3	248.	120.
132.8	56.	171.	77.2	210.	98.9	249.	120.6
133.	56.1	172.	77.8	210.2	99.	249.8	121.

Power Plant Engineers Guide

Table 30-2. Circumferences and Areas of Circles

Diam.	Circum.	Area	Diam.	Circum.	Area	Diam.	Circum.	Area	Diam	Circum.	Area
1/64	0.0491	0.0002	3	9.4248	7.0686	8	25.1327	50.265	16	50.2655	201.06
1/32	0.0982	0.0008	1/16	9.6211	7.3662	1/8	25.5254	51.849	1/8	50.6582	204.22
1/16	0.1964	0.0031	1/8	9.8175	7.6699	1/4	25.9181	53.456	1/4	51.0509	207.39
3/32	0.2945	0.0059	3/16	10.0138	7.9798	3/8	26.3108	55.088	3/8	51.4436	210.60
1/8	0.3927	0.0123	1/4	10.2102	8.2958	1/2	26.7035	56.745	1/2	51.8363	213.82
5/32	0.4909	0.0192	5/16	10.4065	8.6179	5/8	27.0962	58.426	5/8	52.2290	217.08
3/16	0.5890	0.0276	3/8	10.6029	8.9462	3/4	27.4889	60.132	3/4	52.6217	220.35
7/32	0.6872	0.0376	7/16	10.7992	9.2806	7/8	27.8816	61.862	7/8	53.0144	223.65
1/4	0.7854	0.0491	1/2	10.9956	9.6211	9	28.2743	63.617	17	53.4071	226.98
9/32	0.8836	0.0621	9/16	11.1919	9.9678	1/8	28.6670	65.397	1/8	53.7998	230.33
5/16	0.9817	0.0767	5/8	11.3883	10.321	1/4	29.0597	67.201	1/4	54.1925	233.71
11/32	1.0799	0.0928	11/16	11.5846	10.680	3/8	29.4524	69.029	3/8	54.5852	237.10
3/8	1.1781	0.1105	3/4	11.7810	11.045	1/2	29.8451	70.882	1/2	54.9779	240.53
13/32	1.2763	0.1296	13/16	11.9773	11.416	5/8	30.2378	72.760	5/8	55.3706	243.98
7/16	1.3745	0.1503	7/8	12.1737	11.793	3/4	30.6305	74.662	3/4	55.7633	247.45
15/32	1.4726	0.1726	15/16	12.3700	12.177	7/8	31.0232	76.589	7/8	56.1560	250.95
1/2	1.5708	0.1964	4	12.5664	12.566	10	31.4159	78.540	18	56.5487	254.47
17/32	1.6690	0.2217	1/16	12.7627	12.962	1/8	31.8086	80.516	1/8	56.9414	258.02
9/16	1.7672	0.2485	1/8	12.9591	13.364	1/4	32.2013	82.516	1/4	57.3341	261.59
19/32	1.8653	0.2769	3/16	13.1554	13.772	3/8	32.5940	84.541	3/8	57.7268	265.18
5/8	1.9635	0.3068	1/4	13.3518	14.185	1/2	32.9867	86.590	1/2	58.1195	268.80
21/32	2.0617	0.3382	5/16	13.5481	14.607	5/8	33.3794	88.664	5/8	58.5122	272.45
11/16	2.1598	0.3712	3/8	13.7445	15.033	3/4	33.7721	90.763	3/4	58.9049	276.12
23/32	2.2580	0.4057	7/16	13.9408	15.466	7/8	34.1648	92.886	7/8	59.2976	279.81
3/4	2.3562	0.4418	1/2	14.1372	15.904	11	34.5575	95.033	19	59.6903	283.53
25/32	2.4544	0.4794	9/16	14.3335	16.349	1/8	34.9502	97.205	1/8	60.0830	287.27
13/16	2.5525	0.5185	5/8	14.5299	16.800	1/4	35.3429	99.402	1/4	60.4757	291.04
27/32	2.6507	0.5591	11/16	14.7262	17.257	3/8	35.7356	101.62	3/8	60.8684	294.83
7/8	2.7489	0.6013	3/4	14.9226	17.721	1/2	36.1283	103.87	1/2	61.2611	298.65
29/32	2.8471	0.6450	13/16	15.1189	18.190	5/8	36.5210	106.14	5/8	61.6538	302.49
15/16	2.9452	0.6903	7/8	15.3153	18.665	3/4	36.9137	108.43	3/4	62.0465	306.35
31/32	3.0434	0.7371	15/16	15.5116	19.147	7/8	37.3064	110.75	7/8	62.4392	310.24
1	3.1416	0.7854	5	15.7080	19.685	12	37.6991	113.10	20	62.8319	314.16
1/16	3.3379	0.8866	1/16	15.9043	20.129	1/8	38.0918	115.47	1/8	63.2246	318.10
1/8	3.5343	0.9940	1/8	16.1007	20.629	1/4	38.4845	117.86	1/4	63.6173	322.06
3/16	3.7306	1.1075	3/16	16.2970	21.135	3/8	38.8772	120.28	3/8	64.0100	326.05
1/4	3.9270	1.2272	1/4	16.4934	21.648	1/2	39.2699	122.72	1/2	64.4026	330.06
5/16	4.1233	1.3530	5/16	16.6897	22.166	5/8	39.6626	125.19	5/8	64.7953	334.10
3/8	4.3197	1.4849	3/8	16.8861	22.691	3/4	40.0553	127.68	3/4	65.1880	338.16
7/16	4.5160	1.6230	7/16	17.0824	23.221	7/8	40.4480	130.19	7/8	65.5807	342.25
1/2	4.7124	1.7671	1/2	17.2788	23.758	13	40.8407	132.73	21	65.9734	346.36
9/16	4.9087	1.9175	9/16	17.4751	24.301	1/8	41.2334	135.30	1/8	66.3661	350.50
5/8	5.1051	2.0739	5/8	17.6715	24.850	1/4	41.6261	137.89	1/4	66.7588	354.66
11/16	5.3014	2.2365	11/16	17.8678	25.406	3/8	42.0188	140.50	3/8	67.1515	358.84
3/4	5.4978	2.4053	3/4	18.0642	25.967	1/2	42.4115	143.14	1/2	67.5442	363.05
13/16	5.6941	2.5802	13/16	18.2605	26.535	5/8	42.8042	145.80	5/8	67.9369	367.28
7/8	5.8905	2.7612	7/8	18.4569	27.109	3/4	43.1969	148.49	3/4	68.3296	371.54
15/16	6.0868	2.9483	15/16	18.6532	27.688	7/8	43.5896	151.20	7/8	68.7223	375.83
2	6.2832	3.1416	6	18.8496	28.274	14	43.9823	153.94	22	69.1150	380.13
1/16	6.4795	3.3410	1/8	19.2423	29.465	1/8	44.3750	156.70	1/8	69.5077	384.46
1/8	6.6759	3.5466	1/4	19.6350	30.680	1/4	44.7677	159.48	1/4	69.9004	388.82
3/16	6.8722	3.7583	3/8	20.0277	31.919	3/8	45.1604	162.30	3/8	70.2931	393.20
1/4	7.0686	3.9761	1/2	20.4204	33.183	1/2	45.5531	165.13	1/2	70.6858	397.61
5/16	7.2649	4.2000	5/8	20.8131	34.472	5/8	45.9458	167.99	5/8	71.0785	402.04
3/8	7.4613	4.4301	3/4	21.2058	35.785	3/4	46.3385	170.87	3/4	71.4712	406.49
7/16	7.6576	4.6664	7/8	21.5984	37.122	7/8	46.7312	173.78	7/8	71.8639	410.97
1/2	7.8540	4.9087	7	21.9911	38.485	15	47.1239	176.71	23	72.2566	415.48
9/16	8.0503	5.1572	1/8	22.3838	39.871	1/8	47.5166	179.67	1/8	72.6493	420.00
5/8	8.2467	5.4119	1/4	22.7765	41.282	1/4	47.9093	182.65	1/4	73.0420	424.56
11/16	8.4430	5.6727	3/8	23.1692	42.718	3/8	48.3020	185.66	3/8	73.4347	429.13
3/4	8.6394	5.9396	1/2	23.5619	44.179	1/2	48.6947	188.69	1/2	73.8274	433.74
13/16	8.8357	6.2126	5/8	23.9546	45.664	5/8	49.0874	191.75	5/8	74.2201	438.36
7/8	9.0321	6.4918	3/4	24.3473	47.173	3/4	49.4801	194.83	3/4	74.6128	443.01
15/16	9.2284	6.7771	7/8	24.7400	48.707	7/8	49.8728	197.93	7/8	75.0055	447.69

Table 30-2—*Continued*

Diam.	Circum.	Area	Diam.	Circum.	Area	Diam.	Circum.	Area	Diam.	Circum.	Area
24	75.3982	452.39	32	100.531	804.25	40	125.664	1256.6	48	150.796	1809.6
1/8	75.7909	457.11	1/8	100.924	810.54	1/8	126.056	1264.5	1/8	151.189	1819.0
1/4	76.1836	461.86	1/4	101.316	816.86	1/4	126.449	1272.4	1/4	151.582	1828.5
3/8	76.5763	466.64	3/8	101.709	823.21	3/8	126.842	1280.3	3/8	151.975	1837.9
1/2	76.9690	471.44	1/2	102.102	829.58	1/2	127.235	1288.2	1/2	152.367	1847.5
5/8	77.3617	476.26	5/8	102.494	835.97	5/8	127.627	1296.2	5/8	152.760	1857.0
3/4	77.7544	481.11	3/4	102.887	842.39	3/4	128.020	1304.2	3/4	153.153	1866.5
7/8	78.1471	485.98	7/8	103.280	848.83	7/8	128.413	1312.2	7/8	153.545	1876.1
25	78.5398	490.87	33	103.673	855.30	41	128.805	1320.3	49	153.938	1885.7
1/8	78.9325	495.79	1/8	104.065	861.79	1/8	129.198	1328.3	1/8	154.331	1895.4
1/4	79.3252	500.74	1/4	104.458	868.31	1/4	129.591	1336.4	1/4	154.723	1905.0
3/8	79.7179	505.71	3/8	104.851	874.85	3/8	129.983	1344.5	3/8	155.116	1914.7
1/2	80.1106	510.71	1/2	105.243	881.41	1/2	130.376	1352.7	1/2	155.509	1924.4
5/8	80.5033	515.72	5/8	105.636	888.00	5/8	130.769	1360.8	5/8	155.902	1934.2
3/4	80.8960	520.77	3/4	106.029	894.62	3/4	131.161	1369.0	3/4	156.294	1943.9
7/8	81.2887	525.84	7/8	106.421	901.26	7/8	131.554	1377.2	7/8	156.687	1953.7
26	81.6814	530.93	34	106.814	907.92	42	131.947	1385.4	50	157.080	1963.5
1/8	82.0741	536.05	1/8	107.207	914.61	1/8	132.340	1393.7	1/8	157.472	1973.3
1/4	82.4668	541.19	1/4	107.600	921.32	1/4	132.732	1402.0	1/4	157.865	1983.2
3/8	82.8595	546.35	3/8	107.992	928.06	3/8	133.125	1410.3	3/8	158.258	1993.1
1/2	83.2522	551.55	1/2	108.385	934.82	1/2	133.518	1418.6	1/2	158.650	2003.0
5/8	83.6449	556.76	5/8	108.778	941.61	5/8	133.910	1427.0	5/8	159.043	2012.9
3/4	84.0376	562.00	3/4	109.170	948.42	3/4	134.303	1435.4	3/4	159.436	2022.8
7/8	84.4303	567.27	7/8	109.563	955.25	7/8	134.696	1443.8	7/8	159.829	2032.8
27	84.8230	572.56	35	109.956	962.11	43	135.088	1452.2	51	160.221	2042.8
1/8	85.2157	577.87	1/8	110.348	969.00	1/8	135.481	1460.7	1/8	160.614	2052.8
1/4	85.6084	583.21	1/4	110.741	975.91	1/4	135.874	1469.1	1/4	161.007	2062.9
3/8	86.0011	588.57	3/8	111.134	982.84	3/8	136.267	1477.6	3/8	161.399	2073.0
1/2	86.3938	593.96	1/2	111.527	989.80	1/2	136.659	1486.2	1/2	161.792	2083.1
5/8	86.7865	599.37	5/8	111.919	996.87	5/8	137.052	1494.7	5/8	162.185	2093.2
3/4	87.1792	604.81	3/4	112.312	1003.8	3/4	137.445	1503.3	3/4	162.577	2103.3
7/8	87.5719	610.27	7/8	112.705	1010.8	7/8	137.837	1511.9	7/8	162.970	2113.5
28	87.965	615.75	36	113.097	1017.9	44	138.230	1520.5	52	163.363	2123.7
1/8	88.357	621.26	1/8	113.490	1025.0	1/8	138.623	1529.2	1/8	163.756	2133.9
1/4	88.750	626.80	1/4	113.883	1032.1	1/4	139.015	1537.9	1/4	164.148	2144.2
3/8	89.143	632.36	3/8	114.275	1039.2	3/8	139.408	1546.6	3/8	164.541	2154.5
1/2	89.535	637.94	1/2	114.668	1046.3	1/2	139.801	1555.3	1/2	164.934	2164.8
5/8	89.928	643.55	5/8	115.061	1053.5	5/8	140.194	1564.0	5/8	165.326	2175.1
3/4	90.321	649.18	3/4	115.454	1060.7	3/4	140.586	1572.8	3/4	165.719	2185.4
7/8	90.713	654.84	7/8	115.846	1068.0	7/8	140.979	1581.6	7/8	166.112	2195.8
29	91.106	660.52	37	116.239	1075.2	45	141.372	1590.4	53	166.504	2206.2
1/8	91.499	666.23	1/8	116.632	1082.5	1/8	141.764	1599.3	1/8	166.897	2216.6
1/4	91.892	671.96	1/4	117.024	1089.8	1/4	142.157	1608.2	1/4	167.290	2227.0
3/8	92.284	677.71	3/8	117.417	1097.1	3/8	142.550	1617.0	3/8	167.683	2237.5
1/2	92.677	683.49	1/2	117.810	1104.5	1/2	142.942	1626.0	1/2	168.075	2248.0
5/8	93.070	689.30	5/8	118.202	1111.8	5/8	143.335	1634.9	5/8	168.468	2258.5
3/4	93.462	695.13	3/4	118.596	1119.2	3/4	143.728	1643.9	3/4	168.861	2269.1
7/8	93.855	700.98	7/8	118.988	1126.7	7/8	144.121	1652.9	7/8	169.253	2279.6
30	94.248	706.86	38	119.381	1134.1	46	144.513	1661.9	54	169.646	2290.2
1/8	94.640	712.70	1/8	119.773	1141.0	1/8	144.906	1670.9	1/8	170.039	2300.8
1/4	95.033	718.69	1/4	120.166	1149.1	1/4	145.299	1680.0	1/4	170.431	2311.5
3/8	95.426	724.64	3/8	120.559	1156.6	3/8	145.691	1689.1	3/8	170.824	2322.1
1/2	95.819	730.62	1/2	120.951	1164.2	1/2	146.084	1698.2	1/2	171.217	2332.8
5/8	96.211	736.62	5/8	121.344	1171.7	5/8	146.477	1707.4	5/8	171.609	2343.5
3/4	96.604	742.64	3/4	121.737	1179.3	3/4	146.869	1716.5	3/4	172.002	2354.3
7/8	96.997	748.69	7/8	122.129	1186.9	7/8	147.262	1725.7	7/8	172.395	2365.0
31	97.389	754.77	39	122.522	1194.6	47	147.655	1734.9	55	172.788	2375.8
1/8	97.782	760.87	1/8	122.915	1202.3	1/8	148.048	1744.2	1/8	173.180	2386.6
1/4	98.175	766.99	1/4	123.308	1210.0	1/4	148.440	1753.5	1/4	173.573	2397.5
3/8	98.567	773.14	3/8	123.700	1217.7	3/8	148.833	1762.7	3/8	173.966	2408.3
1/2	98.960	779.31	1/2	124.093	1225.4	1/2	149.226	1772.1	1/2	174.358	2419.2
5/8	99.353	785.51	5/8	124.486	1233.2	5/8	149.618	1781.4	5/8	174.751	2430.1
3/4	99.746	791.73	3/4	124.878	1241.0	3/4	150.011	1790.8	3/4	175.144	2441.1
7/8	100.138	797.98	7/8	125.271	1248.8	7/8	150.404	1800.1	7/8	175.536	2452.0

Table 30-3. Properties of Saturated Steam

Vacuum (Inches of Hg.)	Absolute Pressure (psia)	Temperature (°F)	Total Heat Above 32°F		Latent Heat (H-h Heat Units)	Volume (ft.³/lb.)	Weight (lbs./ft.³)	Entropy of the Water	Entropy of Evaporation
			In the Water (h Heat Units)	In the Steam (H Heat Units)					
29.74	0.0886	32	0.00	1073.4	1073.4	3294	0.000304	0.0000	2.1832
29.67	0.1217	40	8.05	1076.9	1068.9	2438	0.000410	0.0162	2.1394
29.56	0.1780	50	18.08	1081.4	1063.3	1702	0.000587	0.0361	2.0865
29.40	0.2562	60	28.08	1085.9	1057.8	1208	0.000828	0.0555	2.0358
29.18	0.3626	70	38.06	1090.3	1052.3	871	0.001148	0.0745	1.9868
28.89	0.505	80	48.03	1094.8	1046.7	636.8	0.001570	0.0932	1.9398
28.50	0.696	90	58.00	1099.2	1041.2	469.3	0.002131	0.1114	1.8944
28.00	0.846	100	67.97	1103.6	1035.6	350.8	0.002851	0.1295	1.8505
27.88	1	101.83	69.8	1104.4	1034.6	333.0	0.00300	0.1327	1.8427
25.85	2	126.15	94.0	1115.0	1021.0	173.5	0.00576	0.1749	1.7431
23.81	3	141.52	109.4	1121.6	1012.3	118.5	0.00845	0.2008	1.6840
21.78	4	153.01	120.9	1126.5	1005.7	90.5	0.01107	0.2198	1.6416
19.34	5	162.28	130.1	1130.5	1000.3	73.33	0.01364	0.2348	1.6084
17.70	6	170.06	137.9	1133.7	995.8	61.89	0.01616	0.2471	1.5814
15.67	7	176.85	144.7	1136.5	991.8	53.56	0.01867	0.2579	1.5582
13.63	8	182.86	150.8	1139.0	988.2	47.27	0.02115	0.2673	1.5380
11.60	9	188.27	156.2	1141.1	985.0	42.36	0.02361	0.2756	1.5202
9.56	10	193.22	161.1	1143.1	982.0	38.38	0.02606	0.2832	1.5042
7.52	11	197.75	165.7	1144.9	979.2	35.10	0.02849	0.2902	1.4895
5.49	12	201.96	169.9	1146.5	976.6	32.36	0.03090	0.2967	1.4760
3.45	13	205.87	173.8	1148.0	974.2	30.03	0.03330	0.3025	1.4639
1.42	14	209.55	177.5	1149.4	971.9	28.02	0.03569	0.3081	1.4523
Psig									
0.0	14.70	212	180.0	1150.4	970.4	26.79	0.03732	0.3118	1.4447
0.3	15	213.0	181.0	1150.7	969.7	26.27	0.03806	0.3133	1.4416
1.3	16	216.3	184.4	1152.0	967.6	24.79	0.04042	0.3183	1.4311
2.3	17	219.4	187.5	1153.1	965.6	23.38	0.04277	0.3229	1.4215
3.3	18	222.4	190.5	1154.2	963.7	22.16	0.04512	0.3273	1.4127
4.3	19	225.2	193.4	1155.2	961.8	21.07	0.04746	0.3315	1.4045
5.3	20	228.0	196.1	1156.2	960.0	20.08	0.04980	0.3355	1.3965
6.3	21	230.6	198.8	1157.1	958.3	19.18	0.05213	0.3393	1.3887
7.3	22	233.1	201.3	1158.0	956.7	18.37	0.05445	0.3430	1.3811
8.3	23	235.5	203.8	1158.8	955.1	17.62	0.05676	0.3465	1.3739
9.3	24	237.8	206.1	1159.6	953.5	16.93	0.05907	0.3499	1.3670
10.3	25	240.1	208.4	1160.4	952.0	16.30	0.0614	0.3532	1.3604
11.3	26	242.2	210.6	1161.2	950.6	15.72	0.0636	0.3564	1.3542
12.3	27	244.4	212.7	1161.9	949.2	15.18	0.0659	0.3594	1.3483
13.3	28	246.4	214.8	1162.6	947.8	14.67	0.0682	0.3623	1.3425
14.3	29	248.4	216.8	1163.2	946.4	14.19	0.0705	0.3652	1.3367
15.3	30	250.3	218.8	1163.9	945.1	13.74	0.0728	0.3680	1.3311
16.3	31	252.2	220.7	1164.5	943.8	13.32	0.0751	0.3707	1.3257
17.3	32	254.1	222.6	1165.1	942.5	12.93	0.0773	0.3733	1.3205
18.3	33	255.8	224.4	1165.7	941.3	12.57	0.0795	0.3759	1.3155
19.3	34	257.6	226.2	1166.3	940.1	12.22	0.0818	0.3784	1.3107
20.3	35	259.3	227.9	1166.8	938.9	11.89	0.0841	0.3808	1.3060
21.3	36	261.0	229.6	1167.3	937.7	11.58	0.0863	0.3832	1.3014

Table 30-3.—Continued

Vacuum (Inches of Hg.)	Absolute Pressure (psia)	Temperature (°F)	Total Heat Above 32°F		Latent Heat (H-h Heat Units)	Volume (ft.^{3}/lb.)	Weight (lbs./ft.^{3})	Entropy of the Water	Entropy of Evaporation
			In the Water (h Heat Units)	In the Steam (H Heat Units)					
22.3	37	262.6	231.3	1167.8	936.6	11.29	0.0886	0.3855	1.2969
23.3	38	264.2	232.9	1168.4	935.5	11.01	0.0908	0.3877	1.2925
24.3	39	265.8	234.5	1168.9	934.4	10.74	0.0931	0.3899	1.2882
25.3	40	267.3	236.1	1169.4	933.3	10.49	0.0953	0.3920	1.2841
26.3	41	268.7	237.6	1169.8	932.2	10.25	0.0976	0.3941	1.2800
27.3	42	270.2	239.1	1170.3	931.2	10.02	0.0998	0.3962	1.2759
28.3	43	271.7	240.5	1170.7	930.2	9.80	0.1020	0.3982	1.2720
29.3	44	273.1	242.0	1171.2	929.2	9.59	0.1043	0.4002	1.2681
30.3	45	274.5	243.4	1171.6	928.2	9.39	0.1065	0.4021	1.2644
31.3	46	275.8	244.8	1172.0	927.2	9.20	0.1087	0.4040	1.2607
32.3	47	277.2	246.1	1172.4	926.3	9.02	0.1109	0.4059	1.2571
33.3	48	278.5	247.5	1172.8	925.3	8.84	0.1131	0.4077	1.2536
34.3	49	279.8	248.8	1173.2	924.4	8.67	0.1153	0.4095	1.2502
35.3	50	281.0	250.1	1173.6	923.5	8.51	0.1175	0.4113	1.2468
36.3	51	282.3	251.4	1174.0	922.6	8.35	0.1197	0.4130	1.2435
37.3	52	283.5	252.6	1174.3	921.7	8.20	0.1219	0.4147	1.2402
38.3	53	284.7	253.9	1174.7	920.8	8.05	0.1241	0.4164	1.2370
39.3	54	285.9	255.1	1175.0	919.9	7.91	0.1263	0.4180	1.2339
40.3	55	287.1	256.3	1175.4	919.0	7.78	0.1285	0.4196	1.2309
41.3	56	288.2	257.5	1175.7	918.2	7.65	0.1307	0.4212	1.2278
42.3	57	289.4	258.7	1176.0	917.4	7.52	0.1329	0.4227	1.2248
43.3	58	290.5	259.8	1176.4	916.5	7.40	0.1350	0.4242	1.2218
44.3	59	291.6	261.0	1176.7	915.7	7.28	0.1372	0.4257	1.2189
45.3	60	292.7	262.1	1177.0	914.9	7.17	0.1394	0.4272	1.2160
46.3	61	293.8	263.2	1177.3	914.1	7.06	0.1416	0.4287	1.2132
47.3	62	294.9	264.3	1177.6	913.3	6.95	0.1438	0.4302	1.2104
48.3	63	295.9	265.4	1177.9	912.5	6.85	0.1460	0.4316	1.2077
49.3	64	297.0	266.4	1178.2	911.8	6.75	0.1482	0.4330	1.2050
50.3	65	298.0	267.5	1178.5	911.0	6.65	0.1503	0.4344	1.2024
51.3	66	299.0	268.5	1178.8	910.2	6.56	0.1525	0.4358	1.1998
52.3	67	300.0	269.6	1179.0	909.5	6.47	0.1547	0.4371	1.1972
53.3	68	301.0	270.6	1179.3	908.7	6.38	0.1569	0.4385	1.1946
54.3	69	302.0	271.6	1179.6	908.0	6.29	0.1590	0.4398	1.1921
55.3	70	302.9	272.6	1179.8	907.2	6.20	0.1612	0.4411	1.1896
56.3	71	303.9	273.6	1180.1	906.5	6.12	0.1634	0.4424	1.1872
57.3	72	304.8	274.5	1180.4	905.8	6.04	0.1656	0.4437	1.1848
58.3	73	305.8	275.5	1180.6	905.1	5.96	0.1678	0.4449	1.1825
59.3	74	306.7	276.5	1180.9	904.4	5.89	0.1699	0.4462	1.1801
60.3	75	307.6	277.4	1181.1	903.7	5.81	0.1721	0.4474	1.1778
61.3	76	308.5	278.3	1181.4	903.0	5.74	0.1743	0.4487	1.1755
62.3	77	309.4	279.3	1181.6	902.3	5.67	0.1764	0.4499	1.1732
63.3	78	310.3	280.2	1181.8	901.7	5.60	0.1786	0.4511	1.1710
64.3	79	311.2	281.1	1182.1	901.0	5.54	0.1808	0.4523	1.1687
65.3	80	312.0	282.0	1182.3	900.3	5.47	0.1829	0.4535	1.1665
66.3	81	312.9	282.9	1182.5	899.7	5.41	0.1851	0.4546	1.1644
67.3	82	313.8	283.8	1182.8	899.0	5.34	0.1873	0.4557	1.1623
68.3	83	314.6	284.6	1183.0	898.4	5.28	0.1894	0.4568	1.1602
69.3	84	315.4	285.5	1183.2	897.7	5.22	0.1915	0.4579	1.1581
70.3	85	316.3	286.3	1183.4	897.1	5.16	0.1937	0.4590	1.1561

Table 30-3—Continued

Vacuum (Inches of Hg.)	Absolute Pressure (psia)	Temperature (°F)	Total Heat Above 32°F		Latent Heat (H-h Heat Units)	Volume (ft.³/lb.)	Weight (lbs./ft.³)	Entropy of the Water	Entropy of Evaporation
			In the Water (h Heat Units)	In the Steam (H Heat Units)					
71.3	86	317.1	287.2	1183.6	896.4	5.10	0.1959	0.4601	1.1540
72.3	87	317.9	288.0	1183.8	895.8	5.05	0.1980	0.4612	1.1520
73.3	88	318.7	288.9	1184.0	895.2	5.00	0.2001	0.4623	1.1500
74.3	89	319.5	289.7	1184.2	894.6	4.94	0.2023	0.4633	1.1481
75.3	90	320.3	290.5	1184.4	893.9	4.89	0.2044	0.4644	1.1461
76.3	91	321.1	291.3	1184.6	893.3	4.84	0.2065	0.4654	1.1442
77.3	92	321.8	292.1	1184.8	892.7	4.79	0.2087	0.4664	1.1423
78.3	93	322.6	292.9	1185.0	892.1	4.74	0.2109	0.4674	1.1404
79.3	94	323.4	293.7	1185.2	891.5	4.69	0.2130	0.4684	1.1385
80.3	95	324.1	294.5	1185.4	890.9	4.65	0.2151	0.4694	1.1367
81.3	96	324.9	295.3	1185.6	890.3	4.60	0.2172	0.4704	1.1348
82.3	97	325.6	296.1	1185.8	889.7	4.56	0.2193	0.4714	1.1330
83.3	98	326.4	296.8	1186.0	889.2	4.51	0.2215	0.4724	1.1312
84.3	99	327.1	297.6	1186.2	888.6	4.47	0.2237	0.4733	1.1295
85.3	100	327.8	298.3	1186.3	888.0	4.429	0.2258	0.4743	1.1277
87.3	102	329.3	299.8	1186.7	886.9	4.347	0.2300	0.4762	1.1242
89.3	104	330.7	301.3	1187.0	885.8	4.268	0.2343	0.4780	1.1208
91.3	106	332.0	302.7	1187.4	884.7	4.192	0.2336	0.4798	1.1174
93.3	108	333.4	304.1	1187.7	883.6	4.118	0.2429	0.4816	1.1141
95.3	110	334.8	305.5	1188.0	882.5	4.047	0.2472	0.4834	1.1108
97.3	112	336.1	306.9	1188.4	881.4	3.978	0.2514	0.4852	1.1076
99.3	114	337.4	308.3	1188.7	880.4	3.912	0.2556	0.4869	1.1045
101.3	116	338.7	309.6	1189.0	879.3	3.848	0.2599	0.4886	1.1014
103.3	118	340.0	311.0	1189.3	878.3	3.786	0.2641	0.4903	1.0984
105.3	120	341.3	312.3	1189.6	877.2	3.726	0.2683	0.4919	1.0954
107.3	122	342.5	313.6	1189.8	876.2	3.668	0.2726	0.4935	1.0924
109.3	124	343.8	314.9	1190.1	875.2	3.611	0.2769	0.4951	1.0895
111.3	126	345.0	316.2	1190.4	874.2	3.556	0.2812	0.4967	1.0865
113.3	128	346.2	317.4	1190.7	873.3	3.504	0.2854	0.4982	1.0837
115.3	130	347.4	318.6	1191.0	872.3	3.452	0.2897	0.4998	1.0809
117.3	132	348.5	319.9	1191.2	871.3	3.402	0.2939	0.5013	1.0782
119.3	134	349.7	321.1	1191.3	870.4	3.354	0.2981	0.5028	1.0755
121.3	136	350.8	322.3	1191.7	869.4	3.308	0.3023	0.5043	1.0728
123.3	138	352.0	323.4	1192.6	868.5	3.263	0.3065	0.5057	1.0702
125.3	140	353.1	324.6	1192.2	867.6	3.219	0.3107	0.5072	1.0675
127.3	142	354.2	325.8	1192.5	866.7	3.175	0.3150	0.5086	1.0649
129.3	144	355.3	326.9	1192.7	865.8	3.133	0.3192	0.5100	1.0624
131.3	146	356.3	328.0	1192.9	864.9	3.092	0.3234	0.5114	1.0599
133.3	148	357.4	329.1	1193.2	864.0	3.052	0.3276	0.5128	1.0574
135.3	150	358.5	330.2	1193.4	863.2	3.012	0.3320	0.5142	1.0550
137.3	152	359.5	331.4	1193.6	862.3	2.974	0.3362	0.5155	1.0525
139.3	154	360.5	332.4	1193.8	861.4	2.938	0.3404	9.5169	1.0501
141.3	156	361.6	333.5	1194.1	860.6	2.902	0.3446	0.5182	1.0477
143.3	158	362.6	334.6	1194.3	859.7	2.868	0.3488	0.5195	1.0454
145.3	160	363.6	335.6	1194.5	858.8	2.834	0.3529	0.5208	1.0431
147.3	162	364.6	336.7	1194.7	858.0	2.801	0.3570	0.5220	1.0409
149.3	164	365.6	337.7	1194.9	857.2	2.769	0.3612	0.5233	1.0387
151.3	166	366.5	338.7	1195.1	856.4	2.737	0.3654	0.5245	1.0365
153.3	168	367.5	339.7	1195.3	855.5	2.706	0.3696	0.5257	1.0343

Table 30-3.—Continued

Vacuum (Inches of Hg.)	Absolute Pressure (psia)	Temperature (°F)	Total Heat Above 32°F		Latent Heat (H−h Heat Units)	Volume (ft.³/lb.)	Weight (lbs./ft.³)	Entropy of the Water	Entropy of Evaporation
			In the Water (h Heat Units)	In the Steam (H Heat Units)					
155.3	170	368.5	340.7	1195.4	854.7	2.675	0.3738	0.5269	1.0321
157.3	172	369.4	341.7	1195.6	853.9	2.645	0.3780	0.5281	1.0300
159.3	174	370.4	342.7	1195.8	853.1	2.616	0.3822	0.5293	1.0278
161.3	176	371.3	343.7	1196.0	852.3	2.588	0.3864	0.5305	1.0257
163.3	178	372.2	344.7	1196.2	851.5	2.560	0.3906	0.5317	1.0235
165.3	180	373.1	345.6	1196.4	850.8	2.533	0.3948	0.5328	1.0215
167.3	182	374.0	346.6	1196.6	850.0	2.507	0.3989	0.5339	1.0195
169.3	184	374.9	347.6	1196.8	849.2	2.481	0.4031	0.5351	1.0174
171.3	186	375.8	348.5	1196.9	848.4	2.455	0.4073	0.5362	1.0154
173.3	188	376.7	349.4	1197.1	847.7	2.430	0.4115	0.5373	1.0134
175.3	190	377.6	350.4	1197.3	846.9	2.406	0.4157	0.5384	1.0114
177.3	192	378.5	351.3	1197.4	846.1	2.381	0.4199	0.5395	1.0095
179.3	194	379.3	352.2	1197.6	845.4	2.358	0.4241	0.5405	1.0076
181.3	196	380.2	353.1	1197.8	844.7	2.335	0.4283	0.5416	1.0056
183.3	198	381.0	354.0	1197.9	843.9	2.312	0.4325	0.5426	1.0038
185.3	200	381.9	354.9	1198.1	843.2	2.290	0.437	0.5437	1.0019
190.3	205	384.0	357.1	1198.5	841.4	2.237	0.447	0.5463	0.9973
195.3	210	386.0	359.2	1198.8	839.6	2.187	0.457	0.5488	0.9928
200.3	215	388.0	361.4	1199.2	837.9	2.138	0.468	0.5513	0.9885
205.3	220	389.9	363.4	1199.6	836.2	2.091	0.478	0.5538	0.9841
210.3	225	319.9	365.5	1199.9	834.4	2.046	0.489	0.5562	0.9799
215.3	230	393.8	367.5	1200.2	832.8	2.004	0.499	0.5586	0.9758
220.3	235	395.6	369.4	1200.6	831.1	1.964	0.509	0.5610	0.9717
225.3	240	397.4	371.4	1200.9	829.5	1.924	0.520	0.5633	0.9676
230.3	245	399.3	373.3	1201.2	827.9	1.887	0.530	0.5655	0.9638
235.3	250	401.1	375.2	1201.5	826.3	1.850	0.541	0.5676	0.9600
245.3	260	404.5	378.9	1202.1	823.1	1.782	0.561	0.5719	0.9525
255.3	270	407.9	382.5	1202.6	820.1	1.718	0.582	0.5760	0.9454
265.3	280	411.2	386.0	1203.1	817.1	1.658	0.603	0.5800	0.9385
275.3	290	414.4	389.4	1203.6	814.2	1.602	0.624	0.5840	0.9316
285.3	300	417.5	392.7	1204.1	811.3	1.551	0.645	0.5878	0.9251
295.3	310	420.5	395.9	1204.5	808.5	1.502	0.666	0.5915	0.9187
305.3	320	423.4	399.1	1204.9	805.8	1.456	0.687	9.5951	0.9125
315.3	330	426.3	402.2	1205.3	803.1	1.413	0.708	0.5986	0.9065
325.3	340	429.1	405.3	1205.7	800.4	1.372	0.729	0.6020	0.9006
335.3	350	431.9	408.2	1206.1	797.8	1.334	0.750	0.6053	0.8949
345.3	360	434.6	411.2	1206.4	795.3	1.298	0.770	0.6085	0.8894
355.3	370	437.2	414.0	1206.8	792.8	1.264	0.791	0.6116	0.8840
365.3	380	439.8	416.8	1207.1	790.3	1.231	0.812	0.6147	0.8788
375.3	390	442.3	419.5	1207.4	787.9	1.200	0.833	0.6178	0.8737
385.3	400	444.8	422	1208	786	1.17	0.86	0.621	0.868
435.3	450	456.5	435	1209	774	1.04	0.96	0.635	0.844
485.3	500	467.3	448	1210	762	0.93	1.08	0.648	0.822
535.3	550	477.3	459	1210	751	0.83	1.20	0.659	0.801
585.3	600	486.6	469	1210	741	0.76	1.32	0.670	0.783
Source	684	500	484	1209	725	0.66	1.52	0.686	0.755
	1062	550	542	1200	658	0.42	2.36	0.743	0.650
	1574	600	604	1176	572	0.27	3.75	0.799	0.540
	2265	650			441	0.16	6.2		0.396
	2974	689				0.05			
	3075	700							
	4300.2	752							
	5017.1	779							
	5659.9	810.6							

SOURCE: *Steam Tables and Diagrams*, by Marks and Davis (Longmans, Green and Co., 1909), with permission.

Table 30-4. Properties of Superheated Steam

Press. Abs. Lbs. per Sq. In.	Temp. Sat. Steam	Degrees of Superheat.									
			0	50	100	150	200	250	300	400	500	600
20	228.0	v	20.08	21.69	23.25	24.80	26.33	27.85	29.37	32.39	35.40	38.40
		h	1156.2	1179.9	1203.5	1227.1	1250.6	1274.1	1297.6	1344.8	1392.2	1440.0
		n	1.7320	1.7652	1.7961	1.8251	1.8524	1.8781	1.9026	1.9479	1.9893	2.0275
40	267.3	v	10.49	11.33	12.13	12.93	13.70	14.48	15.25	16.78	18.30	19.80
		h	1169.4	1194.0	1218.4	1242.4	1266.4	1290.3	1314.1	1361.6	1409.3	1457.4
		n	1.6761	1.7089	1.7392	1.7674	1.7940	1.8189	1.8427	1.8867	1.9271	1.9646
60	292.7	v	7.17	7.75	8.30	8.84	9.36	9.89	10.41	11.43	12.45	13.46
		h	1177.0	1202.6	1227.6	1252.1	1276.4	1300.4	1324.3	1372.2	1420.0	1468.2
		n	1.6432	1.6761	1.7062	1.7342	1.7603	1.7849	1.8081	1.8511	1.8908	1.9279
80	312.0	v	5.47	5.92	6.34	6.75	7.17	7.56	7.95	8.72	9.49	10.24
		h	1182.3	1208.8	1234.3	1259.0	1283.6	1307.8	1331.9	1379.8	1427.9	1476.2
		n	1.6200	1.6532	1.6833	1.7110	1.7368	1.7612	1.7840	1.8265	1.8658	1.9025
100	327.8	v	4.43	4.79	5.14	5.47	5.80	6.12	6.44	7.07	7.69	8.31
		h	1186.3	1213.8	1239.7	1264.7	1289.4	1313.6	1337.8	1385.9	1434.1	1482.5
		n	1.6020	1.6358	1.6655	1.6933	1.7188	1.7428	1.7656	1.8079	1.8468	1.8829
120	341.3	v	3.73	4.04	4.33	4.62	4.89	5.17	5.44	5.96	6.48	6.99
		h	1189.6	1217.9	1244.1	1269.3	1294.1	1318.4	1342.7	1391.0	1439.4	1487.8
		n	1.5873	1.6216	1.6517	1.6789	1.7041	1.7280	1.7505	1.7924	1.8311	1.8669
140	353.1	v	3.22	3.49	3.75	4.00	4.24	4.48	4.71	5.16	5.61	6.06
		h	1192.2	1221.4	1248.0	1273.3	1298.2	1322.6	1346.9	1395.4	1443.8	1492.4
		n	1.5747	1.6096	1.6395	1.6666	1.6916	1.7152	1.7376	1.7792	1.8177	1.8533
160	363.6	v	2.83	3.07	3.30	3.53	3.74	3.95	4.15	4.56	4.95	5.34
		h	1194.5	1224.5	1251.3	1276.8	1301.7	1326.2	1350.6	1399.3	1447.9	1496.6
		n	1.5639	1.5993	1.6292	1.6561	1.6810	1.7043	1.7266	1.7680	1.8063	1.8418
180	373.1	v	2.53	2.75	2.96	3.16	3.35	3.54	3.72	4.09	4.44	4.78
		h	1196.4	1227.2	1254.3	1279.9	1304.8	1329.5	1353.9	1402.7	1451.4	1500.3
		n	1.5543	1.5904	1.6201	1.6468	1.6716	1.6948	1.7169	1.7581	1.7962	1.8316
200	381.9	v	2.29	2.49	2.68	2.86	3.04	3.21	3.38	3.71	4.03	4.34
		h	1198.1	1229.8	1257.1	1282.6	1307.7	1332.4	1357.0	1405.9	1454.7	1503.7
		n	1.5456	1.5823	1.6120	1.6385	1.6632	1.6862	1.7082	1.7493	1.7872	1.8225
220	389.9	v	2.09	2.28	2.45	2.62	2.78	2.94	3.10	3.40	3.69	3.98
		h	1199.6	1232.2	1259.6	1285.2	1310.3	1335.1	1359.8	1408.8	1457.7	1506.8
		n	1.5379	1.5753	1.6049	1.6312	1.6558	1.6787	1.7005	1.7415	1.7792	1.8145
240	397.4	v	1.92	2.09	2.26	2.42	2.57	2.71	2.85	3.13	3.40	3.67
		h	1200.9	1234.3	1261.7	1287.6	1312.8	1337.6	1362.3	1411.5	1460.5	1509.8
		n	1.5309	1.5690	1.5985	1.6246	1.6492	1.6720	1.6937	1.7344	1.7721	1.8072
260	404.5	v	1.78	1.94	2.10	2.24	2.39	2.52	2.65	2.91	3.16	3.41
		h	1202.1	1236.4	1264.1	1289.9	1315.1	1340.0	1364.7	1414.0	1463.2	1512.5
		n	1.5244	1.5631	1.5926	1.6186	1.6430	1.6658	1.6874	1.7280	1.7655	1.8005
280	411.2	v	1.66	1.81	1.95	2.09	2.22	2.35	2.48	2.72	2.95	3.19
		h	1203.1	1238.4	1266.2	1291.9	1317.2	1342.2	1367.0	1416.4	1465.7	1515.1
		n	1.5185	1.5580	1.5873	1.6133	1.6375	1.6603	1.6818	1.7223	1.7597	1.7945
300	417.5	v	1.55	1.69	1.83	1.96	2.09	2.21	2.33	2.55	2.77	2.99
		h	1204.1	1240.3	1268.2	1294.0	1319.3	1344.3	1369.2	1418.6	1468.0	1517.6
		n	1.5129	1.5530	1.5824	1.6082	1.6323	1.6550	1.6765	1.7168	1.7541	1.7889
350	431.9	v	1.33	1.46	1.58	1.70	1.81	1.92	2.02	2.22	2.41	2.60
		h	1206.1	1244.6	1272.7	1298.7	1324.1	1349.3	1374.3	1424.0	1473.7	1523.5
		n	1.5002	1.5423	1.5715	1.5971	1.6210	1.6436	1.6650	1.7052	1.7422	1.7767
400	444.8	v	1.17	1.28	1.40	1.50	1.60	1.70	1.79	1.97	2.14	2.30
		h	1207.7	1248.6	1276.9	1303.0	1328.6	1353.9	1379.1	1429.0	1478.9	1528.9
		n	1.4894	1.5336	1.5625	1.5880	1.6117	1.6342	1.6554	1.6955	1.7323	1.7666
450	456.5	v	1.04	1.14	1.25	1.35	1.44	1.53	1.61	1.77	1.93	2.07
		h	1209	1252	1281	1307	1333	1358	1383	1434	1484	1534.0
		n	1.479	1.526	1.554	1.580	1.603	1.626	1.647	1.687	1.723	1.758
500	467.3	v	0.93	1.03	1.13	1.22	1.31	1.39	1.47	1.62	1.76	1.89
		h	1210	1256	1285	1311	1337	1362	1388	1438	1489	1539
		n	1.470	1.519	1.548	1.573	1.597	1.619	1.640	1.679	1.715	1.750

Source: *Steam Tables and Diagrams*, by Marks and Davis (Longmans, Green and Co., 1909) with permission.

Notes: v is the specific volume in ft^3 per pound; h is total heat, from water at 32°F., in Btu per pound; n is entropy from water at 32°F.

Table 30-5. Specific Heat of Superheated Steam

Lb. per sq. in. Temp. sat. °F.		14.2 210	28.4 248	56.9 289	85.3 316	113.3 336	142.2 350	170.6 368	199.1 381	227.5 392	256.0 403	284.4 412
°F.	°C.											
212	100	0.463										
302	150	.462	0.478	0.515								
392	200	.462	.475	.502	0.530	0.560	0.597	0.635	0.677			
482	250	.463	.474	.495	.514	.532	.552	.570	.588	0.609	0.635	0.664
572	300	.464	.475	.492	.505	.517	.530	.541	.550	.561	.572	.585
662	350	.468	.477	.492	.503	.512	.522	.529	.536	.543	.550	.557
752	400	.473	.481	.494	.504	.512	.520	.526	.531	.537	.542	.547

Power Plant Engineers Guide

Table 30-6. Factors of Evaporation

Pressure in Pounds Per Square Inch Above the Atmosphere

Temp. Feed Water, °F.	0	5.	15.	25.	35.	45.	55.	65.	75.	85.	95.	105.	115.	125.	135.	145.	155.	165.	175.	185.	200.	Temp. Feed Water, °F.
Deg. 32	1.187	1.192	1.199	1.204	1.209	1.212	1.216	1.218	1.221	1.223	1.226	1.228	1.230	1.231	1.233	1.235	1.236	1.238	1.239	1.240	1.241	32
35	1.184	1.189	1.196	1.201	1.206	1.209	1.213	1.215	1.218	1.220	1.223	1.225	1.227	1.228	1.230	1.232	1.233	1.235	1.236	1.237	1.238	35
40	1.179	1.184	1.191	1.196	1.201	1.204	1.208	1.210	1.213	1.215	1.218	1.220	1.222	1.223	1.225	1.227	1.228	1.230	1.231	1.232	1.233	40
45	1.173	1.178	1.185	1.190	1.195	1.198	1.202	1.204	1.207	1.209	1.212	1.214	1.216	1.217	1.219	1.221	1.222	1.224	1.225	1.226	1.227	45
50	1.168	1.173	1.180	1.185	1.190	1.193	1.197	1.199	1.202	1.204	1.207	1.209	1.211	1.212	1.214	1.216	1.217	1.219	1.220	1.221	1.222	50
55	1.163	1.168	1.175	1.180	1.185	1.188	1.192	1.194	1.197	1.199	1.202	1.204	1.206	1.207	1.209	1.211	1.212	1.214	1.215	1.216	1.217	55
60	1.158	1.163	1.170	1.175	1.180	1.183	1.187	1.189	1.192	1.194	1.197	1.199	1.201	1.202	1.204	1.206	1.207	1.209	1.210	1.211	1.212	60
65	1.153	1.158	1.165	1.170	1.175	1.178	1.182	1.184	1.187	1.189	1.192	1.194	1.196	1.197	1.199	1.201	1.202	1.204	1.205	1.206	1.207	65
70	1.148	1.153	1.160	1.165	1.170	1.173	1.177	1.179	1.182	1.184	1.187	1.189	1.191	1.192	1.194	1.196	1.197	1.199	1.200	1.201	1.202	70
75	1.143	1.148	1.155	1.160	1.165	1.168	1.172	1.174	1.177	1.179	1.182	1.184	1.186	1.187	1.189	1.191	1.192	1.194	1.195	1.196	1.197	75
80	1.137	1.142	1.149	1.154	1.159	1.162	1.166	1.168	1.171	1.173	1.176	1.178	1.180	1.181	1.183	1.185	1.186	1.188	1.189	1.190	1.191	80
85	1.132	1.137	1.144	1.149	1.154	1.157	1.161	1.163	1.166	1.168	1.171	1.173	1.175	1.176	1.178	1.180	1.181	1.183	1.184	1.185	1.186	85
90	1.127	1.132	1.139	1.144	1.149	1.152	1.156	1.158	1.161	1.163	1.166	1.168	1.170	1.171	1.173	1.175	1.176	1.178	1.179	1.180	1.181	90
95	1.122	1.127	1.134	1.139	1.144	1.147	1.151	1.153	1.156	1.158	1.161	1.163	1.165	1.166	1.168	1.170	1.171	1.173	1.174	1.175	1.176	95
100	1.117	1.122	1.129	1.134	1.139	1.142	1.146	1.148	1.151	1.153	1.156	1.158	1.160	1.161	1.163	1.165	1.166	1.168	1.169	1.170	1.171	100
105	1.111	1.116	1.123	1.128	1.133	1.136	1.140	1.142	1.145	1.147	1.150	1.152	1.154	1.155	1.157	1.159	1.160	1.162	1.163	1.164	1.165	105
110	1.106	1.111	1.118	1.123	1.128	1.131	1.135	1.137	1.140	1.142	1.145	1.147	1.149	1.150	1.152	1.154	1.155	1.157	1.158	1.159	1.160	110
115	1.101	1.106	1.113	1.118	1.123	1.126	1.130	1.132	1.135	1.137	1.140	1.142	1.144	1.145	1.147	1.149	1.150	1.152	1.153	1.154	1.155	115
120	1.096	1.101	1.108	1.113	1.118	1.121	1.125	1.127	1.130	1.132	1.135	1.137	1.139	1.140	1.142	1.144	1.145	1.147	1.148	1.149	1.150	120
125	1.091	1.096	1.103	1.108	1.113	1.116	1.120	1.122	1.125	1.127	1.130	1.132	1.134	1.135	1.137	1.139	1.140	1.142	1.143	1.144	1.145	125
130	1.085	1.090	1.097	1.102	1.107	1.110	1.114	1.116	1.119	1.121	1.124	1.126	1.128	1.129	1.131	1.133	1.134	1.136	1.137	1.138	1.139	130
135	1.080	1.085	1.092	1.097	1.102	1.105	1.109	1.111	1.114	1.116	1.119	1.121	1.123	1.124	1.126	1.128	1.129	1.131	1.132	1.133	1.134	135
140	1.075	1.080	1.087	1.092	1.097	1.100	1.104	1.106	1.109	1.111	1.114	1.116	1.118	1.119	1.121	1.123	1.124	1.126	1.127	1.128	1.129	140
145	1.070	1.075	1.082	1.087	1.092	1.095	1.099	1.101	1.104	1.106	1.109	1.111	1.113	1.114	1.116	1.118	1.119	1.121	1.122	1.123	1.124	145
150	1.065	1.070	1.077	1.082	1.087	1.090	1.094	1.096	1.099	1.101	1.104	1.106	1.108	1.109	1.111	1.113	1.114	1.116	1.117	1.118	1.119	150
155	1.059	1.064	1.071	1.076	1.081	1.084	1.088	1.090	1.093	1.095	1.098	1.100	1.102	1.103	1.105	1.107	1.108	1.110	1.111	1.112	1.113	155
160	1.054	1.059	1.066	1.071	1.076	1.079	1.083	1.085	1.088	1.090	1.093	1.095	1.097	1.098	1.100	1.102	1.103	1.105	1.106	1.107	1.108	160
165	1.049	1.054	1.061	1.066	1.071	1.074	1.078	1.080	1.083	1.085	1.088	1.090	1.092	1.093	1.095	1.097	1.098	1.100	1.101	1.102	1.103	165
170	1.044	1.049	1.056	1.061	1.066	1.069	1.073	1.075	1.078	1.080	1.083	1.085	1.087	1.088	1.090	1.092	1.093	1.095	1.096	1.097	1.098	170
175	1.039	1.044	1.051	1.056	1.061	1.064	1.068	1.070	1.073	1.075	1.078	1.080	1.082	1.083	1.085	1.087	1.088	1.090	1.091	1.092	1.093	175
180	1.033	1.038	1.045	1.050	1.055	1.058	1.062	1.064	1.067	1.069	1.072	1.074	1.076	1.077	1.079	1.081	1.082	1.084	1.085	1.086	1.087	180
185	1.028	1.033	1.040	1.045	1.050	1.053	1.057	1.059	1.062	1.064	1.067	1.069	1.071	1.072	1.074	1.076	1.077	1.079	1.080	1.081	1.082	185
190	1.023	1.028	1.035	1.040	1.045	1.048	1.052	1.054	1.057	1.059	1.062	1.064	1.066	1.067	1.069	1.071	1.072	1.074	1.075	1.076	1.077	190
195	1.018	1.023	1.030	1.035	1.040	1.043	1.047	1.049	1.052	1.054	1.057	1.059	1.061	1.062	1.064	1.066	1.067	1.069	1.070	1.071	1.072	195
200	1.013	1.018	1.025	1.030	1.035	1.038	1.042	1.044	1.047	1.049	1.052	1.054	1.056	1.057	1.059	1.061	1.062	1.064	1.065	1.066	1.067	200
205	1.001	1.012	1.019	1.024	1.029	1.032	1.036	1.038	1.041	1.043	1.046	1.048	1.050	1.051	1.053	1.055	1.056	1.058	1.059	1.060	1.062	205
210	1.002	1.007	1.014	1.019	1.024	1.027	1.031	1.033	1.036	1.038	1.041	1.043	1.045	1.046	1.048	1.050	1.051	1.053	1.054	1.055	1.057	210
212	1.000	1.005	1.012	1.017	1.022	1.025	1.029	1.031	1.034	1.036	1.039	1.041	1.043	1.044	1.046	1.048	1.049	1.051	1.052	1.053	1.056	212

TABLES AND DATA

Table 30-7. Specific Heat of Various Substances

Solid	Specific Heat	Liquid	Specific Heat
Brass	0.0939	Alcohol	0.7000
Cast iron	0.1315	Ether	0.5034
Copper	0.0951	Kerosene	0.5001
Glass	0.1937	Mercury	0.0333
Graphite	0.2014	Naptha	0.3100
Lead	0.0314	Machine oil	0.4000
Steel, soft	0.1165	Sulfuric acid	0.3354
Steel, hard	0.1175	Water	1.0000

Gas	Constant-Pressure Specific Heat	Constant-Volume Specific Heat
Air	0.2375	0.1684
Alcohol	0.4534	0.3990
Ammonia	0.5080	0.2990
Hydrogen	3.4090	2.4122
Nitrogen	0.2438	0.1727
Oxygen	0.2175	0.1550

Table 30-8. Maximum Allowable Stress for Carbon and Alloy Steel

Spec. No. Grade	Spec. minimum tensile	For metal temperatures not exceeding deg. F.							
		20 to 650	700	750	800	850	900	950	1000
Carbon Steel									
S-1	55000	11000	10400	9500	8000	6300	4400	2600	1350
S-2 .... A	45000	9000	8800	8400	6900	5700	4400	2600	1350
S-2 .... B	50000	10000	9600	9000	7500	6000	4400	2600	1350
S-42 .... A	55000	11000	10400	9500	8500	7200	5600	3800	2000
S-42 .... B	60000	12000	11400	10400	9100	7400	5600	3800	2000
S-55 .... A	65000	13000	12300	11100	9400	7600	5600	3800	2000
S-55 .... B	70000	14000	13300	11900	10000	7800	5600	3800	2000
Low-Alloy Steels									
S-28 .... A	75000	15000	14100	12400	10100	7800	5600	3800	2000
S-28 .... B	85000	15000	14100	12400	10100	7800	5600	3800	2000
S-43 .... A	65000	13000	12300	11100	9400	7800	5600	3800	2000
S-43 .... B	70000	14000	13300	11900	10000	7800	5600	3800	2000
S-43 C	75000	15000	14100	12400	10100	7800	5600	3800	2000
S-44 .... A	65000	13000	13000	13000	12500	11500	10000	8000	5000
S-44 .... B	70000	14000	14000	14000	13500	12000	10200	8000	5000
S-44 C	75000	15000	15000	15000	14400	12700	10400	8000	5000

Table 30-9. Efficiency of Riveted Joints

Type of Joint	Thickness of Plate (T) in Ins.	Diam. of Rivet Holes (D) in Ins.	Pitch of Rivets (P) in Ins.	Efficiency (E) %
Single-riveted lap joint	5/16	3/4	2	56.6
	3/8	7/8	2 3/8	54.0
	1/2	1 1/16	2 1/2	56.7
	5/8	1 1/8	2 5/8	48.5
Double-riveted lap joint	5/16	3/4	2 3/8	68.4
	3/8	7/8	2 1/2	65.0
	1/2	1 1/16	2 7/8	63.00
	5/8	1 1/8	3 1/4	65.4
Double-riveted butt joint	5/16	3/4	2 5/8	71.4
	3/8	13/16	2 7/8	71.7
	1/2	15/16	3 1/4	71.1
	5/8	1	3 3/4	73.3
Treble-riveted butt joint	5/16	3/4	2 5/8	85.7
	3/8	13/16	2 7/8	85.3
	1/2	1	3 1/4	84.6
	5/8	1 1/16	3 3/4	85.8

TABLES AND DATA

Table 30-10. Properties of Standard Lap Welded Boiler Tubes

External Diameter	Internal Diameter	Standard Thickness		Internal Circumference	External Circumference	Internal Area	External Area	Length of Tube per Sq. Ft. of Inside Surface	Length of Tube per Sq. Ft. of Outside Surface	Weight per Lineal Foot
ins.	ins.	ins.	B.W.G.	ins.	ins.	sq. ins.	sq. ins.	ft.	ft.	lbs.
1	.810	.095	13	2.545	3.142	.515	.785	4.479	3.820	.90
1¼	1.060	.095	13	3.330	3.927	.882	1.227	3.604	3.056	1.15
1½	1.310	.095	13	4.115	4.712	1.348	1.767	2.916	2.547	1.40
1¾	1.560	.095	13	4.901	5.498	1.911	2.405	2.448	2.183	1.679
2	1.810	.095	13	5.686	6.283	2.573	3.142	2.110	1.910	1.932
2½	2.282	.109	12	7.169	7.854	4.090	4.909	1.674	1.528	2.783
3	2.782	.109	12	8.740	9.425	6.079	7.069	1.373	1.273	3.365
3½	3.260	.120	11	10.242	10.996	8.347	9.621	1.172	1.091	4.331
4	3.732	.134	10	11.724	12.566	10.939	12.566	1.024	.995	5.532
4½	4.232	.134	10	13.295	14.137	14.066	15.904	.903	.849	6.248
5	4.704	.148	9	14.778	15.708	17.379	19.635	.812	.764	7.669
6	5.670	.165	8	17.813	18.850	25.250	28.274	.674	.637	10.282

Table 30-11. Circumference, Surface, Area and Volume

External diameter inches	Circumference inches	Surface per lineal foot		Lineal feet of tube per square foot of surface	Transverse area square inches	Volume or displacement per lineal foot		
		Square inches	Square feet			Cubic inches	Cubic feet, also area in square feet	United States gallons
1.000	3.1416	37.699	.2618	3.8197	.7854	9.4248	.0055	.0408
1.250	3.9270	47.124	.3272	3.0558	1.2272	14.726	.0085	.0637
1.500	4.7124	56.549	.3927	2.5465	1.7671	21.206	.0123	.0918
1.750	5.4978	65.973	.4581	2.1827	2.4053	28.863	.0167	.1249
2.000	6.2832	75.398	.5236	1.9099	3.1416	37.699	.0218	.1632
2.250	7.0686	84.823	.5890	1.6977	3.9761	47.713	.0276	.2065
2.500	7.8540	94.248	.6545	1.5279	4.9087	58.905	.0341	.2550
2.750	8.6394	103.67	.7199	1.3890	5.9396	71.275	.0412	.3085
3.000	9.4248	113.10	.7854	1.2732	7.0686	84.823	.0491	.3672
3.250	10.210	122.52	.8508	1.1753	8.2958	99.549	.0576	.4309
3.500	10.996	131.95	.9163	1.0913	9.6211	115.45	.0668	.4998
4.000	12.566	150.80	1.0472	.9549	12.566	150.80	.0873	.6528
4.500	14.137	169.65	1.1781	.8488	15.904	190.85	.1104	.8262
5.000	15.708	188.50	1.3090	.7639	19.635	235.62	.1364	1.0200
5.500	17.279	207.35	1.4399	.6945	23.758	285.10	.1650	1.2342
6.000	18.850	226.19	1.5708	.6366	28.274	339.29	.1963	1.4688
6.500	20.420	245.04	1.7017	.5876	33.183	398.20	.2304	1.7238
7.000	21.991	263.89	1.8326	.5457	38.485	461.81	.2673	1.9992
7.500	23.562	282.74	1.9635	.5093	44.179	530.14	.3068	2.2950
8.000	25.133	301.59	2.0944	.4775	50.265	603.19	.3491	2.6112
8.500	26.704	320.44	2.2253	.4494	56.745	680.94	.3941	2.9478
9.000	28.274	339.29	2.3562	.4244	63.617	763.41	.4418	3.3048
9.500	29.845	358.14	2.4871	.4021	70.882	850.59	.4922	3.6822
10.000	31.416	376.99	2.6180	.3820	78.540	942.48	.5454	4.0800

Table 30-12. Standard Wrought Pipe

Size Inches	Size Milli-meters	Diameters External Inches	Diameters Approximate Internal Inches	Nominal Thickness Inches	Circumference External Inches	Circumference Internal Inches	Transverse Areas External Sq. Ins.	Transverse Areas Internal Sq. Ins.	Transverse Areas Metal Sq. Ins.	Length of Pipe Per Sq. foot of External Surface Feet	Length of Pipe Per Sq. foot of Internal Surface Feet	Length of Pipe Containing One Cubic Foot Feet	Nominal Weight Per Foot Plain Ends	Nominal Weight Per Foot Threaded and Coupled	Number of Threads per Inch of Screw
⅛	3	.405	.269	.068	1.272	.845	.129	.057	.072	9.431	14.199	2533.775	.244	.245	27
¼	6	.540	.364	.088	1.696	1.144	.229	.104	.125	7.073	10.493	1383.789	.424	.425	18
⅜	10	.675	.493	.091	2.121	1.549	.358	.191	.167	5.658	7.747	754.360	.567	.568	18
½	13	.840	.622	.109	2.639	1.954	.554	.304	.250	4.547	6.141	473.906	.850	.852	14
¾	19	1.050	.824	.113	3.299	2.589	.866	.533	.333	3.637	4.635	270.034	1.130	1.134	14
1	25	1.315	1.049	.133	4.131	3.296	1.358	.864	.494	2.904	3.641	166.618	1.678	1.684	11½
1¼	32	1.660	1.380	.140	5.215	4.335	2.164	1.495	.669	2.301	2.767	96.275	2.272	2.281	11½
1½	38	1.900	1.610	.145	5.969	5.058	2.835	2.036	.799	2.010	2.372	70.733	2.717	2.731	11½
2	50	2.375	2.067	.154	7.461	6.494	4.430	3.355	1.075	1.608	1.847	42.913	3.652	3.678	11½
2½	64	2.875	2.469	.203	9.032	7.757	6.492	4.788	1.704	1.328	1.547	30.077	5.793	5.819	8
3	76	3.500	3.068	.216	10.996	9.638	9.621	7.393	2.228	1.091	1.245	19.479	7.575	7.616	8
3½	90	4.000	3.548	.226	12.566	11.146	12.566	9.886	2.680	.954	1.076	14.565	9.109	9.202	8
4	100	4.500	4.026	.237	14.137	12.648	15.904	12.730	3.174	.848	.948	11.312	10.790	10.889	8
4½	113	5.000	4.506	.247	15.708	14.156	19.635	15.947	3.688	.763	.847	9.030	12.538	12.642	8
5	125	5.563	5.047	.258	17.477	15.856	24.306	20.006	4.300	.686	.756	7.198	14.617	14.810	8
6	150	6.625	6.065	.280	20.813	19.054	34.472	28.891	5.581	.576	.629	4.984	18.974	19.185	8
7	175	7.625	7.023	.301	23.955	22.063	45.664	38.738	6.926	.500	.543	3.717	23.544	23.769	8
8	200	8.625	8.071	.277	27.096	25.356	58.426	51.161	7.265	.442	.473	2.815	24.696	25.000	8
8	200	8.625	7.981	.322	27.096	25.073	58.426	50.027	8.399	.442	.478	2.878	28.554	28.809	8
9	225	9.625	8.941	.342	30.238	28.089	72.760	62.746	9.974	.396	.427	2.294	33.907	34.188	8
10	250	10.750	10.192	.279	33.772	32.019	90.763	81.585	9.178	.355	.374	1.765	31.201	32.000	8
10	250	10.750	10.136	.307	33.772	31.843	90.763	80.691	10.072	.355	.376	1.785	34.240	35.000	8
10	250	10.750	10.020	.365	33.772	31.479	90.763	78.855	11.908	.355	.381	1.826	40.483	41.132	8
11	275	11.750	11.000	.375	36.914	34.558	108.434	95.033	13.401	.325	.347	1.545	45.557	46.247	8
12	300	12.750	12.090	.330	40.055	37.982	127.676	114.800	12.876	.299	.315	1.254	43.773	45.000	8
12	300	12.750	12.000	.375	40.055	37.699	127.676	113.097	14.579	.299	.318	1.273	49.562	50.706	8

Table 30-13. Properties of Extra Strong* Pipe

Diameter			Circumference		Transverse Areas			Length of Pipe per Square Foot of		Length of Pipe Containing One Cubic Foot	Nominal Weight per Foot Plain Ends
Nominal Internal Inches	External Inches	Approximate Internal Diameter Inches	Nominal Thickness Inches	External Inches	Internal Inches	External Sq. Inches	Internal Sq. Inches	Metal Sq. Inches	External Surface Feet	Internal Surface Feet	Feet	Pounds
1/8	.405	.213	.095	1.272	.675	.129	.036	.093	9.434	17.766	3966.392	.314
1/4	.540	.302	.119	1.696	.949	.229	.072	.157	7.073	12.648	2010.290	.535
3/8	.675	.423	.126	2.121	1.329	.358	.141	.217	6.658	9.030	1024.689	.738
1/2	.840	.546	.147	2.639	1.715	.554	.234	.320	4.547	6.995	615.017	1.087
3/4	1.050	.742	.154	3.299	2.331	.866	.433	.433	3.637	5.147	333.016	1.473
1	1.315	.957	.179	4.131	3.007	1.358	.719	.639	2.904	3.991	200.193	2.171
1 1/4	1.660	1.278	.191	5.215	4.015	2.164	1.283	.881	2.301	2.988	112.256	2.996
1 1/2	1.900	1.500	.200	5.969	4.712	2.835	1.767	1.068	2.010	2.546	81.487	3.631
2	2.375	1.939	.218	7.461	6.092	4.430	2.953	1.477	1.603	1.969	48.766	5.022
2 1/2	2.875	2.323	.276	9.032	7.298	6.492	4.238	2.254	1.328	1.644	33.976	7.661
3	3.500	2.900	.300	10.996	9.111	9.621	6.605	3.016	1.091	1.317	21.801	10.252
3 1/2	4.000	3.364	.318	12.566	10.568	12.566	8.888	3.678	.954	1.135	16.202	12.505
4	4.500	3.826	.337	14.137	12.020	15.901	11.497	4.407	.848	.998	12.525	14.983
4 1/2	5.000	4.290	.355	15.708	13.477	19.635	14.455	5.180	.763	.890	9.962	17.611
5	5.563	4.813	.375	17.477	15.120	24.306	18.194	6.112	.686	.793	7.915	20.778
6	6.625	5.761	.432	20.813	18.099	34.472	26.067	8.405	.576	.663	5.524	28.573
7	7.625	6.625	.500	23.955	20.813	45.664	34.472	11.192	.500	.576	4.177	38.048
8	8.625	7.625	.500	27.096	23.955	58.426	45.663	12.763	.442	.500	3.154	43.388
9	9.625	8.625	.500	30.238	27.096	72.760	58.426	14.334	.396	.442	2.464	48.728
10	10.750	9.750	.500	33.772	30.631	90.763	74.662	16.101	.355	.391	1.929	54.735
11	11.750	10.750	.500	36.914	33.772	108.434	90.763	17.671	.325	.355	1.587	60.075
12	12.750	11.750	.500	40.055	36.914	127.676	108.434	19.242	.299	.325	1.328	65.415

Table 30-14. Double Extra Strong* Pipe

Nominal Internal Inches	Diameter		Nominal Thickness Inches	Circumferences		Transverse Areas			Length of Pipe per Square Foot of		Length of Pipe Containing One Cubic Foot Feet	Nominal Weight per Foot Plain Ends Pounds
	External Inches	Approximate Internal Diameter Inches		External Inches	Internal Inches	External Sq. Inches	Internal Sq. Inches	Metal Sq. Inches	External Surface Feet	Internal Surface Feet		
½	.840	.252	.294	2.639	.792	.554	.050	.504	4.547	15.157	2887.164	1.714
¾	1.050	.434	.308	3.299	1.363	.866	.148	.718	3.637	8.801	973.404	2.440
1	1.345	.599	.358	4.131	1.882	1.358	.282	1.076	2.904	6.376	510.998	3.659
1¼	1.660	.896	.382	5.215	2.815	2.164	.630	1.534	2.301	4.263	228.379	5.214
1½	1.900	1.100	.400	5.969	3.456	2.835	.950	1.885	2.010	3.472	151.526	6.408
2	2.375	1.503	.436	7.461	4.722	4.430	1.774	2.656	1.608	2.541	81.162	9.029
2½	2.875	1.771	.552	9.032	5.664	6.492	2.464	4.028	1.328	2.156	58.457	13.695
3	3.500	2.300	.600	10.996	7.226	9.621	4.155	5.466	1.091	1.660	34.659	18.583
3½	4.000	2.728	.636	12.566	8.570	12.566	5.845	6.721	.954	1.400	24.637	22.850
4	4.500	3.152	.674	14.137	9.902	15.904	7.803	8.101	.848	1.211	18.454	27.541
4½	5.000	3.580	.710	15.708	11.247	19.635	10.066	9.569	.763	1.066	14.306	32.530
5	5.563	4.063	.750	17.477	12.764	24.306	12.966	11.340	.686	.940	11.107	38.552
6	6.625	4.897	.864	20.813	15.384	34.472	18.835	15.637	.576	.780	7.646	53.160
7	7.625	5.875	.875	23.955	18.457	45.664	27.109	18.555	.500	.650	5.312	63.079
8	8.625	6.875	.875	27.096	21.598	58.426	37.122	21.304	.442	.555	3.879	72.424

Note.—Sizes 3½ inch and larger are made by telescoping.

NOTE.—The word *heavy* was formerly used in place of *strong*.

NOTE.—Pipe practice or customs of the trade: Orders for pipe larger than 12 inches should specify the actual outside diameter of the pipe and the thickness of the wall. Standard weight pipe is listed extra strong, and will be carried in stock, threaded and coupled, and will be shipped unless order specified otherwise. Extra strong, double extra strong, hydraulic, and large o.d. pipe is listed in plain ends only, and will be so shipped unless order specifies otherwise. An extra charge is made for threads and couplings on these weights. For pipe smoothed on the inside, known as reamed and drifted, an extra charge is made. Such pipe is furnished in random lengths 20 feet and shorter. Random lengths of extra strong and double extra strong pipe are considered to be 12 to 22 feet, dealer to have privilege of supplying not to exceed 5% of total order in lengths 6 to 12 feet. For cut lengths of any size, an extra charge above random lengths will be made. For galvanized or asphalted pipe, an extra charge above black will be made. Sizes 8, 10, and 12 inch standard pipe are listed in several weights and orders or inquiries should specify the weight required.

Table 30-15. American Standard Bolt Sizes, Course Thread Series

Dia. of Bolt	Dia. at root of Thread	Dia. of Tap Drill	Threads per Inch	Nuts Full Height	Nuts Width across flats	Thickness of Bolt Head
1/4"	.1850	3/16"	20	7/32"	7/16"	11/64"
5/16	.2403	1/4	18	17/64	9/16	13/64
3/8	.2938	19/64	16	21/64	5/8	1/4
7/16	.3447	23/64	14	3/8	3/4	19/64
1/2	.4001	27/64	13	7/16	13/16	21/64
9/16	.4542	31/64	12	1/2	7/8	3/8
5/8	.5069	17/32	11	35/64	1	27/64
3/4	.6201	21/32	10	21/32	1 1/8	1/2
7/8	.7307	49/64	9	49/64	1 5/16	19/32
1	.8376	7/8	8	7/8	1 1/2	21/32
1 1/8	.9394	63/64	7	1	1 11/16	3/4
1 1/4	1.0644	1 7/64	7	1 3/32	1 7/8	27/32
1 3/8	1.1585	1 7/32	6	1 13/64	2 1/16	29/32
1 1/2	1.2835	1 11/32	6	1 5/16	2 1/4	1
1 5/8	1.3868	1 29/64	5 1/2	1 27/64	2 7/16	1 3/32
1 3/4	1.4902	1 9/16	5	1 17/32	2 5/8	1 5/32
1 7/8	1.6152	1 11/16	5	1 41/64	2 13/16	1 1/4
2	1.7113	1 25/32	4 1/2	1 3/4	3	1 11/32

Table 30-16. Physical Properties of Common Refrigerants

		Water	Freon 11 F-11	Freon 12 F-12	Ammonia	Freon 22 F-22	Carbon-Dioxide
1.	Common Name	Water	Freon 11 F-11	Freon 12 F-12	Ammonia	Freon 22 F-22	Carbon-Dioxide
2.	Chemical Symbol	H_2O	CCl_3F	CCl_2F_2	NH_3	$CHClF_2$	CO_2
3.	Boiling Point @ Atm.	212F	74.7F	−21.7F	−28.F	−41.4F	−109.3F
4.	Cond. Pres. @ 86F psig	28.67″	3.6	93.2	154.5	169.1	1024
5.	Evap. Pres. @ 5F psig	29.67″	23.9″	11.8	19.6	28.4	319.7
6.	Latent Heat @ 5F Btu/#	1071	84	69.5	565	93.4	115.3
7.	Volume @ 5F cu. ft/#	2444	12.3	1.48	8.15	1.24	.267
8.	Displ. @ 5F 86F cfm/ton	477	36.3	5.81	3.44	3.57	.943
9.	Critical Temp. °F	706.1	388.4	232.7	271.2	204.8	87.8
10.	Freezing Point °F	32	−168	−252.4	−107.9	−256	−69.9
11.	Least Detect. Odor	None	20%	20%	53ppm	20%	None
12.	Irritability	None	None	None	700ppm	None	None
13.	Toxicity	None	10% 2 hr.	30% 2 hr.	.5% ½ hr.	--	30% 1 hr.
14.	Flammability	None	None	None	13 to 27%	None	None
15.	Toxic in Flame	None	Yes	Yes	None	Yes	None
16.	Sp. Gr. Liquid 5F H_2O = 1	1.00	1.57	1.44	0.66	1.34	0.985
17.	Sp. Gr. Vapor Atm. Air = 1	--	4.85	5.2	0.74	4.0	2.35
18.	Power 5F, 86F	1.12	0.98	1.03	0.99	0.98	1.78
19.	Compression Temp. 5F & 86F	282F	113F	102F	210F	131F	160F
20.	Net refrig. effect (86F to 5F) Btu per lb	1025.3	67.54	51.07	474.4	69.3	55.5
21.	Heat content liquid leaving 86F condenser, Btu per lb		25.34	27.72	138.9		83.0
22.	Heat content saturated vapor in 5F evaporator, Btu per lb		92.88	78.79	613.3		138.7
23.	Refrigerant flow, (86F to 5F) lbs. per min. per ton	0.195	2.96	3.91	0.421	2.89	3.61
24.	Liquid refrig. flow (86F to 5F) cu. in. per min. per ton	5.4	56.0	83.8	19.6	68.1	167.0

Table 30-17. Pressure-Temperature Relations For Refrigerants

Temperature in degrees Fahrenheit. Pressures in pounds per square inch, (#), vacuum (″)				
Temp. deg.F.	Kind of Refrigerant			
	NH$_3$	F-12	CO$_2$	Carrene
−10	9.0#	4.5#	242.6#	28.1″
− 5	12.2	6.7	266.0	27.8″
0	15.7	9.2	290.8	27.5″
5	19.6	11.8	219.7	27.0″
10	23.8	14.7	345.5	26.7″
15	28.4	17.7	375.3	26.2″
20	33.5	21.1	407.1	25.6″
25	39.0	24.6	440.6	25.0″
30	45.0	28.5	476.1	24.3″
35	51.6	32.6	504.6	23.5″
40	58.6	37.0	553.1	22.6″
45	65.8	41.7	595.0	21.7″
50	74.5	46.7	638.9	20.7″
55	83.4	52.0	685.3	19.5″
60	92.9	57.7	733.9	18.2″
65	103.1	63.7	785.3	16.7″
70	115.1	70.1	833.7	15.1″
75	125.8	76.9	895.1	13.4″
80	138.3	84.1	954.0	11.5″
86	154.5	93.2	1024.3	9.2″
90	165.9	99.6		7.3″
95	181.0	108.0		5.0″
100	197.2	116.9		2.4″
105	214.2	126.2		0.2#
110	232.2	136.0		1.6
115	251.5	146.3		3.1
120	271.7	157.1		4.8

Table 30-18. Pressure-Temperature Relations for "Freons"

Temp.	F-11	F-12	F-13	F-22	F-113	F-114	F-502
-50	28.9"	15.4"	57.0#	6.1"		27.2"	0.0#
-45	28.7	13.3		2.7		26.7	2.0
-40	28.4	11.0	72.7	0.5=		26.1	4.3
-35	28.1	8.4		2.5		25.5	6.7
-30	27.8	5.5	90.9	4.8	29.3"	24.7	9.4
-25	27.4	2.3		7.3	29.2	23.9	12.3
-20	27.0	0.6#	111.7	10.1	29.1	22.9	15.5
-15	26.5	2.4		13.1	28.9	21.8	19.0
-10	26.0	4.5	135.4	16.4	28.7	20.6	22.8
-5	25.4	6.7		20.0	28.5	19.3	26.9
0	24.7	9.2	162.2	23.9	28.2	17.8	31.2
+5	24.0	11.8		28.1	27.9	16.1	36.0
10	23.1	14.8	192.2	32.7	27.6	14.3	41.1
15	22.1	17.7		37.7	27.2	12.3	46.6
20	21.1	21.0	225.8	43.0	26.8	10.1	52.4
25	19.9	24.6		48.7	26.3	7.6	58.7
30	18.6	28.5	263.3	54.8	25.8	5.0	65.4
35	17.2	32.6		61.4	25.2	2.1	72.6
40	15.6	37.0	305.0	68.5	24.5	0.5#	80.2
45	13.9	41.7		76.0	23.8	2.2	87.7
50	12.0	46.7	351.2	84.0	22.9	4.0	96.9
55	10.0	52.0		92.5	22.1	6.0	109.7
60	7.7	57.7	402.4	101.6	21.0	8.1	115.6
65	5.3	63.8		111.2	19.9	10.4	125.8
70	2.6	70.2	458.8	121.4	18.7	12.9	136.6
75	0.1#	77.0		132.2	17.3	15.5	147.9
80	1.6	84.2	521.0	143.6	15.9	18.3	159.9
85	3.2	91.8		155.6	14.3	21.4	172.5
90	5.0	99.8		168.4	12.5	24.6	185.8
95	6.8	108.3		181.8	10.6	28.0	199.7
100	8.9	117.2		195.9	8.6	31.7	214.4
105	11.1	126.6		210.7	6.4	35.6	229.7
110	13.4	136.4		226.3	4.0	39.7	245.8
115	15.9	146.8		242.7	1.4	44.1	266.1
120	18.5	157.7		259.9	0.7#	48.7	280.3
125	21.3	169.1		277.9	2.2	53.7	298.7
130	24.3	181.0		296.8	3.7	58.8	318.0
135	27.4	193.5		316.5	5.4	64.3	338.1
140	30.8	206.6		337.2	7.2	70.1	359.2

APPENDIX A

Rules And Regulations of the Environmental Protection Agency (EPA)

These standards of performance limit emissions of sulfur dioxide (SO_2), particulate matter, and nitrogen oxides (NO_x) from new, modified, and reconstructed electric utility steam generating units capable of combusting more than 73 megawatts (MW) heat input (250 million Btu/hour) of fossil fuel. A new reference method for determining continuous compliance with SO_2 and NO_x standards is also established. The Clean Air Act Amendments of 1977 require EPA to revise the current standards of performance for fossil-fuel-fired stationary sources. The intended effect of this regulation is to require new, modified, and reconstructed electric utility steam generating units to use the best demonstrated technological system of continuous emission reduction and to satisfy the requirements of the Clean Air Act Amendments of 1977. The effective date of this regulation was June 11, 1979. (Appendix A in these Rules refers to material at the end of Appendix 1.)

Power Plant Engineers Guide
SUMMARY OF STANDARDS

Applicability

The standards apply to electric utility steam generating units capable of firing more than 73 MW (250 million Btu/hour) heat input of fossil fuel, for which construction commenced after September 18, 1978. Industrial cogeneration facilities that sell less than 25 MW of electricity, or less than one-third of their potential electrical output capacity, are not covered. For electric utility combined cycle gas turbines, applicability of the standards is determined on the basis of the fossil-fuel fired to the steam generator exclusive of the heat input and electrical power contribution of the gas turbine.

SO_2 Standards

The SO_2 standards are as follows:

(1) Solid and solid-derived fuels (except solid solvent refined coal): SO_2 emissions to the atmosphere are limited to 520 ng/J (1.20 lb/million Btu) heat input, and a 90% reduction in potential SO_2 emissions is required at all times except when emissions to the atmosphere are less than 260 ng/J (0.60 lb/million Btu) heat input. When SO_2 emissions are less than 260 ng/J (0.60 lb/million Btu) heat input, a 70% reduction in potential emissions is required. Compliance with the emission limit and percent reduction requirements is determined on a continuous basis by using continuous monitors to obtain a 30-day rolling average. The percent reduction is computed on the basis of overall SO_2 removed by all types of SO_2 and sulfur removal technology, including flue gas desulfurization (FGD) systems and fuel pretreatment systems (such as coal cleaning, coal gasification, and coal liquefaction). Sulfur removed by a coal pulverizer or in bottom ash and fly ash may be included in the computation.

(2) Gaseous and liquid fuels not derived from solid fuels: SO_2 emissions into the the atmosphere are limited to 340 ng/J (0.80 lb/million Btu) heat input, and a 90% reduction in potential SO_2 emissions is required. The percent reduction requirement does not apply if SO_2 emissions into the atmosphere are less than 86 ng/J (0.20 lb/million Btu) heat input. Compliance with the SO_2 emission

limitation and percent reduction is determined on a continuous basis by using continuous monitors to obtain a 30-day rolling average.

(3) Anthracite coal: Electric utility steam generating units firing anthracite coal alone are exempt from the percentage reduction requirement of the SO_2 standard but are subject to the 520 ng/J (1.20 lb/million Btu) heat input emission limit on a 30-day rolling average, and all other provisions of the regulations including the particulate matter and NO_x standards.

(4) Noncontinental areas: Electric utility steam generating units located in noncontinental areas (State of Hawaii, the Virgin Islands, Guam, American Samoa, the Commonwealth of Puerto Rico, and the Northern Mariana Islands) are exempt from the percentage reduction requirement of the SO_2 standard but are subject to the applicable SO_2 emission limitation and all other provisions of the regulations including the particulate matter and NO_x standards.

(5) Resource recovery facilities: Resource recovery facilities that fire less than 25% fossil-fuel on a quarterly (90-day) heat input basis are not subject to the percentage reduction requirements but are subject to the 520 ng/J (1.20 lb/million Btu) heat input emission limit. Compliance with the emission limit is determined on a continuous basis using continuous monitoring to obtain a 30-day rolling average. In addition, such facilities must monitor and report their heat input by fuel type.

(6) Solid solvent refined coal: Electric utility steam generating units firing solid solvent refined coal (SRC I) are subject to the 520 ng/J (1.20 lb/million Btu) heat input emission limit (30-day rolling average) and all requirements under the NO_x and particulate matter standards. Compliance with the emission limit is determined on a continuous basis using a continuous monitor to obtain a 30-day rolling average. The percentage reduction requirement for SRC I, which is to be obtained at the refining facility itself, is 85% reduction in potential SO_2 emissions on a 24-hour (daily) averaging basis. Compliance is to be determined by Method 19 (covered later). Initial full scale demonstration facilities may be granted a commercial demonstration permit establishing a requirement of 80% reduction in potential emissions on a 24-hour (daily) basis.

Particulate Matter Standards

The particulate matter standard limits emissions to 13 ng/J (0.03 lb/million Btu) heat input. The opacity standard limits the opacity of emission to 20% (6-minute average). The standards are based on the performance of a well-designed and operated baghouse or electrostatic precipitator (ESP).

NO_x Standards

The NO_x standards are based on combustion modification and vary according to the fuel type. The standards are:

(1) 86 ng/J (0.20 lb/million Btu) heat input from the combustion of any gaseous fuel, except gaseous fuel derived from coal;

(2) 130 ng/J (0.30 lb/million Btu) heat input from the combustion of any liquid fuel, except shale oil and liquid fuel derived from coal;

(3) 210 ng/J (0.50 lb/million Btu) heat input from the combustion of subbituminous coal, shale oil, or any solid, liquid, or gaseous fuel derived from coal;

(4) 340 ng/J (0.80 lb/million Btu) heat input from the combustion in a slag tap furnace of any fuel containing more than 25%, by weight, lignite which has been mined in North Dakota, South Dakota or Montana;

(5) Combustion of a fuel containing more than 25%, by weight, coal refuse is exempt from the NO_x standards and monitoring requirements; and

(6) 260 ng/J (0.60 lb/million Btu) heat input from the combustion of any solid fuel not specified under (3), (4), or (5).

Continuous compliance with the NO_x standards is required, based on a 30-day rolling average. Also, percent reductions in uncontrolled NO_x emission levels are required. The percent reductions are not controlling, however, and compliance with the NO_x emission limits will assure compliance with the percent reduction requirements.

Emerging Technologies

The standards include provisions which allow the Administrator to grant commercial demonstration permits to allow less stringent requirements for the initial full-scale demonstration plants of certain technologies. The standards include the following provisions:

(1) Facilities using SRC I would be subject to an emission limitation of 520 ng/J (1.20 lb/million Btu) heat input, based on a 30-day rolling average, and an emission reduction requirement of 85%, based on a 24-hour average. However, the percentage reduction allowed under a commercial demonstration permit for the initial full-scale demonstration plants, using SRC I would be 80% (based on a 24-hour average). The plant producing the SRC I would monitor to ensure that the required percentage reduction (24-hour average) is achieved and the power plant using the SRC I would monitor to ensure that the 520 ng/J heat input limit (30-day rolling average) is achieved.

(2) Facilities using fluidized bed combustion (FBC) or coal liquefaction would be subject to the emission limitation and percentage reduction requirement of the SO_2 standard and to the particulate matter and NO_x standards. However, the reduction in potential SO_2 emissions allowed under a commercial demonstration permit for the initial full-scale demonstration plants using FBC would be 85% (based on a 30-day rolling average). The NO_x emission limitation allowed under a commercial demonstration permit for the initial full-scale demonstration plants using coal liquefaction would be 300 ng/J (0.70 lb/million Btu) heat input, based on a 30-day rolling average.

(3) No more than 15,000 MW equivalent electrical capacity would be allotted for the purpose of commercial demonstration permits. The capacity will be allocated as follows:

Technology	Pollutant	Equivalent electrical capacity MW
Solid solvent-refined coal	SO_2	5,000–10,000
Fluidized bed combustion (atmospheric)	SO_2	400–3,000
Fluidized bed combustion (pressurized)	SO_2	200–1,200
Coal liquefaction	NO_x	750–10,000

Compliance Provisions

Continuous compliance with the SO_2 and NO_x standards is required and is to be determined with continuous emission monitors. Reference methods or other approved procedures must be used to supplement the emission data when the continuous emission monitors malfunction, to provide emissions data for at least 18 hours of each day for at least 22 days out of any 30 successive days of boiler operation.

A malfunctioning FGD system may be bypassed under emergency conditions. Compliance with the particulate standard is determined through performance tests. Continuous monitors are required to measure and record the opacity of emissions. This data is to be used to identify excess emissions to ensure that the particulate matter control system is being properly operated and maintained.

PART 60—STANDARDS OF PERFORMANCE FOR NEW STATIONARY SOURCES

In 40 CFR Part 60, § 60.8 of Subpart A is revised, the heading and § 60.40 of Subpart D are revised, a new Subpart Da is added, and a new reference method is added to Appendix A as follows:

1. Section 60.8(d) and § 60.8(f) are revised as follows:

§ 60.8 Performance tests.

* * * * *

(d) The owner or operator of an affected facility shall provide the Administrator at least 30 days prior notice of any performance test, except as specified under other subparts, to afford the Administrator the opportunity to have an observer present.

* * * * *

(f) Unless otherwise specified in the applicable subpart, each performance test shall consist of three separate runs using the applicable test method. Each run shall be conducted for the time and under the conditions specified in the applicable standard. For the purpose of determining compliance with an applicable standard, the arithmetic means of results of the three runs shall apply. In the event that a sample is accidentally lost or conditions occur in which one of the three runs must be discontinued because of forced shutdown, failure of an irreplaceable portion of the sample train, extreme meteorological conditions, or other circumstances,

Rules and Regulations of the EPA

beyond the owner or operator's control, compliance may, upon the Administrator's approval, be determined using the arithmetic mean of the results of the two other runs.

2. The heading for Subpart D is revised to read as follows:

Subpart D—Standards of Performance for Fossil-Fuel-Fired Steam Generators for Which Construction Is Commenced After August 17, 1971.

3. Section 60.40 is amended by adding paragraph (d) as follows:

§ 60.40 Applicability and designation of affected facility.

* * * * *

(d) Any facility covered under Subpart Da is not covered under this subpart.

(Sec. 111, 301(a) of the Clean Air Act as amended (42 U.S.C. 7411, 7601(a)).)

4. A new Subpart Da is added as follows:

Subpart Da—Standards of Performance for Electric Utility Steam Generating Units for Which Construction Is Commenced After September 18, 1978.

Authority: Sec. 111, 301(a) of the Clean Air Act as amended (42 U.S.C. 7411, 7601(a)), and additional authority as noted below.

§ 60.40a Applicability and designation of affected facility.

(a) The affected facility to which this subpart applies is each electric utility steam generating unit:

(1) That is capable of combusting more than 73 megawatts (250 million Btu/hour) heat input of fossil fuel (either alone or in combination with any other fuel); and

(2) For which construction or modification is commenced after September 18, 1978.

(b) This subpart applies to electric utility combined cycle gas turbines that are capable of combusting more than 73 megawatts (250 million Btu/hour) heat input of fossil fuel in the steam generator. Only emissions resulting from combustion of fuels in the steam generating unit are subject to this subpart. (The gas turbine emissions are subject to Subpart GG.)

(c) Any change to an existing fossil-fuel-fired steam generating unit to accomodate the use of combustible materials, other than fossil fuels, shall not bring that unit under the applicability of this subpart

(d) Any change to an existing steam generating unit originally designed to fire gaseous or liquid fossil fuels, to accommodate the use of any other fuel (fossil or nonfossil) shall not bring that unit under the applicability of this subpart.

§60.41a Definitions.

As used in this subpart, all terms not defined herein shall have the meaning given them in the Act and in Subpart A of this part.

"Steam generating unit" means any furnace, boiler, or other device used for combusting fuel for the purpose of producing steam (including fossil-fuel-fired steam generators associated with combined cycle gas turbines; nuclear steam generators are not included).

"Electric utility steam generating unit" means any steam electric generating unit that is constructed for the purpose of supplying more than one-third of its potential electric output capacity and more than 25 MW electrical output to any utility power distribution system for sale. Any steam supplied to a steam distribution system for the purpose of providing steam to a steam-electric generator that would produce electrical energy for sale is also considered in determining the electrical energy output capacity of the affected facility.

"Fossil fuel" means natural gas, petroleum, coal, and any form of solid, liquid, or gaseous fuel derived from such material for the purpose of creating useful heat.

"Subbituminous coal" means coal that is classified as subbituminous A, B, or C according to the American Society of Testing and Materials' (ASTM) Standard Specification for Classification of Coals by Rank D388-66.

"Lignite" means coal that is classified as lignite A or B according to the American Society of Testing and Materials' (ASTM) Standard Specification for Classification of Coals by Rank D388-66.

"Coal refuse" means waste products of coal mining, physical coal cleaning, and coal preparation operations (e.g. culm, gob, etc.) containing coal, matrix material, clay, and other organic and inorganic material.

"Potential combustion concentration" means the theoretical emissions (ng/J, lb/million Btu heat input) that would result from combustion of a fuel in an uncleaned state 9 without emission control systems) and:

(a) For particulate matter is:
(1) 3,000 ng/J (7.0 lb/million Btu) heat input for solid fuel; and
(2) 75 ng/J (0.17 lb/million Btu) heat input for liquid fuels.
(b) For sulfur dioxide is determined under §60.48a(b).
(c) For nitrogen oxides is:
(1) 290 ng/J (0.67 lb/million Btu) heat input for gaseous fuels;
(2) 310 ng/J (0.72 lb/million Btu) heat input for liquid fuels; and
(3) 990 ng/J (2.30 lb/million Btu) heat input for solid fuels.

"Combined cycle gas turbine" means a stationary turbine combustion system where heat from the turbine exhaust gases is recovered by a steam generating unit.

"Interconnected" means that two or more electric generating units are electrically tied together by a network of power transmission lines, and other power transmission equipment.

"Electric utility company" means the largest interconnected organization, business, or governmental entity that generates electric power for sale (e.g., a holding company with operating subsidiary companies).

"Principal company" means the electric utility company or companies which own the affected facility.

"Neighboring company" means any one of those electric utility companies with one or more electric power interconnections to the principal company and which have geographically adjoining service areas.

"Net system capacity" means the sum of the net electric generating capability (not necessarily equal to rated capacity) of all electric generating equipment owned by an electric utility company (including steam generating units, internal combustion engines, gas turbines, nuclear units, hydroelectric units, and all other electric generating equipment) plus firm contractual purchases that are interconnected to the affected facility that has the malfunctioning flue gas desulfurization system. The electric generating capability of equipment under multiple ownership is prorated based on ownership unless the proportional entitlement to electric output is otherwise established by contractual arrangement.

"System load" means the entire electric demand of an electric utility company's service area interconnected with the affected facility that has the malfunctioning flue gas desulfurization system plus firm contractual sales to other electric utility companies. Sales to other electric utility companies (e.g., emergency power) not on a firm contractual basis may also be included in the system load when no available system capacity exists in the electric utility company to which the power is supplied for sale.

"System emergency reserves" means an amount of electric generating capacity equivalent to the rated capacity of the single largest electric generating unit in the electric utility company (including steam generating units, internal combustion engines, gas turbines, nuclear units, hydroelectric units, and all other electric generating equipment) which is interconnected with the affected facility that has the malfunctioning flue gas desulfurization system. The electric generating capability of equipment under multiple ownership is prorated based on ownership unless the proportional entitlement to electric output is otherwise established by contractual arrangement.

"Available system capacity" means the capacity determined by subtracting the system load and the system emergency reserves from the net system capacity.

"Spinning reserve" means the sum of the unutilized net generating capability of all units of the electric utility company that are synchronized to the power distribution system and that are capable of immediately accepting additional load. The electric generating capability of equipment under multiple ownership is prorated based on ownership unless the proportional entitlement to electric output is otherwise established by contractual arrangement.

"Available purchase power" means the lesser of the following:

(a) The sum of available system capacity in all neighboring companies.

(b) The sum of the rated capacities of the power interconnection devices between the principal company and all neighboring companies, minus the sum of the electric power load on these interconnections.

(c) The rated capacity of the power transmission lines between the power interconnection devices and the electric generating units (the unit in the principal company that has the malfunctioning flue gas desulfurization system and the unit(s) in the neighboring company supplying replacement electrical power) less the electric power load on these transmission lines.

"Spare flue gas desulfurization system module" means a separate system of sulfur dioxide emission control equipment capable of treating an amount of flue gas equal to the total amount of flue gas generated by an affected facility when operated at maximum capacity divided by the total number of nonspare flue gas desulfurization modules in the system.

"Emergency condition" means that period of time when:

(a) The electric generation output of an affected facility with a malfunctioning flue gas desulfurization system cannot be reduced or electrical output must be increased because:

(1) All available system capacity in the principal company interconnected with the affected facility is being operated, and

(2) All available purchase power interconnected with the affected facility is being obtained, or

(b) The electric generation demand is being shifted as quickly as possible from an affected facility with a malfunctioning flue gas desulfurization system to one or more electrical generating units held in reserve by the principal company or by a neighboring company, or

(c) An affected facility with a malfunctioning flue gas desulfurization system becomes the only available unit to maintain a part or all of the principal company's system emergency reserves and the unit is operated in spinning reserve at the lowest practical electric generation load consistent with not causing significant physical damage to the unit. If the unit is operated at a higher load to meet load demand, an emergency condition would not exist unless the conditions under (a) of this definition apply.

"Electric utility combined cycle gas turbine" means any combined cycle gas turbine used for electric generation that is constructed for the purpose of supplying more than one-third of its potential electric output capacity and more than 25 MW electrical output to any utility power distribution system for sale. Any steam distribution system that is constructed for the purpose of providing steam to a steam electric generator that would produce electrical power for sale is also considered in determining the electrical energy output capacity of the affected facility.

"Potential electrical output capacity" is defined as 33% of the maximum design heat input capacity of the steam generating unit (e.g., a steam generating unit with a 100-MW (340 million Btu/hr) fossil-fuel heat input capacity would have a 33-MW potential electrical output capacity. For electric utility combined cycle gas turbines the potential electrical output capacity is determined on the basis of the fossil-fuel firing capacity of the steam generator exclusive of the heat input and electrical power contribution by the gas turbine.

"Anthracite" means coal that is classified as anthracite according to the American Society of Testing and Materials' (ASTM) Standard Specification for Classification of Coals by Rank D388-66.

"Solid-derived fuel" means any solid, liquid, or gaseous fuel derived from solid fuel for the purpose of creating useful heat and includes, but is not limited to, solvent refined coal, liquefied coal, and gasified coal.

"24-hour period" means the period of time between 12:01 a.m. and 12:00 midnight.

"Resource recovery unit" means a facility that combusts more than 75% nonfossil fuel on a quarterly (calendar) heat input basis.

"Noncontinental area" means the State of Hawaii, the Virgin Islands, Guam, American Samoa, the Commonwealth of Puerto Rico, or the Northern Mariana Islands.

"Boiler operating day" means a 24-hour period during which fossil fuel is combusted in a steam generating unit for the entire 24 hours.

§60.42a Standard for particulate matter.

(a) On and after the date on which the performance test required to be conducted under § 60.8 is completed, no owner or operator subject to the provisions of this subpart shall cause to be discharged into the atmosphere from any affected facility any gases which contain particulate matter in excess of:

(1) 13 ng/J (0.03 lb/million Btu) heat input derived from the combustion of solid, liquid, or gaseous fuel;

(2) 1% of the potential combustion concentration (99% reduction) when combusting solid fuel; and

(3) 30% of potential combustion concentration (70% reduction) when combusting liquid fuel.

(b) On and after the date the particulate matter performance test required to be conducted under § 60.8 is completed, no owner or operator subject to the provisions of this subpart shall cause to be discharged into the atmosphere from any affected facility any gases which exhibit greater than 20% opacity (6-minute average), except for one 6-minute period per hour of not more than 27% opacity.

§60.43a Standard for sulfur dioxide.

(a) On and after the date on which the initial performance test required to be conducted under § 60.8 is completed, no owner or operator subject to the provisions of this subpart shall cause to be discharged into the atmosphere from any affected facility which combusts solid fuel, or solid-derived fuel, except as provided under paragraphs (c), (d), (f) and (h) of this section, any gases which contain sulfur dioxide in excess of:

(1) 520 ng/J (1.20 lb/million Btu) heat input and 10% of the potential combustion concentration (90% reduction), or

(2) 30% of the potential combustion concentration (70% reduction), when emissions are less than 260 ng/J (0.60 lb/million Btu) heat input.

(b) On and after the date on which the initial performance test required to be conducted under § 60.8 is completed, no owner or operator subject to the provisions of this subpart shall cause to be discharged into the atmosphere from any affected facility which combusts liquid or gaseous fuels (except for liquid or gaseous fuels derived from solid fuels and as provided under paragraphs (e) and

(h) of this section), any gases which contain sulfur dioxide in excess of:

(1) 340 ng/J (0.80 lb/million Btu) heat input and 10% of the potential combustion concentration (90% reduction), or

(2) 100% of the potential combustion concentration (zero% reduction) when emissions are less than 86 ng/J (0.20 lb/million Btu) heat input.

(c) On and after the date on which the initial performance test required to be conducted under § 60.8 is complete, no owner or operator subject to the provisions of this subpart shall cause to be discharged into the atmosphere from any affected facility which combusts solid solvent refined coal (SRC-I) any gases which contain sulfur dioxide in excess of 520 ng/J (1.20 lb/million Btu) heat input and 15% of the potential combustion concentration (85% reduction) except as provided under paragraph (f) of this section; compliance with the emission limitation is determined on a 30-day rolling average basis and compliance with the percent reduction requirement is determined on a 24-hour basis.

(d) Sulfur dioxide emissions are limited to 520 ng/J (1.20 lb/million Btu) heat input from any affected facility which:

(1) Combusts 100% anthracite,

(2) Is classified as a resource recovery facility, or

(3) Is located in a noncontinental area and combusts solid fuel or solid-derived fuel.

(e) Sulfur dioxide emissions are limited to 340 ng/J (0.80 lb/million Btu) heat input from any affected facility which is located in a noncontinental area and combusts liquid or gaseous fuels (excluding solid-derived fuels).

(f) The emission reduction requirements under this section do not apply to any affected facility that is operated under an SO_2 commercial demonstration permit issued by the Administrator in accordance with the provisions of § 60.45a.

(g) Compliance with the emission limitation and percent reduction requirements under this section are both determined on a 30-day rolling average basis except as provided under paragraph (c) of this section.

(h) When different fuels are combusted simultaneously, the applicable standard is determined by proration using the following formula:

(1) If emissions of sulfur dioxide to the atmosphere are greater than 260 ng/J (0.60 lb/million Btu) heat input

$$E_{SO_2} = [340\ x + 520\ y]/100 \text{ and}$$
$$P_{SO_2} = 10\%$$

(2) If emissions of sulfur dioxide into the atmosphere are equal to or less than 260 ng/J (0.60 lb/million Btu) heat input:

$$E_{SO_2} = [340\ x + 520\ y]/100 \text{ and}$$
$$P_{SO_2} = [90\ x + 70\ y]/100$$

where:

E_{SO_2} is the prorated sulfur dioxide emission limit (ng/J heat input).

P_{SO_2} is the percentage of potential sulfur dioxide emission allowed (percent reduction required = $100-P_{so2}$).

x is the percentage of total heat input derived from the combustion of liquid or gaseous fuels (excluding solid-derived fuels).

y is the percentage of total heat input derived from the combustion of solid fuel (including solid-derived fuels).

§60.44a Standard for nitrogen oxides.

(a) On and after the date on which the initial performance test required to be conducted under §60.8 is completed, no owner or operator subject to the provisions of this subpart shall cause to be discharged into the atmosphere from any affected facility, except as provided under paragraph (b) of this section, any gases which contain nitrogen oxides in excess of the following emission limits, based on a 30-day rolling average.

(1) NO_x Emission Limits—

Fuel type	Emmission limit ng/J (lb/million) Btu input heat	
Gaseous Fuels;		
Coal-derived fuels	210	(0.50)
All other fuels	86	(0.20)
Liquid Fuels;		
Coal-derived fuels	210	(0.50)
Shade oil	210	(0.50)
All other fuels	130	(0.30)
Solid Fuels;		
Coal-derived fuels	210	(0.50)
Any fuel containing more than 25%, by weight, coal refuse.....	Exempt from NO_x standards and NO_x monitoring requirements	
Any fuel containing more than 25%, by weight, lignite if the lignite is mined in North Dakota, South Dakota, or Montana, and is combusted in a slag tap furnace...........	340	(0.80)
Lignite not subject to the 340 ng/J heat input emission limit	260	(0.60)
Subbituminous coal	210	(0.50)
Bituminous coal	260	(0.60)
Anthracite coal	260	(0.60)
All other fuels ················	260	(0.60)

(2) NO_x reduction requirements—

Fuel type	Percent reduction of potential combustible concentration
Gaseous fuels	25%
Liquid fuels.....................	30%
Solid fuels	65%

(b) The emission limitations under paragraph (a) of this section do not apply to any affected facility which is combusting coal-derived liquid fuel and is operating under a commercial demonstration permit issued by the Administrator in accordance with the provisions of §60.45a.

(c) When two or more fuels are combusted simultaneously, the applicable standard is determined by proration using the following formula:

$$E_{NO_2} = [86\,w + 130\,x + 210\,y + 260\,z]/100$$

where:

E_{NO_2} is the applicable standard for nitrogen oxides when multiple fuels are combusted simultaneously (ng/J heat input);

w is the percentage of total heat input derived from the combustion of fuels subject to the 86 ng/J heat input standard;

x is the percentage of total heat input derived from the combustion of fuels subject to the 130 ng/J heat input standard;

y is the percentage of total heat input derived from the combustion of fuels subject to the 210 ng/J heat input standard; and

z is the percentage of total heat input derived from the combustion of fuels subject to the 260 ng/J heat input standard.

§60.45a Commercial demonstration permit.

(a) An owner or operator of an affected facility proposing to demonstrate an emerging technology may apply to the Administrator for a commercial demonstration permit. The Administrator will issue a commercial demonstration permit in accordance with paragraph (e) of this section. Commercial demonstration permits may be issued only by the Administrator, and this authority will not be delegated.

(b) An owner or operator of an affected facility that combusts solid solvent refined coal (SRC—I) and who is issued a commercial demonstration permit by the Administrator is not subject to the SO_2 emission reduction requirements under §60.43a(c) but must, as a minimum, reduce SO_2 emissions to 20% of the potential combustion concentration (80% reduction) for each 24-hour period of steam generator operation and to less than 520 ng/J (1.20 lb/million Btu) heat input on a 30-day rolling average basis.

(c) An owner or operator of a fluidized bed combustion electric utility steam generator (atmospheric or pressurized) who is issued a commercial demonstration permit by the Administrator is not subject to the SO_2 emission reduction requirements under §60.43a(a) but must, as a minimum, reduce SO_2 emissions to 15% of the potential combustion concentration (85% reduction) on a 30-

day rolling average basis and to less than 520 ng/J (1.20 lb/million Btu) heat input on a 30-day rolling average basis.

(d) The owner or operator of an affected facility that combusts coal-derived liquid fuel and who is issued a commercial demonstration permit by the Administrator is not subject to the applicable NO_x emission limitation and percent reduction under §60.44a(a) but must, as a minimum, reduce emissions to less than 300 ng/J (0.70 lb/million Btu) heat input on a 30-day rolling average basis.

(e) Commercial demonstration permits may not exceed the following equivalent MW electrical generation capacity for any one technology category, and the total equivalent MW electrical generation capacity for all commercial demonstration plants may not exceed 15,000 MW.

Technology	Pollutant	Equivalent electrical capacity (MW electrical output)
Solid solvent refined coal (SRC 1)....................	SO_2	6,000–10,000
Fluidized bed combustion (atmospheric)..............	SO_2	400–3,000
Fluidized bed combustion (pressurized)	SO_2	400–1,200
Coal liquefaction	NO_x	750–10,000
Total allowable for all technologies		15,000

§60.46a Compliance provisions.

(a) Compliance with the particulate matter emission limitation under §60.42a(a)(1) constitutes compliance with the percent reduction requirements for particulate matter under §60.42a(a)(2) and (3).

(b) Compliance with the nitrogen oxides emission limitation under §60.44a(a) constitutes compliance with the percent reduction requirements under §60.44a(a)(2).

(c) The particulate matter emission standards under §60.42a and the nitrogen oxides emission standards under §60.44a apply at all times except during periods of startup, shutdown, or malfunction. The sulfur dioxide emission standards under §60.43a apply at all times except during periods of startup, shutdown, or when both

Rules and Regulations of the EPA

emergency conditions exist and the procedures under paragraph (d) of this section are implemented.

(d) During emergency conditions in the principal company, an affected facility with a malfunctioning flue gas desulfurization system may be operated if sulfur dioxide emissions are minimized by:

(1) Operating all operable flue gas desulfurization system modules, and bringing back into operation any malfunctioned module as soon as repairs are completed,

(2) Bypassing flue gases around only those flue gas desulfurization system modules that have been taken out of operation because they were incapable of any sulfur dioxide emission reduction or which would have suffered significant physical damage if they had remained in operation, and

(3) Designing, constructing, and operating a spare flue gas desulfurization system module for an affected facility larger than 365 MW (1,250 million Btu/hr) heat input (approximately 125 MW electrical output capacity). The Administrator may at his discretion require the owner or operator within 60 days of notification to demonstrate spare module capability. To demonstrate this capability, the owner or operator must demonstrate compliance with the appropriate requirements under paragraph (a), (b), (d), (e), and (i) under §60.43a for any period of operation lasting from 24 hours to 30 days when:

(i) Any one flue gas desulfurization module is not operated,

(ii) The affected facility is operating at the maximum heat input rate,

(iii) The fuel fired during the 24-hour to 30-day period is representative of the type and average sulfur content of fuel used over a typical 30-day period, and

(iv) The owner or operator has given the Administrator at least 30 days notice of the date and period of time over which the demonstration will be performed.

(e) After the initial performance test required under §60.8, compliance with the sulfur dioxide emission limitations and percentage reduction requirements under §60.43a and the nitrogen oxides emission limitations under §60.44a is based on the average emission rate for 30 successive boiler operating days. A separate performance test is completed at the end of each boiler operating day after the initial performance test, and a new 30 day average emission rate for both sulfur dioxide and nitrogen oxides and a new percent reduction for sulfur dioxide are calculated to show compliance with the standards.

(f) For the initial performance test required under §60.8, compliance with the sulfur dioxide emission limitations and percent reduction requirements under §60.43a and the nitrogen oxides emission limitation under §60.44a is based on the average emission rates for sulfur-dioxide, nitrogen oxides, and percent reduction for sulfur dioxide for the first 30 successive boiler operating days. The initial performance test is the only test in which at least 30 days prior notice is required unless otherwise specified by the Administrator. The initial performance test is to be scheduled so that the first boiler operating day of the 30 successive boiler operating days is completed within 60 days after achieving the maximum production rate at which the affected facility will be operated, but not later than 180 days after initial startup of the facility.

(g) Compliance is determined by calculating the arithmetic average of all hourly emission rates for SO_2 and NO_x for the 30 successive boiler operating days, except for data obtained during startup, shutdown, malfunction (NO_x only), or emergency conditions (SO_2 only). Compliance with the percentage reduction requirement for SO_2 is determined based on the average inlet and average outlet SO_2 emission rates for the 30 successive boiler operating days.

(h) If an owner or operator has not obtained the minimum quantity of emission data as required under §60.47a of this subpart, compliance of the affected facility with the emission requirements under §§60.43a and 60.44a of this subpart for the day on which the 30-day period ends may be determined by the Administrator by following the applicable procedures in sections 6.0 and 7.0 of Reference Method 19 (Appendix A).

§60.47a Emission monitoring.

(a) The owner or operator of an affected facility shall install, calibrate, maintain, and operate a continuous monitoring system, and record the output of the system, for measuring the opacity of emissions discharged to the atmosphere, except where gaseous

Rules and Regulations of the EPA

fuel is the only fuel combusted. If opacity interference due to water droplets exists in the stack (for example, from the use of an FGD system), the opacity is monitored upstream of the interference (at the inlet to the FGD system). If opacity interference is experienced at all locations (both at the inlet and outlet of the sulfur dioxide control system), alternate parameters indicative of the particulate matter control system's performance are monitored (subject to the approval of the Administrator).

(b) The owner or operator of an affected facility shall install, calibrate, maintain, and operate a continuous monitoring system, and record the output of the system, for measuring sulfur dioxide emissions, except where natural gas is the only fuel combusted, as follows:

(1) Sulfur dioxide emissions are monitored at both the inlet and outlet of the sulfur dioxide control device.

(2) For a facility which qualifies under the provisions of § 60.43a(d), sulfur dioxide emissions are only monitored as discharged to the atmosphere.

(3) An "as-fired" fuel monitoring system (upstream of coal pulverizers) meeting the requirements of Method 19 (Appendix A) may be used to determine potential sulfur dioxide emissions in place of a continuous sulfur dioxide emission monitor at the inlet to the sulfur dioxide control device as required under paragraph (b)(1) of this section.

(c) The owner or operator of an affected facility shall install, calibrate, maintain, and operate a continuous monitoring system, and record the output of the system, for measuring nitrogen oxides emissions discharged to the atmosphere.

(d) The owner or operator of an affected facility shall install, calibrate, maintain, and operate a continuous monitoring system, and record the output of the system, for measuring the oxygen or carbon dioxide content of the flue gases at each location where sulfur dioxide or nitrogen oxides emissions are monitored.

(e) The continuous monitoring systems under paragraphs (b), (c), and (d) of this section are operated and data recorded during all periods of operation of the affected facility including periods of startup, shutdown, malfunction or emergency conditions, except for continuous monitoring system breakdowns, repairs, calibration checks, and zero and span adjustments.

(f) When emission data are not obtained because of continuous monitoring system breakdowns, repairs, calibration checks and zero and span adjustments, emission data will be obtained by using other monitoring systems as approved by the Administrator or the reference methods as described in paragraph (h) of this section to provide emission data for a minimum of 18 hours in at least 22 out of 30 successive boiler operating days.

(g) The 1-hour averages required under paragraph § 60.13(h) are expressed in ng/J (lbs/million Btu) heat input and used to calculate the average emission rates under § 60.46a. The 1-hour averages are calculated using the data points required under § 60.13(b). At least two data points must be used to calculate the 1-hour averages

(h) Reference methods used to supplement continuous monitoring system data to meet the minimum data requirements in paragraph § 60.47a(f) will be used as specified below or otherwise approved by the Administrator.

(1) Reference Methods 3, 6, and 7, as applicable, are used. The sampling location(s) are the same as those used for the continuous monitoring system.

(2) For Method 6, the minimum sampling time is 20 minutes and the minimum sampling volume is 0.02 dscm (0.71 dscf) for each sample. Samples are taken at approximately 60-minute intervals. Each sample represents a 1-hour average.

(3) For Method 7, samples are taken at approximately 30-minute intervals. The arithmetic average of these two consecutive samples represent a 1-hour average.

(4) For Method 3, the oxygen or carbon dioxide sample is to be taken for each hour when coutinuous SO_2 and NO_x data are taken or when Methods 6 and 7 are required. Each sample shall be taken for a minimum of 30 minutes in each hour using the integrated bag method specified in Method 3. Each sample represents a 1-hour average.

(5) For each 1-hour average, the emissions expressed in ng/J (lb/million Btu) heat input are determined and used as needed to achieve the minimum data requirements of paragraph (f) of this section.

(i) The following procedures are used to conduct monitoring

Rules and Regulations of the EPA

system performance evaluations under §60.13(c) and calibration checks under §60.13(d).

(1) Reference method 6 or 7, as applicable, is used for conducting performance evaluations of sulfur dioxide and nitrogen oxides continuous monitoring systems.

(2) Sulfur dioxide or nitrogen oxides, as applicable, is used for preparing calibration gas mixtures under performance specification 2 of Appendix B to this part.

(3) For affected facilities burning only fossil fuel, the span value for a continuous monitoring system for measuring opacity is between 60 and 80% and for a continuous monitoring system measuring nitrogen oxides is determined as follows:

Fossil fuel	Span value for nitrogen oxides (ppm)
Gas	500
Liquid	500
Solid	1,000
Combination ...	500 (x + y) + 1,000z

where:

x is the fraction of total heat input derived from gaseous fossil fuel,

y is the fraction of total heat input derived from liquid fossil fuel, and

z is the fraction of total heat input derived from solid fossil fuel.

(4) All span values computed under paragraph (b)(3) of this section for burning combinations of fossil fuels are rounded to the nearest 500 ppm.

(5) For affected facilities burning fossil fuel, alone or in combination with nonfossil fuel, the span value of the sulfur dioxide continuous monitoring system at the inlet to the sulfur dioxide control device is 125% of the maximum estimated hourly potential emissions of the fuel fired, and the outlet of the sulfur dioxide control device is 50% of maximum estimated hourly potential emissions of the fuel fired.

(Sec. 114, Clean Air Act as amended (42 U.S.C. 7414).)

§60.48a Compliance determination procedures and methods.

(a) The following procedures and reference methods are used to determine compliance with the standards for particulate matter under §60.42a.

(1) Method 3 is used for gas analysis when applying Method 5 or Method 17.

(2) Method 5 is used for determining particulate matter emissions and associated moisture content. Method 17 may be used for stack gas temperatures less than 160°C. (320°F.).

(3) For Methods 5 or 17, Method 1 is used to select the sampling site and the number of traverse sampling points. The sampling time for each run is at least 120 minutes and the minimum sampling volume is 1.7 dscm (60 dscf) except that smaller sampling times or volumes, when necessitated by process variables or other factors, may be approved by the Administrator.

(4) For Method 5, the probe and filter holder heating system in the sampling train is set to provide a gas temperature no greater than 160°C. (320°F.).

(5) For determination of particulate emissions, the oxygen or carbon-dioxide sample is obtained simultaneously with each run of Methods 5 or 17 by traversing the duct at the same sampling location. Method 1 is used for selection of the number of traverse points except that no more than 12 sample points are required.

(6) For each run using Methods 5 or 17, the emission rate expressed in ng/J heat input is determined using the oxygen or carbon-dioxide measurements and particulate matter measurements obtained under this section, the dry basis F_c-factor and the dry basis emission rate calculation procedure contained in Method 19 (Appendix A).

(7) Prior to the Administrator's issuance of a particulate matter reference method that does not experience sulfuric acid mist interference problems, particulate matter emissions may be sampled prior to a wet flue gas desulfurization system.

(b) The following procedures and methods are used to determine compliance with the sulfur dioxide standards under §60.43a.

(1) Determine the percent of potential combustion concentration (percent PCC) emitted to the atmosphere as follows:

(i) Fuel Pretreatment (% R_f):
Determine the percent reduction achieved by any fuel pretreatment using the procedures in Method 19 (Appendix A). Calculate the average percent reduction for fuel pretreatment on a quarterly basis using fuel analysis data. The determination of percent R_f to calculate the percent of potential combustion concentration emitted to the atmosphere is optional. For purposes of determining compliance with any percent reduction requirements under § 60.43a, any reduction in potential SO_2 emissions resulting from the following processes may be credited:

(A) Fuel pretreatment (physical coal cleaning, hydrodesulfurization of fuel oil, etc.),
(B) Coal pulverizers, and
(C) Bottom and fly ash interactions.

(ii) Sulfur Dioxide Control System (% R_g): Determine the percent sulfur dioxide reduction achieved by any sulfur dioxide control system using emission rates measured before and after the control system, following the procedures in Method 19 (Appendix A); or, a combination of an "as-fired" fuel monitor and emission rates measured after the control system, following the procedures in Method 19 (Appendix A). When the "as-fired" fuel monitor is used, the percent reduction is calculated using the average emission rate from the sulfur dioxide control device and the average SO_2 input rate from the "as-fired" fuel analysis for 30 successive boiler operating days.

(iii) Overall percent reduction (% R_o): Determine the overall percent reduction using the results obtained in paragraphs (b)(1) (i) and (ii) of this section following the procedures in Method 19 (Appendix A). Results are calculated for each 30-day period using the quarterly average percent sulfur reduction determined for fuel pretreatment from the previous quarter and the sulfur dioxide reduction achieved by a sulfur dioxide control system for each 30-day period in the current quarter.

(iv) Percent emitted (% PCC): Calculate the percent of potential combustion concentration emitted to the atmosphere using the following equation: Percent PCC = 100 Percent R_o

(2) Determine the sulfur dioxide emission rates following the procedures in Method 19 (Appendix A).

(c) The procedures and methods outlined in Method 19 (Appendix A) are used in conjunction with the 30-day nitrogen-oxides emission data collected under §60.47a to determine compliance with the applicable nitrogen oxides standard under §60.44.

(d) Electric utility combined cycle gas turbines are performance tested for particulate matter, sulfur dioxide, and nitrogen oxides using the procedures of Method 19 (Appendix A). The sulfur dioxide and nitrogen oxides emission rates from the gas turbine used in Method 19 (Appendix A) calculations are determined when the gas turbine is performance tested under Subpart GG. The potential uncontrolled particulate matter emission rate from a gas turbine is defined as 17 ng/J (0.04 lb/million Btu) heat input.

§60.49a Reporting requirements.

(a) For sulfur dioxide, nitrogen oxides, and particulate matter emissions, the performance test data from the initial performance test and from the performance evaluation of the continuous monitors (including the transmissometer) are submitted to the Administrator.

(b) For sulfur dioxide and nitrogen oxides the following information is reported to the Administrator for each 24-hour period.

(1) Calendar date.

(2) The average sulfur dioxide and nitrogen oxide emission rates (ng/J or lb/million Btu) for each 30 successive boiler operating days, ending with the last 30-day period in the quarter; reasons for non-compliance with the emission standards; and, description of corrective actions taken.

(3) Percent reduction of the potential combustion concentration of sulfur dioxide for each 30 successive boiler operating days, ending with the last 30-day period in the quarter; reasons for non-compliance with the standard; and, description of corrective actions taken.

(4) Identification of the boiler operating days for which pol-

Rules and Regulations of the EPA

lutant or dilutent data have not been obtained by an approved method for at least 18 hours of operation of the facility; justification for not obtaining sufficient data; and description of corrective actions taken.

(5) Identification of the times when emissions data have been excluded from the calculation of average emission rates because of startup, shutdown, malfunction (NO_x only), emergency conditions (SO_2 only), or other reasons, and justification for excluding data for reasons other than startup, shutdown, malfunction, or emergency conditions.

(6) Identification of "F" factor used for calculations, method of determination, and type of fuel combusted.

(7) Identification of times when hourly averages have been obtained based on manual sampling methods.

(8) Identification of the times when the pollutant concentration exceeded full span of the continuous monitoring system.

(9) Description of any modifications to the continuous monitoring system which could affect the ability of the continuous monitoring system to comply with Performance Specifications 2 or 3.

(c) If the minimum quantity of emission data as required by §60.47a is not obtained for any 30 successive boiler operating days, the following information obtained under the requirements of §60.46a(h) is reported to the Administrator for that 30-day period:

(1) The number of hourly averages available for outlet emission rates (n_o) and inlet emission rates (n_i) as applicable.

(2) The standard deviation of hourly averages for outlet emission rates (s_o) and inlet emission rates (s_i) as applicable.

(3) The lower confidence limit for the mean outlet emission rate (E_o^*) and the upper confidence limit for the mean inlet emission rate (E_i^*) as applicable.

(4) The applicable potential combustion concentration.

(5) The ratio of the upper confidence limit for the mean outlet emission rate (E_o^*) and the allowable emission rate (E_{std}) as applicable.

(d) If any standards under §60.43a are exceeded during emergency conditions because of control system malfunction, the owner or operator of the affected facility shall submit a signed statement:

(1) Indicating if emergency conditions existed and requirements under §60.46a(d) were met during each period, and

(2) Listing the following information:

(i) Time periods the emergency condition existed;

(ii) Electrical output and demand on the owner or operator's electric utility system and the affected facility;

(iii) Amount of power purchased from interconnected neighboring utility companies during the emergency period;

(iv) Percent reduction in emissions achieved;

(v) Atmospheric emission rate (ng/J) of the pollutant discharged; and

(vi) Actions taken to correct control system malfunction.

(e) If fuel pretreatment credit toward the sulfur dioxide emission standard under §60.43a is claimed, the owner or operator of the affected facility shall submit a signed statement:

(1) Indicating what percentage cleaning credit was taken for the calendar quarter, and whether the credit was determined in accordance with the provisions of §60.48a and Method 19 (Appendix A); and

(2) Listing the quantity, heat content, and date each pretreated fuel shipment was received during the previous quarter; the name and location of the fuel pretreatment facility; and the total quantity and total heat content of all fuels received at the affected facility during the previous quarter.

(f) For any periods for which opacity, sulfur dioxide or nitrogen oxides emissions data are not available, the owner or operator of the affected facility shall submit a signed statement indicating if any changes were made in operation of the emission control system during the period of data unavailability. Operations of the control system and affected facility during periods of data unavailability are to be compared with operation of the control system and affected facility before and following the period of data unavailability.

(g) The owner or operator of the affected facility shall submit a signed statement indicating whether:

(1) The required continuous monitoring system calibration, span, and drift checks or other periodic audits have or have not been performed as specified.

(2) The data used to show compliance was or was not obtained

Rules and Regulations of the EPA

in accordance with approved methods and procedures of this part and is representative of plant performance.

(3) The minimum data requirements have or have not been met; or, the minimum data requirements have not been met for errors that were unavoidable.

(4) Compliance with the standards has or has not been achieved during the reporting period.

(h) For the purposes of the reports required under §60.7, periods of excess emissions are defined as all 6-minute periods during which the average opacity exceeds the applicable opacity standards under §60.42a(b). Opacity levels in excess of the applicable opacity standard and the date of such excesses are to be submitted to the Administrator each calendar quarter.

(i) The owner or operator of an affected facility shall submit the written reports required under this section and Subpart A to the Administrator for every calendar quarter. All quarterly reports shall be postmarked by the 30th day following the end of each calendar quarter.

(Sec. 114, Clean Air Act as amended (42 U.S.C. 7414).)

4. Appendix A to part 60 is amended by adding new reference Method 19 as follows:

APPENDIX A—REFERENCE METHODS

* * * * *

Method 19. Determination of Sulfur Dioxide Removal Efficiency and Particulate, Sulfur Dioxide and Nitrogen Oxides Emission Rates From Electric Utility Steam Generators

1. Principle and Applicability

1.1 Principle.

1.1.1. Fuel samples from before and after fuel pretreatment systems are collected and analyzed for sulfur and heat content, and the percent sulfur dioxide (ng/Joule, lb/million Btu) reduction is calculated on a dry basis. (Optional Procedure.)

1.1.2 Sulfur dioxide and oxygen or carbon dioxide concentration data obtained from sampling emissions upstream and downstream of sulfur dioxide control devices are used to calculate sulfur dioxide removal efficiencies. (Minimum Requirement.) As an alternative to sulfur dioxide monitoring upstream of sulfur dioxide control devices, fuel samples may be collected in an as-fired condition and analyzed for sulfur and heat content. (Optional Procedure.)

1.1.3 An overall sulfur dioxide emission reduction efficiency is calculated from the effeciency of fuel pretreatment systems and the efficiency of sulfur dioxide control devices.

1.1.4 Particulate, sulfur dioxide, nitrogen oxides, and oxygen or carbon dioxide concentraion data obtained from sampling emissions downstream from sulfur dioxide control devices are used along with F factors to calculate particulate, sulfur dioxide, and nitrogen oxides emission rates. F factors are values relating combustion gas volume to the heat content of fuels.

1.2 Applicability.

This method is applicable for determining sulfur removal efficiencies of fuel pretreatment and sulfur dioxide control devices and the overall reduction of potential sulfur dioxide emissions from electric utility steam generators. This method is also applicable for the determination of particulate, sulfur dioxide, and nitrogen oxides emission rates.

2. Determination of Sulfur Dioxide Removal Efficiency of Fuel Pretreatment Systems

2.1 Solid Fossil Fuel.

2.1.1 *Sample Increment Collection.* Use ASTM D 2234[1], Type I, conditions A, B, or C, and systematic spacing. Determine the number and weight of increments required per gross sample

---

[1] Use the most recent revision or designation of the ASTM procedure specified.

representing each coal lot according to Table 2 or Paragraph 7.1.5.2 of ASTM D 2234[1]. Collect one gross sample for each raw coal lot and one gross sample for each product coal lot.

2.1.2 *ASTM Lot Size.* For the purpose of Section 2.1.1, the product coal lot size is defined as the weight of product coal produced from one type of raw coal. The raw coal lot size is the weight of raw coal used to produce one product coal lot. Typically, the lot size is the weight of coal processed in a 1-day (24 hours) period. If more than one type of coal is treated and produced in 1 day, then gross samples must be collected and analyzed for each type of coal. A coal lot size equaling the 90-day quarterly fuel quantity for a specific power plant may be used if representative sampling can be conducted for the raw coal and product coal.

Note.—Alternate definitions of fuel lot sizes may be specified subject to prior approval of the Administrator.

2.1.3 *Gross Sample Analysis.* Determine the percent sulfer content (%S) and gross calorific value (GCV) of the solid fuel on a dry basis for each gross sample. Use ASTM 2013[1] for sample preparation. ASTM D 3177[1] for sulfer analysis, and ASTM D 3173[1] for moisture analysis. Use ASTM D 3176[1] for gross calorific value determination.

2.2 Liquid Fossil Fuel

2.2.1 *Sample Collection.* Use ASTM D 270[1] following the practices outlined for continuous sampling for each gross sample representing each fuel lot.

2.2.2 *Lot Size.* For the purposes of Section 2.2.1, the weight of product fuel from one pretreatment facility and intended as one shipment (ship load, barge load, etc.) is defined as one product fuel lot. The weight of each crude liquid fuel type used to produce one product fuel lot is defined as one inlet fuel lot.

Note.—Alternate definitions of fuel lot sizes may be specified subject to prior approval of the Administrator.

---

[1]Use the most recent revision or designation of the ASTM procedure specified.

Note.—For the purposes of this method, raw or inlet fuel (coal or oil) is defined as the fuel delivered to the desulfurization pretreatment facility or to the steam generating plant. For pretreated oil the input oil to the oil desulfurization process (e.g. hydrotreatment emitted) is sampled.

2.2.3 *Sample Analysis.* Determine the percent sulfur content (%S) and gross calorific value (GCV). Use ASTMD 240[1] for the sample analysis. This value can be assumed to be on a dry basis.

2.3 Calculation of Sulfur Dioxide Removal Efficiency Due to Fuel Pretreatment.

Calculate the percent sulfur dioxide reaction due to fuel pretreatment using the following equation:

$$\%R_f = 100 \left[1 - \frac{\%S_o/GCV_o}{\%S_i/GCV_i}\right]$$

Where:

$\%R_f$ = Sulfur dioxide removal efficiency due pretreatment; percent.
$\%S_o$ = Sulfur content of the product fuel lot on a dry basis; weight percent.
$\%S_1$ = Sulfur content of the inlet fuel lot on a dry basis; weight percent.
GCV_o = Gross calorific value for the outlet fuel lot on a dry basis; kJ/kg (Btu/lb).
GCV_i = Gross calorific value for the inlet fuel lot on a dry basis; kJ/kg (Btu/lb).

Note.—If more than one fuel type is used to produce the product fuel, use the following equation to calculate the sulfur contents per unit of heat content of the total fuel lot, %S/GCV:

$$\%S/GCV = \sum_{k=1}^{n} Y_k(\%S_k/GCV_k)$$

---

[1] Use the most recent revision or designation of the ASTM procedure specified.

Where:

Y_k = The fraction of total mass input derived from each type, k, of fuel.

$\%S_k$ = Sulfur content of each fuel type, k, on a dry basis; weight percent.

GCV_k = Gross calorific value for each fuel type, k, on a dry basis; kJ/kg (Btu/lb).

n = The number of different types of fuels.

3. Determination of Sulfur Removal Efficiency of the Sulfur Dioxide Control Device

3.1 Sampling.

Determine SO_2 emission rates at the inlet and outlet of the sulfur dioxide control system according to methods specified in the applicable subpart of the regulations and the procedures specified in Section 5. The inlet sulfur dioxide emission rate may be determined through fuel analysis (Optional, see Section 3.3.)

3.2 Calculation.

Calculate the percent removal efficiency using the following equation:

$$\%R_{g(m)} = 100 \times \left(1.0 - \frac{E_{SO_2o}}{E_{SO_2i}}\right)$$

Where:

$\%R_g$ = Sulfur dioxide removal effeciency of the sulfur dioxide control system using inlet and outlet monitoring data; percent.

$E_{SO\,o}$ = Sulfur dioxide emission rate from the outlet of the sulfur dioxide control system; ng/J (lb/million Btu).

$E_{SO\,i}$ = Sulfur dioxide emission rate to the outlet of the sulfur dioxide control system; ng/J (lb/million Btu).

3.3 As-Fired Fuel Analysis (Optional Procedure).

If the owner or operator of an electric utility steam generator chooses to determine the sulfur dioxide input rate at the inlet to the sulfur dioxide control device through an as-fired fuel analysis in lieu of data from a sulfur dioxide control system inlet gas monitor, fuel samples must be collected in accordance with applicable paragraph in Section 2. The sampling can be conducted upstream of any fuel processing, e.g., plant coal pulverization. For the purposes of this section, a fuel lot size is defined as the weight of fuel consumed in 1 day (24 hours) and is directly related to the exhaust gas monitoring data at the outlet of the sulfur dioxide control system.

3.3.1 *Fuel Analysis.* Fuel samples must be analyzed for sulfur content and gross calorific value. The ASTM procedures for determining sulfur content are defined in the applicable paragraphs of Section 2.

3.3.2 *Calculation of Sulfur Dioxide Input Rate.* The sulfur dioxide input rate determined from fuel analysis is calculated by:

$$I_s = \frac{2.0\,(\%S_f)}{GCV} \times 10^7 \text{ for SI units.}$$

$$I_s = \frac{2.0\,(\%S_f)}{GCV} \times 10^4 \text{ for English units.}$$

Where:

I = Sulfur dioxide input rate from as-fired fuel analysis, ng/J (lb/millions Btu).
$\%S_f$ = Sulfur content of as-fired fuel, on a dry basis; weight percent.
GCV = Gross calorific value for as-fired fuel, on a dry basis; kJ/kg (Btu/lb.).

3.3.3 *Calculation of Sulfur Dioxide Emission Reduction Using As-Fired Fuel Analysis.* The sulfur dioxide emission reduction efficiency is calculated using the sulfur input rate from paragraph

Rules and Regulations of the EPA

3.3.2 and the sulfur dioxide emission rate, E_{SO_2}, determined in the applicable paragraph of Section 5.3. The equation for sulfur dioxide emission reduction efficiency is:

$$\%R_{g(f)} = 100 \times \left(1.0 - \frac{E_{SO_2}}{I_s}\right)$$

Where:

$\%R_{g(f)}$ = Sulfur dioxide removal efficiency of the sulfur dioxide control system using as-fired fuel analysis data; percent.
E_{SO_2} = Sulfur dioxide emission rate from sulfur dioxide control systems; ng/J (lb/million Btu).
I_s = Sulfur dioxide input rate from as-fired fuel analysis; ng/J (lb/million Btu).

4. Calculation of Overall Reduction in Potential Sulfur Dioxide Emission

4.1 The overall percent sulfur dioxide reduction calculation uses the sulfur dioxide concentration at the inlet to the sulfur dioxide control device as the base value. Any sulfur reduction realized through fuel cleaning is introduced into the equation as an average percent reduction, $\%R_1$.

4.2 Calculate the overall percent sulfur reduction as:

$$\%R_o = 100\left[1.0 - \left(1.0 - \frac{\%R_f}{100}\right)\left(1.0 - \frac{\%R_g}{100}\right)\right]$$

Where:

$\%R_o$ = Overall sulfur dioxide reduction; percent.
$\%R_f$ = Sulfur dioxide removal efficiency of fuel pretreatment from Section 2; percent. Refer to applicable subpart for definition of applicable averaging period. Sulfur dioxide removal efficiency of sulfur dioxide control device either O_2 or CO_2 based calculation or calculated
$\%R_g$ = from fuel analysis and emission data, from Section 3; percent. Refer to applicable subpart for definition of applicable averaging period.

5. Calculation of Particulate, Sulfur Dioxide, and Nitrogen Oxides Emission Rates

5.1 Sampling.

Use the outlet SO_2 or O_2 or CO_2 concentrations data obtained in Section 3.1. Determine the particulate, NO_x, and O_2 or CO_2 concentrations according to methods specified in an applicable subpart of the regulations.

5.2 Determination of an F Factor.

Select an average F factor (Section 5.2.1) or calculate an applicable F factor (Section 5.2.2.). If combined fuels are fired, the selected or calculated F factors are prorated using the procedures in Section 5.2.3. F factors are ratios of the gas volume released during combustion of a fuel divided by the heat content of the fuel. A dry F factor (F_d) is the ratio of the volume of dry flue gases generated to the calorific value of the fuel combusted; a wet F factor (F_w) is the ratio of the volume of wet flue gases generated to the calorific value of the fuel combusted; and the carbon F factor (F_c) is the ratio of the volume of carbon dioxide generated to the calorific value of the fuel combusted. When pollutant and oxygen concentrations have been determined in Section 5.1, wet or dry F factors are used. (F_w) factors and associated emission calculation procedures are not applicable and may not be used after wet scrubbers; (F_c) or (F_d) factors and associated emission calculation precedures are used after wet scrubbers. When pollutant and carbon dioxide concentrations have been determined in Section 5.1, F_c factors are used.

5.2.1 *Average F Factors.* Table 1 shows average F_d, F_w, and F_c factors (scm/J, scf/million Btu) determined for commonly used fuels. For fuels not listed in Table 1, the F factors are calculated according to the procedures outlined in Section 5.2.2 of this section.

5.2.2 *Calculating an F Factor.* If the fuel burned is not listed in Table 1 or if the owner or operator chooses to determine an F factor rather than use the tabulated data, F factors are calculated

Rules and Regulations of the EPA

using the equations below. The sampling and analysis procedures followed in obtaining data for these calculations are subject to the approval of the Administrator and the Administrator should be consulted prior to data collection.

For SI Units:

$$F_d = \frac{227.0(\%H) + 95.7(\%C) + 35.4(\%S) + 8.6(\%N) - 28.5(\%O)}{GCV}$$

$$F_w = \frac{347.4(\%H) + 95.7(\%C) + 35.4(\%S) + 8.6(\%N) - 28.5(\%O) + 13.0(\%H_2O)^{**}}{GCV_w}$$

$$F_c = \frac{20.0(\%C)}{GVC}$$

For English Units:

$$F_d = \frac{10^6[5.57(\%H) + 1.53(\%C) + 0.57(\%S) + 0.14(\%N) - 0.46(\%O)]}{GCV}$$

$$F_w = \frac{10^6[5.57(\%H) + 1.53(\%C) + 0.57(\%S) + 0.14(\%N) - 0.46(\%O) + 0.21(\%H_2O)^{**}]}{GCV_w}$$

$$F_c = \frac{10^6[0.321(\%C)]}{GCV}$$

Where:

F_d, F_w, and F_c have the units of scm/J, or scf/million Btu; %H, %C, %S, %N, %O, and %H$_2$O are the concentrations by weight (expressed in percent) of hydrogen, carbon, sulfur, nitrogen, oxygen, and water from an ultimate analysis of the fuel; and GCV is the gross calorific value of the fuel in kJ/kg or Btu/lb and consistent with the ultimate analysis. Follow ASTM D 2015* for solid fuels, D 240* for liquid fuels, and D 1826* for gaseous fuels as applicable in determining GCV.

**The %H$_2$O term may be omitted if %H and %O include the unavailable hydrogen and oxygen in the form of H$_2$O.

5.2.3 *Combined Fuel Firing F Factor.* For affected facilities firing combinations of fossil fuels or fossil fuels and wood residue, the F_d, F_w, or F_c factors determined by Sections 5.2.1 or 5.2.2 of this section shall be prorated in accordance with applicable formula as follows:

$$F_d = \sum_{k=1}^{n} x_k F_{dk} \text{ or}$$

$$F_w = \sum_{k=1}^{n} x_k F_{wk} \text{ or}$$

$$F_c = \sum_{k=1}^{n} x_k F_{ck}$$

Where:

x_k = The fraction of total heat input derived from each type of fuel, K.

n = The number of fuels being burned in combination.

5.3 Calculation of Emission Rate

Select from the following paragraphs the applicable calculation procedure and calculate the particulate, SO_2, and NO_x emission rate. The values in the equation are defined as:

E = Pollutant emission rate, ng/J (lb/million Btu).
C = Pollutant concentration, ng/scm (lb/scf).

Note.—It is necessary in some cases to convert measured concentration units to other units for these calculations.

Use the following table for such conversions:

Conversion Factors for Concentration

From—	To—	Multiply by—
g/scm	ng/scm	10^9
mg/scm	ng/scm	10^6
lb/scf	ng/scm	1.602×10^{13}
ppm(SO_2)	ng/scm	2.660×10^6
ppm(NO_x)	ng/scm	1.912×10^6
ppm(SO_2)	lb/scf	1.660×10^{-7}
ppm(NO_x)	lb/scf	1.194×10^{-7}

5.3.1 *Oxygen-Based F Factor Procedure.*

5.3.1.1 *Dry Basis.* When both percent oxygen (%O_{2d}) and the pollutant concentration (C_d) are measured in the flue gas on a dry basis, the following equation is applicable:

$$E = C_d F_d \left[\frac{20.9}{20.9 - \%O_{2d}} \right]$$

5.3.1.2 *Wet Basis.* When both the percent oxygen (%O_{2w}) and the pollutant concentration (C_w) are measured in the flue gas on a wet basis, the following equations are applicable: (Note: F_w factors are not applicable after wet scrubbers.)

$$\text{(a)} \quad E = C_w F_w \left[\frac{20.9}{20.9 (1 - B_{wa}) - \%O_{2w}} \right]$$

Where:

B_{wa} = Proportion by volume of water vapor in the ambient air.

In lieu of actual measurement, B_{wa} may be estimated as follows:

Note.—The following estimating factors are selected to assure that any negative error introduced in the term:

$$\left[\frac{20.9}{20.9 (1 - B_{wa}) - \%O_{2ws}} \right]$$

will not be larger than −1.5 percent. However, positive errors, or overestimation of emissions, of as much as 5% may be introduced depending upon the geographic location of the facility and the associated range of ambient moisture.

(i) B_{wa} = 0.027. This factor may be used as a constant value at any location.

(ii) B_{wa} = Highest monthly average of B_{wa} which occurred within a calendar year at the nearest Weather Service Station.

(iii) B_{wa} = Highest daily average of B_{wa} which occurred within a calendar month at the nearest Weather Service Station, calculated from the data for the past 3 years. This factor shall be calculated for each month and may be used as an estimating factor for the respective calendar month.

$$\text{(b)} \quad E = C_w F_d \left[\frac{20.9}{20.9(1 - B_{ws}) - \%O_{2w}} \right]$$

Where:

B_{wa} = Proportion by volume of water vapor in the stack gas.

5.3.1.3 *Dry/Wet Basis.* When the pollutant concentration (C_w) is measured on a wet basis and the oxygen concentration ($\%O_{2d}$) is measured on a dry basis, the following equation is applicable:

$$E = \left[\frac{C_w F_d}{(1 - B_{ws})} \right] \left[\frac{20.9}{20.9 - \%O_{2d}} \right]$$

When the pollutant concentration (C_d) is measured on a dry basis and the oxygen concentration ($\%O_{2d}$) is measured on a wet basis, the following equation is applicable:

$$E = C_d F_d \frac{20.9}{20.9 - \dfrac{\%O_{2w}}{(1 - B_{ws})}}$$

5.3.2 *Carbon Dioxide-Based F Factor Procedure.*

5.3.2.1 *Dry Basis.* When both the percent carbon dioxide ($\%CO_{2d}$) and the pollutant concentration (C_d) are measured in the flue gas on a dry basis, the following equation is applicable:

$$E = C_d F_c \left(\frac{100}{\%CO_{2d}} \right)$$

5.3.2.2 *Wet Basis.* When both the percent carbon dioxide ($\%CO_{2w}$) and the pollutant concentration (C_w) are measured on a wet basis, the following equation is applicable:

$$E = C_w F_c \left(\frac{100}{\%CO_{2w}} \right)$$

5.3.2.3 *Dry/Wet Basis.* When the pollutant concentration (C_w)

is measured on a wet basis and the percent carbon dioxide ($\%CO_{2d}$) is measured on a dry basis, the following equation is applicable:

$$E = \left[\frac{C_w F_c}{(1 - B_{ws})}\right] \left[\frac{100}{\%CO_{2d}}\right]$$

When the pollutant concentration (C_d) is measured on a dry basis and the percent carbon dioxide ($\%CO_{2w}$) is measured on a wet basis, the following equation is applicable:

$$E = C_d (1 - B_{ws}) F_c \left(\frac{100}{\%CO_{2w}}\right)$$

5.4 Calculation of Emission Rate from Combined Cycle-Gas Turbine Systems

For gas turbine-steam generator combined cycle systems, the emissions from supplemental fuel fired to the steam generator or the percentage reduction in potential (SO_2) emissions cannot be determined directly. Using measurements from the gas turbine exhaust (performance test, subpart GG) and the combined exhaust gases from the steam generator, calculate the emission rates for these two points following the appropriate paragraphs in Section 5.3.

Note.—F_w factors shall not be used to determine emission rates from gas turbines because of the injection of steam nor to calculate emission rates after wet scrubbers; F_d or F_c factor and associated calculation procedures are used to combine effluent emissions according to the procedure in Paragraph 5.2.3.

The emission rate from the steam generator is calculated as:

$$E_{sg} = \frac{E_c - X_{gt} E_{gt}}{X_{sg}}$$

Where:

E_{sg} = Pollutant emission rate from steam generator effluent, ng/J (lb/million Btu).

E_c = Pollutant emission rate in combined cycle effluent; ng/J (lb/million Btu).

POWER PLANT ENGINEERS GUIDE

E_{gt} = Pollutant emission rate from gas turbine effluent; ng/J (lb/million Btu).

X_{sg} = Fraction of total heat input from supplemental fuel fired to the steam generator.

X_{gt} = Fraction of total heat input from gas turbine exhaust gases.

Note.—The total heat input to the steam generator is the sum of the heat input from supplemental fuel fired to the steam generator and the heat input to the steam generator from the exhaust gases from the gas turbine.

5.5 Effect of Wet Scrubber Exhaust, Direct-Fired Reheat Fuel Burning

Some wet scrubber systems require that the temperature of the exhaust gas be raised above the moisture dew-point prior to the gas entering the stack. One method used to accomplish this is direct firing of an auxiliary burner into the exhaust gas. The heat required for such burners is from 1 to 2% of total heat input of the steam generating plant. The effect of this fuel burning on the exhaust gas components will be less than ±1.0% and will have a similar effect on emission rate calculations. Because of this small effect, a determination of effluent gas constituents from direct-fired reheat burners for correction of stack gas concentrations is not necessary.

F Factors for Various fuels[c]

Fuel type	F_d		F_w		F_c	
	dscm / J	dscf / 10^6 Btu	wscm / J	wscf / 10^6	scm / J	scf / 10^6
Coal;						
Anthracite[a]	2.71×10^{-7}	(10540)	2.83×10^{-7}	(10540)	0.530×10^{-7}	(1970)
Bituminous[a]	2.63×10^{-7}	(9780)	2.86×10^{-7}	(10640)	0.484×10^{-7}	(1800)
Lignite	2.65×10^{-7}	(9860)	3.21×10^{-7}	(11950)	0.513×10^{-7}	(1910)
Oil[b]	2.47×10^{-7}	(9190)	2.77×10^{-7}	(10320)	0.383×10^{-7}	(1420)
Gas;						
Natural	2.43×10^{-7}	(8710)	2.85×10^{-7}	(10610)	0.287×10^{-7}	(1040)
Propane	2.34×10^{-7}	(8710)	2.74×10^{-7}	(10200)	0.321×10^{-7}	(1190)
Butane	2.34×10^{-7}	(8710)	2.74×10^{-7}	(10390)	0.337×10^{-7}	(1250)
Wood	2.46×10^{-7}	(9240)	....	......	0.492×10^{-7}	(1830)
Wood Bark	2.58×10^{-7}	(9600)	....	......	0.497×10^{-7}	(1850)

[a] As classified according to ASTM D 388-66.
[b] Crude, residual, or distillate.
[c] Determined at standard conditions 20° C(68° F) and 760 mm Hg (29.92 in. Hg.)

6. Calculation of Confidence Limits for Inlet and Outlet Monitoring Data

6.1 Mean Emission Rates

Calculate the mean emission rates using hourly averages in ng/J (lb/million Btu) for SO_2 and NO_x outlet data and, if applicable, SO_2 inlet data using the following equations:

$$E_o = \frac{\Sigma x_o}{n_o}$$

$$E_i = \frac{\Sigma x_i}{n_i}$$

Where:

E_o = Mean outlet emission rate; ng/J (lb/million Btu).

E_i = Mean inlet emission rate; ng/J (lb/million Btu).

x_o = Hourly average outlet emission rate; ng/J (lb/million Btu).

x_i = Hourly average inlet emission rate; ng/J (lb/million Btu).

n_o = Number of outlet hourly averages available for the reporting period.

n_i = Number of inlet hourly averages available for reporting period.

6.2 Standard Deviation of Hourly Emission Rates

Calculate the standard deviation of the available outlet hourly average emission rates for SO_2 and NO_x and, if applicable, the available inlet hourly average emission rates for SO_2 using the following equations:

$$S_o = \left(\sqrt{\frac{1}{n_o} - \frac{1}{720}}\right) \left(\sqrt{\frac{\Sigma (E_o - x_o)^2}{n_o - 1}}\right)$$

$$S_i = \left(\sqrt{\frac{1}{n_i} - \frac{1}{720}}\right)\left(\sqrt{\frac{\Sigma (E_i - x_i)^2}{n_i - 1}}\right)$$

Where:

S_o = Standard deviation of the average outlet hourly average emission rates for the reporting period; ng/J (lb/million Btu).

S_i = Standard deviation of the average inlet hourly average emission rates for the reporting period; ng/J (lb/million Btu).

6.3 Confidence Limits

Calculate the lower confidence limit for the mean outlet emission rates for SO_2 and NO_x and, if applicable, the upper confidence limit for the mean inlet emission rate for SO_2 using the following equations:

$$E_o^* = E_o - t_{0.95} S_o$$
$$E_i^* = E_i + t_{0.95} S_i$$

Where:

E_o^* = The lower confidence limit for the mean outlet emission rates; ng/J (lb/million Btu).

E_i^* = The upper confidence limit for the mean inlet emission rate; ng/J (lb/million Btu).

$t_{0.95}$ = Values shown below for the indicated number of available data points (n):

Values for $t_{0.95}$

n	$t_{0.95}$
2	6.31
3	2.42
4	2.35
5	2.13
6	2.02

Values for $t_{0.95}$

n	$t_{0.95}$
7	1.94
8	1.89
9	1.86
10	1.83
11	1.81
12–16	1.77
17–21	1.73
22–26	1.71
27–31	1.70
32–51	1.68
52–91	1.67
92–151	1.66
152 or more	1.65

The values of this table are corrected for n-1 degrees of freedom. Use n equal to the number of hourly average data points.

7. Calculation to Demonstrate Compliance When Available Monitoring Data Are Less Than the Required Minimum

7.1 Determine Potential Combustion Concentration (PCC) for SO_2.

7.1.1 When the removal efficiency due to fuel pretreatment (% R_f) is included in the overall reduction in potential sulfur dioxide emissions (% R_o) and the "as-fired" fuel analysis is not used, the potential combustion concentration (PCC) is determined as follows:

$$PCC = E_i^\circ + 2\left(\frac{\%S_i}{GCV_i} - \frac{\%S_o}{GCV_o}\right) 10^7; \text{ng/J}$$

$$PCC = E_i^\circ + 2\left(\frac{\%S_i}{GCV_i} - \frac{\%S_o}{GCV_o}\right) 10^4; \text{lb/million Btu.}$$

Where:

$$\left(\frac{\%S_i}{GCV_i} - \frac{\%S_o}{GCV_o}\right)$$ = Potential emissions removed by the pretreatment process, using the full parameters defined in section 2.3; ng/J (lb/million Btu).

7.1.2 When the "as-fired" fuel analysis is used and the removal efficiency due to fuel pretreatment (% R_f) is not included in the overall reduction in potential sulfur dioxide emissions (% R_o), the potential combustion concentration (PCC) is determined as follows:

$$PCC = I_s$$

Where:

I_s = The sulfur dioxide input rate as defined in section 3.3.

7.1.3 When the "as-fired" fuel analysis is used and the removal efficiency due to fuel pretreatment (% R_f) is included in the overall reduction (% R_o), the potential combustion concentration (PCC) is determined as follows:

$$PCC = I_s + 2\left(\frac{\%S_i}{GCV_i} - \frac{\%S_o}{GCV_o}\right) 10^7; ng/J$$

$$PCC = I_1 + 2\left(\frac{\%S_i}{GCV_i} - \frac{\%S_o}{GCV_o}\right) 10^4; lb/million\ Btu.$$

7.1.4 When inlet monitoring data are used and the removal efficiency due to fuel pretreatment (% R_f) is not included in the overall reduction in potential sulfur dioxide emissions (% R_o), the potential combustion concentration (PCC) is determined as follows:

$$PCC = E_i^*$$

Where:

E_i^* = The upper confidence limit of the mean inlet emission rate, as determined in section 6.3.

Rules and Regulations of the EPA

7.2 Determine Allowable Emission Rates (E_{std}).

7.2.1 NO_x. Use the allowable emission rates for NO_x as directly defined by the applicable standard in terms of ng/J (lb/million Btu).

7.2.2 SO_2. Use the potential combustion concentration (PCC) for SO_2 as determined in section 7.1, to determine the applicable emission standard. If the applicable standard is an allowable emission rate in ng/J (lb/million Btu), the allowable emission rate is used as E_{std}. If the applicable standard is an allowable percent emission, calculate the allowable emission rate (E_{std}) using the following equation:

$$E_{std} = \% \, PCC/100$$

Where:

%PCC = Allowable percent emission as defined by the applicable standard; percent.

7.3 Calculate E_o^*/E_{std}.

To determine compliance for the reporting period calculate the ratio:

$$E_o^*/E_{std}$$

Where:

E_o^* = The lower confidence limit for the mean outlet emission rates, as defined in section 6.3; ng/J (lb/million Btu).

E_{std} = Allowable emission rate as defined in section 7.2; ng/J (lb/million Btu).

If E_o^*/E_{std} is equal to or less than 1.0, the facility is in compliance; if E_o^*/E_{std} is greater than 1.0, the facility is not in compliance for the reporting period.

For details relating to a specific power plant, contact the nearest regional office of the EPA, listed below.

EPA REGIONAL OFFICES

REGION I
John F. Kennedy Federal Bldg.
Room 2203
Boston, Massachusetts 02203
FTS-223-7210
CML-617-223-7210

REGION II
26 Federal Plaza
Room 1009
New York, New York 10007
FTS-264-2525
CML-212-264-2525

REGION III
Curtis Building
6th & Walnut Streets
Philadelphia, Pennsylvania 19106
FTS-597-9814
CML-215-597-9814

REGION IV
345 Courtland Street, N.E.
Atlanta, Georgia 30308
FTS-257-4727
CML-404-881-4727

REGION V
230 South Dearborn Street
Chicago, Illinois 60604
FTS-353-2000
CML-312-353-2000

REGION VI
First International Building
1201 Elm Street
Dallas, Texas 75270
FTS-729-2600
CML-214-767-2600

REGION VII
324 East 11th Street
Kansas City, Missouri 64106
FTS-758-5493
CML-816374-5493

REGION VIII
1860 Lincoln Street
Denver, Colorado 80203
FTS-327-3895
CML-303-837-3895

REGION IX
215 Fremont Street
San Francisco, California 94105
FTS-556-2320
CML-415-556-2320

REGION X
1200 6th Avenue
Seattle, Washington 98101
FTS-399-1220
CML-206-442-1220

REGIONS
Alabama	IV
Alaska	X
Arizona	IX
Arkansas	VI
California	IX

Colorado	VIII
Connecticut	I
Delaware	III
D.C.	III
Florida	IV
Georgia	IV
Hawaii	IX
Idaho	X
Illinois	V
Indiana	V
Iowa	VII
Kansas	VII
Kentucky	IV
Louisiana	VI
Maine	I
Maryland	III
Massachusettes	I
Michigan	V
Minnesota	V
Mississippi	IV
Missouri	VII
Montana	VIII
Nebraska	VII
Nevada	IX
New Hampshire	I
New Jersey	II
New Mexico	VI
New York	II
North Carolina	IV
North Dakota	VIII
Ohio	V
Oklahoma	VI
Oregon	X
Pennsylvania	III
Rhode Island	I
South Carolina	IV
South Dakota	VII
Tennessee	IV
Texas	VI

POWER PLANT ENGINEERS GUIDE

Utah VII
Vermont I
Virginia III
Washington X
West Virginia III
Wisconsin V
Wyoming VII
American Samoa IX
Guam IX
Puerto Rico II
Virgin Islands II

APPENDIX B

Criteria for Reactor Operator Training and Licensing as Established by the Nuclear Regulatory Commission (NRC)

The following criteria are included for general reference only. Contact the chief engineer of the plant of interest or the Nuclear Regulatory Commission, Washington, D.C. 20555 for specific information.

TRAINING AND LICENSING

A. Eligibility Requirements to be Administered an Examination

1. Experience.

a. Applicants for senior operator licenses shall have 4 years of responsible power plant experience. Responsible power plant experience should be that obtained as control room operator (fossil or nuclear) or as a power plant staff engineer involved in the day-to-day activities of the facility, commencing with the final year of construction. A maximum of 2 years power plant experience may be fulfilled by academic or related technical training, on a one-for-one time basis. Two years shall be nuclear power plant experience. At least 6 months of the nuclear power plant experience shall be at the plant for which he seeks a license. Precritical applicants will be required to meet unique qualifications designed to accommodate the fact that their facility has not yet been in operation.

  b. Applicants for senior operator licenses shall have held an operator's license for 1 year.

2. Training.

  a. Senior operator: Applicants shall have 3 months of shift training as an extra man on shift.

  b. Control room operator: Applicants shall have 3 months training on shift as an extra person in the control room.

In both cases above precritical applicants will be required to meet unique qualifications designed to accommodate the fact that their facility has not yet been in operation.

  c. Training programs shall be modified, as necessary, to provide:

  (1) Training in heat transfer, fluid flow, and thermodynamics.

  (2) Training in the use of installed plant systems to control or mitigate an accident in which the core is severely damaged.

  (3) Increased emphasis on reactor and plant transients.

  d. Training center and facility instructors who teach systems, integrated responses, transient, and simulator courses shall demonstrate their competence to NRC by successful completion of a senior operator examination.

  e. Instructors shall be enrolled in appropriate requalification programs to assure they are cognizant of current operating history, problems, and changes to procedures and administrative limitations.

3. Facility Certifications.

Certifications completed pursuant to Sections 55.10(a) (6) and

OPERATOR TRAINING AND LICENSING

55.33a(4) and (5) of 10 CFR Part 55 (see last part of this Appendix) shall be signed by the highest level of corporate management for plant operation (for example, Vice President for Operations).

B. NRC Examinations

1. Increased Scope of Examinations.

a. A new category shall be added to the operator written examination entitled, "Principles of Heat Transfer and Fluid Mechanics."

b. A new category shall be added to the senior operator written examination entitled, "Theory of Fluids and Thermodynamics."

c. Time limits shall be imposed for completion of the written examinations.
  1. Operator: 9 hours.
  2. Senior Operator: 7 hours.

d. The passing grade for the written examination shall be 80% overall and 70% in each category.

e. All applicants for senior operator licenses shall be required to be administered an operating test as well as the written examination.

f. Applicants will grant permission to NRC to inform their facility management regarding the results of the examinations for purposes of enrollment in requalification programs.

C. Requalification Programs

1. Content of the licensed operator requalification programs shall be modified to include instruction in heat transfer, fluid flow, thermodynamics and mitigation of accidents involving a degraded core.

2. The criteria for requiring a licensed individual to participate in accelerated requalification shall be modified to be consistent with the new passing grade for issuance of a license; 80% overall and 70% each category.

3. Programs should be modified to require the control manipulations listed under this heading in this Appendix. Normal control manipulations, such as plant or reactor startups, must be performed. Control manipulations during abnormal or emergency operations must be walked through with, and evaluated by, a

member of the training staff at a minimum. An appropriate simulator may be used to satisfy the requirements for control manipulations.

D. Long Range Criteria and/or Requirements

The following require additional staff work and/or rulemaking prior to their implementation.

1. Qualifications.

a. Shift supervisors shall have an engineering degree or equivalent qualifications.

b. Senior operators shall have successfully completed a course in appropriate engineering and scientific subject equal to 60 credit hours of college level subjects.

2. Training.

a. All applicants shall attend simulator training programs. Required control manipulations and exercises to be performed shall be the same for "cold" and "hot" applicants.

b. Eligibility requirements shall be developed for instructors, in addition to that listed in A.2 above.

3. NRC Examinations.

a. NRC shall administer the certification examinations that are presently administered at the conclusion of the off-site portion of the cold training programs.

b. All applicants shall be required to be administered a simulator examination in addition to the written examinations and plant oral tests.

c. NRC shall administer the requalification program annual examination.

4. Requalification Programs.

All licensees shall participate in simulator programs as part of the requalification programs. Control manipulations shall be performed pursuant to standard established under this heading (see later this Appendix).

TRAINING IN HEAT TRANSFER, FLUID FLOW AND THERMODYNAMICS

1. Basic Properties of Fluids and Matter.

This section should cover a basic introduction to matter and its properties. This section should include such concepts as temperature measurements and effects, density and its effects, specific weight, buoyancy, viscosity, and other properties of fluids. A working knowledge of steam tables should also be included. Energy movement should be discussed including such fundamentals as heat exchange, specific heat, latent heat of vaporization and sensible heat.

2. Fluid Statics.

This section should cover the pressure, temperature, and volume effects on fluids. Examples of these parametric changes should be illustrated by the instructor and related calculations should be performed by the students and discussed in the training sessions. Causes and effects of pressure and temperature changes in the various components and systems should be discussed as applicable to the facility with particular emphasis on safety significant features. The characteristics of force and pressure, pressure in liquids at rest, principles of hydraulics, saturation pressure and temperature, and subcooling should also be included.

3. Fluid Dynamics.

This section should cover the flow of fluids and such concepts as Bernoulli's principle, energy in moving fluids, flow measure theory and devices, and pressure losses due to friction and orificing. Other concepts and terms to be discussed in this section are NPSH, carry over, carry under, kinetic energy, head-loss relationships, and two-phase flow fundamentals. Practical applications relating to the reactor coolant system and steam generators should also be included.

4. Heat Transfer by Conduction, Convection and Radiation.

This section should cover the fundamentals of heat transfer by conductions. This section should include discussions on such concepts and terms as specific heat, heat flux, and atomic action. Heat transfer characteristics of fuel rods and heat exchangers should be included in this section.

This section should cover the fundamentals of heat transfer by convection. Natural and forced circulation should be discussed as applicable to the various systems at the facility. The convection current patterns created by expanding fluids in a confined area should be included in this section. Heat transport and fluid flow reductions or stoppage should be discussed due to steam and/or noncondensible gas formation during normal and accident conditions.

This section should cover the fundamentals of heat transfer by thermal radiation in the form of radiant energy. The electromagnetic energy emitted by a body as a result of its temperature should be discussed and illustrated by the use of equations and sample calculations. Comparisons should be made of a black body absorber and a white body emitter.

5. Change of Phase—Boiling.

This section should include descriptions of the state of matter, their inherent characteristics, and thermodynamic properties such as enthalpy and entropy. Calculations should be performed involving steam quality and void fraction properties. The types of boiling should be discussed as applicable to the facility during normal evolutions and accident conditions.

6. Burnout and Flow Instability.

This section should cover descriptions and mechanisms for calculating such terms as critical flux, critical power, DNB ratio, and hot channel factors. This section should also include instructions for preventing and monitoring for clad or fuel damage and flow instabilities. Sample calculations should be illustrated by the instructor and calculations should be performed by the students and discussed in the training sessions. Methods and procedures for using the plant computer to determine quantitative values of various factors during plant operation and plant heat balance determination should also be covered in this section.

7. Reactor Heat Transfer Limits.

This section should include a discussion of heat transfer limits by examining fuel rod and reactor design and limitations. The basis for the limits should be covered in this section along with recommended methods to ensure that limits are not approached or exceeded. This section should cover discussions of peaking factors, radial and axial power distributions, and changes of these factors due to the influence of other variables such as moderator temperature, xenon, and control rod position.

Suggested References:

Collier, J. G. *Convection Boiling and Condensation.* New York: McGraw-Hill, 1972.
Eckert, E. R. G. and Drake, R. M., Jr. *Analysis of Heat and Mass Transfer.* New York: McGraw-Hill, 1973.
El-Wakil, M. M. *Nuclear Heat Transport.* Scranton, PA: International, 1971.
Gebhart, B. *Heat Transfer.* 2nd ed. New York: McGraw-Hill, 1971.
Mooney, D. *Mechanical Engineering Thermodynamics.* Prentice Hall, 1953.

TRAINING CRITERIA FOR MITIGATING CORE DAMAGE

A program is to be developed to ensure that all operating personnel are training in the use of installed plant systems to control or mitigate an accident in which the core is severely damaged. The training program should include the following topics.

A. Incore Instrumentation

1. Use of fixed or movable incore detectors to determine extent of core damage and geometry changes.

2. Use of thermocouples in determining peak temperatures; methods for extended range readings; methods for direct readings at terminal junctions.

3. Methods for calling up (printing) incore data from the plant computer.

B. Excore Nuclear Instrumentation (NIS)

1. Use of NIS for determination of void formation; void location basis for NIS response as a function of core temperatures and density changes.

C. Vital Instrumentation

1. Instrumentation response in an accident environment; failure sequence (time to failure, method of failure); indication reliability (actual vs indicated level).

2. Alternative methods for measuring flows, pressures, levels, and temperatures.

    a. Determination of pressurizer level if all level transmitters fail.

    b. Determination of letdown flow with a clogged filter (low flow).

    c. Determination of other Reactor Coolant System parameters if the primary method of measurement has failed.

D. Primary Chemistry

1. Expected chemistry results with severe core damage; consequences of transferring small quantities of liquid outside containment; importance of using leak tight systems.

2. Expected isotopic breakdown for core damage; for clad damage.

3. Corrosion effects of extended immersion in primary water; time to failure.

E. Radiation Monitoring

1. Response of Process and Area Monitors to severe damages; behavior of detectors when saturated; method for detecting radiation readings by direct measurement at detector output (over

ranged detector); expected accuracy of detectors at different locations; use of detectors to determine extent of core damage.
2. Methods of determining dose rate inside containment from measurements taken outside containment.

CONTROL MANIPULATIONS

The following control manipulations and plant evolutions where applicable to the plant design are acceptable for meeting the reactivity control manipulations required by the rules. The starred items shall be performed on an annual basis; all other items shall be performed on a two-year cycle. However, the requalification programs shall contain a commitment that each individual shall perform or participate in a combination of reactivity control manipulations based on the availability of plant equipment and systems. Those control manipulations which are not performed at the plant may be performed on a simulator. The use of the Technical Specifications should be maximized during the simulator control manipulations. Personnel with senior licenses are credited with these activities if they direct or evaluate control manipulations as they are performed.

* (1) Plant or reactor startups to include a range that reactivity feedback from nuclear heat addition is noticeable and heatup rate is established.
 (2) Plant shutdown.
* (3) Manual control of steam generators and/or feed water during startup and shutdown.
 (4) Boration and/or dilution during power operation.
* (5) Any significant (>10%) power changes in manual rod control or recirculation flow.
 (6) Any reactor power change of 10% or greater where load change is performed with load limit control or where flux, temperature, or speed control is on manual (for HDGR).
* (7) Loss of coolant including:
  1. significant PWR steam generator leaks
  2. inside and outside primary containment
  3. large and small, including leak-rate determination

4. saturated Reactor Coolant response (PWR).
(8) Loss of instrument air (if simulated plant specific).
(9) Loss of electrical power (and/or degraded power sources.
°(10) Loss of core coolant flow/natural circulation.
(11) Loss of condenser vacuum.
(12) Loss of service water if required for safety.
(13) Loss of shutdown cooling.
(14) Loss of component cooling system or cooling to an individual component.
(15) Loss of normal feed water or normal feed-water system failure.
°(16) Loss of all feed water (normal and emergency).
(17) Loss of protective system channel.
(18) Mispositioned control rod or rods (or rod drops).
(19) Inability to drive control rods.
(20) Conditions requiring use of emergency boration or standby liquid control system.
(21) Fuel cladding failure or high activity in reactor coolant or offgas.
(22) Turbine or generator trip.
(23) Malfunction of automatic control system(s) which affect reactivity.
(24) Malfunction of reactor coolant pressure/volume control system.
(25) Reactor trip.
(26) Main steam line break (inside or outside containment).
(27) Nuclear instrumentation failure(s).

PART 55: RULES AND REGULATIONS: OPERATOR'S LICENSES

General Provisions

§55.1 Purpose.

The regulations in this part establish procedures and criteria for the issuance of licenses to operators, including senior operators, of facilities licensed pursuant to the Atomic Energy Act of 1954, as

OPERATOR TRAINING AND LICENSING

amended, or section 202 of the Energy Reorganization Act of 1974, and provide for the terms and conditions upon which the Commission will issue these licenses.

§55.2 Scope.

The regulations contained in this part apply to any individual who manipulates the controls of any facility licensed to any individual designated by a facility licensee to be responsible for directing the licensed activities of licensed operators.

§55.3 License Requirements.

(a) No person may perform the function of an operator as defined in this part except as authorized by a license issued by the Commission;
(b) No person may perform the function of a senior operator as defined in this part except as authorized by a license issued by the Commission.

§55.4 Definitions.

As used in this part:
(a) "Act" means the Atomic Energy Act of 1954 including any amendments thereto.
(b) "Commission" means the Nuclear Regulatory Commission or its duly authorized representatives.
(c) "Facility" means any production facility or utilization facility as defined by the Rules.
(d) "Operator" is any individual who manipulates a control of a facility. An individual is deemed to manipulate a control if he directs another to manipulate a control.
(e) "Senior operator" is any individual designated by a facility licensee to direct the licensed activities of licensed operators.
(f) "Controls" when used with respect to a nuclear reactor means apparatus and mechanisms the manipulation of which directly affect the reactivity or power level of the reactor. "Controls" when used with respect to any other facility means apparatus and mechanisms the manipulation of which could affect the chemical, physical, metallurgical, or nuclear process of the facility

in such a manner as to affect the protection of health and safety against radiation.

(g) "United States" when used in a geographical sense, includes all territories and possessions of the United States, and Puerto Rico.

§55.5 Communications.

Except where otherwise specified, all communications and reports concerning the regulations in this part, and applications filed under them should be addressed to the Director of Nuclear Reactor Regulation or the Director of Nuclear Material Safety and Safeguards, as appropriate, U.S. Nuclear Regulatory Commission, Washington, D.C. 20555. Communications, reports, and applications may be delivered in person at the Commission's offices at 1717 H Street, N.W., Washington, D.C. or at 7920 Norfolk Avenue, Bethesda, Md.

§55.6 Interpretations.

Except as specifically authorized by the Commission in writing, no interpretation of the meaning of the regulations in this part by any officer or employee of the Commission other than a written interpretation by the General Counsel will be recognized to be binding upon the Commission.

Exemptions

§55.7 Specific exemptions.

The Commission may, upon application by an interested person, or upon its own initiative, grant such exemptions from the requirements of the regulations in this part as it determines are authorized by law and will not endanger life or property and are otherwise in the public interest.

§55.8 Additional requirements.

The Commission may, by rule, regulation, or order, impose upon any licensee such requirements in addition to those established in the regulations in this part, as it deems appropriate or necessary to protect health and to minimize danger to life or property.

§55.9 Exemptions from license.

Nothing in this part shall be deemed to require a license for:

(a) An individual who manipulates the controls of a research or training reactor as part of his training as a student in a nuclear engineering course under the direction and in the presence of a licensed operator or senior operator;

(b) An individual who manipulates the controls of a facility as a part of his training to qualify for an operator license under this part under the direction and in the presence of a licensed operator or senior operator.

License Applications

§55.10 Contents of applications.

(a) Applications for licenses should be filed in triplicate, except for the report of medical examination, with the Director of Nuclear Reactor Regulation or the Director of Nuclear Material Safety and Safeguards, as appropriate, U.S. Nuclear Regulatory Commission, Washington, D.C. 20555. Communications, reports, and applications may be delivered in person at the Commission's offices at 1717 H Street N.W.; Washington, D.C., or 7920 Norfolk Avenue, Bethesda, Md.

Each application for a license shall contain the following information:

(1) The full name, citizenship, age, address and present employment of the applicant;

(2) The education and pertinent experience of the applicant, including detailed information on the extent and nature of responsibility;

(3) Serial numbers of any operator and senior operator license issued by the Commission to the applicant and the expiration date of each;

(4) The specific facility for which the applicant seeks an operator or senior operator license;

(5) The written request of an authorized representative of the facility license that the operating test be administered to the applicant of the facility.

(6) Evidence that the applicant has learned to operate the controls in a competent and safe manner and has need for an operator or a senior operator license in the performance of his duties. The Commission may accept as proof of this a certification of an authorized representative of the facility licensee where the applicant's services will be utilized. This certification shall include details on courses of instruction administered by the facility licensee, number of course hours, number of hours of training and nature of training received at the facility, and for reactors, the startup and shutdown experience received.

(7) A report of a medical examination by a licensed medical practitioner, in one copy in the form prescribed in §55.60.

(a) The Commission may at any time after the filing of the original application, and before the expiration of the license, require further information in order to enable it to determine whether the application should be granted or denied or whether a license should be revoked, modified or suspended.

(b) An applicant whose application has been denied because of his physical condition or general health may submit a further report of medical examination at any time as a supplement to his original application.

(c) Each application and statement shall contain complete and accurate disclosure as to all matters and things required to be disclosed. All applications and statements, other than the matters required by items 5, 6, and 7 of paragraph (a) of this section shall be signed by the applicant.

§55.11 Requirements for the approval of application.

An application for a license pursuant to the regulations in this part will be approved if the Commission finds that

(a) The physical condition and the general health of the applicant are not such as might cause operational errors endangering public health and safety.

(1) Epilepsy, insanity, diabetes, hypertension, cardiac disease, fainting spells, defective hearing or vision or any other physical or

Operator Training and Licensing

mental condition which might cause impaired judgment or motor coordination may constitute sufficient cause for denial of an application.

(2) If an applicant's vision, hearing and general physical condition do not meet the minimum standards normally considered necessary, the Commission may approve the application and include conditions in the license to accommodate the physical defect. The Commission will consider the recommendations of the facility licensee or holder of an authorization and of the examining physician on Form NRC-396 in arriving at its decision.

(b) The applicant has passed a written examination and operating test as may be prescribed by the Commission to determine that he has learned to operate and, in the case of a senior operator, to operate and to direct the licensed activities of licensed operators in a competent and safe manner.

(c) The applicant's service as a licensed operator or senior operator will be utilized on the facility for which he seeks a license or on a similar facility within the United States.

§55.12 Re-applications.

(a) Any applicant whose application for a license has been denied because of failure to pass the written examination or operating test or both may file a new application for license two months after the date of denial. Any new application shall be accompanied by a statement signed by an authorized representative of the facility licensee by whom the applicant will be employed, stating in detail the extent of additional training which the applicant has received and certifying that he is ready for reexamination. An applicant may file a third application six months after the date of denial of his second application, and may file further successive applications two years after the date of denial of each prior application.

(b) An applicant who has passed either the written examination or operating test and failed the other may request in a new application that he be excused from reexamination on the examination or test which he has passed. The Commission may in its descretion grant the request if it determines that sufficient justification is presented under all the circumstances.

Written Examinations and Operating Tests

§55.20 Scope of examinations.

The written examination and operating test for a license as an operator or a senior operator are designed to test the applicant's understanding of the facility design and his familiarity with the controls and operating procedures of the facility. The written examination is based in part on information in the final safety analysis report, operating manuals, and license for the facility.

§55.21 Content of operator written examination.

The operator written examination, to the extent applicable to the facility, will include questions on:

(a) Fundamentals of reactor theory, including fission process, neutron multiplication, source effects, control rod effects, and criticality indications.

(b) General design features of the core, including core structure, fuel elements, control rods, core instrumentation, and coolant flow.

(c) Mechanical design features of the reactor primary system.

(d) Auxiliary systems which affect the facility.

(e) General operating characteristics, including causes and effects of temperature, pressure and reactivity changes, effects of load changes, and operating limitations and reasons for them.

(f) Design, components and functions of reactivity control mechanisms and instrumentation.

(g) Design, components and functions of safety systems, including instrumentation, signals, interlocks, automatic and manual features.

(h) Components, capacity and functions of reserve and emergency systems.

(i) Shielding, isolation and containment design features, including access limitation.

(j) Standard and emergency operating procedures for the facility and plant.

(k) Purpose and operation of radiation monitoring system, including alarm and survey equipment.

(l) Radiological safety principles and procedures.

§55.22 Content of senior operator written examination.

The senior operator written examination to the extent applicable to the facility, will include questions on the items specified in §55.21 and in addition on the following:

(a) Conditions and limitations in the facility license.

(b) Design and operating limitations in the technical specifications for the facility.

(c) Facility licensee procedures required to obtain authority for design and operating changes in the facility.

(d) Radiation hazards which may arise during the performance of experiments, shielding alterations, maintenance activities and various contamination conditions.

(e) Reactor theory, including details of fission process, neutron multiplication, source effects, control rod effects, and criticality indications.

(f) Specific operating characteristics, including coolant chemistry and causes and effects of temperature, pressure and reactivity changes.

(g) Procedures and limitations involved in initial core loading, alterations in initial core configuration, control rod programming, determination of various internal and external effects on core reactivity.

(h) Fuel handling facilities and procedures.

(i) Procedures and equipment available for handling and disposal of radioactive materials and effluents.

§55.23 Scope of operator and senior operator operating tests.

The operating tests administered to applicants for operator and senior operator licenses are generally similar in scope. The operating test, to the extent applicable to the facility, requires the applicant to demonstrate an understanding of:

(a) Prestart-up procedures for the facility, including associated plant equipment which could affect reactivity.

(b) Required manipulation of console controls to bring the facility from shutdown to designated power levels.

(c) The source and significance of annunciator signals and condition-indicating signals and remedial action responsive thereto.

(d) The instrumentation system and the source and significance of reactor instrument readings.

(e) The behavior characteristics of the facility.

(f) The control manipulation required to obtain desired operating results during normal, abnormal and emergency situations.

(g) The operation of the facility's heat removal systems, including primary coolant, emergency coolant, and decay heat removal systems, and the relation of the proper operation of these systems to the operation of the facility.

(h) The operation of the facility's auxiliary systems which could affect reactivity.

(i) The use and function of the facility's radiation monitoring systems, including fixed radiation monitors and alarms, portable survey instruments, and personnel monitoring equipment.

(j) The significance of radiation hazards, including permissible levels of radiation, levels in excess of those authorized and procedures to reduce excessive levels of radiation and to guard against personnel exposure.

(k) The emergency plan for the facility, including the operator's or senior operator's responsibility to decide whether the plan should be executed and the duties assigned under the plan.

(l) The necessity for a careful approach to the responsibility associated with the safe operation of the facility.

§55.24 Waiver of examination and test requirements.

On application, the Commission may waive any or all of the requirements for a written examination and operating test if it finds that the applicant:

(a) Has had extensive actual operating experience at a comparable facility within two years prior to the date of application.

(b) Has discharged his responsibilities competently and safely and is capable of continuing to do so. The Commission may accept as proof of the applicant's past performance a certification of an authorized representative of the facility licensee or holder of an authorization by which the applicant was previously employed. The certification shall contain a description of the applicant's operating experience, including an approximate number of hours the applicant operated the controls of the facility, the duties performed, and the extent of his responsibility.

(c) Has learned the operating procedure for and is qualified to operate competently and safely the facility designated in his application. The Commission may accept as proof of the applicant's qualifications a certification of an authorized representative of the facility licensee or holder of an authorization where the applicant's services will be utilized.

§55.25 Administration of operating test prior to initial criticality.

The Commission may administer a simulated operating test to an applicant for a license to operate a reactor prior to its initial criticality if a written request by an authorized representative of the facility licensee is sufficient for the Commission to find that:
 (a) There is an immediate need for the applicant's services.
 (b) The applicant has had extensive actual operating experience at a comparable reactor.
 (c) The applicant has a thorough knowledge of the reactor control system, instrumentation and operating procedues under normal, abnormal, and emergency conditions.
 (d) The reactor control mechanism and instrumentation are in such condition as determined by the Commission to permit effective administration of a simulated operating test.

Licenses

§55.30 Issuance of licenses.

On determining that an application meets the requirements of the Act and the regulations of the Commission, the Commission will issue a license in such form and containing such conditions and limitations as it deems appropriate and necessary.

§55.31 Conditions of the licenses.

Each license shall contain and is subject to the following conditions, whether stated in the license or not:
 (a) Neither the license nor any right under the license shall be assigned or otherwise transferred.
 (b) The license is limited to the facility for which it is issued.

(c) The license is limited to those controls of the facility specified in the license.

(d) The license is subject to, and the licensee shall observe, all applicable rules, regulations and orders of the Commission.

(e) If a licensee has not been actively performing the functions of an operator or senior operator for a period of four months or longer, he shall, prior to resuming activities licensed pursuant to this part, demonstrate to the Commission that the knowledge and understanding of facility operation and administration are satisfactory. The Commission may accept as evidence, a certification by an authorized representative of the facility licensee by which the licensee has been employed.

(f) Such other conditions as the Commission may impose to protect health or to minimize danger to life or property.

§55.32 Expiration.

Each operator and senior operator license shall expire two years after the date of issuance.

§55.33 Renewal of licenses.

(a) Application for renewal of a license shall be signed by the applicant and shall contain the following information:

(1) The full name, citizenship, address and present employment of the applicant;

(2) The serial number of the license for which renewal is sought;

(3) The experience of the applicant under his existing license, including the approximate number of hours during which he has operated the facility;

(4) A statement that during the effective term of his current license the applicant has satisfactorily completed the requalification program for the facility for which operator or senior operator license renewal is sought. In the case of an application for license renewal filed within two years after September 17, 1973, if the facility licensee has not implemented the requalification program requirements in time for the applicant to complete an approved requalification program before the effective term of his current license expires, the applicant shall submit a statement showing his current enrollment in an approved requalification program and

OPERATOR TRAINING AND LICENSING

describing those portions of the program which he had completed by the date of his application for license renewal;

(5) Evidence that the licensee has discharged his license responsibilities competently and safely. The Commission may accept as evidence of this a certificate of an authorized representative of the facility licensee or holder of an authorization by which the licensee has been employed;

(6) A report by a licensed medical practitioner in the form prescribed in § 55.60.

(b) In any case in which a licensee not less than thirty days prior to the expiration of his existing license has filed an application in proper form for renewal or for a new license, the existing license shall not expire until the application for renewal or for a new license has been finally determined by the Commission.

(c) The license will be renewed if the Commission finds that:

(1) The physical condition and the general health of the licensee continue to be such as not to cause operational errors which might endanger public health and safety; and

(2)(i) The licensee has been actively and extensively engaged as an operator or as a senior operator under his existing license, has discharged his responsibilities competently and safely, and is capable of continuing to do so.

(ii) The licensee has completed a requalification program or is presently enrolled in a requalification program if the completion of the requalification program will occur after the expiration of his license as provided in subparagraph (a)(4) of this section.

(iii) If the requirements of paragraph (c)(2)(i) and (ii) of this section are not met, the Commission may require the applicant for renewal to take a written examination or an operating test or both.

(3) There is a continued need for a license to operate or direct operators at the facility designated in the application.

Modification and Revocation of Licenses

§55.40 Modification and revocation of licenses.

(a) The terms and conditions of all licenses shall be subject to amendment, revision, or modification by reason of amendments to the Act, or by reason of rules, regulations or orders issued in accordance with the Act or any amendments thereto.

(b) Any license may be revoked, suspended or modified, in whole or in part, for any material false statement in the application or any statement of fact required under section 182 of the Act, or because of conditions revealed by such application or statement of fact or any report, record, inspection or other means which would warrant the Commission to refuse to grant a license on an original application, or for violation of, or failure to observe any of the terms and conditions of the Act, or the license, or of any rule, regulation or order of the Commission, or any conduct determined by the Commission to be a hazard to safe operation of the facility.

§55.41 Notification of disability.

The licensee shall notify the Director of Nuclear Reactor Regulation or the Director of Nuclear Material Safety and Safeguards, as appropriate, U.S. Nuclear Regulatory Commission, Washington, D.C. 20555, within fifteen (15) days after its occurrence of any disability referred to in § 55.11(a)(1) which occurs after the submission of his medical examination form.

Enforcement

§55.50 Violations.

An injunction or other court order may be obtained prohibiting any violation of any provision of the Atomic Energy Act of 1954, as amended, or Title II of the Energy Reorganization Act of 1974, or any regulation or order issued thereunder. A court order may be obtained for the payment of a civil penalty imposed pursuant to section 234 of the Act for violation of section 53, 57, 62, 63, 81, 82, 101, 103, 104, 107, or 109 of the Act, or section 206 of the Energy Reorganization Act of 1974, or any rule, regulation, or order issued thereunder, or any term, condition, or limitation of any license issued thereunder, or for any violation for which a license may be revoked under section 186 of the Act. Any person who willfully violates any provision of the Act or any regulation or order issued thereunder may be guilty of a crime and, upon conviction, may be punished by fine or imprisonment or both, as provided by law.

Certificate of Medical Examination

§55.60 Examination form.

(a) An applicant shall complete and sign Form NRC-396, "Certificate of Medical Examination."

(b) The examining physician shall complete and sign Form NRC-396 and shall mail the completed form to the Director of Nuclear Reactor Regulation or Director of Nuclear Material Safety and Safeguards, as appropriate, U.S. Nuclear Regulatory Commission, Washington, D.C. 20555.

Note.—Copies of Form NRC-396 may be obtained by writing to the Director of Nuclear Reactor Regulation or Director of Nuclear Material Safety and Safeguards, as appropriate, U.S. Nuclear Regulatory Commission, Washington, D.C. 20555.

APPENDIX A: REQUALIFICATION PROGRAMS FOR LICENSED OPERATORS OF PRODUCTION AND UTILIZATION FACILITIES

Introduction

Section 50.54 of 10 CFR Part 50 requires that individuals who manipulate controls of production and utilization facilities be licensed as operators by the Commission and that individuals who direct the licensed activities of licensed operators be licensed as senior operators in accordance with 10 CFR Part 55. Section 55.33 of 10 CFR Part 55 requires that each licensed individual demonstrate his continued competence every two years in order for his license to be renewed. Competence may be demonstrated, in lieu of reexamination, by satisfactory completion of a requalification program which has been reviewed and approved by the Commission.

Periodic requalification for all operators and senior operators of production and utilization facilities is necessary for the personnel to maintain competence, particularly to respond to abnormal and emergency situations. The complexity of design and operating modes of production and utilization facilities require that ongoing comprehensive requalification programs be conducted for all licensed operators and senior operators as a matter of sound principle and practice.

Licensed operators and senior operators of production and utilization facilities who have been actively and extensively engaged as operators or as senior operators shall participate in requalification programs meeting the requirements of this Appendix. Individuals who maintain operator or senior operator licenses for the purpose of providing backup capability to the operating staff shall participate in the requalification programs except to the extent that their normal duties preclude the need for specific retraining in particular areas. Licensed operators or senior operators whose licenses are conditioned to permit manipulation of specific controls only shall participate in those portions of the requalification program appropriate to the duties they perform.

The requalification program requirements involving manipulation of controls may be performed on the facility for which the operator is licensed. However, the use of a simulator as specified in paragraphs 3e and 4d of this appendix is permissible and such use is encouraged.

Requalification Program Requirements

1. Schedule.

The requalification program shall be conducted for a continuous period not to exceed two years, and upon conclusion shall be promptly followed, pursuant to a continuous schedule, by successive requalification programs.

2. Lectures.

The requalification program shall include preplanned lectures on a regular and continuing basis throughout the license period in those areas where annual operator and senior operator written examinations indicate that emphasis in s(~pe and depth of coverage is needed in the following subjects:
  a. Theory and principles of operation.
  b. General and specific plant operating characteristics.

OPERATOR TRAINING AND LICENSING

c. Plant instrumentation and control systems.
d. Plant protection systems.
e. Engineered safety systems.
f. Normal, abnormal, and emergency operating procedures.
g. Radiation control safety.
h. Technical specifications.
i. Applicable portions of Title 10, Chapter I, Code of Federal Regulations.

Other training techniques including films, videotapes and other effective training aids may also be used.

Individual study on the part of each operator shall be encouraged. However, a requalification program based solely upon the use of films, videotapes and on individual study is not an acceptable substitute for a lecture series.

3. On-the job training.

The requalification program shall include on-the-job training so that:

a. Each licensed operator of a production or utilization facility manipulates the plant controls and each licensed senior operator either manipulates the controls or directs the activities of individuals during plant control manipulations during the term of their licenses. For reactor operators and senior operators, these manipulations shall consist of at least 10 reactivity control manipulations in any combination of reactor startups, reactor shutdowns or other control manipulations which demonstrate skill and/or familiarity with reactivity control systems.

b. Each licensed operator and senior operator has demonstrated satisfactory understanding of the operation of all apparatus and mechanisms and knows the operating procedures in each area for which he is licensed.

c. Each licensed operator and senior operator is cognizant of facility design changes, procedure changes, and facility license changes.

d. Each licensed operator and senior operator reviews the contents of all abnormal and emergency procedures on a regularly scheduled basis.

e. A simulator may be used in meeting the requirements of

paragraphs 3a* and 3b* if the simulator reproduces the general operating characteristics of the facility involved, and the arrangement of the instrumentation and controls of the simulator is similar to that of the facility involved.

4. Evaluation.

The requalification program shall include:

a. Annual written examinations which determine areas in which retraining is needed to upgrade licensed operator and senior operator knowledge.

b. Written examinations which determine licensed operators' and senior operators' knowledge of subjects covered in the requalification program and provide a basis for evaluating their knowledge of abnormal and emergency procedures.

c. Systematic observation and evaluation of the performance and competency of licensed operators and senior operators by supervisors and/or training staff members including evaluation of actions taken or to be taken during actual or simulated abnormal and emergency conditions.

d. Simulation of emergency or abnormal conditions that may be accomplished by using the control panel of the facility involved or by using a simulator. Where the control panel of the facility is used for simulation, the actions taken or to be taken for the emergency or abnormal condition shall be discussed; actual manipulation of the plant controls is not required. If a simulator is used in meeting the requirements of paragraph 4c, the simulator shall accurately reproduce the operating characteristics of the facility involved and the arrangement of the instrumentation and controls of the simulator shall closely parallel that of the facility involved.

e. Provisions for each licensed operator and senior operator to participate in an accelerated requalification program where performance evaluations conducted pursuant to paragraphs 4a through 4d clearly indicate the need.

5. Records.

a. Records of the requalification program shall be maintained for a period of two years from the date of the recorded event to

Operator Training and Licensing

document the participation of each licensed operator and senior operator in the requalification program. The records shall contain copies of written examinations administered, the answers given by the licensee, results of evaluations and documentation of any additional training administered in areas in which an operator or senior operator has exhibited deficiencies.

b. Records which must be maintained pursuant to this Appendix may be the original or a reproduced copy or microform if such reproduced copy or microform is duly authenticated by authorized personnel and the microform is capable of producing a clear and legible copy after storage for the period specified by Commission regulations.

c. If there is a conflict between the Commission's regulations in this part, license condition, or other written Commission approval or authorization pertaining to the retention period for the same type of record, the retention period specified in the regulations in this part for such records shall apply unless the Commission, pursuant to § 55.7 has granted a specific exemption from the record retention requirements specified in the regulations in this part.

6. Alternative training programs.

The requirements of this Appendix may be met by requalification programs conducted by persons other than the facility licensee if such requalification programs are similar to the programs described in paragraphs 1 through 5, and the alternative program has been approved by the Commission.

7. Applicability to research and test reactors and non-reactor facilities.

To accomodate specialized modes of operation and differences in control, equipment, and operator skills and knowledge, the requalification program for each licensed operator and senior operator of a research or test reactor or of a non-reactor facility shall conform generally but need not be identical to the requalification program outlined in paragraphs 1 through 6 of this Appendix. However, significant deviations from the requirements of this Appendix shall be permitted only if supported by written justification and approved by the Commission.

Index

A

Air compressors, 677
  aftercoolers, 695, 703
  axial flow, 685
  centrifugal, 683
  cooling, 688
  direct-connected, 685
  double-acting, 682
  helical, 684
  intercoolers, 695, 698
  liquid-piston, 684
  operation of, 712
    running, 716
    startup, 712
  packaged systems, 692
  pressure controls, 704
  rotary, 684
  single-acting, 682
  single stage medium pressure, 683
  speed controls, 704
  two-impeller straight-lobe, 684
  two-stage high pressure, 683
  types of, 681
  unloaders, 707
  valves, 690
Air-evacuation pump, 441
AC (alternating current), 756, 767
Alternators, electrical, 767
Anthracite coal, 33
Area of circle, calculation, 846
Ash, 81
Atmospheric pressure, 95
Atomizing oil burners, 619

B

Back pressure, steam, 439
Barometers, 23
Barometric pressure, 23
Basic principles, 13
Big boilers, 193
  classification, 193
  headers, 194
Bleeder turbine, steam, 662
Blow-down valves, 269
Blow-off tank, 274
Blow-off valve, 269

Boilers
  attachments, 261,
  bent-tube, 115
    disadvantages, 196
  big, 193
  blowdown valves, 269
  blowing-down, 530
  blow-off tank, 274
  blow-off valves, 269
  calculations, 219
    Adamson flues
      (max. pressure), 249
    area of heads to be stayed, 246
    bursting pressure, 222
    circular flues
      (max. pressure), 248
    displacement, 322
    furnace, 252
    furnace grate size, 253
    heating surface, 254
    ligaments, efficiency of, 241
    output, 250
    pump discharge per hour, 323
    pump pressure production, 322
    reinforcement flat surfaces, 243
    riveted joint strength, 228-239
    shell strength, 219
    shell thickness, 227
    stay tubes (area), 248
    working pressure, 222
  caulking, 537
  check valves, 266
  classifications of, 109
  cleaning, 526
  construction
    fire-tube, 131
    lap joints, 133
    riveted joints, 133
    shell, 131
    stays, 138
    water-tube, 165
    welded joints, 132
  conversion factors, 836
  feed pumps, 297
    centrifugal pumps, 324
      turbine driven, 325
    duplex system, 312
      cushion valves, 313

Index

Boilers—cont
  feed pumps
    "D" and "B" type gears,
      319, 320
    "lost motion,", 317
    valve gear, 316
    maintenance, 328
    centrifugal pumps, 338
    reciprocating pumps, 329
    rotary pumps, 340
    simplex system, 300
      auxiliary piston, 303
      auxiliary valve, 303
    types, 297
    variable stroke, 341
  fire-tube boiler, 109
  firing, 497
    alternate front & back spread, 498
    alternate side-spread, 497
    coking method, 500
    even-spread, 497
  fittings, 261
  fixtures, 261
  forced-circulation, 203
  furnace, 294
    pulverized coal burners, 595
  fusible plug, 286
  grooving, 535
  horsepower, calculating, 833
  installation, 485
  laying-up, 528
  low water cut off, 813
  maintenance, 533
    lubricants, 552
      choice of, 559
      classes of, 556
      desirable qualities, 555
      grease, 572
    lubrication, 560
      external, 566
      internal, 561
      splash system, 574
    lubricators, 561
      wick-feed, 568
    tube cleaners, 546
  material strength, 205
  metals
    characteristics, 207
    testing, 210-218
  nonsectional, 112
  oil, elimination of, 524
  oilers, 560-572
  operation, 485
  packaged, 129

Boilers—cont
  parts of, 109
  priming, 516
  repair, 533
    burned plate, 535
    cracks, 534
    gasket, 538
    patch, 536
    tube, 540
  routine operation and care, 518
  safety valves, 264
  scale elimination, 523
  sectional boiler, 112
  soot deposits, 519
  startup, 485
    running, 497
  steam gauge, 281
  steam gauge test, 815
  stoker operation, 588
  stop valves, 266
  through-tube, 114
  tubes, 124, 151
  types of, 109
    miscellaneous, 186
  vertical, 105-6
  vertical tube, 114
  waste-heat recovery, 126
  water, 510
  water-tube boiler, 109
  watergauge, 510
  zinc, as anti-corrosive, 522
Boiling, 90
Boiling point, 95-98
Bolt sizes, American standard
  coarse thread series, sizes 862
Boyle's law, application, 678
British thermal unit (Btu), 89
Burners
  automatic control
    pulverizers, 607
  dual-fuel systems, 631
  gas, 613
  multi-fuel systems, 631
  oil, 613
    atomizing, 619
    classifications of, 617
    gravity feed, 614
    gun-type, 627
    pot-type, 627
    pressurized, 615
    proportioning burners,
      low-pressure, 625
    rotary oil burners, 621
    spray burners, 623

INDEX

Burners—cont
  oil
    vaporizing, 617
    pulverized coal, 602
    split flame burner, 610
    waste-fuel, 613, 634

C

Calculations
  Adamson flues (max. pressure), 249
  basic math, 823-832
  boiler horsepower, 833
  boiler output, 250
  boiler pressure, 222
  circle, 834
  circular flues (max. pressure), 248
  cylindrical, 833
  discharge per hour, pump, 323
  displacement, 322
  engine horsepower, 834
  furnace grate size, 253
  heating surface (boiler), 254
  inscribed circle, 835
  kinetic energy, 660
  parallelogram, 832
  pump pressure production, 322
  rectangle, 832
  reinforcement of flat surfaces (boilers), 243
  shell thickness, 227
  spherical, 835
  square, 832
  stay tubes (area), 248
  strength of riveted joints, 228-239
  trapezoid, 833
  triangle, 833
  water, 835
Capacitance, electrical, 771
Capacitor, 758
Carbon, 66
Carbon dioxide (CO_2), 78
Caulking, boiler, 551
Cells, electrical, 760
Centrifugal pumps, 324
  turbine driven, 325
Charles' law, applications, 678
Chemical substances in feed-water characteristics of, 419
Chemical terms, rel. to feed-water, 410
Circuits, electrical, 756
Circumference of circle, 846
Class 1 condensers, 444
Class 2 condensers, 451

Clyde boiler, 118
CO_2, 78
Coal, 32
  classifications of, 32
  combustion of, 596
  evaporation of, 36
  furnace, 67
  pulverized, 35
  pulverizers, 598
  sizes of
    anthracite, 32
Cocks, 275
  gauge, 277
  Mississippi, 278
Coke, 35
  combustion of, 506
Collector, steam, 291
Combustible material, 66
Combustion, 65, 596
  complete, 69, 76
  controls, 67
  experiments with, 82-84
  incomplete, 69, 76
  products, 68
Compound turbines, steam, 651
Compressed air, 677
Compressor, refrigerant, 726
Condensation, 24, 94
Condensers, 439
  direct-contact (Class 1), 444
    jet condenser, 444
      barometric, 450
    modern condensate systems, 458
    operation of, 456
    spray-flow deaerators, 460
    steam-flow deaerator, 461
  surface (Class 2), 451
    tubes, 454
    vacuum breaker, 447
      air admission, 447
      reduced contact surface, 447
    water and steam circuits, 455
Condensing, economy of, 440
Conversion factors, boiler, 836
Conversions
  Celsius to Fahrenheit, 844
  English-to-English, 841
  English-to-metric, 840
  Fahrenheit to Celsius, 844
  metric-to-English, 837
  miscellaneous, 842
Cooling ponds, 465
  operation of, 466
  spray pond, 467

947

Index

Cooling towers, 465, 472
  efficiency of, calculations, 473
  hyperbolic, 476, 481, 482
Cooling water problems, 480
Cooling water treatment, 480
Cornish boiler, 121
Corrosion, boiler, 520
Crow-foot stay, 149
Crown stay, 151
Current, electrical, 754
  measuring, 772

D

DC (direct current), 756, 764
Deaerators, feed-water, 369, 391
  spray-flow, 460
  steam-flow, 461
Defrosting systems, 744
Diagonal stay, 147-148
Diesel engine principles, 779
  combustion, 790
  compression pressures, 782
  four-cycle, 779
  fuel injection methods, 783
  fuel-injection system, 795
  service, 800
  timing, 803
  two-cycle, 801
Discharge rate calculation, 323
Displacement, calculation, 322
Displacement regulators,
  feed-water, 371
Draft, 69
  forced, 70
  natural, 69

E

Economizers, feed-water, 362
  calculation for gain with, 367
Efficiency, 26
  riveted joints, 856
Ejectors, feed-water, 369, 397
  applications, 398
  classifications, 398
  maintenance, 402
  operation, 399
  steam-jet air, 397
Electric circuits, 756
Electrical generators, 763
Electrical symbols, 759
Electricity, basics, 751
  current, 754

Electricity, basics—cont
  measurement, 760
  resistance, 754
  voltage, 754
Electromagnet, 754
Energy, 20
Engine horsepower, calculating, 834
English-to-metric conversion, 840
English to English conversion, 841
Environmental P.A. (EPA)
  rules and regulations, 867
Evaporation, factor of, 100

F

Farenheit to Celsius conversion, 54
Feed pumps, boiler, 297
  centrifugal pumps, 324
  maintenance, 328
  centrifugal pumps, 338
  rotary pumps, 340
  variable stroke, 341
Feed water
  chemicals in, characteristics of, 412
  chemical terms, related, 410
  deaerators, 369, 391
  economizers, 362
    calculation for gain with, 367
  ejectors, 369, 397
    applications, 398
    classifications, 398
    maintenance, 402
    operation, 399
    steam-jet air, 397
  hardness, 409
  heaters, 349
    classifications, 350
    closed heaters, 355
    fuel saving, 350
    open heaters, 351
  heating savings calculation, 35
  injectors, 369, 376
    calculation for correct size, 385
    connection of, 387
    failures, cause of, 389
    maintenance, 386
    operation, automatic, 386
    selection of, 383
  precipitation, 410
  regulators, 369
    classifications, 370
    displacement, 371
    evaporation, 373
    expansion, 374

948

INDEX

Feed water—cont
 saturation point, 409
 testing, 405
  simple tests, 413
 treatment, 405, 419
  cold lime-soda, 421
  corrosion prevention, 438
  evaporators, 419
  hot lime-soda, 423
  hydrogen zeolite process, 431
  phosphate treatment, 428
  scale and mud, 407
  silica removal, 435
  sodium zeolite process, 430
  softening by ion exchange, 432
  zeolite, 429
Feed-water connection, 281
Firebox boiler, 116
Firing
 burners, pulverized coal, 602
 mechanical stokers, 575
  (see Mechanical stokers)
 pulverized coal, 597
Fire-tube boiler, 109
 horizontal, 158
Flame detectors, 815
Forced circulation boilers, 203
Foster Wheeler delayed
 coking system, 34
Friction, 25
Fuel oils, 613
 classification of, 613
Fuels, 31
 gaseous, 41
 liquid, 39
 solid, 32
Furnace, 294
Fusible plug, 286

G

Galloway tubes, 118, 121
Gas burners, 613
Gaseous fuels, 41
Gasket, repair, boilers, 538
Gauge cock, 277
Generators, electrical, 763
Generators, steam
 demiflash, 125
Geothermal energy, 48
Governors, steam turbine, 649
Gravity-feed oil burners, 614
Grease cups, bearings, 572

Gun-type oil burners, 627
Gusset stays, 147

H

Heat, 15, 53
 experiments with, 62-64
 latent, 89
  external, 89
  internal, 89
 measurement, 53
 specific, 62
 specific, table, 855
Heat pumps, 735
 air-to-air, 736
 water-to-water, 738
Heaters, feed water, 349
Heating surface (boiler)
 calculations, 254
Helical-flow turbine, steam, 662
Horizontal return tube boiler
 (HRT), 117, 119
Horsepower
 boiler, calculations, 833
 engine horsepower, 834
Hydrocarbons, 68
Hydrogen, 67
Hyperbolic cooling tower, 476, 481, 482

I

Impact-and-attrition pulverization, 600
Impact mill, 600
Impulse, in steam turbines, 642
Inductance, electrical, 771
Inductor, electrical, 758
Injectors, feed-water, 369, 376
 calculation for correct size, 385
 connection of, 387
 failures, cause of, 389
 maintenance, 386
 operation, automatic, 386
 selection of, 383
Ion exchange, softening by, 432

J

Jaw stay, 149-150

K

Kidwell boiler, 115
Kinetic energy, 660

949

Index

L

Lap points
 strength of, 228-239
Laying up, boiler, 528
Ligaments, 239
Licensing criteria
 reactor operator, 917
Liquid fuels, 39
Lubricants, boilers, 552
 choice of, 559
 classes of, 556
 desirable qualities, 555
 temperature range, 556
Lubrication, 560
 splash system, 574
 steam turbine, 668

M

Magnetic flux, 753
Magnetism, 752
Maintenance, boilers, 533
Math, basic calculations, 823
Matter, 14
Measurement, metric, 837
Mechanical stokers, 575
 operation of, 588
 overfeed stokers, 576
 rotary stokers, 583
 spreader stokers, 586
 sprinkler stokers, 583, 586
 traveling-grate stokers
 types of, 576
 underfeed stokers, 577
Metals
 characteristics of, 26
 expansion by heat, 57
 testing, 210-218
 used in boilers, 205
Metric measurement, 837
 metric to English conversion, 837
Mississippi cock, 278
Morrison corrugated boiler, 118
Motors, electrical, 773
Multistage turbines, steam, 652

N

Nitrogen, 73
Nuclear energy, 44
Nuclear power plant, 47
Nuclear reactor
 criteria for operator training

Nuclear reactor—cont
 criteria
 and licensing (NRC), 917
Nuclear reactor core, 48

O

Oil, 39
Oil burners, 613
 atomizing, 619
 classifications of, 617
 gravity-feed, 614
 gun-type burners, 627
 pot-type burners, 627
 pressurized, 616
 proportioning burners, 625
 rotary, 621
 spray burners, 623
 vaporizing, 617
Oilers, boilers, 560-572
Output (boiler), 250
Overfeed stokers, 576
Oxidation, 65

P

Packaged boilers, 129
Palm stay, 144-145
Patches, boiler, 536
pH, definition, 428
Phosphate treatment, feed-water, 428
Pipe
 double extra strong, 861
 extra strong, properties of, 860
 standard wrought, table, 859
Plant safety, 805
Porcupine boiler, 121, 122
Pot-type oil burners, 627
Power, 20
Pressure, 21
 barometric, 23
 calculations (boiler), 222
 pump, production of, 322
Priming, 93
Pulverized coal burners, 595
Pulverizers, 598
 attrition (bowl), 598
 automatic control, 607
 burners, 602
 impact-and-attrition, combined, 600
 impact mill, 600
Pumps
 air-evacuation, 441
 dry-air, 446

INDEX

Pumps—cont
  feed, 514
  oil, boilers, 565
  pressure production, 322
  water removal, 447
  wet-air-removal, 446

R

Radiant heat, 595
Reaction, in steam turbines, 641
Refrigerant compressors
  centrifugal, 731
  compound, 727
  horizontal, 728
  reciprocating, 728
  rotary, 730
  V/W, 729
Refrigerants, 722, 724
  freons
    pressure-temperature
      relations, 865
  physical properties of common, 863
  pressure-temperature relations, 864
Refrigeration, 28, 721
  absorption method, 733
  basics, 721
  cascade system, 735
  controls, 738
  defrosting systems, 744
  heat-pumps, 735
  maintenance, 748
  repair, 748
  steam-jet system, 734
  system components, 721
Regulations,
  Environmental Protection
    Agency (EPA), 867
Regulators, feed-water, 369
Reinforcement, flat surfaces, 243
Resistance, electric, 754
Resistance, measuring, 772
Riveted joints, efficiency of, 856
Rotary air-compressor, 684
Rotary oil burners, 621
Rotary stokers, 583
Rules
  Environmental Protection
    Agency (EPA), 867

S

SAE viscosity numbers, 559

Safety guidelines, 817
Safety valves, 262, 264, 805
  boiler, 264
  pop, 808
Saturation point, 409
Saturated steam, properties of, 848
Scale, 407, 408
  eliminating from boilers, 523
  removal, calculations for, 418
  treatment, 408
Scotch boiler, 118, 120
Screw stay, 141
  riveted, 142
Separators, 289
Shell thickness, 227
Shell boiler, 160
  openings, 160
Shell strength, 219
Sizes
  American standard bolt,
    course thread series, 862
Smoke, 75
  classifications of, 77
  density, 77-78
Smoke and flue gases, 74
Socket stay, 144
Solar energy, 49
Solar-fossil fuel power plant, 51
Solid fuels, 32
Soot deposits, boiler, 519
Specific heat, 62
  various substances, 855
Split flame burner, controlled flow, 610
Spray burners, oil, 623
Spray-flow deaerator, 460
Spreader stokers, 586
Sprinkler stokers, 583, 586
Standard wrought pipe, table, 859
Stay bolt, 140, 141
  thread, 141
Stay-rod, 146
Stay tubes, calculation of area, 248
Stays, 138
  methods of fastening, 157
Steam, 23
  boiling point, 95-98
  classifications of (by pressure), 88
  condensation, 24, 94
  dry, 87
  expansion, 25
  formation of, 88
  gauge test, 815
  how made by boiler, 85
  nature of, 85

951

INDEX

Steam—cont
  saturated, 86, 91-92
    properties of, 848
  superheated, 93, 98
    properties of, 852
    specific heat of, 99
    volume of, 99
  tables, 96
  types of, 86
  under pressure, 95
  wet, 87
Steam dome, 163
Steam-flow deaerator, 461
Steam-gauge, 283
Steam generators, 125
  flash, 125
Steam loop, 292
Steam traps, 7
Steam turbines, 639
  bleeder, 662
  classifications of, 640
  compound turbines, 651
  erection of, 663
  governors, 649
  helical-flow, 662
  impulse, 641
  low-pressure, 662
  lubrication of, 668
  maintenance of, 669
  mixed pressure, 662
  multistage turbines, 652
  operation of, 666
  reaction, 641
  troubleshooting, 673
  unequal-pressure (reaction), 656
  velocity ("impulse"), 645
Steel
  angle stay, 149
  stress table, 855
  temperature by color, 57
Sterling boiler, 115
Straight tubes, 196
Stress calculation, 219
Superheated steam, properties of, 852
Surface (class 2) condensers, 451

T

Temperature, 18
  Celsius to Fahrenheit, 844
  Fahrenheit to Celsius, 844
Tests, simple feed-water, 413
Thermocirculation, 104
Thermocouple, 55

Transformers, electrical, 770
Traveling-grate stokers, 583
Tube cleaners, boiler, 546
Tube repair, boilers, 540
  condenser, 454

U

Underfeed stokers, 577
Unequal-pressure (reaction)
  turbines, 656
Unloaders, air compressors, 707

V

Vacuum, 21
Vacuum breaker, 447
Valves
  air compressor, 690
  blow-down valves, 269
  blow-off, boiler, 269
  check, boiler, 266
  safety, 805
  stop, boiler, 266
Vaporizing oil burners, 617
Vertical boiler, 106
Vertical tubular boilers, 153
Voltage, 754
  measuring, 772

W

Water-cooled furnace, 201
Water gauge, 279
Water gauge, boiler, 510
Water and steam circuits,
  condenser, 455
Water tubes, 172
Water-tube boilers, 58, 109, 168
  sectional and
    nonsectional, 179
  section headers, 169
  types, 165
Waste fuel, 39
Waste-fuel burners, 613
Waste heat, 44
Welded construction, 239
Wood, 37
Work, 19

Z

Zinc, as an anti-corrosive, 522

**Over a Century of Excellence
for the Professional
and
Vocational Trades and the Crafts**

**Order now from your local bookstore
or use the convenient order form at
the back of this book.**

AUDEL

These fully illustrated, up-to-date guides and manuals mean a better job done for mechanics, engineers, electricians, plumbers, carpenters, and all skilled workers.

Contents

Electrical	II
Machine Shop and Mechanical Trades	III
Plumbing	IV
Heating, Ventilating and Air Conditioning	IV
Pneumatics and Hydraulics	V
Carpentry and Construction	V
Woodworking	VI
Maintenance and Repair	VI
Automotive and Engines	VII
Drafting	VII
Hobbies	VII
Macmillan Practical Arts Library	VIII

Electrical

House Wiring sixth edition
Roland E. Palmquist
5½ x 8¼ Hardcover 256 pp. 150 illus.
ISBN: 0-672-23404-1 $13.95

Rules and regulations of the current National Electrical Code® for residential wiring, fully explained and illustrated: • basis for load calculations • calculations for dwellings • services • nonmetallic-sheathed cable • underground feeder and branch-circuit cable • metal-clad cable • circuits required for dwellings • boxes and fittings • receptacle spacing • mobile homes • wiring for electric house heating.

Practical Electricity fourth edition
Robert G. Middleton; revised by L. Donald Meyers
5½ x 8¼ Hardcover 504 pp. 335 illus.
ISBN: 0-672-23375-4 $14.95

Complete, concise handbook on the principles of electricity and their practical application: • magnetism and electricity • conductors and insulators • circuits • electromagnetic induction • alternating current • electric lighting and lighting calculations • basic house wiring • electric heating • generating stations and substations.

II

Guide to the 1984 Electrical Code®
Roland E. Palmquist
5½ x 8¼ Hardcover 664 pp. 225 illus.
ISBN: 0-672-23398-3 $19.95

Authoritative guide to the National Electrical Code® for all electricians, contractors, inspectors, and homeowners: • terms and regulations for wiring design and protection • wiring methods and materials • equipment for general use • special occupancies • special equipment and conditions • and communication systems. Guide to the 1987 NEC® will be available in mid-1987.

Mathematics for Electricians and Electronics Technicians
Rex Miller
5½ x 8¼ Hardcover 312 pp. 115 illus.
ISBN: 0-8161-1700-4 $14.95

Mathematical concepts, formulas, and problem solving in electricity and electronics: • resistors and resistance • circuits • meters • alternating current and inductance • alternating current and capacitance • impedance and phase angles • resonance in circuits • special-purpose circuits. Includes mathematical problems and solutions.

Fractional Horsepower Electric Motors
Rex Miller and Mark Richard Miller
5½ x 8¼ Hardcover 436 pp. 285 illus.
ISBN: 0-672-23410-6 $15.95

Fully illustrated guide to small-to-moderate-size electric motors in home appliances and industrial equipment: • terminology • repair tools and supplies • small DC and universal motors • split-phase, capacitor-start, shaded pole, and special motors • commutators and brushes • shafts and bearings • switches and relays • armatures • stators • modification and replacement of motors.

Electric Motors
Edwin P. Anderson; revised by Rex Miller
5½ x 8¼ Hardcover 656 pp. 405 illus.
ISBN: 0-672-23376-2 $14.95

Complete guide to installation, maintenance, and repair of all types of electric motors: • AC generators • synchronous motors • squirrel-cage motors • wound rotor motors • DC motors • fractional-horsepower motors • magnetic contractors • motor testing and maintenance • motor calculations • meters • wiring diagrams • armature windings • DC armature rewinding procedure • and stator and coil winding.

Home Appliance Servicing fourth edition
Edwin P. Anderson; revised by Rex Miller
5½ x 8¼ Hardcover 640 pp. 345 illus.
ISBN: 0-672-23379-7 $15.95

Step-by-step illustrated instruction on all types of household appliances: • irons • toasters • roasters and broilers • electric coffee makers • space heaters • water heaters • electric ranges and microwave ovens • mixers and blenders • fans and blowers • vacuum cleaners and floor polishers • washers and dryers • dishwashers and garbage disposals • refrigerators • air conditioners and dehumidifiers.

Television Service Manual
fifth edition
Robert G. Middleton; revised by Joseph G. Barrile
5½ x 8¼   Hardcover   512 pp.   395 illus.
ISBN: 0-672-23395-9 $15.95

Practical up-to-date guide to all aspects of television transmission and reception, for both black and white and color receivers: • step-by-step maintenance and repair • broadcasting • transmission • receivers • antennas and transmission lines • interference • RF tuners • the video channel • circuits • power supplies • alignment • test equipment.

Electrical Course for Apprentices and Journeymen
second edition
Roland E. Palmquist
5½ x 8¼   Hardcover   478 pp.   290 illus.
ISBN: 0-672-23393-2 $14.95

Practical course on operational theory and applications for training and re-training in school or on the job: • electricity and matter • units and definitions • electrical symbols • magnets and magnetic fields • capacitors • resistance • electromagnetism • instruments and measurements • alternating currents • DC generators • circuits • transformers • motors • grounding and ground testing.

Questions and Answers for Electricians Examinations
eighth edition
Roland E. Palmquist
5½ x 8¼   Hardcover   320 pp.   110 illus.
ISBN: 0-672-23399-1 $12.95

Based on the current National Electrical Code®, a review of exams for apprentice, journeyman, and master, with explanations of principles underlying each test subject: • Ohm's Law and other formulas • power and power factors • lighting • branch circuits and feeders • transformer principles and connections • wiring • batteries and rectification • voltage generation • motors • ground and ground testing.

Machine Shop and Mechanical Trades

Machinists Library
fourth edition   3 vols
Rex Miller
5½ x 8¼   Hardcover   1,352 pp.   1,120 illus.
ISBN: 0-672-23380-0 $38.95

Indispensable three-volume reference for machinists, tool and die makers, machine operators, metal workers, and those with home workshops.
Volume I, Basic Machine Shop
5½ x 8¼   Hardcover   392 pp.   375 illus.
ISBN: 0-672-23381-9 $14.95

• Blueprint reading • benchwork • layout and measurement • sheet-metal hand tools and machines • cutting tools • drills • reamers • taps • threading dies • milling machine cutters, arbors, collets, and adapters.
Volume II, Machine Shop
5½ x 8¼   Hardcover   528 pp.   445 illus
ISBN: 0-672-23382-7 $14.95

• Power saws • machine tool operations • drilling machines • boring • lathes • automatic screw machine • milling • metal spinning.
Volume III, Toolmakers Handy Book
5½ x 8¼   Hardcover   432 pp.   300 illus.
ISBN: 0-672-23383-5 $14.95

• Layout work • jigs and fixtures • gears and gear cutting • dies and diemaking • toolmaking operations • heat-treating furnaces • induction heating • furnace brazing • cold-treating process.

Mathematics for Mechanical Technicians and Technologists
John D. Bies
5½ x 8¼   Hardcover   392 pp.   190 illus.
ISBN: 0-02-510620-1 $17.95

Practical sourcebook of concepts, formulas, and problem solving in industrial and mechanical technology: • basic and complex mechanics • strength of materials • fluidics • cams and gears • machine elements • machining operations • management controls • economics in machining • facility and human resources management.

Millwrights and Mechanics Guide
third edition
Carl A. Nelson
5½ x 8¼   Hardcover   1,040 pp.   880 illus.
ISBN: 0-672-23373-8 $22.95

Most comprehensive and authoritative guide available for millwrights and mechanics at all levels of work or supervision: • drawing and sketching • machinery and equipment installation • principles of mechanical power transmission • V-belt drives • flat belts • gears • chain drives • couplings • bearings • structural steel • screw threads • mechanical fasteners • pipe fittings and valves • carpentry • sheet-metal work • blacksmithing • rigging • electricity • welding • pumps • portable power tools • mensuration and mechanical calculations.

Welders Guide
third edition
James E. Brumbaugh
5½ x 8 ¼   Hardcover   960 pp.   615 illus.
ISBN: 0-672-23374-6 $23.95

Practical, concise manual on theory, operation, and maintenance of all welding machines: • gas welding equipment, supplies, and process • arc welding equipment, supplies, and process • TIG and MIG welding • submerged-arc and other shielded-arc welding processes • resistance, thermit, and stud welding • solders and soldering • brazing and braze welding • welding plastics • safety and health measures • symbols and definitions • testing and inspecting welds. Terminology and definitions as standardized by American Welding Society.

Welder/Fitters Guide
John P. Stewart
8½ x 11   Paperback   160 pp.   195 illus.
ISBN: 0-672-23325-8 $7.95

Step-by-step instruction for welder/fitters during training or on the job: • basic assembly tools and aids • improving blueprint reading skills • marking and alignment techniques • using basic tools • simple work practices • guide to fabricating weldments • avoiding mistakes • exercises in blueprint reading • clamping devices • introduction to using hydraulic jacks • safety in weld fabrication plants • common welding shop terms.

Sheet Metal Work
John D. Bies
5½ x 8¼   Hardcover   456 pp.   215 illus.
ISBN: 0-8161-1706-3 $17.95

On-the-job sheet metal guide for manufacturing, construction, and home workshops: • mathematics for sheet metal work • principles of drafting • concepts of sheet metal drawing • sheet metal standards, specifications, and materials • safety practices • layout • shear cutting • holes • bending and folding • forming operations • notching and clipping • metal spinning • mechanical fastening • soldering and brazing • welding • surface preparation and finishes • production processes.

III

Power Plant Engineers Guide
third edition
Frank D. Graham; revised by Charlie Buffington
5½ x 8¼ Hardcover 960 pp. 530 illus.
ISBN: 0-672-23329-0 $16.95

All-inclusive question-and-answer guide to steam and diesel-power engines: • fuels • heat • combustion • types of boilers • shell or fire-tube boiler construction • strength of boiler materials • boiler calculations • boiler fixtures, fittings, and attachments • boiler feed pumps • condensers • cooling ponds and cooling towers • boiler installation, startup, operation, maintenance and repair • oil, gas, and waste-fuel burners • steam turbines • air compressors • plant safety.

Mechanical Trades Pocket Manual
second edition
Carl A. Nelson
4 x 6 Paperback 364 pp. 255 illus.
ISBN: 0-672-23378-9 $10.95

Comprehensive handbook of essentials, pocket-sized to fit in the tool box or • mechanical and isometric drawing • machinery installation and assembly • belts • drives • gears • couplings • screw threads • mechanical fasteners • packing and seals • bearings • portable power tools • welding • rigging • piping • automatic sprinkler systems • carpentry • stair layout • electricity • shop geometry and trigonometry.

Plumbing

Plumbers and Pipe Fitters Library
third edition 3 vols
Charles N. McConnell; revised by Tom Philbin
5½ x 8¼ Hardcover 952 pp. 560 illus.
ISBN: 0-672-23384-3 $34.95

Comprehensive three-volume set with up-to-date information for master plumbers, journeymen, apprentices, engineers, and those in building trades.

Volume 1, Materials, Tools, Roughing-In
5½ x 8¼ Hardcover 304 pp. 240 illus.
ISBN: 0-672-23385-1 $12.95

• Materials • tools • pipe fitting • pipe joints • blueprints • fixtures • valves and faucets.

Volume 2, Welding, Heating, Air Conditioning
5½ x 8¼ Hardcover 384 pp. 220 illus.
ISBN: 0-672-23386-X $13.95

• Brazing and welding • planning a heating system • steam heating systems • hot water heating systems • boiler fittings • fuel-oil tank installation • gas piping • air conditioning.

Volume 3, Water Supply, Drainage, Calculations
5½ x 8¼ Hardcover 264 pp. 100 illus.
ISBN: 0-672-23387-8 $12.95

• Drainage and venting • sewage disposal • soldering • lead work • mathematics and physics for plumbers and pipe fitters.

Home Plumbing Handbook
third edition
Charles N. McConnell
8½ x 11 Paperback 200 pp. 100 illus.
ISBN: 0-672-23413-0 $10.95

Clear, concise, up-to-date fully illustrated guide to home plumbing installation and repair: • repairing and replacing faucets • repairing toilet tanks • repairing a trip-lever bath drain • dealing with stopped-up drains • working with copper tubing • measuring and cutting pipe • PVC and CPVC pipe and fittings • installing a garbage disposals • replacing dishwashers • repairing and replacing water heaters • installing or resetting toilets • caulking around plumbing fixtures and tile • water conditioning • working with cast-iron soil pipe • septic tanks and disposal fields • private water systems.

The Plumbers Handbook
seventh edition
Joseph P. Almond, Sr.
4 x 6 Paperback 352 pp. 170 illus.
ISBN: 0-672-23419-X $10.95

Comprehensive, handy guide for plumbers, pipe fitters, and apprentices that fits in the tool box or pocket: • plumbing tools • how to read blueprints • heating systems • water supply • fixtures, valves, and fittings • working drawings • roughing and repair • outside sewage lift station • pipes and pipelines • vents, drain lines, and septic systems • lead work • silver brazing and soft soldering • plumbing systems • abbreviations, definitions, symbols, and formulas.

Questions and Answers for Plumbers Examinations
second edition
Jules Oravetz
5½ x 8¼ Paperback 256 pp. 145 illus.
ISBN: 0-8161-1703-9 $9.95

Practical, fully illustrated study guide to licensing exams for apprentice, journeyman, or master plumber: • definitions, specifications, and regulations set by National Bureau of Standards and by various state codes • basic plumbing installation • drawings and typical plumbing system layout • mathematics • materials and fittings • joints and connections • traps, cleanouts, and backwater valves • fixtures • drainage, vents, and vent piping • water supply and distribution • plastic pipe and fittings • steam and hot water heating.

HVAC

Air Conditioning: Home and Commercial
second edition
Edwin P. Anderson; revised by Rex Miller
5½ x 8¼ Hardcover 528 pp. 180 illus.
ISBN: 0-672-23397-5 $15.95

Complete guide to construction, installation, operation, maintenance, and repair of home, commercial, and industrial air conditioning systems, with troubleshooting charts: • heat leakage • ventilation requirements • room air conditioners • refrigerants • compressors • condensing equipment • evaporators • water-cooling systems • central air conditioning • automobile air conditioning • motors and motor control.

Heating, Ventilating and Air Conditioning Library
second edition 3 vols
James E. Brumbaugh
5½ x 8¼ Hardcover 1,840 pp. 1,275 illus.
ISBN: 0-672-23388-6 $42.95

Authoritative three-volume reference for those who install, operate, maintain, and repair HVAC equipment commercially, industrially, or at home. Each volume fully illustrated with photographs, drawings, tables and charts.

Volume I, Heating Fundamentals, Furnaces, Boilers, Boiler Conversions
5½ x 8¼ Hardcover 656 pp. 405 illus.
ISBN: 0-672-23389-4 $16.95

• Insulation principles • heating calculations • fuels • warm-air, hot water, steam, and electrical heating systems • gas-fired, oil-fired, coal-fired, and electric-fired furnaces • boilers and boiler fittings • boiler and furnace conversion.

Volume II, Oil, Gas and Coal Burners, Controls, Ducts, Piping, Valves
5½ x 8¼ Hardcover 592 pp. 455 illus.
ISBN: 0-672-23390-8 $15.95

• Coal firing methods • thermostats and humidistats • gas and oil controls and other automatic controls •

ducts and duct systems • pipes, pipe fittings, and piping details • valves and valve installation • steam and hot-water line controls.

Volume III, Radiant Heating, Water Heaters, Ventilation, Air Conditioning, Heat Pumps, Air Cleaners
5½ x 8¼ Hardcover 592 pp. 415 illus.
ISBN: 0-672-23391-6 $14.95

• Radiators, convectors, and unit heaters • fireplaces, stoves, and chimneys • ventilation principles • fan selection and operation • air conditioning equipment • humidifiers and dehumidifiers • air cleaners and filters.

Oil Burners fourth edition
Edwin M. Field
5½ x 8¼ Hardcover 360 pp. 170 illus.
ISBN: 0-672-23394-0 $15.95

Up-to-date sourcebook on the construction, installation, operation, testing, servicing, and repair of all types of oil burners, both industrial and domestic: • general electrical hookup and wiring diagrams of automatic control systems • ignition system • high-voltage transportation • operational sequence of limit controls, thermostats, and various relays • combustion chambers • drafts • chimneys • drive couplings • fans or blowers • burner nozzles • fuel pumps.

Refrigeration: Home and Commercial second edition
Edwin P. Anderson; revised by Rex Miller
5½ x 8¼ Hardcover 768 pp. 285 illus.
ISBN: 0-672-23396-7 $17.95

Practical, comprehensive reference for technicians, plant engineers, and homeowners on the installation, operation, servicing, and repair of everything from single refrigeration units to commercial and industrial systems: • refrigerants • compressors • thermoelectric cooling • service equipment and tools • cabinet maintenance and repairs • compressor lubrication systems • brine systems • supermarket and grocery refrigeration • locker plants • fans and blowers • piping • heat leakage • refrigeration-load calculations.

Pneumatics and Hydraulics

Hydraulics for Off-the-Road Equipment second edition
Harry L. Stewart; revised by Tom Philbin
5½ x 8¼ Hardcover 256 pp. 175 illus.
ISBN: 0-8161-1701-2 $13.95

Complete reference manual for those who own and operate heavy equipment and for engineers, designers, installation and maintenance technicians, and shop mechanics: • hydraulic pumps, accumulators, and motors • force components • hydraulic control components • filters and filtration, lines and fittings, and fluids • hydrostatic transmissions • maintenance • troubleshooting.

Pneumatics and Hydraulics fourth edition
Harry L. Stewart; revised by Tom Philbin
5½ x 8¼ Hardcover 512 pp. 315 illus.
ISBN: 0-672-23412-2 $15.95

Practical guide to the principles and applications of fluid power for engineers, designers, process planners, tool men, shop foremen, and mechanics: • pressure, work and power • general features of machines • hydraulic and pneumatic symbols • pressure boosters • air compressors and accessories • hydraulic power devices • hydraulic fluids • piping • air filters, pressure regulators, and lubricators • flow and pressure controls • pneumatic motors and tools • rotary hydraulic motors and hydraulic transmissions • pneumatic circuits • hydraulic circuits • servo systems.

Pumps fourth edition
Harry L. Stewart; revised by Tom Philbin
5½ x 8¼ Hardcover 508 pp. 360 illus.
ISBN: 0-672-23400-9 $15.95

Comprehensive guide for operators, engineers, maintenance workers, inspectors, superintendents, and mechanics on principles and day-to-day operations of pumps: • centrifugal, rotary, reciprocating, and special service pumps • hydraulic accumulators • power transmission • hydraulic power tools • hydraulic cylinders • control valves • hydraulic fluids • fluid lines and fittings.

Carpentry and Construction

Carpenters and Builders Library
fifth edition 4 vols
John E. Ball; revised by Tom Philbin
5½ x 8¼ Hardcover 1,224 pp. 1,010 illus.
ISBN: 0-672-23369-x $39.95
Also available in a new boxed set at no extra cost:
ISBN: 0-02-506450-9 $39.95

These profusely illustrated volumes, available in a handsome boxed edition, have set the professional standard for carpenters, joiners, and woodworkers.
Volume 1, Tools, Steel Square, Joinery
5½ x 8¼ Hardcover 384 pp. 345 illus.
ISBN: 0-672-23365-7 $10.95

• Woods • nails • screws • bolts • the workbench • tools • using the steel square • joints and joinery • cabinetmaking joints • wood patternmaking • and kitchen cabinet construction.
Volume 2, Builders Math, Plans, Specifications
5½ x 8¼ Hardcover 304 pp. 205 illus.
ISBN: 0-672-23366-5 $10.95

• Surveying • strength of timbers • practical drawing • architectural drawing • barn construction • small house construction • and home workshop layout.
Volume 3, Layouts, Foundations, Framing
5½ x 8¼ Hardcover 272 pp. 215 illus.
ISBN: 0-672-23367-3 $10.95

• Foundations • concrete forms • concrete block construction • framing, girders and sills • skylights • porches and patios • chimneys, fireplaces, and stoves • insulation • solar energy and paneling.
Volume 4, Millwork, Power Tools, Painting
5½ x 8¼ Hardcover 344 pp. 245 illus.
ISBN: 0-672-23368-1 $10.95

• Roofing, miter work • doors • windows, sheathing and siding • stairs • flooring • table saws, band saws, and jigsaws • wood lathes • sanders and combination tools • portable power tools • painting.

Complete Building Construction
second edition
John Phelps; revised by Tom Philbin
5½ x 8¼ Hardcover 744 pp. 645 illus.
ISBN: 0-672-23377-0 $19.95

Comprehensive guide to constructing a frame or brick building from the

v

footings to the ridge: • laying out building and excavation lines • making concrete forms and pouring fittings and foundation • making concrete slabs, walks, and driveways • laying concrete block, brick, and tile • building chimneys and fireplaces • framing, siding, and roofing • insulating • finishing the inside • building stairs • installing windows • hanging doors.

Complete Roofing Handbook
James E. Brumbaugh
5½ x 8¼   Hardcover   536 pp.   510 illus.
ISBN: 0-02-517850-4 $29.95

Authoritative text and highly detailed drawings and photographs, on all aspects of roofing: • types of roofs • roofing and reroofing • roof and attic insulation and ventilation • skylights and roof openings • dormer construction • roof flashing details • shingles • roll roofing • built-up roofing • roofing with wood shingles and shakes • slate and tile roofing • installing gutters and downspouts • listings of professional and trade associations and roofing manufacturers.

Complete Siding Handbook
James E. Brumbaugh
5½ x 8¼   Hardcover   512 pp.   450 illus.
ISBN: 0-02-517880-6 $23.95

Companion to *Complete Roofing Handbook*, with step-by-step instructions and drawings on every aspect of siding: • sidewalls and siding • wall preparation • wood board siding • plywood panel and lap siding • hardboard panel and lap siding • wood shingle and shake siding • aluminum and steel siding • vinyl siding • exterior paints and stains • refinishing of siding, gutter and downspout systems • listings of professional and trade associations and siding manufacturers.

Masons and Builders Library
second edition   2 vols
Louis M. Dezettel; revised by Tom Philbin
5½ x 8¼   Hardcover   688 pp.   500 illus.
ISBN: 0-672-23401-7 $23.95

Two-volume set on practical instruction in all aspects of materials and methods of bricklaying and masonry: • brick • mortar • tools • bonding • corners, openings, and arches • chimneys and fireplaces • structural clay tile and glass block • brick walls, floors, and terraces • repair and maintenance • plasterboard and plaster • stone and rock masonry • reading blueprints.

Volume 1, Concrete, Block, Tile, Terrazzo
5½ x 8¼   Hardcover   304 pp.   190 illus.
ISBN: 0-672-23402-5 $13.95

Volume 2, Bricklaying, Plastering, Rock Masonry, Clay Tile
5½ x 8¼   Hardcover   384 pp.   310 illus.
ISBN: 0-672-23403-3 $12.95

Woodworking

Woodworking and Cabinetmaking
F. Richard Boller
5½ x 8¼   Hardcover   360 pp.   455 illus.
ISBN: 0-02-512800-0 $16.95

Compact one-volume guide to the essentials of all aspects of woodworking: • properties of softwoods, hardwoods, plywood, and composition wood • design, function, appearance, and structure • project planning • hand tools • machines • portable electric tools • construction • the home workshop • and the projects themselves – stereo cabinet, speaker cabinets, bookcase, desk, platform bed, kitchen cabinets, bathroom vanity.

Wood Furniture: Finishing, Refinishing, Repairing second edition
James E. Brumbaugh
5½ x 8¼   Hardcover   352 pp.   185 illus.
ISBN: 0-672-23409-2 $12.95

Complete, fully illustrated guide to repairing furniture and to finishing and refinishing wood surfaces for professional woodworkers and do-it-yourselfers: • tools and supplies • types of wood • veneering • inlaying • repairing, restoring, and stripping • wood preparation • staining • shellac, varnish, lacquer, paint and enamel, and oil and wax finishes • antiquing • gilding and bronzing • decorating furniture.

Maintenance and Repair

Building Maintenance second edition
Jules Oravetz
5½ x 8¼   Hardcover   384 pp.   210 illus.
ISBN: 0-672-23278-2 $9.95

Complete information on professional maintenance procedures used in office, educational, and commercial buildings: • painting and decorating • plumbing and pipe fitting

• concrete and masonry • carpentry • roofing • glazing and caulking • sheet metal • electricity • air conditioning and refrigeration • insect and rodent control • heating • maintenance management • custodial practices.

Gardening, Landscaping and Grounds Maintenance
third edition
Jules Oravetz
5½ x 8¼   Hardcover   424 pp.   340 illus.
ISBN: 0-672-23417-3 $15.95

Practical information for those who maintain lawns, gardens, and industrial, municipal, and estate grounds: • flowers, vegetables, berries, and house plants • greenhouses • lawns • hedges and vines • flowering shrubs and trees • shade, fruit and nut trees • evergreens • bird sanctuaries • fences • insect and rodent control • weed and brush control • roads, walks, and pavements • drainage • maintenance equipment • golf course planning and maintenance.

Home Maintenance and Repair: Walls, Ceilings and Floors
Gary D. Branson
8½ x 11   Paperback   80 pp.   80 illus.
ISBN: 0-672-23281-2 $6.95

Do-it-yourselfer's step-by-step guide to interior remodeling with professional results: • general maintenance • wallboard installation and repair • wallboard taping • plaster repair • texture paints • wallpaper techniques • paneling • sound control • ceiling tile • bath tile • energy conservation.

Painting and Decorating
Rex Miller and Glenn E. Baker
5½ x 8¼   Hardcover   464 pp.   325 illus.
ISBN: 0-672-23405-x $18.95

Practical guide for painters, decorators, and homeowners to the most up-to-date materials and techniques: • job planning • tools and equipment needed • finishing materials • surface preparation • applying paint and stains · decorating with coverings • repairs and maintenance • color and decorating principles.

Tree Care *second edition*
John M. Haller
8½ x 11   Paperback   224 pp.   305 illus.
ISBN: 0-02-062870-6 $9.95

New edition of a standard in the field, for growers, nursery owners, foresters, landscapers, and homeowners: • planting • pruning • fertilizing • bracing and cabling • wound repair • grafting • spraying • disease and insect management • coping with environmental damage • removal • structure and physiology • recreational use.

Upholstering
updated
James E. Brumbaugh
5½ x 8¼   Hardcover   400 pp.   380 illus.
ISBN: 0-672-23372-x $12.95

Essentials of upholstering for professional, apprentice, and hobbyist: • furniture styles • tools and equipment • stripping • frame construction and repairs • finishing and refinishing wood surfaces • webbing • springs • burlap, stuffing, and muslin • pattern layout • cushions • foam padding • covers • channels and tufts • padded seats and slip seats • fabrics • plastics • furniture care.

Automotive and Engines

Diesel Engine Manual *fourth edition*
Perry O. Black; revised by William E. Scahill
5½ x 8¼   Hardcover   512 pp.   255 illus.
ISBN: 0-672-23371-1 $15.95

Detailed guide for mechanics, students, and others to all aspects of typical two- and four-cycle engines: • operating principles • fuel oil • diesel injection pumps • basic Mercedes diesels • diesel engine cylinders • lubrication • cooling systems • horsepower • engine-room procedures • diesel engine installation • automotive diesel engine • marine diesel engine • diesel electrical power plant • diesel engine service.

Gas Engine Manual *third edition*
Edwin P. Anderson; revised by Charles G. Facklam
5½ x 8¼   Hardcover   424 pp.   225 illus.
ISBN: 0-8161-1707-1 $12.95

Indispensable sourcebook for those who operate, maintain, and repair gas engines of all types and sizes: • fundamentals and classifications of engines · engine parts • pistons • crankshafts • valves • lubrication, cooling, fuel, ignition, emission control and electrical systems • engine tune-up • servicing of pistons and piston rings, cylinder blocks, connecting rods and crankshafts, valves and valve gears, carburetors, and electrical systems.

Small Gasoline Engines
Rex Miller and Mark Richard Miller
5½ x 8¼   Hardcover   640 pp.   525 illus.
ISBN: 0-672-23414-9 $16.95

Practical information for those who repair, maintain, and overhaul two- and four-cycle engines – with emphasis on one-cylinder motors – including lawn mowers, edgers, grass sweepers, snowblowers, emergency electrical generators, outboard motors, and other equipment up to ten horsepower: • carburetors, emission controls, and ignition systems • starting systems • hand tools • safety • power generation • engine operations • lubrication systems • power drivers • preventive maintenance • step-by-step overhauling procedures • troubleshooting • testing and inspection • cylinder block servicing.

Truck Guide Library 3 vols
James E. Brumbaugh
5½ x 8¼   Hardcover   2,144 pp.   1,715 illus.
ISBN: 0-672-23392-4 $45.95

Three-volume comprehensive and profusely illustrated reference on truck operation and maintenance.

Volume 1, Engines
5½ x 8¼   Hardcover   416 pp.   290 illus.
ISBN: 0-672-23356-8 $16.95

• Basic components · engine operating principles • troubleshooting • cylinder blocks • connecting rods, pistons, and rings • crankshafts, main bearings, and flywheels • camshafts and valve trains • engine valves.

Volume 2, Engine Auxiliary Systems
5½ x 8¼   Hardcover   704 pp.   520 illus.
ISBN: 0-672-23357-6 $16.95

• Battery and electrical systems • spark plugs • ignition systems, charging and starting systems • lubricating, cooling, and fuel systems • carburetors and governors • diesel systems • exhaust and emission-control systems.

Volume 3, Transmissions, Steering, and Brakes
5½ x 8¼   Hardcover   1,024 pp.   905 illus.
ISBN: 0-672-23406-8 $16.95

• Clutches • manual, auxiliary, and automatic transmissions • frame and suspension systems • differentials and axles, manual and power steering • front-end alignment • hydraulic, power, and air brakes • wheels and tires • trailers.

Drafting

Answers on Blueprint Reading
fourth edition
Roland E. Palmquist; revised by Thomas J. Morrisey
5½ x 8¼   Hardcover   320 pp.   275 illus.
ISBN: 0-8161-1704-7 $12.95

Complete question-and-answer instruction manual on blueprints of machines and tools, electrical systems, and architecture: • drafting scale • drafting instruments • conventional lines and representations • pictorial drawings • geometry of drafting • orthographic and working drawings • surfaces • detail drawing • sketching • map and topographical drawings • graphic symbols • architectural drawings • electrical blueprints • computer-aided design and drafting. Also included is an appendix of measurements • metric conversions • screw threads and tap drill sizes • number and letter sizes of drills with decimal equivalents • double depth of threads • tapers and angles.

Hobbies

Complete Course in Stained Glass
Pepe Mendez
8½ x 11   Paperback   80 pp.   50 illus.
ISBN: 0-672-23287-1 $8.95

Guide to the tools, materials, and techniques of the art of stained glass, with ten fully illustrated lessons: • how to cut glass • cartoon and pattern drawing • assembling and cementing • making lamps using various techniques • electrical components for completing lamps • sources of materials • glossary of terminology and techniques of stained glasswork.

Macmillan Practical Arts Library
Books for and by the Craftsman

World Woods in Color
W.A. Lincoln
7 × 10   Hardcover   300 pages
300 photos
ISBN: 0-02-572350-2 $39.95

Large full-color photographs show the natural grain and features of nearly 300 woods: • commercial and botanical names • physical characteristics, mechanical properties, seasoning, working properties, durability, and uses • the height, diameter, bark, and places of distribution of each tree • indexing of botanical, trade, commercial, local, and family names • a full bibliography of publications on timber study and identification.

The Woodworker's Bible
Alf Martensson
8 × 10   Paperback   288 pages   900 illus.
ISBN: 0-02-011940-2 $12.95

For the craftsperson familiar with basic carpentry skills, a guide to creating professional-quality furniture, cabinetry, and objects d'art in the home workshop: • techniques and expert advice on fine craftsmanship whether tooled by hand or machine • joint-making • assembling to ensure fit • finishes. Author, who lives in London and runs a workshop called Woodstock, has also written *The Book of Furnituremaking*.

Cabinetmaking: The Professional Approach
Alan Peters
8½ × 11   Hardcover   208 pages   175 illus. (8 pp. color)
ISBN: 0-02-596200-0 $29.95

A unique guide to all aspects of professional furniture making, from an English master craftsman: • the Cotswold School and the birth of the furniture movement • setting up a professional shop • equipment • finance and business efficiency • furniture design • working to commission • batch production, training, and techniques • plans for nine projects.

The Woodturner's Art: Fundamentals and Projects
Ron Roszkiewicz
8 × 10   Hardcover   256 pages   300 illus.
ISBN: 0-02-605250-4 $24.95

A master woodturner shows how to design and create increasingly difficult projects step-by-step in this book suitable for the beginner and the more advanced student: • spindle and faceplate turning • tools • techniques • classic turnings from various historical periods • more than 30 types of projects including boxes, furniture, vases, and candlesticks • making duplicates • projects using combinations of techniques and more than one kind of wood. Author has also written *The Woodturner's Companion*.

Cabinetmaking and Millwork
John L. Feirer
7⅛ × 9½   Hardcover   992 pages   2,350 illus. (32 pp. in color)
ISBN: 0-02-537350-1 $47.50

The classic on cabinetmaking that covers in detail all of the materials, tools, machines, and processes used in building cabinets and interiors, the production of furniture, and other work of the finish carpenter and millwright: • fixed installations such as paneling, built-ins, and cabinets • movable wood products such as furniture and fixtures • which woods to use, and why and how to use them in the interiors of homes and commercial buildings • metrics and plastics in furniture construction.

Carpentry and Building Construction
John L. Feirer and Gilbert R. Hutchings
7½ × 9½   hardcover   1,120 pages   2,000 photos (8 pp. in color)
ISBN: 0-02-537360-9 $50.00

A classic by Feirer on each detail of modern construction: • the various machines, tools, and equipment from which the builder can choose • laying of a foundation • building frames for each part of a building • details of interior and exterior work • painting and finishing • reading plans • chimneys and fireplaces • ventilation • assembling prefabricated houses.